# PRINT READING FOR ENGINEERING AND MANUFACTURING TECHNOLOGY

## Second Edition

D0707392

# PRINT READING FOR ENGINEERING AND MANUFACTURING TECHNOLOGY

## Second Edition

# David A. Madsen

*Faculty Emeritus, Former Department Chair,*
*Drafting Technology, Clackamas Community College;*
*Former Member, Board of Directors,*
*American Drafting and Design Association*

**THOMSON**

**DELMAR LEARNING**

Australia • Canada • Mexico • Singapore • Spain • United Kingdom • United States

# Print Reading for Engineering and Manufacturing Technology
## Second Edition

David A. Madsen

Vice President, Technology
and Trades SBU:
**Alar Elken**
Editorial Director:
**Sandy Clark**
Senior Acquisitions Editor:
**James Devoe**
Senior Development Editor
**John Fisher**

Marketing Director:
**Dave Garza**
Channel Manager:
**Fair Huntoon**
Marketing Coordinator:
**Casey Bruno**
Editorial Assistant:
**Katherine Bevington**
Production Director:
**Mary Ellen Black**

Production Manager:
**Andrew Crouth**
Production Editor:
**Stacy Masucci**
Technology Project Manager:
**Kevin Smith**
Technology Project Specialist
**Linda Verde**
Production Services
**TIPS Technical Publishing, Inc.**

Library of Congress Cataloging-in-
Publication Data

Madsen, David A.

Print reading for engineering and
manufacturing technology / David A.
Madsen.— 2nd ed.

p. cm. — (Delmar learning
blueprint reading series)
Includes index.

ISBN 1-4018-5163-0
(pbk.; perforated : alk. paper)

1. Photographic reproduction of plans,
drawings, etc. I. Title. II. Series.

TR920.M33 2004
670—dc22

2004005063

## Notice to the Reader

# TABLE OF CONTENTS

# PREFACE

*Print Reading for Engineering and Manufacturing Technology* is a practical comprehensive workbook that is easy to use and understand. You can use this content as presented, following a logical sequence of learning activities for engineering and manufacturing technology print reading; or you can arrange the chapters to accommodate alternate formats for traditional or individualized instruction. This is the only training and reference material you will need for engineering and manufacturing technology print reading.

## Prerequisites

An interest in print reading for engineering and manufacturing technology, and basic arithmetic, written communication, and reading skills are the only prerequisites required. If you begin with an interest in engineering and manufacturing technology, you will end with the knowledge and skills required to read complete sets of working drawings for manufactured projects.

## *Major Features*

### Practical

*Print Reading for Engineering and Manufacturing Technology* provides a practical approach to reading prints as related to current common practices. One excellent and necessary foundation of print reading training is the emphasis of standardization and quality real-world engineering and manufacturing print examples. When you become a professional in the manufacturing industry, this text will go along as a valuable desk reference.

### Realistic

Chapters contain realistic examples, actual prints, illustrations, related tests, and print-reading problems based on real-world engineering prints. The examples demonstrate recommended presentation with actual engineering prints used for reinforcement.

## Practical Approach to Problem Solving

The professional in the manufacturing technology industry is responsible for accurately reading and interpreting engineering prints. This workbook explains how to read actual industry prints and interpret requirements in a knowledge-building format; one concept is learned before the next is introduced. Print-reading problem assignments are presented in order of difficulty and in a manner that provides you with a wide variety of print-reading experiences. The concepts and skills learned from one chapter to the next allow you to read complete sets of working drawings in manufacturing and mechanical engineering. The prints are presented in a manner that is consistent with actual engineering office practices. You must be able to think through the process of print reading with a foundation of how prints and related manufacturing processes are implemented. The goals and objectives of each problem assignment are consistent with recommended evaluation criterion based on the progression of learning activities.

## Prints Prepared Using Computer-Aided Drafting (CAD)

Computer-aided drafting is used in engineering drafting applications. This print reading workbook is written and presented with current CAD technology standards. This is an advantage as you, the students, proceed into a manufacturing career in the twenty-first century. All of the print reading examples and problems are prepared using CAD in a manner that displays the highest industry standards.

## Standards and Manufacturing Techniques

National standards: ANSI (American National Standard Institute, American National Standard Engineering Drawings and Related Documentation Practices, published by The American Society of Mechanical Engineers), are used throughout the text as related to specific instruction and applications. This text acknowledges the difference in manufacturing methods and introduces you to the format used to read prints for each method. The print-reading problem assignments are designed to provide drawings that involve a variety of alternatives.

## Fundamental Through Advanced Coverage

You can use this text in the Engineering Drafting Technology curriculum that covers the basics of mechanical engineering drawing. In this application, you can use the chapters directly associated with reading actual industry prints. The balance of the text can remain as reference for future study or as a valuable desk reference. Use the workbook in the comprehensive Manufacturing Technology program where manufacturing skills and theory are required. In this application you can expand on the primary objective of reading prints followed by or along with actual manufacturing projects. This workbook also includes:

1. Introduction to prints and how they are made.
2. Communicating with a sketch.
3. Using scales and precision measuring equipment.
4. Basic manufacturing math.
5. Manufacturing processes.
6. Welding processes and representations.
7. Fasteners and springs.
8. Geometric tolerancing.
9. Pictorial drawings.
10. Precision sheet metal fabrication prints.
11. Electrical diagrams and schematics.
12. Gears, cams, and bearings.

## Course Plan

Engineering and manufacturing technology print reading is a primary emphasis of many technical drafting and manufacturing technology curriculums, while other programs offer only an exploratory course in this field. This text is appropriate for either application. The content of this text reflects the common elements in a comprehensive engineering and manufacturing technology curriculum that you can use in part or totally.

## Section Length

Chapters are presented in individual learning segments that begin with elementary concepts and build until each chapter provides complete coverage of each topic. Instructors can select to present lectures in short 15-minute discussions, or divide each chapter into 40–50-minute lectures.

## Applications

Special emphasis has been placed on providing realistic print-reading examples and problems. The examples and problems have been supplied by manufacturing companies, drafting departments, drafters, engineers, and engineering designers. Problems are given in order of complexity so that you can be exposed to a variety of print reading experiences. Problems require you to go through the same thinking process that a professional manufacturing worker is faced with daily; including reading symbols, notes, finding and interpreting information, and many other activities. Chapter tests provide a complete review of each chapter to help you evaluate your progress or to use as study questions.

## Instructor's Manual

The instructor's manual contains learning objectives, problem and test solutions, and transparency masters.

## Online Companion

Visit *www.delmarlearning.com/companions/* to locate this text's online companion containing Appendices A–E.

## Acknowledgements

We give special thanks and acknowledgement to the many professionals who reviewed the manuscript for this text in an effort to help us publish the best *Print Reading for Engineering and Manufacturing Technology* text.

## List of Reviewers

Gilbert Atkins, Mercer Country Vocational Technical Center, Princeton, West Virginia

Samuel Barnes, A-B Technical School, Asheville, North Carolina

Charles Case, ITT Technical Institute, Indianapolis, Indiana

Victor Marshall, Washington-Holmes Vocational Technical Center, Chipley, Florida

Michael Price, Walters State Community College, Morristown, Tennessee

Ray Schaeffer, East Central College, Washington, Missouri

Bruce Whipple, Trident Technical College, Charleston, South Carolina

The quality of this text is also enhanced by the support and contributions from designers, engineers, and vendors. The list of contributors is extensive and acknowledgement is given at each figure illustration; however, the following individuals and companies gave an extraordinary amount of support:

Tom Pearce, Stanley Hydraulic Tools—A Division of the Stanley Works

Mike Stegmann, HYSTER Company

Paul D. Young, ENOCH Manufacturing Company

Greg Lanz, Brod & McClung—PACE Co.

Advance Machine Technology

Manufacturers Tool Service/Clackamas, Oregon

David George

Jim B. MacDonald

Richard Clouser

## Using This Text

This Print Reading for Engineering and Manufacturing Technology workbook is designed for you, the student. The development and format presentation has been tested in actual conventional and individualized classroom instruction. The information presented is based on engineering and manufacturing standards, drafting room practice, and trends in the manufacturing industry. This text is designed to be a comprehensive coverage of engineering and manufacturing technology print reading. Use the text as a learning tool while in school and take it along as a desk reference when you enter the profession. The amount of written text is complete, but kept to a minimum. Examples and illustrations are used extensively. Many students learn best by observation of examples. Here are a few helpful hints:

1. Read the text. The text content is intentionally designed to make easy reading. It gives the facts in as few, easy to understand, words as possible. Don't pass up the reading, because the content helps you clearly understand the prints that you will read.

2. Look carefully at the examples. The figure examples are presented in a manner that is consistent with engineering and manufacturing technology standards. Look at the examples carefully in an attempt to understand the intent of specific applications. If you can understand why something is done a certain way, it is easier to read the actual prints. Print reading is often like doing a puzzle; you may need to carefully search the print to find the desired information.

3. Use the text as a reference. Few professionals know everything about standard practices, techniques, and concepts; so always be ready to use the reference if you need to verify how specific applications are handled. Become familiar with the use and definitions of technical terms. It will be difficult to memorize everything in this text but, after considerable use of the concepts, print reading applications should become second nature.

4. Learn each concept and skill before you continue to the next. The text is presented in a logical learning sequence. Each chapter is designed for learning development, and chapters are sequenced so print reading knowledge grows from one chapter to the next. Print reading problem assignments are presented in the same learning sequence as the chapter content, and also reflect progressive levels of difficulty.

5. Do the chapter tests. Each chapter has a test at the end. Answering these test questions gives you an opportunity to review the material that you just studied. This reinforcement helps you fully learn the material.

6. Do the print reading problems. There are several print reading problems following each chapter test. These problems require that you answer questions as you read actual engineering and manufacturing technology prints. There is no substitute for reading actual prints. The practice of print reading becomes easier when you have had a chance to read several prints. By the time you complete the print-reading problems, the content covered in the preceding chapter will be easier for you to identify on future prints.

# Introduction to Engineering and Manufacturing Technology Print Reading

## LEARNING OBJECTIVES

After completing this chapter you will be able to:

■ Answer questions and identify terminology related to the blueprint, blue-line, diazo, photocopy, microfilm, and computer-aided design and drafting (CADD) print processes.

■ Answer questions related to the use of CADD in the engineering and manufacturing technology industry.

■ Identify and properly fold standard print sizes.

■ Define tolerance and answer related questions.

■ Identify and read print scales.

■ Answer questions and read items on prints related to zoning, title block information, and the drawing field.

---

This text deals with reading prints that are prepared for mechanical engineering and the manufacturing technology industry. *Mechanical engineering* is the art and science of designing manufactured products. The term "manufactured," as used here, refers to any mechanical product from the family toaster to the 18-wheeler that rolls down the highway to the airliner that carries people around the world. The proper reading of prints for the manufacturing industry requires a basic knowledge of manufacturing materials and processes; standard drafting practices; fabrication methods, such as welding, fastening, and precision sheet metal; gearing; and electrical diagrams.

## ABOUT PRINTS

This text refers to the reproduction of manufacturing drawings as *prints*. The term "blueprint" is an old term that is generally used in the manufacturing industry when referring to prints. *Blueprinting* is an old method that results in a print with a dark blue background and white lines. You may find some of these in a museum or in a company's drawing archives. The reproduction methods commonly used today are diazo, photocopy, printer, and plotter.

### The Diazo Print Process

*Diazo prints* are also known as *blue-line prints* because the resulting print generally consists of blue lines on a white background. The diazo process is an inexpensive way to make copies from any type of translucent paper, film, or cloth material. The diazo print is made by a print process that uses an ultraviolet light passing through a translucent original drawing to expose a chemically coated paper or print material underneath. The light does not go through the lines of the drawing; thus, the chemical coating beneath the lines remains unexposed. The print material is then exposed to ammonia vapor, which activates the remaining chemical coating to produce the blue lines. Diazo materials are also available that make black or brown lines. The diazo print process is becoming outdated as it is being replaced by other print processes. The diazo print process is demonstrated in Figure 1–1.

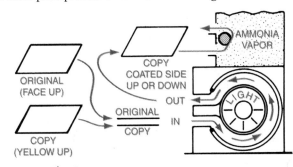

Figure 1–1 The Diazo print process.

### Photocopy Prints

The photocopy process is gaining wide use for copying engineering and architectural drawings. These copy machines are much like the copier found in the traditional office setting. The only difference is the capacity that allows the copying of large drawings.

The photocopy process is popular because there is no need for coated copy materials or exposure to the possible hazards of ammonia. The photocopy process also copies drawings or other documents from any type of original and onto most types of copy material, although using brand name products is normally recommended. Some photocopiers allow you to reduce or enlarge drawings to suit specific needs.

## Microfilm

*Microfilm* is a photographic reproduction on film of a drawing or other document, which is highly reduced for ease in storage and sending from one place to another. The microfilm used in mechanical engineering is generally prepared as one frame or drawing attached to an aperture card as shown in Figure 1–2. Although the aperture card is more popular, some microfilm users prefer the roll film, shown in Figure 1–3. The use of microfilm is rapidly becoming outdated as industry converts to computer-aided design and drafting (CADD). CADD drawings are stored in computer files for easy access and distribution.

As a manufacturing print reader, you probably do not really care what process is used to make the print, because your main interest is that the print is clear and easy to read. The end result of any quality drawing and proper reproduction of that drawing is a print that you can easily read. This discussion about print processes is provided so you understand some of the basic terminology when you talk to an engineer or drafter in your industry.

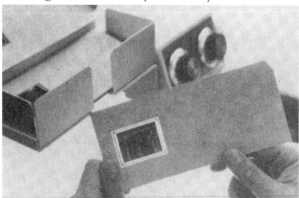

Figure 1–2 The aperture card and microfilm. *Courtesy of the 3M Company.*

Figure 1–3 Roll microfilm. Courtesy of the 3M Company.

## Computer-Aided Design and Drafting Reproduction

Computer-aided design and drafting is used extensively in the mechanical engineering industry. Drawings that are created using CADD can be reproduced on a printer or plotter. The resulting print can be sent directly to manufacturing, or copies can be made using the diazo or photocopy process previously discussed. Figure 1–4 shows a drawing on the computer screen and the plotter that reproduces the drawing. The equipment used to reproduce the drawing from the computer is called an *output device*. The quality of a CADD drawing is greatly influenced by the reproduction method and the type of output device.

The quality of a CADD drawing made using liquid ink pens and a plotter is excellent.

Some companies keep their CADD drawings on 8-1/2 x 11-inch sheets and are then able to reproduce them using a quality laser printer like the one shown in Figure 1–5. Larger format laser printers are also available.

*Inkjet printers*, also called *inkjet plotters*, produce high-quality plots very fast, and are generally less expensive than laser printers. An inkjet printer sprays droplets of ink onto the paper to form dot-matrix images. Print quality is measured in *dots per inch* (dpi).

Figure 1–4 A computer system being used in an architectural office. *Photo courtesy of Hewlett-Packard Co.*

Inkjet printers are available for the reproduction of small- or large-format drawings.

Another type of quality reproduction is made using an electrostatic plotter. While more expensive than other output devices, the electrostatic plotter is gaining in popularity because of its high speed and quality performance. This plotter works by attaching permanent ink to electrically charged dots. An electrostatic plotter is shown in Figure 1–6.

Figure 1–5 A laser printer. *Courtesy of the Hewlett-Packard Company.*

## THE INFLUENCE OF CADD IN TODAY'S ENGINEERING AND MANUFACTURING TECHNOLOGY PRINT-READING WORLD

Traditional pencil-on-paper drafting is referred to as *manual drafting*. The full range of design and drafting using a computer is called *computer-aided design and drafting*. CADD is the predominant method used for creating drawings in industry today. All of this is not too important to you as a print reader, except that prints from CADD drawings are often of better quality than manually drawn prints. This is not always true, because many traditional drafters make drawings that compare with the quality of CADD prints. However, if you have a chance to look at the prints from several different companies, you will probably find some poorly created prints that were reproduced from manual drawings. It is possible for a drafter to make a technically incorrect CADD drawing, but normally CADD drawings are easy to read. Properly prepared CADD drawings are also extremely accurate, have consistent quality, and display proper drafting standards. These are some of the advantages of CADD over manual drafting:

Figure 1–6 An electrostatic plotter. *Courtesy Versatec*

■ Some minor differences can be found, but generally all CADD drawings produced in a single company look the same. You do not have to get used to reading prints with different drafting styles.

■ When properly used, CADD is extremely accurate. A complete drawing must contain all of the dimensions needed for you to manufacture the part. Trying to determine dimensions by using a scale (ruler) on the print should *never* be done.

■ CADD lettering is easy to read. All numbers and letters are clear and you do not need to worry about trying to read different personal styles of manual lettering.

■ Prints made from CADD drawings normally have sharp, clear lines and lettering, depending on the method used to plot or print the drawing. Most professional drafters can make quality lines and lettering manually. However, you are likely to be asked to create drawings for immediate turnaround, and the end result can be lines and letters that are difficult to read. This is not the case with CADD. Even when the CADD drafter is in a hurry, the end result is always of the same quality.

■ Although less important to print readers, in many cases CADD increases drafting productivity over manual drafting. CADD is especially fast when a drawing requires changes. This aspect of CADD can be helpful if you are waiting for a print to be revised.

■ A huge advantage of CADD is that the engineer can create the design on the computer screen and then prepare the final drawings for manufacturing or send the preliminary design to the drafting department to complete the drawings. A team of engineers and drafters can all share drawings among their respective computers. This is not new to manual drafting, because something similar has been done for a long time with pencil and paper. However, with CADD, the process is much easier, and once a change is made on one part of the design, this change can reach every team member's computer almost instantly. Another exciting feature is that each part of the design is extremely accurate. This allows one part to check the next as the design goes together.

## 3D AND SOLID MODELING

Drafters using traditional manual drafting practices or standard two-dimensional (2D) computer-aided design and drafting programs use 2D drawings to design, develop, and document an object. The 2D drawings are then used to manufacture the item. For example, the typical process of documenting a part design for manufacturing is to create a 2D production-part drawing that contains enough views to fully describe the part. The 2D drawing provides the necessary information to manufacture the physical three-dimensional (3D) part. In contrast to the 2D drafting approach, drafters using 3D CADD programs generate 3D models. A *model* is an exact, computer-generated, 3D representation of an object that is used to develop and visualize objects, as shown in Figure 1–7.

Models allow you to explore part- and assembly-design options before components are ever manufactured. Many 3D CADD programs allow you to generate solid models that are often more realistic and effective than wire frame or surface models, because a *solid model* contains accurate mass characteristics to help you analyze physical properties and better understand the model. A *wire frame model* contains only information about object edges. A *surface model* contains information about object faces, planes, and edges, but does not recognize solid mass. A number of 3D CADD programs also contain 2D drawing capabilities that allow you to quickly generate 2D production drawings from an existing 3D model. Using the previous example, the process for documenting a part design for manufacturing using a 3D CADD program involves creating a 3D model of the actual part, not just 2D part views. Once the initial model is generated, you can redefine the part characteristics, and if necessary, develop additional models and interconnected assemblies, study model properties, and further interact with the model, almost as if the model is a real object. Then, if required, 2D drawing views are quickly taken from the model, dimensions and notes are added, and the physical 3D part is manufactured. The 2D drawing, shown in Figure 1–8, is created from the model in Figure 1–7.

Many 3D CADD programs function on the concept of parametric solid modeling. *Parametric solid modeling* programs allow you to create mechanical part and assembly models that contain parametric relationships. *Parametric relationships*, or *parametrics*, function on the idea of associations between model parameters and components. When you define the size and position of model geometry using specific parameter information, you need to modify predefined size and position specifications for model changes to occur. As a result, if model-parameter modifications conflict with existing geometry, typically the model cannot be redefined without errors. For example, a line and the line's dimension have a simple parametric relationship. In this example, if the line is an inch long, the dimension you place on the line is one inch. If you change the size of the line to half an inch, the dimension *automatically* changes to half an inch. Parametric relationships are often identified as intelligence, because parametric modeling is the result of the program's ability to effectively store and manage model information in a database. The database contains information about each model parameter, including geometric sizes and positions, and model history. A *database* is a collection of information stored in computerized form.

The intent of this book about print reading is to show you the best-quality industrial drawings available. The drawings for the print-reading exercises are produced using CADD. You will be

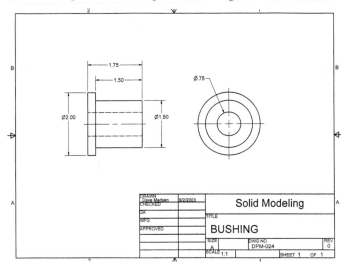

Figure 1–8 The 2D drawing created using the 3D model in Figure 1–7.

introduced to proper standards and print preparation techniques throughout this book. While the drawings selected are the best professional work available, some companies have standards that are slightly different from the national standards used as the basis for instruction in this book. Your ability to effectively read prints may often be influenced by your flexibility. In other words, things may not always be the way they are supposed to be.

## SHEET SIZES, TITLE BLOCKS, AND BORDERS

*ASME/ANSI* standard sheet sizes and format are specified in the documents ANSI Y14.1–1980 (Reaffirmed 1987) Drawing Sheet Size and Format, and ASME Y14.1M–1992 Metric Drawing Sheet Size and Format.

All professional drawings have title blocks. Standards have been developed for the information put into the title block and on the

Figure 1–7 An example of a three-dimensional computer-generated model.

surrounding sheet next to the border, so the drawing is easier to read and file than drawings that do not follow a standard format.

## Sheet Sizes

ANSI Y14.1 specifies sheet size specifications in inches as follows:

| SHEET SIZES IN INCHES | | |
|---|---|---|
| DESIGNATION | VERTICAL | HORIZONTAL |
| A | 8-1/2 | 11 |
| B | 11 | 17 |
| C | 17 | 22 |
| D | 22 | 34 |
| E | 34 | 44 |
| F | 28 | 40 |
| A0 | 841 | 1189 |
| A1 | 594 | 841 |
| A2 | 420 | 594 |
| A3 | 297 | 420 |
| A4 | 210 | 297 |

There are four additional size designations (G, H, J, and K); these apply to roll sizes.

The M in the title of the document Y14.1M means all specifications are given in metric measurements. Standard metric drawing sheet sizes are designed as follows:

Longer lengths are referred to as elongated and extra-elongated drawing sizes. These are available in multiples of the short side of the sheet size.

Standard inch sheet sizes are shown in Figure 1–9(a), and metric sheet sizes are shown in Figure 1–9(b).

## Zoning

Some companies use a system of numbers along the top and bottom margins and letters along the left and right margins called *zoning*. Notice in Figure 1–9(a) that numbered and lettered zoning begins on C-size drawing sheets, and in Figure 1–9(b) zoning is used on A3 sheet sizes and larger. Zoning allows the drawing to read like a road map. For example, you can refer to the location of a specific item as D4, which means that the item can be found at or near the intersection of D across and 4 up or down.

## Title Blocks

Companies generally have title blocks and borders preprinted on drawing sheets or in CADD files to reduce drafting time and cost. Drawing sheet sizes and sheet format items, such as borders, title blocks, zoning, revision columns, and general note locations, have

been standardized so the same general relationship exists between engineering drawings internationally. The ANSI Y14.1 and ASME Y14.1M documents specify the exact size and location for each item found on the drawing sheet. It is recommended that standard sheet sizes and format be followed to improve readability, handling, filing, and reproduction. Each company may use a slightly different design, although the following basic information is located in approximately the same place on most engineering drawings:

1. Title block. Lower-right corner.

   ■ Company name
   ■ Confidential statement
   ■ Unspecified dimensions and tolerances
   ■ Sheet size
   ■ Drawing number
   ■ Part name
   ■ Material
   ■ Scale
   ■ Drafter signature
   ■ Checker signature
   ■ Engineer signature

2. Revision column. Upper-right corner, over or next to the title block.

   ■ Revision symbol, number or letter
   ■ Description
   ■ Drafter
   ■ Date

3. Border line.

   ■ Without zoning
   ■ With zoning

Figure 1–10 shows a sample title block.

## Title Block Definitions

*Dimension.* A dimension is a numerical value indicated on a drawing and in documents to define the size, shape, location, geometric characteristic, or surface texture of a feature.

*Tolerances.* Tolerances are discussed in detail in Chapter 7, although it is important to know that a tolerance is a given amount of acceptable variation in a size or location dimension. All dimensions have a tolerance.

*Millimeters and inches.* All dimensions are in millimeters (mm) or inches (in.) unless otherwise specified.

*Unless otherwise specified.* This means that, in general, all of the features or dimensions on a drawing have the relationship or specifications given in the title block, unless a specific note or dimensional tolerance is provided at a particular location in the drawing.

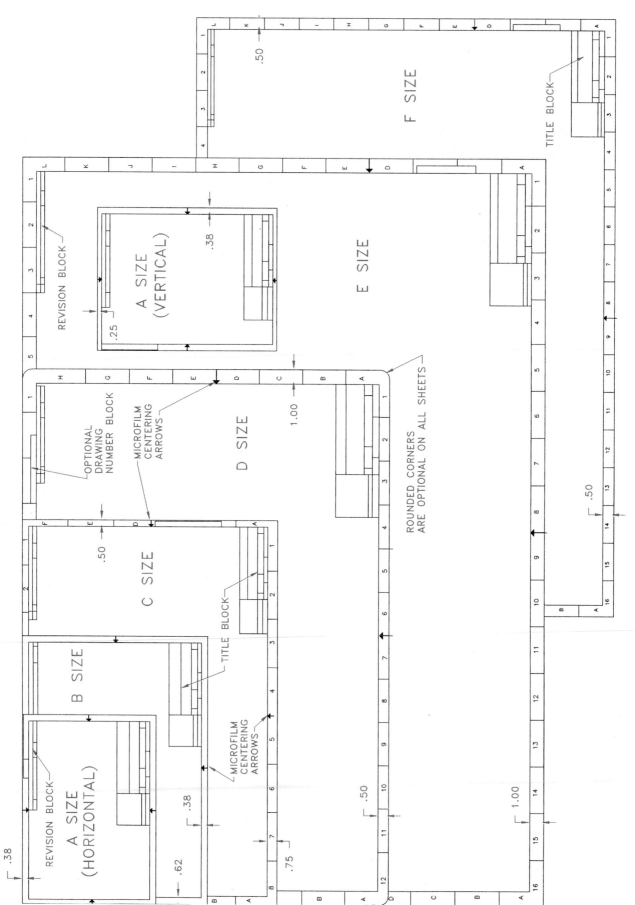

Figure 1–9(a) Standard inch drawing sheet sizes.

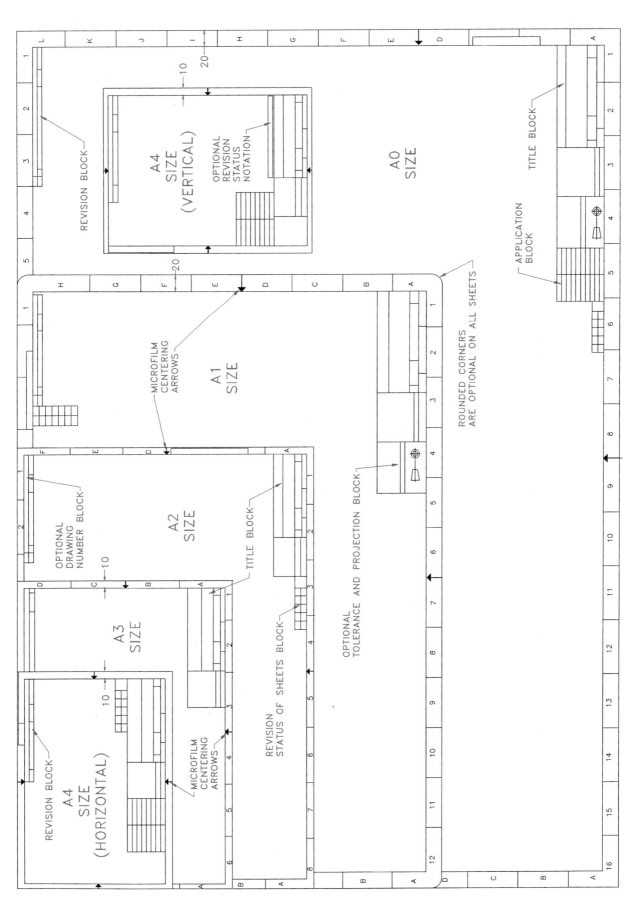

Figure 1–9(b) Standard metric drawing sheet sizes.

| HUNTER | HUNTER FAN COMPANY |
|---|---|
| SINCE 1886 | 2500 FRISCO AVE., MEMPHIS, TENN. 38114 |

THIS DOCUMENT AND THE INFORMATION IT DISCLOSES IS THE EXCLUSIVE PROPERTY OF HUNTER FAN COMPANY. ANY REPRODUCTION OR USE OF THIS DRAWING, IN PART OR IN WHOLE, WITHOUT THE EXPRESS CONSENT OF THE PROPRIETOR ARE PROHIBITED.

| TOLERANCES: (UNLESS OTHERWISE SPECIFIED) | DATE: 5/21/96 | DRAWN BY: DICK PEARCE | DEPARTMENTAL APPROVALS | |
|---|---|---|---|---|
| DECIMAL: | SCALE: FULL | CHK'D BY: | R&D: | Q.A.: |
| .XX = ±_____ | FIRST USE: AIR FORCE FAN | | ENG: | MFG: |
| .XXX = ±_____ | REFERENCE: | | MRKTG: | I.D.: |
| FRACTIONAL: ±_____ | PART NAME | | PART NO. | |
| ANGULAR: ±_____ | BLADE RING | | 92678 | |

Figure 1–10 Sample title block. *Courtesy Hunter Fan Company, Memphis, TN.*

*Unspecified tolerances.* Unspecified tolerance refers to any dimension on the drawing that does not have a tolerance specified. This is when the dimensional tolerance required is the same as the general tolerance shown in the title block.

*Revisions.* When parts are redesigned or altered for any reason, the drawing will be changed. All drawing changes are commonly documented and filed for future reference. When this happens, you should reference the documentation on the drawing so users can identify that a change has been made. Before any revision can be made, the drawing must be released for manufacturing.

## Title and Revision Block Instructions

The title block displayed in Figure 1–11(a) shows most of the common elements found in industrial title blocks. The revision block, Figure 1–11(b), found in this format is located in the upper-right corner of the drawing sheet. Some companies use other placement.

The large numbers shown on Figure 1–11 refer to the following items normally found in title and revision blocks.

1. DRAWN BY: Identifies the drafter using initials, such as JRM, DAM, JLT, unless otherwise specified.

2. SCALE: Shows examples of drawing scales to fill in this block are FULL or 1:1, HALF or 1:2, DBL or 2:1, QTY or 1:4; NONE.

3. DATE: This gives the date in the order of day, month, and year, such as 18 NOV 03, or numerically with month, day, and year, such as 11/18/03.

4. APVD: This block is to be used by the checker that approves the drawing by initialing it.

5. MATERIAL: Describes the material used to make the part, for example, BRONZE, CAST IRON, SAE 4320.

6. PART NAME: The name of the part, such as COVER, HOUSING.

7. B: Gives sheet sizes shown in Figure 1–8.

8. PART NO.: Most companies have their own part numbering systems. While numbering systems differ, they are often keyed to categories, such as the disposition of the drawing (casting, machining, assembly), materials used, related department within the company, or a numerical classification of the part.

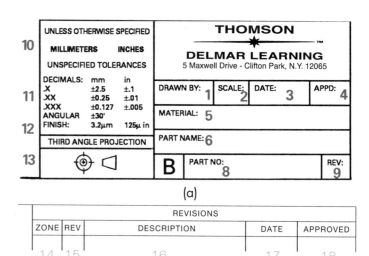

(a)

| REVISIONS | | | | |
|---|---|---|---|---|
| ZONE | REV | DESCRIPTION | DATE | APPROVED |
| 14 | 15 | 16 | 17 | 18 |

(b)

Figure 1–11 (a) Title block elements. (b) Revision block elements.

9. REV.: The revision letter of the part or drawing. A new or original drawing is — (dash) or 0 (zero). The first time a drawing is revised, the — or 0 changes to an A. For the second drawing change, a B is placed here, and so on. The letters *I, O, S, X,* and *Z* are not used. When all of the available letters A through Y have been used, double letters, such as *AA* and *AB,* or *BA* and *BB,* are used. Some companies use revision numbers rather than letters.

10. UNLESS OTHERWISE SPECIFIED: Unless otherwise specified, drawing dimensions are given in millimeters or inches.

11. UNSPECIFIED TOLERANCES: The INCHES is blacked out or left off the title block if the drawing dimension values are in millimeters. *Unless otherwise specified* means that values on the drawing that have no specific tolerance identification relate back to the information given in the title block. All of the features or dimensions on the drawing have the relationship or specifications given in the title block unless a specific note or dimensional tolerance is provided in a particular location on the drawing. For example, if INCHES is specified in the title block, a value on the drawing, such as 2.625, is in inches and has the tolerance of a three-place decimal, as given in the title block. The term *unspecified tolerance* refers to any dimension on the drawing that does not have a tolerance specified, and the tolerance is given in the title block or in a general note. A *specified tolerance* designates a dimension on a drawing that has its own tolerance that is different from the standard tolerance given in the title block. One-, two-, and three-place decimals are established as MILLIMETERS or INCHES in a manner the same as item 10. Angular tolerances for unspecified angular dimensions are ±30°. These applications depend on company practice.

12. FINISH: This block gives the unspecified surface finish, for surfaces that are identified for finishing without a specific callout.

13. THIRD-ANGLE PROJECTION: Third-angle projection view presentation is common in the United States. A complete discussion of this topic is found in Chapter 5.

14. ZONE: Zone refers to the place on the drawing where the change is located. Remember from an earlier discussion, the zoning allows the drawing to read like a road map; numbers along the top and bottom and letters along the sides direct you to a specific location on the drawing. If the location of the change is D4, D4 is placed in the ZONE column for this change. The ZONE column is used only if the drawing has zoning.

15. REV: A *revision* is a change that is formally made to a drawing. The revision letter or number is found in this block, such as A, B, C, etc. Succeeding letters are to be used for each Engineering Change Notice (ECN) or group of related ECNs regardless of quantity of changes on an ECN. (An ECN is a numbered document that provides information about the change. ECNs are discussed in Chapter 13.)

Note: the REV block in the title block must be changed to agree with the last REV letter in the revision block. Some companies use revision numbers.

16. DESCRIPTION: This block has the ECN letter covering the Engineering Change Request (ECR) or group of ECRs that requires the drawing to be revised. Another option is a short description of the change.

17. DATE: The day, month, and year on which the ECN package was ready for release to production, such as 6 APR 03, or use month, day, and year numbers, such as 4/6/03.

18. APPROVED: This column contains the initials of the person approving the change, and optional date.

Drafting revisions are completed in chronological order by adding horizontal columns and extending the vertical column lines.

The previous example showed you a typical title block and revision block. Each company has its own design and format that is similar, and most CADD programs have predefined sheet-layout options with borders and title blocks. These are normally based on the ASME standard title blocks and revision blocks. The document ANSI Y14.1 gives the recommended title block and revision block dimensions shown in Figure 1–12. The title block shown in Figure 1–12 is the recommendation for A-, B-, C-, and G-size sheets.

The recommended title block for D-, E-, F-, H-, J-, and K-size sheets is bigger, measuring 7.62 inches long and 2.50 inches high. The revision block is recommended for the upper-right corner of the drawing border, but some companies place it in other locations, such as over the title block or to the left of the title block. Look again at Figure 1–12. Notice the large letters A and B. These are items found in the ANSI Y14.1 examples that are not displayed in Figure 1–11. These are described as follows:

A   This is the FSCM number, which is optional. *FSCM* stands for *Federal Supply Code for Manufacturers*. The FSCM number is a five-digit numerical code used on all drawings that produce items used by the Federal government.

B.   The SHEET compartment is for the identification of sheet numbers. This is used by companies that anticipate having the drawing for a particular application occupy

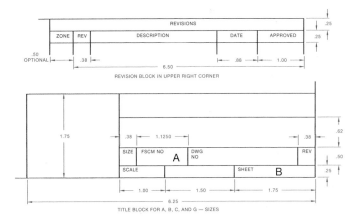

Figure 1–12 Title block and revision block dimensions recommended by ANSI Y14.1.

more than one sheet. 1 OF 1 is placed here if the drawing is on one sheet. If the drawing is on two sheets, the first has 1 OF 2 and the second has 2 OF 2 in this compartment.

Notice that the ANSI title block in Figure 1–12 has DWG NO rather than PART NO, as in Figure 1–11. This is a company preference.

## HOW DO YOU FIND INFORMATION ON PRINTS?

Print reading is basically finding information on prints. Information can be displayed on a print in the form of lines, symbols, and notes. These items are either located in the title block or in the *field* of the drawing. The field of the drawing is anywhere within the border lines outside the title block. As you will learn later, many items on the drawing are made up of symbols. If you know that the information you are looking for is usually displayed as a symbol, you need to look for that symbol in a location where it is often found. Information about symbols is discussed as they relate to specific applications throughout this book. For now, here are some general rules to follow when reading information on prints:

■ Scan the entire drawing while looking at the general layout. This helps you become familiar with the major parts of the drawing.

■ Look at the title block to find information that is related to the drawing, such as part name and number, material, tolerances, number of revisions, engineer's name, drafter's name, and sheet number, if more than one sheet.

■ Look at the views on the print to get a quick grasp of what is included. For example, look at the views, determine if sectioning is used, and get a general idea of the overall size of the part. Each of these concepts is covered in this book.

■ Quickly read the general notes to become familiar with the information that relates to the entire drawing. *General notes* provide written information related to the entire drawing. General notes are usually found in the lower-left corner, upper-left corner, or above the title block. *Specific* or *local notes* are notes that relate to a specific feature or features on the drawing and are found where they apply. Notes are discussed in detail in Chapter 7.

■ Now that you know the major features found on the drawings, take more time to look for the specific information that you seek.

If you follow these basic guidelines, you will normally find it easy to determine what the print includes, and you will be able to find the information you need. Sometimes, however, the drawing can be very complex and the information you need is difficult to find. When this happens, print reading can be time consuming. After you study this text and when you gain experience reading a variety of prints, you will routinely be able to read a print quickly and interpret the information given.

## HOW TO PROPERLY FOLD PRINTS

As you have already learned, prints can come in a variety of sizes ranging from a small 8-1/2 x 11 inches to 34 x 44 inches or larger. It is easy to file the 8-1/2 x 11 size prints, because standard file cabinets are designed to hold this size. There are file cabinets available called flat files that can be used to store full-size unfolded prints. However, many companies use standard file cabinets. Larger prints must be properly folded before they can be filed in a standard file cabinet. It is also important to properly fold a print if it is to be mailed. Folding large prints is much like folding a road map. Folding is done in a pattern of bends that results in the title block and sheet identification ending up on the front. This is desirable for easy identification in the file cabinet. The proper method used to fold prints also aids in unfolding or refolding prints. Look at Figure 1–13 to see how large prints are properly folded.

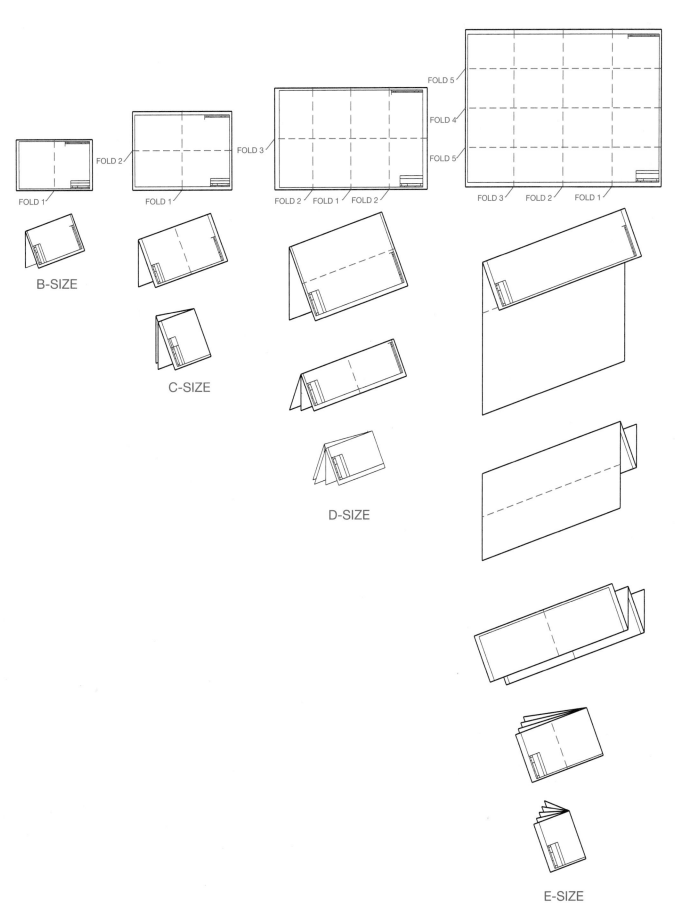

Figure 1–13 How to properly fold B-size, C-size, D-size, and E-size prints.

# CHAPTER 1 TEST

Multiple choice: Respond to the following by selecting a, b, c, or d to best answer the question or complete the statement.

1. This is an old term that is generally used in the manufacturing industry when referring to print.
   a. blueprint
   b. diazo
   c. photocopy
   d. microfilm

2. Blue-line prints are also known as:
   a. blueprint
   b. diazo
   c. photocopy
   d. microfilm

3. This is a full-size print process where there is no need for coated copy materials or the possible hazards of ammonia.
   a. blueprint
   b. diazo
   c. photocopy
   d. microfilm

4. This print process copies drawings or other documents from any type of original and onto most types of copy material.
   a. blueprint
   b. diazo
   c. photocopy
   d. microfilm

5. Some machines that use this print process allow the user to reduce or enlarge drawings to suit specific needs.
   a. blueprint
   b. diazo
   c. photocopy
   d. microfilm

6. This process produces a reduced reproduction of a drawing or other document for ease of storage and for convenience in sending from one place to another.
   a. blueprint
   b. diazo
   c. photocopy
   d. microfilm

7. The microfilm used in mechanical engineering is generally prepared as one frame or drawing attached to this type of card.
   a. photocopy
   b. aperture
   c. roll
   d. microfilm

8. The letters CADD refer to:
   a. computer-aided drawing and design
   b. computer-assisted design and drawing
   c. computer-aided design and drafting
   d. common applications for drawing and distribution

9. A standard A-size print is:
   a. 22" x 34"
   b. 17" x 22"
   c. 11" x 17"
   d. 8.5" x 11"

10. A standard B-size print is:
    a. 22" x 34"
    b. 17" x 22"
    c. 11" x 17"
    d. 8.5" x 11"

11. A standard C-size print is:
    a. 22" x 34"
    b. 17" x 22"
    c. 11" x 17"
    d. 8.5" x 11"

12. A standard D-size print is:
    a. 22" x 34"
    b. 17" x 22"
    c. 11" x 17"
    d. 8.5" x 11"

13. A given amount of acceptable variation in a size or location dimension is:
    a. units
    b. basic
    c. tolerance
    d. limits

14. Manufacturing engineering drawings are generally drawn in these units.
    a. feet and inches
    b. inches or millimeters
    c. scientific units
    d. engineering units

15. This term means that all of the features or dimensions on the drawing have the relationship or specifications given in the title block, unless a specific note or dimensional tolerance is provided in a particular location in the drawing.
    a. bilateral tolerance
    b. unspecified tolerance
    c. specified tolerance
    d. unless otherwise specified

16. This term refers to any dimension on the drawing that does not have a tolerance specified, and the tolerance is given in the title block or in a general note.
    a. bilateral tolerance
    b. unspecified tolerance
    c. specified tolerance
    d. unless otherwise specified

17. This term designates a dimension on a drawing that has its own tolerance which is different from the standard tolerance given in the title block.
    a. bilateral tolerance
    b. unspecified tolerance
    c. specified tolerance
    d. unless otherwise specified

18. A system of numbers along the top and bottom margins and letters along the left and right margins that allow a drawing to be read like a map is called:
    a. zoning
    b. referencing
    c. mapping
    d. margin reference

19. The area inside the border lines and outside the title block of a drawing is referred to as:
    a. drawing area
    b. zone
    c. revision area
    d. field

20. Prints should be properly folded for:
    a. easy identification in the file cabinet
    b. placing in a file cabinet
    c. mailing
    d. all of the above

21. Drawings that are created using CADD can be reproduced on:
    a. a printer
    b. a plotter
    c. both A and B
    d. neither A nor B

22. The quality of a CADD drawing is greatly influenced by the reproduction method and the type of output device.
    a. true
    b. false

23. This is a CADD output device that uses liquid ink pens to plot a drawing.
    a. laser printer
    b. electrostatic plotter
    c. pen plotter
    d. impact printer

24. This high-speed output device achieves results of excellent quality by attaching permanent ink to electrically charged dots.
    a. laser printer
    b. electrostatic plotter
    c. pen plotter
    d. dot-matrix printer

25. This output device sprays droplets of ink onto the paper to form dot-matrix images.
    a. laser printer
    b. electrostatic plotter
    c. inkjet printer
    d. dot-matrix printer

## CHAPTER 1 PROBLEMS

**PROBLEM 1–1** Given the title block and revision block shown on this page, describe the elements located at each number in the space provided.

1. _____

2. _____

3. _____

4. _____

5. _____

6. _____

7. _____

8. _____

9. _____

10. _____

11. _____

12. _____

13. _____

14. _____

15. _____

16. _____

B

| 13 | | 14 | | 15 | 16 |
|---|---|---|---|---|---|
| REV. | | DESCRIPTION | | REVISED BY | DATE |

**REVISIONS**

THIS PRINT AND ITS CONTENTS ARE THE PROPERTY OF INCOM INTERNATIONAL INC., AND THEY ARE LOANED TO YOU SUBJECT TO RETURN UPON INCOM INTERNATIONAL INC.'S DEMAND, AND ARE TO BE MAINTAINED IN CONFIDENCE BY YOU AND ARE NOT TO BE USED IN ANY WAY DIRECTLY OR INDIRECTLY DETRIMENTAL TO INCOM INTERNATIONAL INC.'S INTERESTS.

UNLESS OTHERWISE SPECIFIED — ANGLES ± 1/2°. 2 PLACE DEC. ± .03. 3 PLACE DEC. ± .016 COMMERCIAL TOLERANCES SHALL APPLY TO SIZES OF BAR, ROD, WIRE, SHEET, TUBE, ETC. PLATED PARTS MUST FIT GAUGES AND MEET SPECIFIED TOLERANCES AFTER PLATING.

7

DRAWN BY    DATE
1

CHECKED BY    DATE
2

APPROVED BY    DATE
3

MODEL NO.
4

NEXT ASSY.

**FINCOR**

INCOM INTERNATIONAL INC

**INCOM**

3750 East Market Street
York, Pennsylvania 17402
(717) 757-4641

PART NAME:
8

10

MATERIAL
6

SCALE
5

SHT. 9 OF

DWG. SIZE
C

PART NUMBER:
11

REV.
12

A

2

1

Problem 1–1 Courtesy FINCOR Electronics Div.

**PROBLEM 1–2** Given the title block and revision block shown on this page, answer the following questions using short, complete statements.

1. What is the company name?

   _____

2. Who was the drafter?

   _____

3. What is the scale of this drawing?

   _____

4. What is the date of this drawing?

   _____

5. Give the part name.

   _____

6. What is the part number?

   _____

7. What is the model number?

   _____

8. Give general specifications associated with this drawing.

   _____

9. Identify at least two ways you can tell that the dimensions in this drawing are in millimeters.

   _____

10. What is the tolerance for unspecified one-place decimals?

    _____

11. What is the tolerance for unspecified two-place decimals?

    _____

12. What is the tolerance for unspecified angular dimensions?

    _____

13. What is the tolerance for unspecified whole dimensions?

    _____

14. Give the current drawing revision.

    _____

15. What is the ECN number of the current revision?

    _____

16. Describe the last change.

    _____

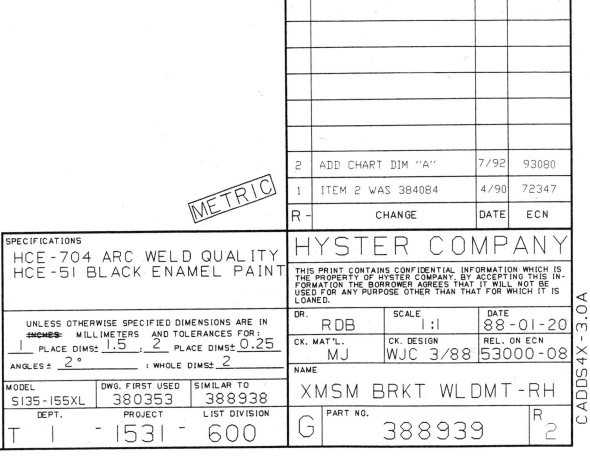

| 2 | ADD CHART DIM "A" | 7/92 | 93080 |
| 1 | ITEM 2 WAS 384084 | 4/90 | 72347 |
| R - | CHANGE | DATE | ECN |

METRIC

SPECIFICATIONS
HCE-704 ARC WELD QUALITY
HCE-51 BLACK ENAMEL PAINT

HYSTER COMPANY

THIS PRINT CONTAINS CONFIDENTIAL INFORMATION WHICH IS THE PROPERTY OF HYSTER COMPANY. BY ACCEPTING THIS INFORMATION THE BORROWER AGREES THAT IT WILL NOT BE USED FOR ANY PURPOSE OTHER THAN THAT FOR WHICH IT IS LOANED.

UNLESS OTHERWISE SPECIFIED DIMENSIONS ARE IN ~~INCHES~~ MILLIMETERS AND TOLERANCES FOR:
1 PLACE DIMS± 1.5 ; 2 PLACE DIMS± 0.25
ANGLES ± 2° ; WHOLE DIMS± 2

| DR. RDB | SCALE 1:1 | DATE 88-01-20 |
| CK. MAT'L. MJ | CK. DESIGN WJC 3/88 | REL. ON ECN 53000-08 |

NAME
XMSM BRKT WLDMT-RH

| MODEL S135-155XL | DWG. FIRST USED 380353 | SIMILAR TO 388938 |

| DEPT. | PROJECT | LIST DIVISION |
| T 1 - 1531 - 600 |

G   PART NO. 388939   R 2

CADDS4X-3.0A

Problem 1–2 Courtesy Hyster Company.

# Communicating with a Sketch

## LEARNING OBJECTIVES

After completing this chapter you will be able to:

- Sketch lines, circles, and arcs.
- Sketch objects using correct proportions.
- Sketch irregular shapes.
- Make basic multiview sketches.
- Prepare isometric sketches of objects.

## SKETCHING

*Sketching* is freehand drawing, that is, drawing without the aid of drafting equipment. Sketching is convenient because all that is needed is paper, pencil, and an eraser. There are a number of advantages and uses for freehand sketching. Sketching is fast visual communication. The ability to create an accurate sketch quickly can often be an asset when communicating with people at work or at home. Especially when technical concepts are the topic of discussion, a sketch can be the best form of communication. A sketch can be a useful form of illustration in technical reports. Sketching is also used in job shops where one-of-a kind products are made. In the job shop, the sketch is often used as a formal production drawing.

The quality of a sketch depends on the intended purpose. Normally a sketch does not have to be of very good quality, as long as it adequately represents what you want to display. *Speed is a key to sketching.* You normally want to prepare the sketch as fast as possible while making it as clear and easy to read as possible. Normally a sketch does not need to have the quality of a formal presentation. The degree of quality can vary depending on the intent of the sketch. A sketch can be used as an artistic impression of a product, or as a one-time detail drawing for manufacturing purposes. However, the sketch is normally used in preliminary planning or to relate an idea to someone very quickly. The quality of your classroom sketches depends on your course objectives. Your instructor may want quality sketches or very quick sketches. You should confirm this in advance. In the professional world, your own judgment determines the nature and desired quality of the sketch.

## TOOLS AND MATERIALS

Sketching equipment is not very elaborate. As mentioned, all you need is paper, pencil, and an eraser. The pencil should have a soft lead; a common number 2 pencil works fine, or an automatic 0.7 mm or 0.9 mm pencil with H, F, or HB lead is also good. The pencil lead should not be sharp. A dull, slightly rounded pencil point is best. Different thicknesses of line, if needed, can be drawn by changing the amount of pressure you apply to the pencil or by the sharpness of the point. A sharper point is good for thin lines, while a duller point is good for thicker lines. The quality of the paper is not critical either. A good sketching paper is newsprint, although almost any kind works. Actually, paper with a surface that is not too smooth is best. Many engineering designs have been created on a napkin at a lunch table. Sketching paper should not be taped to the table. The best sketches are made when you are able to move the paper to the most comfortable drawing position. Some people make horizontal lines better than vertical lines. If this is true for you, move the paper so vertical lines become horizontal. Such movement of the paper may not always be possible, so it does not hurt to keep practicing all forms of lines for best results. Graph paper is also good to use for sketching because it has grid lines that you can use as a guide for your sketch lines.

## SKETCHING STRAIGHT LINES

Lines should be sketched in short, light, connected segments, as shown in Figure 2–1. If you sketch one long stroke in one continuous movement, your arm tends to make the line curved rather than straight, as shown in Figure 2–2. Also, when you make a dark line, you may have to erase if you make an error, but if you draw a light line, often there is no need to erase an error.

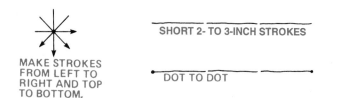

Figure 2–1 Sketching short line segments.

Figure 2–2 Long movements tend to cause a line to curve.

Use the following procedure to sketch a horizontal straight line with the dot-to-dot method:

**STEP 1** Mark the starting and ending positions, as in Figure 2–3. The letters A and B are for instruction only. All you need are the points.

**STEP 2** Without actually touching the paper with the pencil point, make a few trial motions between the marked points to adjust the eye and hand to the anticipated line.

**STEP 3** Sketch very light lines between the points by moving in short light strokes (2 to 3 inches long). Keep one eye directed toward the end point while keeping the other eye directed on the pencil point. With each stroke, try to correct the most obvious defects of the preceding stroke so the finished light lines are relatively straight (see Figure 2–4).

**STEP 4** Darken the finished line with a dark, distinct, uniform line directly on top of the light line. Press the pencil harder to achieve the right darkness (see Figure 2–5).

Figure 2–3 Step 1: use dots to identify both ends of a line.

Figure 2–4 Step 3: use short light strokes.

Figure 2–5 Step 4: darken to finish the line.

You can sketch very long, straight lines by using the edge of the paper or the edge of a table as a guide. To do this, position the paper in a comfortable position with your hand placed along the edge, as shown in Figure 2–6(a). Extend the pencil point out to the location of the line. Next, place one of your fingers or the palm of your hand along the edge of the paper as a guide. Now, move your hand and the pencil continuously along the edge of the paper as shown in Figure 2–6(b). A problem with this method is that it works best if the line is fairly close to the edge of the paper. A sketch does not have to be perfect, but a little practice should help.

## SKETCHING CIRCULAR LINES

Figure 2–7 shows the parts of a circle. There are several sketching techniques to use when making a circle. This text explains the quick freehand method for small circles; and it explains the box method, the centerline method, the hand-compass method, and the trammel method for very large circles

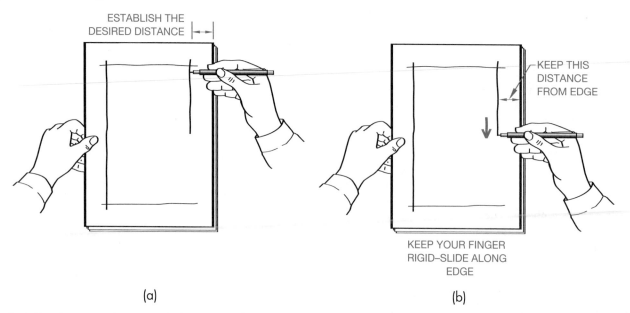

(a)                                              (b)

Figure 2–6 Sketching very long straight lines using the edge of the sheet as a guide. (a) Place your hand along the edge as a guide. (b) Move your hand and the pencil along the edge of the paper, using your finger or palm as a guide to keep the pencil a constant distance from the edge.

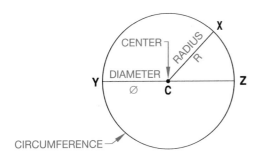

Figure 2–7 The parts of a circle.

## Sketching Quick Small Circles

Small circles are easy to sketch if you treat them just like drawing the letter O. You should be able to do this in two strokes by sketching a half circle on each side as shown in Figure 2–8.

STEP 1          STEP 2

Figure 2–8 Sketching a small circle just like drawing the letter "O."

## Using the Box Method

It is always faster to sketch a circle without first creating other construction guides, but it can be difficult to do so. The box method can help you by providing a square that contains the desired circle. Start this method by very lightly sketching a square box that is equal in size to the diameter of the proposed circle, as shown in Figure 2–9. The very light lines are called *construction lines*. Next, sketch diagonals across the square. This establishes the center and allows you to mark the radius of the circle on the diagonals, as shown in Figure 2–10. Use the sides of the square and the marks on the diagonals as a guide to sketch the circle. Create the circle by drawing arcs that are tangent to the sides of the square and go through the marks on the diagonals, as shown in Figure 2–11. If you have trouble sketching the circle as dark and thick as desired, sketch it very lightly first, and then go back over it to make it dark. You can easily correct very lightly sketched lines, but it is difficult to correct very dark lines. You won't need to erase your construction lines if they are sketched very lightly.

## Using the Centerline Method

The centerline method is similar to the box method, but without a box. This method uses very lightly sketched horizontal, vertical, and two 45° diagonal centerlines, as shown in Figure 2–12. Next, mark the approximate radius of the circle on the centerlines, as shown in Figure 2–13. Create the circle by drawing arcs that go through the marks on the centerline, as shown in Figure 2–14.

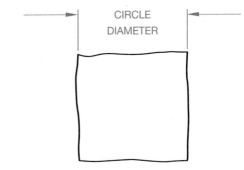

Figure 2–9 Very lightly sketch a square box that is equal in size to the diameter of the proposed circle.

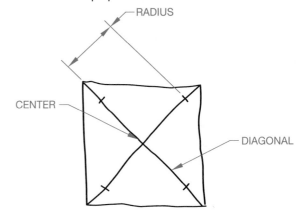

Figure 2–10 Sketch light diagonal lines across the square, and mark the radius on the diagonals.

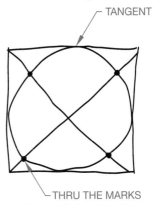

Figure 2–11 Create the circle by sketching arcs that are tangent to the sides of the square and go through the marks on the diagonals.

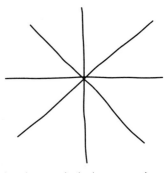

Figure 2–12 Sketch very light horizontal, vertical, and 45° lines that meet at the center of the proposed circle.

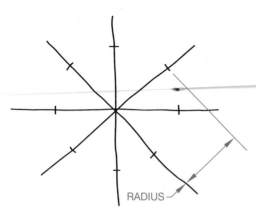

Figure 2–13 Mark the approximate radius of the circle on the centerlines created in Figure 2–12.

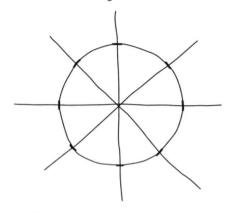

Figure 2–14 Create the circle by sketching arcs that go through the marks on the centerlines.

## Using the Hand-Compass Method

The hand-compass method is a quick and reasonably accurate method of sketching circles, although it is a method that takes some practice.

**STEP 1** Be sure that your paper is free to rotate completely around 360°. Remove anything from the table that might stop this rotation.

**STEP 2** To use your hand and a pencil as a compass, place the pencil in your hand between your thumb and the upper part of your index finger, so your index finger becomes the *compass point* and the pencil becomes the *compass lead*. The other end of the pencil rests in your palm, as shown in Figure 2–15.

**STEP 3** Determine the circle radius by adjusting the distance between your index finger and the pencil point. Now, with the desired approximate radius established, place your index finger on the paper at the proposed center of the circle.

**STEP 4** With the desired radius established, keep your hand and pencil point in one place while rotating the paper with your other hand. Try to keep the radius steady as you rotate the paper (see Figure 2–16).

**STEP 5** You can perform step 4 very lightly and then go back and darken the circle; or, if you have had a lot of practice, you may be able to draw a dark circle as you go.

Figure 2–15 Step 2: holding the pencil in the hand compass.

Figure 2–16 Step 4: rotate the paper under your finger center point.

## Trammel Method

Avoid the trammel method if you are creating a quick sketch, because it takes extra time and materials to set up this technique. Also, the trammel method is intended for large to very large circles that are difficult to sketch when using the other methods. The following examples demonstrate the trammel method to create a small circle. Use the same principles to draw a large circle.

**STEP 1** Make a trammel to sketch a 6-inch diameter circle. Cut or tear a strip of paper approximately 1 inch wide and longer than the radius, 3 inches. On the strip of paper, mark an approximate 3-inch radius with tick marks, such as A and B in Figure 2–17.

**STEP 2** Sketch a straight line representing the circle radius at the place where the circle is to be located. On the sketched line, locate with a dot the center of the circle to be sketched. Use the marks on the trammel to mark the other end of the radius line, as shown in Figure 2–18. With the trammel next to the sketched line, be sure point B on the trammel is aligned with the center of the circle you are about to sketch.

**STEP 3** Pivot the trammel at point B, making tick marks at point A as you go, as shown in Figure 2–19, until the circle is complete.

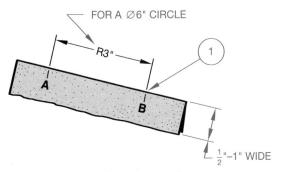

Figure 2–17 Step 1: make a trammel.

**STEP 4** Lightly sketch the circumference over the tick marks to complete the circle, and then darken it (Step 5), as shown in Figure 2–20.

Another, similar, trammel method, generally used to sketch very large circles, is to tie a string between a pencil and a pin. The distance between the pencil and pin is the radius of the circle. Use this method when a large circle is to be sketched, because the other methods may not work as well. Workers at a construction site sometimes use this method by tying the string to a nail and driving the nail at the center location.

## SKETCHING ARCS

Sketching arcs is similar to sketching circles. An arc is part of a circle, as you can see in Figure 2–21. An arc is commonly used as a rounded corner or at the end of a slot. When an arc is a rounded corner, the ends of the arc are typically tangent to adjacent lines. In

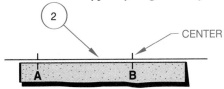

Figure 2–18 Step 2: locate the center of the circle.

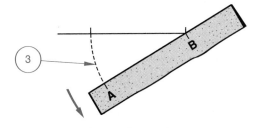

Figure 2–19 Step 3: begin the circle construction.

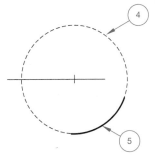

Figure 2–20 Steps 4 and 5: Sketch and darken the circle.

short, *tangent* means that the arc touches the line at only one point and does not cross over the line, as shown in Figure 2–21. An arc is generally drawn with a radius. The most comfortable way to sketch an arc is to move the paper so your hand faces the inside of the arc. Having the paper free to move helps with this practice.

One way to sketch an arc is to create a box at the corner. The box establishes the arc center and radius, as shown in Figure 2–22. You can also sketch a 45° construction line from the center to the outside corner of the box, and mark the radius on the 45° line (see Figure 2–22). Now, sketch the arc by using the tangent points and the mark as a guide, as shown in Figure 2–23. You should generally connect the straight lines to the arc after the arc is created, because it is usually easier to sketch straight lines than it is to sketch arcs.

The same technique can be used to sketch any arc. For example, a full-radius arc is sketched in Figure 2–24. This arc is a half-circle, so using half of the box method or centerline method works well.

## SKETCHING ELLIPSES

If you look directly at a coin, it represents a circle. As you rotate a coin, it takes the shape of an ellipse. Figure 2–25 shows the relationship between a circle and an ellipse, and shows the parts of an ellipse.

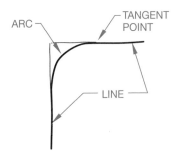

Figure 2–21 An arc is part of a circle. This arc is used to create a rounded corner. Notice that the arc creates a smooth connection at the point of tangency with the straight lines.

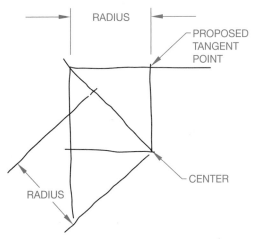

Figure 2–22 The box establishes the center and radius of the arc. The 45° diagonal helps establish the radius.

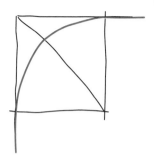

Figure 2–23 Sketch the arc using the tangent points and mark on the diagonal as a guide for the radius.

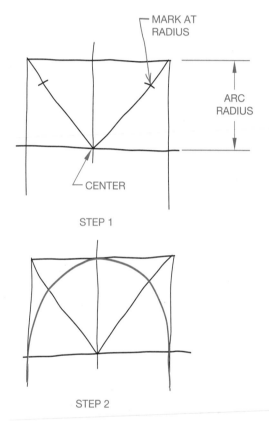

STEP 1

STEP 2

Figure 2–24 Sketching a full-radius arc uses the same method as sketching any arc or circle.

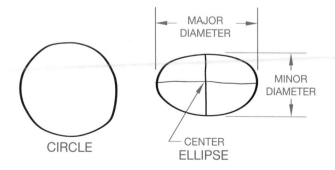

CIRCLE

ELLIPSE

Figure 2–25 The relationship between an ellipse and a circle.

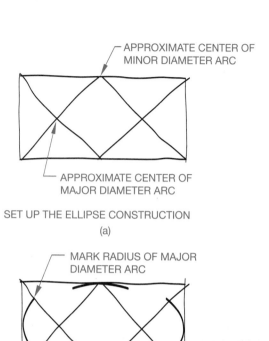

APPROXIMATE CENTER OF MINOR DIAMETER ARC

APPROXIMATE CENTER OF MAJOR DIAMETER ARC

SET UP THE ELLIPSE CONSTRUCTION

(a)

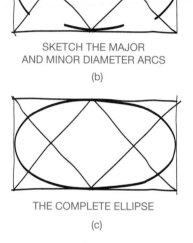

MARK RADIUS OF MAJOR DIAMETER ARC

SKETCH THE MAJOR AND MINOR DIAMETER ARCS

(b)

THE COMPLETE ELLIPSE

(c)

Figure 2–26 Sketching an ellipse. (a) Sketch a light rectangle equal in length and width to the major and minor diameters of the desired ellipse. Sketch crossing lines from the corners of the minor diameter to the midpoint of the major diameter sides. (b) Use the point where the lines cross as the center to sketch the major diameter arcs. Use the midpoint of the minor diameter sides as the center to sketch the minor diameter arcs. (c) Blend in connecting arcs to fill the gaps.

If you can accurately sketch an ellipse without a construction line, do it. If you need help, an ellipse can also be sketched using a box method. To start this technique, sketch a light rectangle equal in length and width to the major and minor diameters of the desired ellipse, as shown in Figure 2–26(a). Next, sketch crossing lines from the corners of the minor diameter to the midpoint of the major diameter sides, as in Figure 2–26(a). Now, using the point where the lines cross as the center, sketch the major diameter arcs (see Figure 2–26(b)). Use the midpoint of the minor diameter sides as the center to sketch the minor diameter arcs, as shown in Figure 2–26(b). Finally, blend in connecting arcs to fill the gaps, as shown in Figure 2–26(c).

## MEASUREMENT LINES AND PROPORTIONS

When sketching objects, all the lines that make up the object are related to each other by size and direction. For a sketch to communicate accurately and completely, it must be drawn in the same proportion as the object. The actual size of the sketch depends on the paper size and how large you want the sketch to be. The sketch should be large enough to be clear, but the proportions of the features are more important than the size of the sketch.

Look at the lines in Figure 2–27. How long is line 1? How long is line 2? Answer these questions without measuring either line, but instead relate each line to the other. For example, line 1 could be stated as being half as long as line 2, or line 2 called twice as long as line 1. Now you know how long each line is in relationship to the other (proportion), but you do not know how long either line is in relationship to a measured scale. No scale is used for sketching, so this is not a concern. Whatever line you decide to sketch first determines the scale of the drawing. This first line sketched is called the *measurement line*. Relate all the other lines in the sketch to that first line. This is one of the secrets of making a sketch look like the object being sketched.

LINE 1 ————————

LINE 2 ————————————

Figure 2–27 Measurement lines.

The second thing you must know about the relationship of the two lines in the above example is the direction and position held relative to each other. For example, do they touch each other, are they parallel, perpendicular, or at some angle to each other? When you look at a line, ask yourself the following questions (for this example, use the two lines given in Figure 2–28).

Figure 2–28 Measurement lines.

1.  How long is the second line?

    ■ Same length as the first line?

    ■ Shorter than the first line? How much shorter?

    ■ Longer than the first line? How much longer?

2.  In what direction and position is the second line related to the first line?

Typical answers to these questions for the lines in Figure 2–28 would be as follows:

1.  The second line is about three times as long as the first line.

2.  Line two touches the lower end of the first line with about a 90° angle between them.

Carrying this concept a step farther, a third line can relate to the first line or the second line and so forth. Again, the first line drawn (measurement line) sets the scale for the whole sketch.

This idea of relationship can also apply to spaces. In Figure 2–29, the location of the square can be determined by space proportions. A typical verbal location for the square in this block might be as follows: the square is located about one-half square width from the top of the object or about two square widths from the bottom, and about one square width from the right side or about three square widths from the left side of the object. All the parts must be related to the whole object.

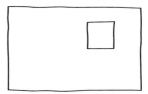

Figure 2–29 Space proportions.

### Introduction to the Block Technique

Any illustration of an object can be surrounded with some sort of an overall rectangle, as shown in Figure 2–30. Before starting a sketch, visualize the object to be sketched inside a rectangle in your mind. Then use the measurement-line technique with the rectangle, or block, to help you determine the shape and proportion of your sketch.

## PROCEDURES IN SKETCHING

**STEP 1** When starting to sketch an object, visualize the object surrounded with an overall rectangle. Sketch this rectangle first with very light lines. Sketch the proper proportion with the measurement-line technique, as shown in Figure 2–31.

**STEP 2** Cut out sections, using proper proportions as measured by eye, using light lines, as in Figure 2–32.

**STEP 3** Finish the sketch by darkening in the desired outlines for the finished sketch (see Figure 2–33).

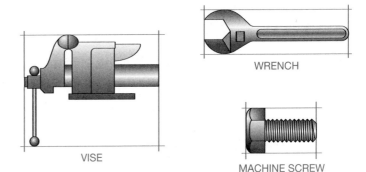

Figure 2–30 Block technique.

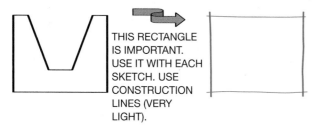

THIS RECTANGLE IS IMPORTANT. USE IT WITH EACH SKETCH. USE CONSTRUCTION LINES (VERY LIGHT).

Figure 2–31 Step 1: Outline the drawing area with a block.

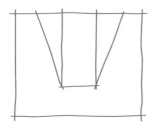

Figure 2—32 Step 2: Draw features to proper proportions.

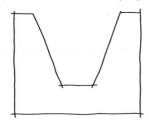

Figure 2–33 Step 3, darken the object lines.

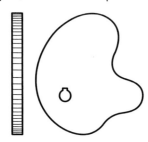

Figure 2–34 A cam.

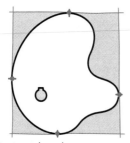

Figure 2–35 Step 1: Box the object.

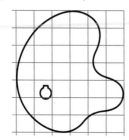

Figure 2–36 Step 2: Evenly spaced grid.

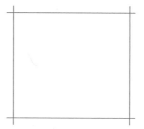

Figure 2–37 Step 3, proportioned box.

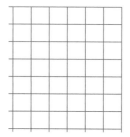

Figure 2–38 Step 4: Regular grid.

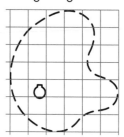

Figure 2–39 Step 5: Sketched shape using the regular grid.

## Sketching Irregular Shapes

By using a frame of reference or an extension of the block method, irregular shapes can be sketched easily to their correct proportions. Follow these steps to sketch the cam shown in Figure 2–34.

**STEP 1** Place the object in a lightly constructed box (see Figure 2–35).

**STEP 2** Draw several equally spaced horizontal and vertical lines, as shown in Figure 2–36. If you are sketching an object already drawn, just draw your reference lines on top of the object's lines to establish a frame of reference. If you are sketching an object directly, you have to visualize these reference lines on the object you sketch.

**STEP 3** On your sketch, correctly locate a proportioned box similar to the one established on the original drawing or object, as shown in Figure 2–37.

**STEP 4** Using the drawn box as a frame of reference, include the grid lines in correct proportion, as seen in Figure 2–38.

**STEP 5** Then, using the grid, sketch the small irregular arcs and lines that match the lines of the original, as in Figure 2–39.

**STEP 6** Darken the outline for a complete proportioned sketch, as shown in Figure 2–40.

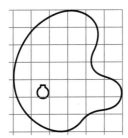

Figure 2–40 Step 6: Completely darken the outline of the object.

## CREATING MULTIVIEW SKETCHES

*Multiview projection* is also known as *orthographic projection.* Chapter 5 provides complete coverage of reading multiview drawings. Multiviews are two-dimensional views of an object that are established by a line of sight that is perpendicular (90°) to the surface of the object. When making multiview sketches, a systematic order should be followed. Most drawings are in the multiview form. Learning to sketch multiview drawings will save you time when making a formal drawing. The *pictorial view* shows the object in a 3D picture, while the multiview shows the object in a 2D representation. Figure 2–41 shows an object in 3D and 2D.

### Multiview Alignment

To keep your drawing in a common form, sketch the front view in the lower-left portion of the paper, the top view directly above the front view, and the right-side view to the right side of the front view. (See Figure 2–41.) The views needed may differ depending on the object. Your ability to visualize between 3D objects and 2D views is very important in understanding how to lay out a multiview sketch and when reading multiview drawings. Multiview arrangement is explained in detail in Chapter 5.

### Multiview Sketching Technique

Steps in sketching:

**STEP 1** Sketch and align the proportional rectangles for the front, top, and right-side views of the pictorial object given in Figure 2–41. Sketch a 5° line to help transfer width dimensions. The 45° line is established by projecting the width from the top view across and the width from the right-side view up until the lines intersect, as shown in Figure 2–42.

**STEP 2** Complete the shapes by cutting out the rectangles, as shown in Figure 2–43.

**STEP 3** Darken the lines of the object as in Figure 2–44. Remember to keep the views aligned for ease of sketching and readability.

**STEP 4** In the views where some of the features are hidden, show these features with hidden lines, which are dashed lines, as shown in Figure 2–45. Start the practice of sketching object lines thick and hidden lines thin.

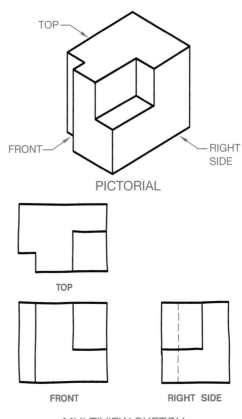

Figure 2–41 Views of objects shown in pictorial view and multiview.

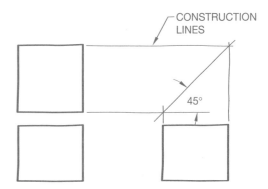

Figure 2–42 Step 1: block out views.

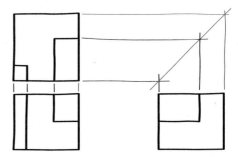

Figure 2–43 Step 2: Block out shapes.

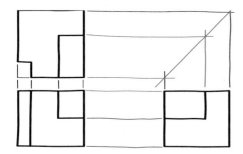

Figure 2–44 Step 3: Darken all object lines.

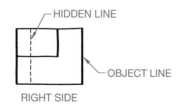

Figure 2–45 Step 4: Draw hidden features.

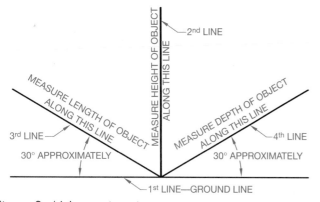

Figure 2–46 Isometric axis.

## Creating Isometric Sketches

*Isometric* sketches provide a three-dimensional pictorial representation of an object. Isometric sketches are easy to create and make a very realistic exhibit of the object. The surface features or the axes of the objects are drawn at equal angles from horizontal. Isometric sketches tend to represent the objects as they appear to the eye. Isometric sketches help in the visualization of an object, because three sides of the object are sketched in a single three-dimensional view.

## Establishing Isometric Axes

In setting up an isometric axis, you need four beginning lines: a horizontal reference line, two 30° angular lines, and one vertical line. Draw them as very light construction lines (see Figure 2–46).

**STEP 1** Sketch a horizontal reference line. (Consider this the ground-level line.)

**STEP 2** Sketch a vertical line perpendicular to the ground-level line and somewhere near its center. The vertical line is used to measure height.

**STEP 3** Sketch two 30° angular lines, each starting at the intersection of the first two lines, as shown at Figure 2–47.

## Making an Isometric Sketch

To make an isometric sketch, follow these steps:

**STEP 1** Select an appropriate view of the object.

**STEP 2** Determine the best position in which to show the object.

**STEP 3** Begin your sketch by setting up the isometric axes (see Figure 2–47).

**STEP 4** By using the measurement-line technique, draw a rectangular box, using correct proportion, which could surround the object to be drawn. Use the object shown in Figure 2–48 for this example. Imagine the rectangular box in your mind. Begin to sketch the box by marking off the width at any convenient length, as in Figure 2–49. This is your measurement line. Next, estimate and mark the length and height as related to the measurement line (see Figure 2–50). Sketch the three-dimensional box by using lines parallel to the original axis lines (see Figure 2–51). Sketching the

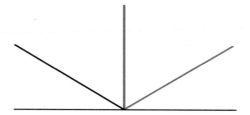

Figure 2–47 Step 3: Sketch the isometric axis.

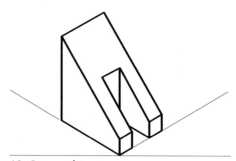

Figure 2–48 Given object.

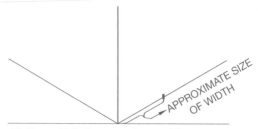

Figure 2–49 Step 4(a): Lay out the width.

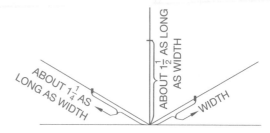

Figure 2–50 Step 4(b): Lay out the length and height.

box is the most critical part of the construction. It must be done correctly; otherwise, your sketch will be out of proportion. All lines drawn in the same direction must be parallel.

**STEP 5** Lightly sketch the slots, holes, insets, and other features that define the details of the object. By estimating distances on the rectangular box, the features of the object are easier to sketch in correct proportion than trying to draw them without the box (see Figure 2–52).

**STEP 6** Finish the sketch, darken all the object lines (outlines), as in Figure 2–53. For clarity, do not show any hidden lines.

## Nonisometric Lines

*Isometric lines* are lines that are on or parallel to the three original isometric axes lines. All other lines are nonisometric lines. Isometric lines can be measured in true length.

*Nonisometric lines* appear either longer or shorter than they actually are (see Figure 2–54). You can measure and draw nonisometric lines by connecting their end points. You can find the end points of the nonisometric lines by measuring along isometric lines. To locate where nonisometyric lines should be placed, you have to relate to an isometric line. Follow these

steps, using the object in Figure 2–55 as an example.

**STEP 1** Develop a proportional box, as in Figure 2–56.

**STEP 2** Sketch in all isometric lines, as shown in Figure 2–57.

**STEP 3** Locate the starting and end points for the nonisometric lines (see Figure 2–58).

**STEP 4** Sketch the nonisometric lines, as shown in Figure 2–59, by connecting the points established in Step 3. Also darken all outlines.

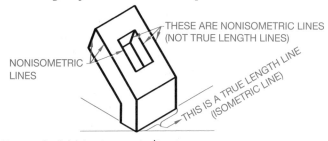

Figure 2–54 Nonisometric lines.

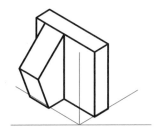

Figure 2–55 Guide.

Figure 2–56 Step 1: Sketch the box.

Figure 2–57 Step 2: Sketch isometric lines.

Figure 2–58 Step 3: Locate nonisometric line end points.

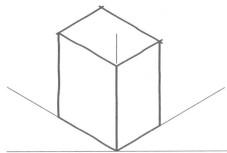

Figure 2–51 Step 4(c): Sketch the 3D box.

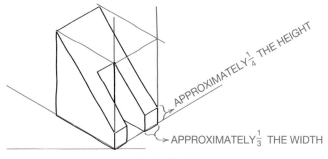

Figure 2–52 Step 5: Sketch the features.

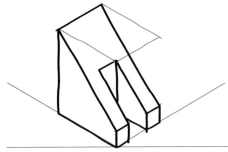

Figure 2–53 Step 6: Darken the outline.

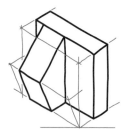

Figure 2–59 Step 4: Complete the sketch and darken all outlines.

## Sketching Isometric Circles

Circles and arcs appear as ellipses in isometric views. To sketch isometric circles and arcs correctly, you need to know the relationship between circles and faces, or planes, of an isometric cube.

Depending on which face the circle is to appear, isometric circles look like one of the ellipses shown in Figure 2–60. The angle the ellipse (isometric circle) slants is determined by the surface on which the circle is to be sketched.

To practice sketching isometric circles, you need isometric surfaces to put them on. The surfaces can be found by first sketching a cube in isometric. A *cube* is a box with six equal sides. Notice, as shown in Figure 2–61, that only three of the sides can be seen in an isometric drawing.

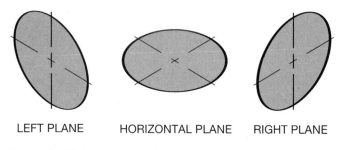

LEFT PLANE      HORIZONTAL PLANE      RIGHT PLANE

Figure 2–60 Isometric circles.

## Four-center Method

The four-center method of sketching an isometric ellipse is easier to perform, but care must be taken to form the ellipse arcs properly so the ellipse does not look distorted.

**STEP 1** Draw an isometric cube similar to Figure 2–61.

**STEP 2** On each surface of the cube, draw line segments that connect the 120° corners to the centers of the opposite sides (see Figure 2–62).

**STEP 3** With points 1 and 2 as centers, sketch arcs that begin and end at the centers of the opposite sides on each isometric surface (see Figure 2–63).

**STEP 4** On each isometric surface, with points 3 and 4 as the centers, complete the isometric ellipses by sketching arcs that meet the arcs sketched in Step 3 (see Figure 2–64).

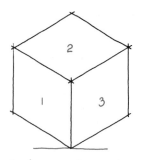

Figure 2–61 Step 1: draw an isometric cube.

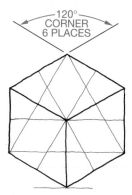

Figure 2–62 Step 2: four-center isometric ellipse construction.

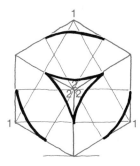

Figure 2–63 Step 3: sketch arcs from points 1 and 2 as centers.

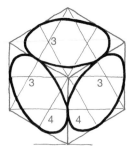

Figure 2–64 Step 4: sketch arcs from points 3 and 4 as centers.

## Sketching Isometric Arcs

Sketching isometric arcs is similar to sketching isometric circles. First, block out the overall configuration of the object, and then establish the centers of the arcs. Finally, sketch the arc shapes, as shown in Figure 2–65. Remember that isometric arcs, just like isometric circles, must lie in the proper plane and have the correct shape.

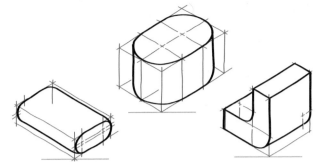

Figure 2–65 Sketching isometric arcs.

# CHAPTER 2 TEST

Multiple choice: Respond to the following by selecting a, b, c, or d to best answer the question or complete the statement.

1. Sketching is also known as:
   a. diazo
   b. freehand drawing
   c. art
   d. production drawing

2. Which of the following is an advantage of sketching?
   a. convenient
   b. fast visual communication
   c. good for communicating technical concepts
   d. all of the above

3. Use this pencil for sketching.
   a. 2
   b. HB
   c. 4H
   d. none of the above

4. The best sketches are made when:
   a. the paper is taped to the table
   b. the paper is placed vertically
   c. the paper is free to move
   d. the paper has a slick surface

5. The distance from the center to the circumference of a circle is called:
   a. radius
   b. diameter
   c. circumference
   d. tangent

6. The distance across a circle through the center is called:
   a. radius
   b. diameter
   c. circumference
   d. tangent

7. The distance all the way around the outside of a circle is known as:
   a. radius
   b. diameter
   c. circumference
   d. tangent

8. When sketching, all of the lines which make up the object are related to each other by size and direction. The first line of the object that you sketch sets the proportion for the rest of the sketch. This line is called the:
   a. reference line
   b. sketch line
   c. measurement line
   d. proportion line

9. The method used to sketch a circle using a strip of paper with the radius marked on the edge is called:
   a. compass method
   b. trammel method
   c. hand-compass method
   d. hand-trammel method

10. A method used to sketch circles that requires only your hand, a pencil, and a piece of paper on which to sketch is called:
    a. compass method
    b. trammel method
    c. hand-compass method
    d. hand-trammel method

11. When you sketch two-dimensional views of an object that are established by a line of sight which is perpendicular to the surface of the object, you are sketching the:
    a. isometric view
    b. pictorial view
    c. view alignment
    d. multiview

12. When these types of sketches provide a three-dimensional representation of an object where the surface features or the axes of the objects are drawn at equal angles (30° from horizontal), you are sketching the:
    a. isometric view
    b. pictorial view
    c. view alignment
    d. multiview

## CHAPTER 2 PROBLEMS

**PROBLEM 2–1** List the length, direction, and proportion of each line numbered in the drawing that follows. Remember to use the measurement-line method described in this chapter. Do not measure the lines with a scale. Example: Line 2 is the same length as line 1, and touches the top of line 1 at a 90° angle.

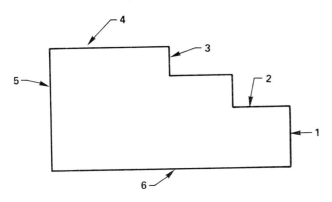

Problem 2–1 Diagram.

Line 1 _____

Line 2 _____

Line 3 _____

Line 4 _____

Line 5 _____

Line 6 _____

**PROBLEM 2–2** Sketch straight lines between the numbered points below.

Problem 2–2 Diagram.

**PROBLEM 2–3** Use the trammel method to sketch circles, given the center points and the radius of each circle that follows. For example, "C1" represents the center, and "R1" is a point establishing the radius for circle 1, "C2" and "R2" for circle 2.

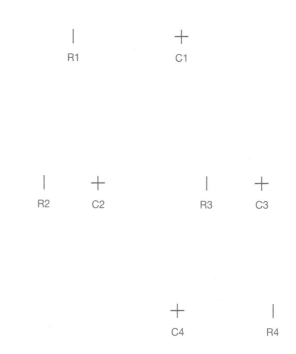

Problem 2–3 Diagram.

PROBLEM 2–4 Use the hand-compass method to sketch circles, given the center points and the radius of each circle following. For example, "C1" represents the center, and "R1" is a point establishing the radius for circle 1, "C2" and "R2" for circle 2.

Problem 2–4 Diagram.

PROBLEM 2–5 Given the following drawing of a cam, resketch it larger in the rectangular grid provided. Use the grid as a guide as discussed in this chapter.

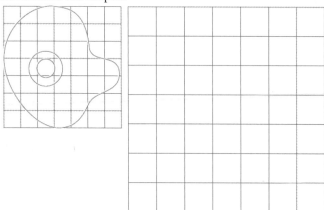

Problem 2–5 Diagram.

PROBLEM 2–6 Given the following drawing of a wrench, resketch it larger in the rectangular grid provided. Use the grid as a guide as discussed in this chapter.

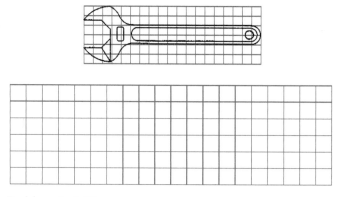

Problem 2–6 Diagram.

PROBLEM 2–7 On a separate piece of paper, make a sketch of the machine screw drawing that follows. Use a frame of reference to make your sketch twice as big as the given sketch.

MACHINE SCREW

Problem 2–7 Machine screw.

PROBLEM 2–8 On a separate piece of paper, make a sketch of the following vise drawing. Use a frame of reference to make your sketch twice as big as the given sketch.

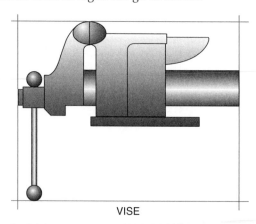

VISE

Problem 2–8 Vise.

**PROBLEM 2–9** Pocket block. Using the pictorial drawing as a guide, sketch the missing lines in the multiviews.

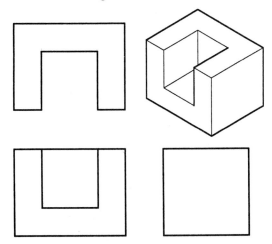

Problem 2–9 Pocket block.

**PROBLEM 2–10** Angle gage. Using the pictorial drawing as a guide, sketch the missing lines in the multiviews.

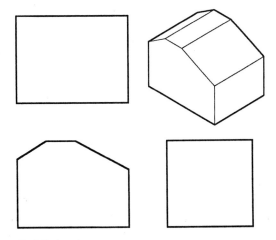

Problem 2–10 Angle gage.

**PROBLEM 2–11** Base. Using the pictorial drawing as a guide, sketch the missing lines in the multiviews.

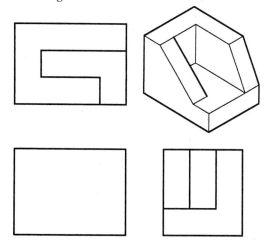

Problem 2–11 Base.

**PROBLEM 2–12** Corner Block. Using the pictorial drawing as a guide, sketch the missing lines in the multiviews.

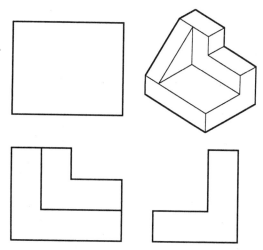

**PROBLEM 2–13** Find a stapler, tape dispenser, coffee cup, or something similar, and sketch a two-dimensional, frontal view using the block technique in the following space. Do not measure the object. Use the measurement-line method to approximate proper proportions.

**PROBLEM 2–14** Using the same object selected for Problem 2–13, sketch an isometric representation in the space following. Do not measure the object; use the measurement-line method to approximate dimensions.

**PROBLEM 2–15** Given the pictorial drawing of the following V-block, sketch a front and right-side view. Use the measurement-line method to approximate dimensions when preparing your sketch in the space below.

**PROBLEM 2–16** Given the pictorial drawing of the guide base that follows, sketch a front, top, and right-side view. Use the measurement-line method to approximate dimensions when preparing your sketch in the space below.

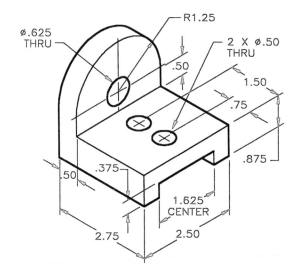

# Scales and Precision Measurement

## SCALES

If you are involved in creating or reading prints, it is a good idea for you to be familiar with the scales used to produce drawings and the tools used to measure manufactured parts. While a working knowledge of these scales is important, it is never appropriate to measure features on a print. A print must be read by finding the necessary information that is presented. Any attempt to measure features from the print can lead to inaccuracies and inappropriate information. Consult with the engineer or drafter if a dimension seems to be missing on a print.

### Scale Shapes

There are four basic scale shapes, as shown in Figure 3–1.

| TWO BEVEL | FOUR BEVEL | OPPOSITE BEVEL | TRIANGULAR |

Figure 3–1 Scale shapes.

### Scale Notation

The scale of a drawing is usually noted in the title block or below the view of an object that differs in scale to that given in the title block. Drawings are scaled so the object represented can be illustrated clearly on standard sizes of paper. It would be difficult, for example, to make a full-size drawing of a Boeing 747, so you must use a scale that reduces the size of such a large object. Machine parts are often drawn full size or even twice, four, or 10 times larger than full size, depending on the actual size of the part.

The following scales and their notation are frequently used on mechanical drawings.

Full scale = FULL or 1:1

Half scale = HALF or 1:2

Quarter scale = QUARTER or 1:4

Twice scale = DOUBLE or 2:1

Four times scale = 4:1

Ten times scale = 10:1

### Metric Scale

ANSI According to the American National Standards Institute, the commonly used SI (International System of Units) linear unit used on engineering drawings is the millimeter. On drawings where all dimensions are either in inches or millimeters, individual identification of units is not required. However, the drawing shall contain a note stating: UNLESS OTHERWISE SPECIFIED, ALL DIMENSIONS ARE IN INCHES (or MILLIMETERS, as applicable). Where some millimeters are shown on an inch-dimensioned drawing, the millimeter value should be followed by the symbol "mm." Where some inches are shown on a millimeter-dimensioned drawing, the inch value should be followed by the abbreviation IN.

Metric symbols are:

*millimeter* = mm

*centimeter* = cm

*decimeter* = dm

*meter* = m

*decameter* = dam

*hectometer* = hm

*kilometer* = km

Some metric-to-metric equivalents are:

10 millimeters = 1 centimeter

10 centimeters = 1 decimeter

10 decimeters = 1 meter

10 meters = 1 decameter

10 decameters = 1 kilometer

Some metric-to-U.S. customary equivalents are:

1 mile = 1.6093 kilometers = 1609.3 meters

1 yard = 914.4 millimeters = .9144 meters

1 foot = 304.8 millimeters = .3048 meters

1 inch = 25.4 millimeters = .0254 meters

To convert inches to millimeters, multiply inches by 25.4 mm.

Figure 3–2 shows the common scale calibrations found on the triangular metric scale.

## Civil Engineer's Scale

The triangular civil engineer's scale contains six scales, one on each of its sides. The civil engineer's scales are calibrated in multiples of 10. Civil engineering scales are commonly used in civil engineering for the design and drawing of maps and other related projects. Some of the scales can be conveniently used in mechanical drafting for manufacturing. The following table, Figure 3–3, shows some of the many scale options available with this scale.

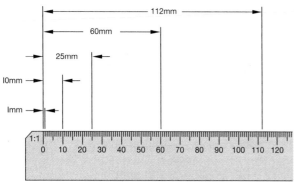

FULL SCALE = 1:1

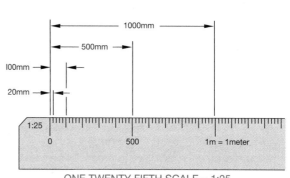

ONE TWENTY FIFTH SCALE = 1:25

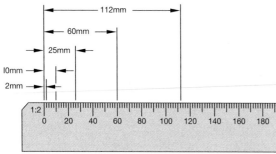

HALF SCALE = 1:2

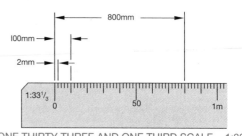

ONE THIRTY THREE AND ONE THIRD SCALE = 1:33 $\frac{1}{3}$

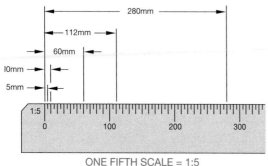

ONE FIFTH SCALE = 1:5

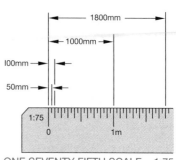

ONE SEVENTY FIFTH SCALE = 1:75

Figure 3–2 Metric scale calibrations.

| Civil Engineer's Scale | | | | | |
|---|---|---|---|---|---|
| Divisions | Ratio | Scales Used with This Division | | | |
| 10 | 1:1 | 1" = 1" | 1" = 1' | 1" = 10' | 1" = 100' |
| 20 | 1:2 | 1" = 2" | 1" = 2' | 1" = 20' | 1" = 200' |
| 30 | 1:3 | 1" = 3" | 1" = 3' | 1" = 30' | 1" = 300' |
| 40 | 1:4 | 1" = 4" | 1" = 4' | 1" = 40' | 1" = 400' |
| 50 | 1:5 | 1" = 5" | 1" = 5' | 1" = 50' | 1" = 500' |
| 60 | 1:6 | 1" = 6" | 1" = 6' | 1" = 60' | 1" = 600' |

Figure 3–3 Civil engineer's scale.

The 10 scale is often used in mechanical drafting as a full, decimal-inch scale, shown in Figure 3–4. Increments of 1/10 (.1) inch can easily be read on the 10 scale. Readings of less than .1 inch require you to approximate the desired amount, as shown in Figure 3–3. Some scales are available that refine the increments to 1/50 of an inch.

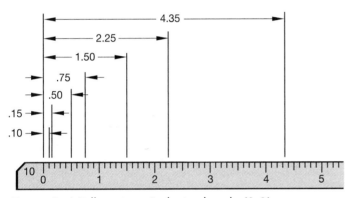

Figure 3–4 Full engineer's decimal scale (1:1).

The 20 scale is commonly used in mechanical drawing to represent dimensions on a drawing at half scale (1:2). Figure 3–5 shows examples of half-scale decimal dimensions.

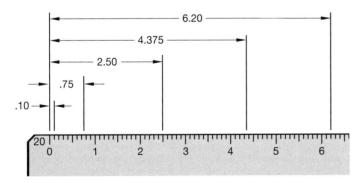

Figure 3–5 Half scale on the engineer's scale (1:2).

The remaining scales on the engineer's scale can be used in a similar fashion, for example, 1 inch = 50 feet, and so on. The symbol for inch is (") and the symbol for feet is ('). These symbols can be used when referring to inches and feet, but are not generally used on a drawing.

## Architect's Scale

The architect's scale is commonly used for architectural design and drafting, but the 1/16 division on the architect's scale has some application in mechanical engineering when related to prints that have fractional-inch values, although this use is not as common as decimal-inch values.

The triangular architect's scale contains 11 different scales. On 10 of them, each inch represents a foot and is subdivided into multiples of 12 parts to represent inches and fractions of an inch. The eleventh scale is the full scale with a 16 in the margin. The 16 means that each inch is divided into 16 parts and each part is equal to 1/16 inch. Look at Figure 3–6 for a comparison between the 10 engineer's scale and the 16 architect's scale. Figure 3–7 shows an example of the full architect's scale, while Figure 3–8 shows the fraction calibrations.

## Mechanical Engineer's Scale

The mechanical engineer's scale is commonly used for mechanical drafting when drawings are in fractional or decimal inches. The mechanical engineer's scale typically has full-scale divisions that are divided into 1/16, 10, and 50. The 1/16 divisions are just like

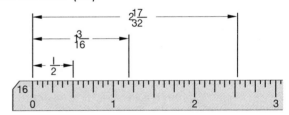

Figure 3–6 Comparison of full engineer's scale (10) and architect's scale (16).

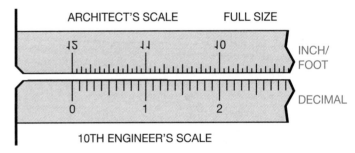

Figure 3–7 Full (1:1), or 12"=1'—0", architect's scale.

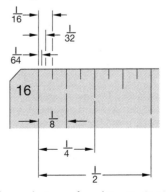

Figure 3–8 Enlarged view of architect's (16) scale.

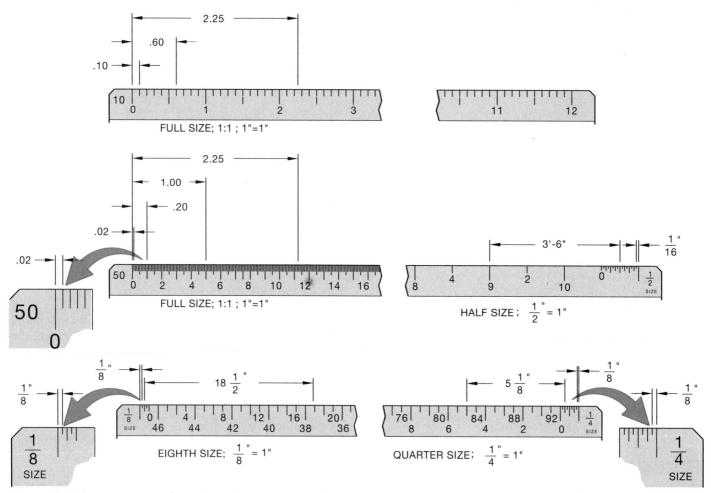

Figure 3–9 A comparison of mechanical engineers' scales.

the 16 architect's scale, where there are 12 inches and each inch is divided into 1/16-inch increments, or sometimes 1/32-inch divisions. The 10 scale is the same as the 10 civil engineer's scale, where each inch is divided into 10 parts, with each division being .10 inch. The 50 scale is for scaling dimensions that require additional accuracy, because each inch has 50 divisions. This makes each increment 1/50 inch or .02 inch (1 ÷ 50 = .02). Figure 3–9 shows a comparison of these scales. The mechanical engineer's scale also has options for scaling down the size of drawings. These scales are half size (1/2 size, 1/2 inch = 1 inch) quarter size (1/4 size, 1/4 inch = 1 inch), and eighth size (1/8 size, 1/8 inch = 1 inch). These scales are shown in Figure 3–9. A drawing that is represented at full scale (1:1), half scale (1:2), and quarter scale (1:4) is shown in Figure 3–10.

## PRECISION MEASURING INSTRUMENTS

You should have a good working knowledge of precision measuring tools, such as the vernier caliper and the micrometer. As a student in drafting, manufacturing, or related fields, you need to be able to take exact measurements. Some practice with both calipers and micrometers can help ensure that your readings are accurate.

### Handling Precision Measuring Instruments

Precision measuring instruments are designed for easy reading and handling, but you need to be careful when using them. The senses of sight and touch are critical for accurate and consistent measurements. Practice using precision measuring instruments to learn their proper feel. Never force or overadjust the tools when making measurements. You can achieve the best practice by measuring several parts that have internal and external features. If readings are consistently the same, or nearly the same, you have achieved good control of the instruments.

### Vernier Calipers

The vernier caliper consists of a stationary bar and a movable vernier slide assembly. The station rule is a hardened graduated bar with a fixed measuring jaw. The movable vernier slide assembly combines a movable jaw, a vernier plate, clamp screws, and an adjusting nut.

The vernier slide assembly moves as a unit along the graduations of the bar to bring the jaws in contact with the work. Readings are taken in thousandths of an inch by reading the position of the vernier plate in relation to the graduations on the stationary bar. Figure 3–11 shows which parts of a vernier caliper are used to take inside, outside, or depth measurements.

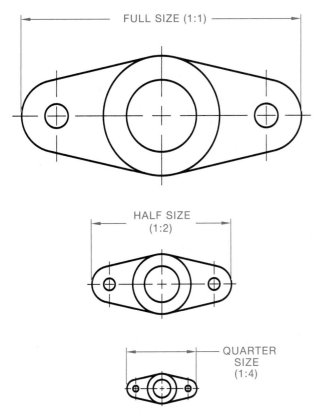

Figure 3–10 A sample drawing represented at full scale, half scale, and quarter scale.

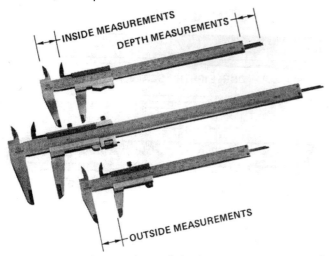

Figure 3–11 Inside, outside, and depth measurement parts of a vernier caliper. *Courtesy Mitutoyo, MTI Corporation.*

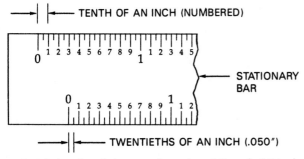

Figure 3–12 A caliper's bar graduated in .10 and .05 inch.

## Reading Vernier Calipers

The following steps should be used when reading vernier calipers.

*U.S. Customary.* On a U.S. customary vernier caliper, the bar is graduated in twentieths of an inch (.05 inch). Every second division represents 1/10 (.10) of an inch and is numbered as seen in Figure 3–12. The vernier plate is divided into 50 parts and numbered 0, 5, 10 … 45, 50. The 50 divisions on the vernier plate occupy the same space as 49 divisions on the bar (see Figure 3–13).

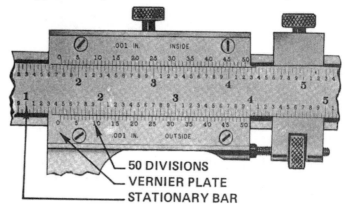

Figure 3–13 A vernier plate calibrated in .001 inch. *Courtesy the L.S. Starrett Company.*

The difference between the width of one of the 50 spaces on the vernier plate and one of the forty-nine spaces on the bar is 1/1000 of an inch (1/50 of 1/2). If the tool is set so the 0 line on the vernier plate coincides with the 0 line on the bar, the line to the right of the 0 on the vernier plate will differ from the line to the right of the 0 on the bar by 1/1000; the second line by 2/1000 and so on. The difference will continue to increase 1/1000 of an inch for each division until the 50 on the vernier coincides with the 49 line on the steel rule.

To read the tool, note how many inches, tenths (or .10), and twentieths (or .05) the 0 mark on the vernier slide is from the bar's 0 mark. Then note the number of divisions on the vernier slide from the 0 to a line which coincides exactly with a line on the bar.

In Figure 3–14, for example, the outside vernier plate has been moved to the right so the 0 mark has passed the one and four-tenths and one-twentieth inches (1.450) measurement, as shown on the bar. The fourteenth line on the vernier slide coincides exactly with a line on the bar. Fourteen-thousandths of an inch, therefore, are to be added to the reading on the bar, making the total reading one and four-hundred and sixty-four thousandths inches (1.464). You add three readings to make your measurement as follows:

**EXAMPLE:**

> 1.000 on the bar
>  .450 also on the bar
> +.014 on the vernier plate (outside)
> ———————————
> 1.464 is your measurement

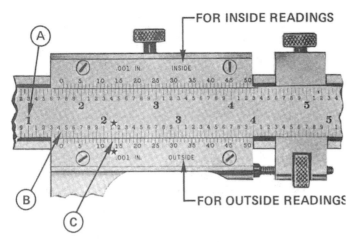

Figure 3–14 How to read a U.S. Customary vernier caliper. *Courtesy the L.S. Starrett Company.*

*Metric.* Each graduation on a metric vernier caliper bar is 0.5 mm. Every twentieth graduation is numbered in sequence, 10, 20, 30 mm, and so on, over the full range of the bar, which provides a direct reading in millimeters. The vernier plate is graduated in 50 parts, each representing 0.02 mm. Every fifth line is numbered in sequence, 0.10, 0.20, 0.30 mm, and so on, which provides a direct reading in hundredths of a millimeter.

To read the gage, first count how many millimeters lie between the 0 line on the bar and the 0 line on the vernier plate. Then, find the graduation on the vernier plate that exactly coincides with a line on a bar and note its value in hundredths of a millimeter. Add the vernier plate reading in hundredths of a millimeter to the number of millimeters counted on the bar for the total measurement (see Figure 3–15).

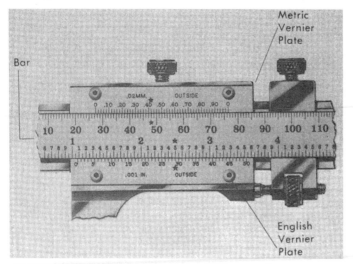

Figure 3–15 How to read a metric vernier caliper. *Courtesy the L.S. Starrett Company.*

**EXAMPLE:**

27.00 mm on the bar

+0.42 mm on the vernier plate

27.42 mm is your measurement

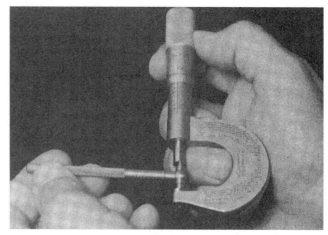

Figure 3–16 Proper holding of a micrometer. *Courtesy the L.S. Starrett Company.*

## Micrometers

A micrometer caliper, or micrometer, as it is commonly called, combines the double contact of a slide caliper with a precision screw adjustment which can be read with great accuracy. For most measurements, you hold the micrometer, as shown in Figure 3–16, with the work held in the hand, while the spindle is turned down on the feature with the thumb and index finger of the other hand. Do not remove the micrometer from the object before taking the reading. If the reading cannot be seen because of the position of the micrometer, set the lock nut before sliding the micrometer off the object.

## Reading Micrometers

The following steps should be used when reading the micrometers.

*U.S. Customary.* U.S. customary micrometers are graduated in thousandths (.001) of an inch. The pitch of the screw thread on the spindle is 1/40 inch or 40 threads per inch. One complete revolution of the thimble advances the spindle face toward or away from the anvil face precisely 1/40 or .025 inch. The reading line on the sleeve is divided into 40 equal parts by vertical lines that correspond to the number of threads on the spindle. Every fourth vertical line is longer than the others and designates hundreds of thousandths of an inch. For example, the line marked 1 represents .100 inch, the line marked 2 represents .200 inch, the line marked 3 represents .300 inch, and so on.

The beveled edge of the thimble is divided into 25 equal parts with each line representing .001 inch. On some micrometers every line is numbered consecutively, while on others every fifth line is numbered. Rotating the thimble from one line to the next moves the spindle longitudinally 1/25 of .025 or .001 inch, rotating two divisions represents .002 inch, and so on. Twenty-five divisions indicate a complete revolution of .025 or 1/40 to an inch.

To read the micrometer in thousandths, multiply the number of vertical divisions visible on the sleeve by .25, and add to this the number of thousandths indicated by the line on the thimble which coincides with the reading line on the sleeve.

**EXAMPLE:**

Refer to Figure 3–17.

| | |
|---|---|
| The one-line sleeve is visible, representing | .100" |
| Three additional lines are visible, each representing .025"; | 3 × .025" = .075" |
| Line 3 on the thimble coincides with the reading line on the sleeve, each line representing .001"; | 3 × .001" = .003" |
| The micrometer reading is: | = .178" |

The beveled edge of the thimble is graduated in 50 divisions, every fifth line being numbered from 0 to 50. Because one revolution of the thimble advances or withdraws the spindle 0.5 mm, each thimble graduation equals 1/50 of 0.5 mm or 0.01 mm. Thus two thimble graduations equal 0.02 mm, three graduations 0.03 mm, and so on.

To read the micrometer, add the number of millimeters and half-millimeters visible on the sleeve to the number of hundredths of a millimeter indicated by the thimble graduation, which coincides with the reading line on the sleeve.

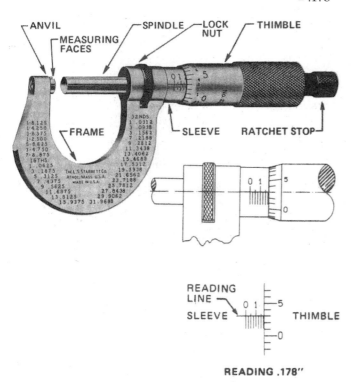

**READING .178"**

Figure 3–17 How to read a micrometer graduated in thousandths (.001) of an inch. *Courtesy the L.S. Starrett Company.*

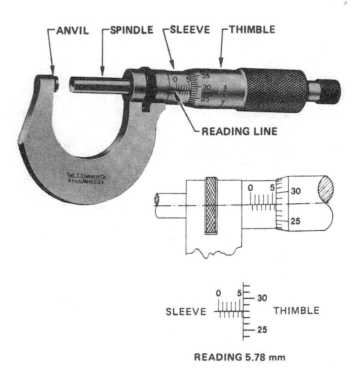

**READING 5.78 mm**

Figure 3–18 How to read a micrometer graduated in hundredths (0.01) of a millimeter. *Courtesy the L.S. Starrett Company.*

An easy way to remember how to read a customary micrometer is to think of the varied units as if you were making change from a $10 bill. Count the figures on the sleeve as dollars, the vertical lines on the sleeve as quarters, and the divisions on the thimble as cents. Add up your change and put a decimal point instead of a dollar sign in front of the figures.

*Metric.* A metric micrometer is graduated in hundredths (0.01) of a millimeter. The pitch of the spindle screw is one-half millimeter (0.5 mm) and one revolution of the thimble advances the spindle toward or away from the anvil the same 0.5 mm distance.

The reading line on the sleeve is graduated in millimeters (1.0 mm), with every fifth millimeter being numbered from 0 to 25. Each millimeter is also divided in half (0.5 mm), and so two revolutions of the thimble are required to advance thespindle 1.0 mm.

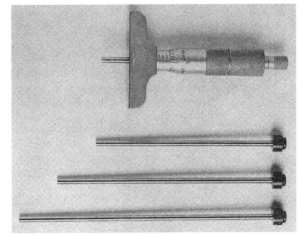

Figure 3–19 A micrometer depth gage. *Courtesy Mitutoyo, MTI Corporation.*

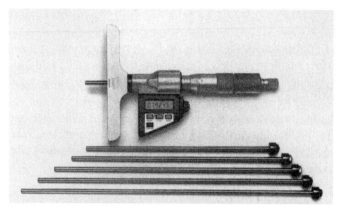

Figure 3–20 Digital micrometer depth gage. *Courtesy Mitutoyo, MTI Corporation.*

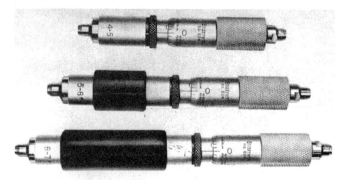

Figure 3–21 Inside micrometer. *Courtesy L.S. Starrett Company.*

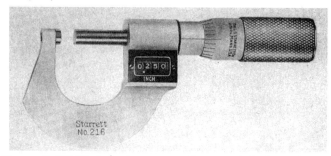

Figure 3–22 Digital micrometer. *Courtesy L.S. Starrett Company.*

Figure 3–23 Bench micrometer. *Courtesy L.S. Starrett Company.*

**EXAMPLE:**

| | |
|---|---:|
| Refer to Figure 3–18. | |
| The 5-mm sleeve graduation is visible | 5.00mm |
| One additional 0.5-mm line is visible on the sleeve | 0.50mm |
| Line 28 on the thimble coincides with the reading line on the sleeve, so 28 X 0.01 mm = | +0.28mm |
| The micrometer reading is | 5.78 mm |

## Other Micrometers

A micrometer depth gage (see Figure 3–19) is used to measure the depth of holes, slots, recesses, keyways, and other internal features. You take the reading the same as with an outside micrometer, except that sleeve graduations run in the opposite direction. Depth gages are also available with digital readouts, as shown in Figure 3–20.

You make inside measurements using an inside micrometer, as shown in Figure 3–21. The sleeve and thimble are between two ends that adjust outward to contact the surfaces to be measured. The same calibrations as found on outside micrometers are used on inside micrometers.

You can read a digital micrometer, shown in Figure 3–22, quicky and easily. The frame-mounted counter saves handling time because it can be read instantly.

The bench micrometer, shown in Figure 3–23, is a fine precision instrument for use in the shop or inspection laboratory. Use it as a comparator measuring to 50 millionths (.000050) of an inch or as a direct measuring tool to ten-thousandths (.0001) of an inch.

Friction thimble micrometers are available that give uniform contact pressure for readings that are independent of feel. The friction mechanism is designed so that the spindle will not turn after more than a given amount of pressure is applied.

## Protractors

A variety of protractors is available for measuring angles. The common half-circle (180°) protractor graduated in degrees can have a rectangular shape, as shown in Figure 3–24, so that any of four edges can be used as a vertical or horizontal reference line. The universal bevel protractor with vernier measures any angle to 5 feet, as shown in Figure 3–25.

## Dial Indicators

One of the most widely used inspection and quality control instruments is the dial indicator. Dial-indicator graduations are available from .001 to .000001 inch. Reading directly from a dial gage provides the accuracy and speed necessary in inspection operations. The dial indicator has been incorporated in all types of gaging equipment and in many machine tools. The dial comparator (see Figure 3–26) is used for inspecting duplicate parts. A hand lever

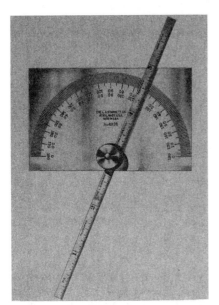

Figure 3–24 Protractor. *Courtesy L.S. Starrett Company.*

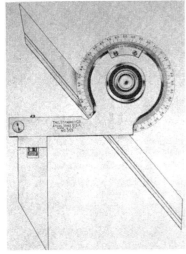

Figure 3–25 Universal bevel protractor. *Courtesy L.S. Starrett Company.*

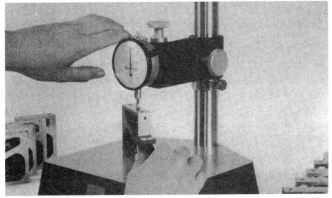

Figure 3–26 Dial comparator. *Courtesy L.S. Starrett Company.*

operates the indicator to contact the work, while size variations are read from the dial.

Dial and digital calipers are also available so that fast, accurate internal, external, and depth readings can be made using a slide

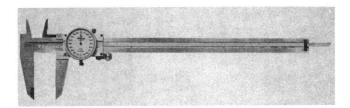

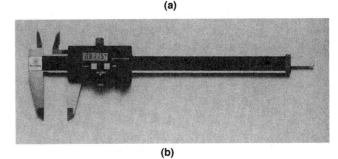

Figure 3–27 (a) Dial caliper; (b) digital caliper. *Courtesy L.S. Starrett Company.*

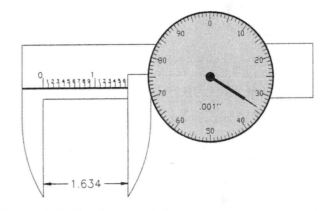

Figure 3–28 Reading the dial caliper.

caliper measuring instrument, as seen in Figure 3–27(a) and 3–27(b).

## Reading the Dial Caliper

The dial caliper, like the vernier caliper, has a movable slide assembly and a stationary bar. The caliper range is determined by the length of slide travel, such as 0 to 6 inches. Dial caliper graduations are .001 inch. When reading the dial caliper, first determine

Figure 3–29 Setting the dial indicator at 0.

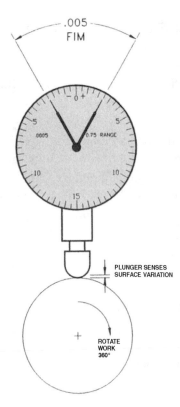

Figure 3–30 The dial indicator full indicator movement (FIM) of .005 inch.

the number of inches and tenths of an inch (.10) that the movable bar has passed on the stationary bar, for example, 1.60, as shown in Figure 3–28. Next, look at the dial. One complete revolution of the dial needle represents .10 inch. Each increment is .001 inch. The 10 on the dial face represents .01 inch. The dial shown in Figure 3–28 has a reading of .034 inch. The total reading is

$$
\begin{array}{r}
1.60'' \\
+ .03'' \\
+ .004'' \\
\hline
= 1.634''
\end{array}
$$

## Reading the Dial Indicator

The dial indicator operates by a plunger that is placed in contact with the work. As the work is moved, the dial indicator registers the difference between the high and low points. The negative (–) side of the dial reads the low points and the positive (+) side reads the high points. The maximum amount of movement between the positive reading and the negative reading is known as the full indicator movement (FIM) or total indicator reading (TIR). When you first set the dial indicator plunger on the work, the dial is set at 0, as shown in Figure 3–29. The dial indicator shown in Figure 3–29 has .0005-inch increments. There are 10 of these increments between the 0 and 5 readings. Therefore, if the needle on the dial moves to 5, this is a reading of

Figure 3–31 Gage block. *Courtesy the L.S. Starrett Company.*

$$
\begin{array}{r}
10.0 \\
\times \quad .0005 \\
\hline
= \quad .005''
\end{array}
$$

Now, with the dial plunger resting on the work and the dial set at 0, the work is either moved laterally, if it is lineal, or rotated between centers, if it is cylindrical. The needle moves as the plunger senses any variation on the surface of the material. While this movement takes place, the operator looks for the total variation between the positive and negative readings on the dial. This is the FIM. Figure 3–30 shows an FIM of .005 inch.

## Gage Blocks

Dimensional quality control is established by gage blocks, which are accurate to millionths of an inch. Precision gage blocks, such as the one shown in Figure 3–31, (.00001) are the primary standard vital to dimensional quality. Both rectangular and angle gage blocks are used for calibrating precision measuring instruments and for establishing comparative dimensions for part inspection.

# CHAPTER 3 TEST

Multiple choice: Respond to the following by selecting a, b, c, or d to best answer the question or complete the statement.

1. The scale of a drawing is usually given in the:
   a. title block or below the view of an object that differs to that given in the title block
   b. general notes
   c. specific notes
   d. engineering change notice

2. A full scale is often noted as:
   a. 1:1
   b. 1:2
   c. 2:1
   d. 1:4

3. A quarter-scale drawing is noted as:
   a. 1:1
   b. 1:2
   c. 2:1
   d. 1:4

4. A half-scale drawing is often noted as:
   a. 1:1
   b. 1:2
   c. 2:1
   d. 1:4

5. A double-scale drawing is often noted as:
   a. 1:1
   b. 1:2
   c. 2:1
   d. 1:4

6. A four-times scale drawing is often noted as:
   a. 1:1
   b. 1:2
   c. 4:1
   d. 1:4

7. According to ANSI, the commonly used SI linear unit used on engineering drawings is the:
   a. inch
   b. millimeter
   c. foot
   d. meter

8. 10 millimeters equals:
   a. 1 cm
   b. 1 dm
   c. 1 m
   d. 1 km

9. 100 millimeters equals:
   a. 1 cm
   b. 1 dm
   c. 1 m
   d. 1 km

10. 1000 millimeters equals:
    a. 1 cm
    b. 1 dm
    c. 1 m
    d. 1 km

11. Convert .50 inch to millimeters.
    a. 25.4 mm
    b. 127 mm
    c. 1.27 mm
    d. 12.7 mm

12. The civil engineer's scale is calibrated in units of:
    a. feet and inches
    b. 10
    c. millimeters
    d. fractions

13. The civil engineer's 10 scale is often used in mechanical drafting as a _____, decimal-inch scale.
    a. full
    b. half
    c. quarter
    d. double

14. The civil engineer's 20 scale is often used in mechanical drafting as a _____, decimal-inch scale.
    a. full
    b. half
    c. quarter
    d. double

15. The architect's 16 scale is the _____ scale.
    a. full
    b. half
    c. quarter
    d. double

16. The architect's 16 scale is equal to what fraction of an inch?
    a. 16
    b. 1/8
    c. 1/16
    d. 1/10

17. The mechanical engineer's scale includes a decimal scale with 50 calibrations per inch, which means that decimal inches can be read directly to what fraction of an inch?
    a. .01
    b. .02
    c. .05
    d. .005

18. Precision measuring instruments require which of the following care and handling for proper use?
    a. Use your senses of sight and touch, which are critical for accurate and consistent measurements.
    b. Practice using precision measuring instruments to learn their proper feel.
    c. Never force or overadjust the tools when making measurements.
    d. All of the above.

19. On a U.S. customary vernier caliper, the bar is graduated in:
    a. .01 inch
    b. .02 inch
    c. .05 inch
    d. .10 inch

20. Each graduation on a metric vernier caliper bar is:
    a. 0.1 mm
    b. 0.5 mm
    c. 1.0 mm
    d. 5.0 mm

21. U.S. customary micrometers are graduated in _____ of an inch.
    a. .001
    b. .002
    c. .005
    d. .010

22. A metric micrometer is graduated in _____ of a millimeter.
    a. 0.001
    b. 0.01
    c. 0.1
    d. 1.0

23. A micrometer depth gage is used to measure the depth of:
    a. holes
    b. slots
    c. recesses
    d. all of the above

24. This type of a micrometer has a frame-mounted counter for quick and easy readings.
    a. display micrometer
    b. digital micrometer
    c. commercial micrometer
    d. friction-thimble micrometer

25. This precision instrument can be used as a comparator measuring to .000050 of an inch.
    a. display micrometer
    b. digital micrometer
    c. commercial micrometer
    d. bench micrometer

26. One of the most widely used inspection and quality control instruments available in graduations from .001 to .000050 inch provides accuracy and speed and is available in all types of gaging equipment and in many machine tools.
    a. digital indicator
    b. bench micrometer
    c. dial indicator
    d. protractor

27. These instruments are accurate to millionths of an inch and are used for calibrating precision measuring instruments and for establishing comparative dimensions for part inspection.
    a. gage blocks
    b. dial indicator
    c. protractor
    d. computer numerical control

28. Use these for measuring angles.
    a. gage blocks
    b. dial indicator
    c. protractor
    d. angle block

29. What should you do if a dimension seems to be missing on the print you are reading?
    a. Use a scale to measure the desired feature.
    b. Use math to calculate the missing dimension from given values.
    c. Use a caliper or micrometer to measure the feature.
    d. Consult with the engineer or drafter to determine the missing dimension.

30. Each division of the 50 scale on the mechanical engineer's scale represents what part of an inch?
    a. .001
    b. .002
    c. .02
    d. .005

# CHAPTER 3 PROBLEMS

**PROBLEM 3–1** Given the following civil engineer's scale, determine the readings at A, B, C, D, and E.

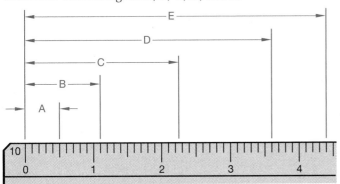

FULL SCALE = 1:1

A _____

B _____

C _____

D _____

E _____

**PROBLEM 3–2** Given the following civil engineer's scale, determine the readings at A, B, C, and D.

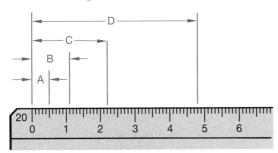

HALF SCALE = 1:2

A _____

B _____

C _____

D _____

**PROBLEM 3–3** Given the following architect's scale, determine the readings at A, B, C, D, E, and F.

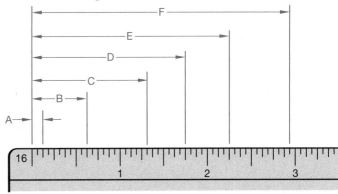

FULL SCALE = 1:1

A _____

B _____

C _____

D _____

E _____

**PROBLEM 3–4** Given the following metric scale, determine the readings at A, B, C, D, and E.

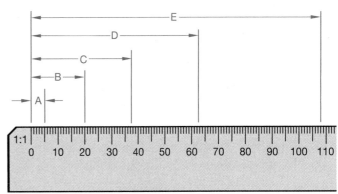

FULL SCALE = 1:1

A _____

B _____

C _____

D _____

E _____

**PROBLEM 3–5** Given the following metric scale, determine the readings at A, B, C, D, and E.

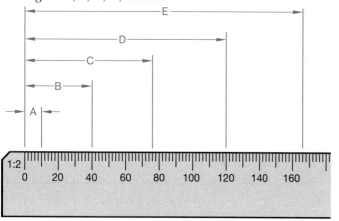

HALF SCALE = 1:2

A _____

B _____

C _____

D _____

E _____

**PROBLEM 3–6** Given the following architect's scale, determine the readings at A, B, C, and D.

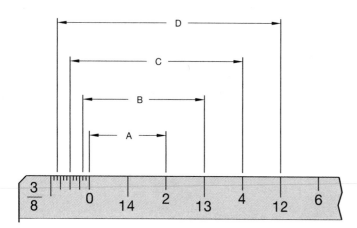

A _____

B _____

C _____

D _____

**PROBLEM 3–7** Given the following mechanical engineer's scale, determine the readings at A, B, C, and D.

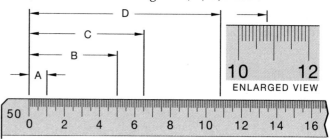

FULL SIZE; 1:1; 1"=1"

A _____

B _____

C _____

D _____

**PROBLEM 3–8** Determine the readings in inches on the following vernier calipers.

**INCH VERNIER CALIPERS**

1.

READING :_____

2.

READING :_____

3.

READING :_____

4.

READING :_____

**PROBLEM 3–9** Determine the readings in millimeters on the following vernier calipers.

### METRIC VERNIER CALIPERS

1.

READING : _____

2.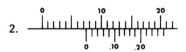

READING : _____

3.

READING : _____

4.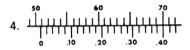

READING : _____

**PROBLEM 3–10** Determine the readings in thousandths of an inch on the following micrometers.

### INCH MICROMETER

1. 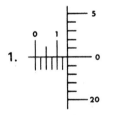     2.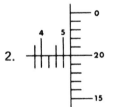

READING : _____     READING : _____

3. 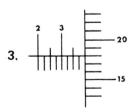     4.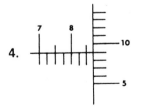

READING : _____     READING : _____

**PROBLEM 3–11** Determine the readings in hundredths of a millimeter (0.01) on the following micrometers.

**PROBLEM 3–12** Determine the readings on the following dial calipers. The inches and tenths are given.

### METRIC MICROMETERS

1. 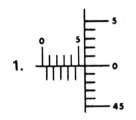     2.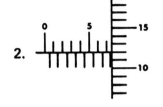

READING : _____     READING : _____

3. 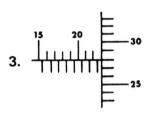     4.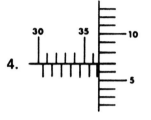

READING : _____     READING : _____

READING: 1.4 _____

READING: 2.5 _____

READING: 1.7 _____

READING: 3.2 _____

**PROBLEM 3–13** Determine the full indicator movement on the following dial indicators.

READING: _____

READING: _____

READING: _____

READING: _____

# *Reading Lettering and Lines on a Drawing*

## LEARNING OBJECTIVES

After completing this chapter you will be able to:

■ Identify the ANSI standard for line conventions and lettering.

■ Locate and read notes on prints.

■ Locate and identify line types on prints.

## *LETTERING ON PRINTS*

Information on drawings that cannot be represented graphically by lines can be presented by lettered dimensions, notes, and titles.

*ASME* The standard for lettering was established in 1935 by the American National Standards Institute (ANSI) and published by The American Society of Mechanical Engineers (ASME). This standard is now conveyed by the document ASME Y14.2M–1992 (revision of ANSI Y14.2M–1979), *Line Conventions and Lettering*. Lettering on drawings is generally uppercase, as shown in Figure 4–1.

ABCDEFGHIJKLMNOP
QRSTUVWXYZ&
1234567890

Figure 4–1 Vertical uppercase letters and numbers.

## *MICROFONT LETTERING*

When adequately sized standard lettering is created with clear, open features, there is generally no problem with microfilm or photocopy reductions, and many companies that use microfilm continue to use standard lettering. However, an alternate lettering design has been developed by the National Micrographics Association that is intended to provide greater legibility when

reduced for microfilm applications. This lettering, known as Microfont, is a style that provides a more open face for better reproduction capabilities (see Figure 4–2).

ABCDEFGHIJKLMNO
PQRSTUVWXYZ
1234567890

Figure 4–2 Microfont alphabet on numerals.

## *OTHER LETTERING STYLES*

Mechanical drafting for manufacturing typically used vertical uppercase lettering as previously discussed. Other types of lettering styles are sometimes used for other disciplines as explained in the following.

### Inclined Lettering

Some companies prefer inclined lettering. Structural drafting is one field where slanted lettering may be commonly found. Figure 4–3 shows slanted uppercase letters.

### Lowercase Lettering

Occasionally, lowercase letters are used; however, they are very uncommon in mechanical drafting. Civil or map drafters use lowercase lettering for some practices. Figure 4–4 shows lowercase lettering styles.

Figure 4-3 Uppercase inclined letters and numbers.

Figure 4-4 Lowercase lettering.

### Architectural Styles

Architectural lettering is much more varied in style than mechanical lettering (see Figure 4-5).

MAIN FLOOR PLAN
1/4" = 1'-0"

LIVING
12⁶ × 14⁰

ABCDEFGHIJKLM
NOPQRSTUVWXYZ
1 2 3 4 5 6 7 8 9 10

Figure 4-5 Architectural lettering.

### Computer-Aided Design and Drafting (CADD) Lettering Styles

Many CADD systems possess a variety of lettering styles, or fonts. The term font refers to a complete assortment of any one style of letters. Figure 4-6 shows some of the styles and sizes of characters that can be used in CADD. The size and style of characters used is dictated by the nature of the drawing.

## NOTES ON A PRINT

The two types of notes found on a print are called *specific* and *general* notes. Specific notes are found anywhere on the drawing where they relate to information about specific features or processes. Specific notes are also often referred to as *local notes*. General notes provide information that relates to the entire drawing. General notes are usually found all together in one location on the print. Look in the lower-left corner, the upper-left corner, or near the title block to find the general notes. Figure 4-7 shows an example of a typical set of general notes. Reading notes is discussed in detail in Chapter 7.

ABCDEFGHIJKLMNOPQRSTUVWXYZ
acbcdefghijklmnopqrstuvwxyz
1234567890

**ABCDEFGHI JKLMNOPQRSTUVWXYZ**
**acbcdefghi jkl mnopqrstuvwxyz**
**1234567890**

ABCDEFGHI JKLMNOPQRSTUVWXYZ
acbcdefghi jkl mnopqrstuvwxyz
1234567890

Figure 4-6 A sample of computer-aided design and drafting (CADD) character font styles.

3. ALL FILLETS AND ROUNDS R.125

2. REMOVE ALL BURRS AND SHARP EDGES

1. INTERPRET DIMENSIONS AND TOLERANCES PER ASME Y14.5M-1994.

# NOTES:

Figure 4-7 Common general notes found on a drawing.

## LINES ON A PRINT

Drafting is a graphic language using lines, symbols, and notes to describe objects to be manufactured or built. Certain lines are drawn thick so that they stand out clearly from other information on the drawing. Other lines are drawn thin. Thin lines are not necessarily less important than thick lines, but they are subordinate for identification purposes.

*ASME* The American Society of Mechanical Engineers recommends two line thicknesses, with bold lines twice as thick as thin lines. Drawings that meet military documentation standards (MIL) require three thicknesses of lines: thick (cutting plane, viewing plane, short break, and object), medium (hidden and phantom), and thin (center, dimension, extension, leader, long break, and section). Figure 4-8 shows widths and types of lines as taken from ASME standard, Line Conventions and Lettering, ASME Y14.2M-1992. Figure 4-9 shows a sample drawing using the various kinds of lines.

## THE KINDS OF LINES FOUND ON A PRINT

Lines on a print are created to aid in the print-reading process, because each type of line has its own meaning. The following discussion provides a general explanation of the function of each line type and gives examples of their use. Additional information is provided in the chapter where each line is discussed in detail.

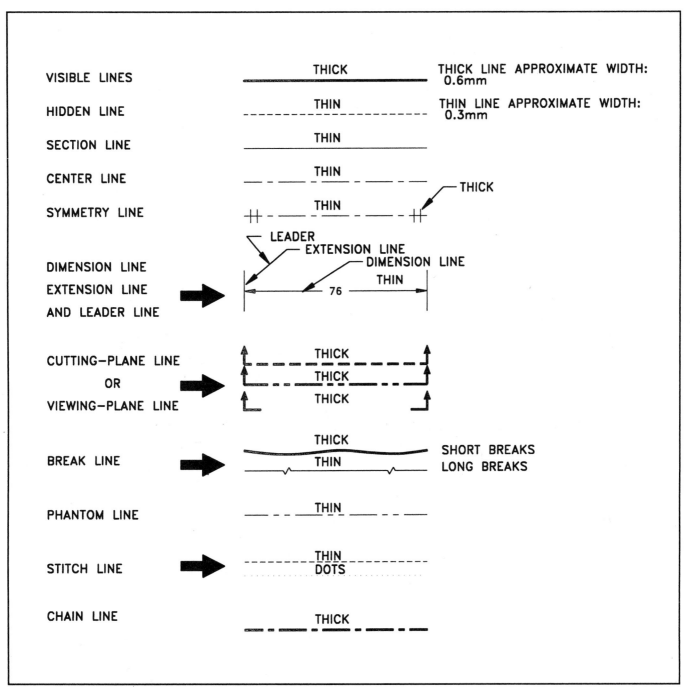

Figure 4–8 Line conventions, width, and type of lines. Based on ASME Y14.2M–1992.

## Object Lines

*Object lines*, also called visible lines, or outlines, describe the visible surface or edge of the object. These are drawn as thick lines, as shown in Figure 4–10.

## Hidden Lines

A *hidden line* represents an invisible edge on an object (see Figure 4–11). Hidden lines are thin lines, half as thick as object lines for contrast.

## Centerlines

*Centerlines* are used to show and locate the centers of circles and arcs. They are also used to represent the center axis of a circular or symmetrical form. Centerlines are thin lines on a drawing (see Figure 4–12).

You can find centerlines for holes in a bolt circle in either of two ways, depending upon how the holes are located, as shown in Figure 4–13.

When a centerline represents symmetry, as in the center plane of an object, the symmetry symbol, shown in Figure 4–14, may be used, or symmetry may be assumed.

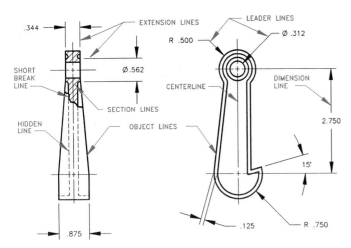

Figure 4–9 Sample drawing with a variety of lines displayed.

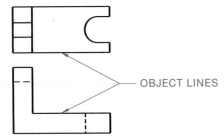

Figure 4–10 Object lines.

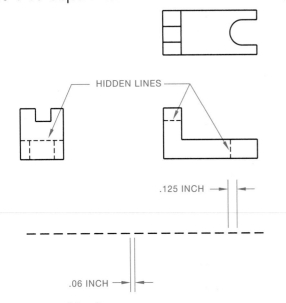

Figure 4–11 Hidden line representation.

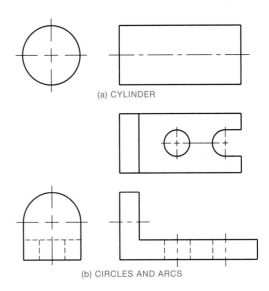

Figure 4–12 Centerline representation and examples.

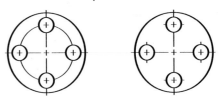

Figure 4–13 Bolt circle centerline options.

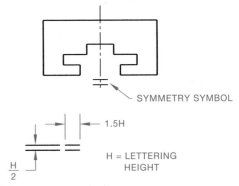

Figure 4–14 Symmetry symbol.

## Extension Lines

*Extension lines* are thin lines used to establish the extent of a dimension. Extension lines begin with a short space from the object and extend to about .125 inch or beyond the last dimension, as shown in Figure 4–15. Circular features, such as holes, are located by their centers in the view where they appear as circles. In this practice, centerlines become extension lines, as shown in Figure 4–16.

## Dimension Lines and Leader Lines

*Dimension lines* and thin lines are capped on the ends with arrowheads and broken along their length to provide a space for the dimension numeral. *Dimension lines* indicate the length of the dimension (see Figure 4–17).

Leaders, or leader lines, are thin lines used to connect a specific note to a feature, as shown in Figure 4–18. The leader has a .25-inch shoulder at one end that begins at the center of the vertical height of the lettering, and an arrowhead at the other end pointing to the feature.

## Arrowheads

*Arrowheads* are used to terminate dimension lines and leaders. Individual company preference dictates if arrowheads are filled in or left open, as shown in Figure 4–19.

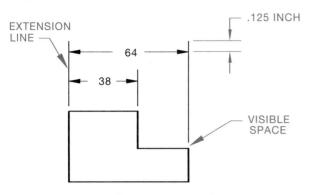

Figure 4–15 Extension lines.

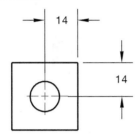

Figure 4–16 The centerline becomes an extension line when used for dimensioning.

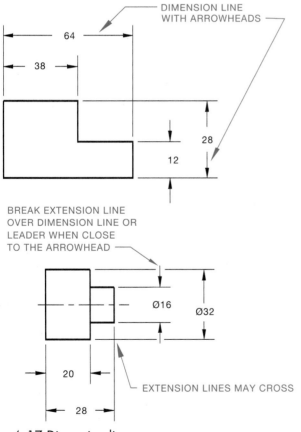

Figure 4–17 Dimension lines.

## Cutting-Plane Lines and Viewing-Plane Lines

*Cutting-plane lines* are thick lines used to identify where a sectional view is taken.

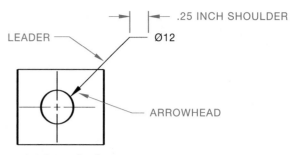

Figure 4–18 Leader line.

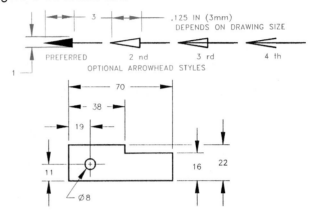

Figure 4–19 Arrowheads.

*Viewing-plane lines* are also thick lines, and they are used to identify where a view is taken for view enlargements or for partial views. Cutting-plane and viewing-plane lines are properly found in either of the two ways shown in Figure 4–20. The cutting-plane line takes precedence over the centerline when used in the place of a centerline.

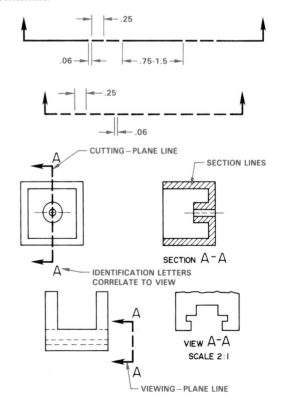

Figure 4–20 Cutting-plane lines and viewing-plane lines.

The scale of the view may be increased or remain the same as the view from the viewing plane, depending upon the clarity of information presented. When the location of the cutting plane or viewing plane is easily understood or if the clarity of the drawing is improved, the portion of the line between the arrowheads may be omitted, as shown in Figure 4–21.

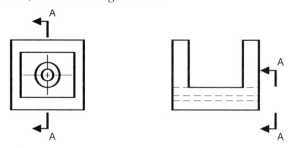

Figure 4–21 Simplified cutting-plane lines and viewing-plane lines.

## Section Lines

*Section lines* are thin lines used in the view of a section to show where the cutting-plane line has cut through material (see Figure 4–22). Section lines are generally found equally spaced at 45°. Any convenient angle can be used to avoid placing section lines parallel or perpendicular to other lines of the object; 30° and 60° are common. Section lines are drawn in opposite directions on adjacent parts (see Figure 4–27). The space between section lines may vary depending upon the size of the object (see Figure 4–23). When a very large area requires section lining, outline section lining can be used, as shown in Figure 4–24.

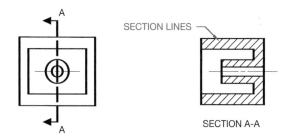

Figure 4–22 Section lines.

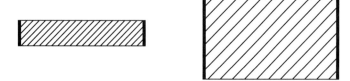

Figure 4–23 Space between section lines.

The section lines shown in Figure 4–22 through 4–24 were all drawn as general section-line symbols. General section lines are used for any material and are specifically used for cast or malleable iron. Coded section lines, as shown in Figure 4–25, can be used when the material must be clearly described, as in the section through an assembly of parts made of different materials (see Figure 4–26). Very thin parts are left unsectioned rather than sec-

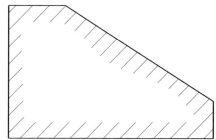

Figure 4–24 Outline section lines.

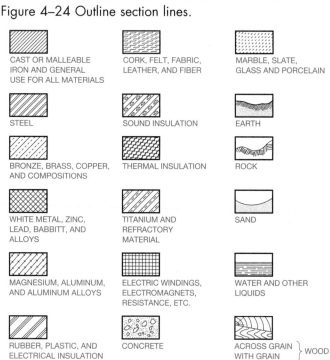

| CAST OR MALLEABLE IRON AND GENERAL USE FOR ALL MATERIALS | CORK, FELT, FABRIC, LEATHER, AND FIBER | MARBLE, SLATE, GLASS AND PORCELAIN |
| STEEL | SOUND INSULATION | EARTH |
| BRONZE, BRASS, COPPER, AND COMPOSITIONS | THERMAL INSULATION | ROCK |
| WHITE METAL, ZINC, LEAD, BABBITT, AND ALLOYS | TITANIUM AND REFRACTORY MATERIAL | SAND |
| MAGNESIUM, ALUMINUM, AND ALUMINUM ALLOYS | ELECTRIC WINDINGS, ELECTROMAGNETS, RESISTANCE, ETC. | WATER AND OTHER LIQUIDS |
| RUBBER, PLASTIC, AND ELECTRICAL INSULATION | CONCRETE | ACROSS GRAIN / WITH GRAIN } WOOD |

Figure 4–25 Coded section lines.

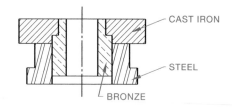

Figure 4–26 Coded section lines in assembly.

tion lined. This option is often used for a gasket (see Figure 4–27). Previous practice allowed thin parts to be blackened.

## Break Lines

There are two types of break lines: the short break line and the long break line. The thick, short break is very common on detail drawings, although the thin, long break can be used for breaks of long distances as the drafter chooses (see Figure 4–28). Other conventional breaks may be used for cylindrical features, as in Figure 4–29.

## Phantom Lines

*Phantom lines* are thin lines made of one long and two short dashes alternately spaced. Phantom lines are used to identify

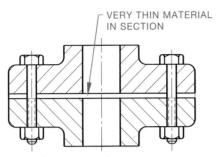

Figure 4–27 Very thin material in section.

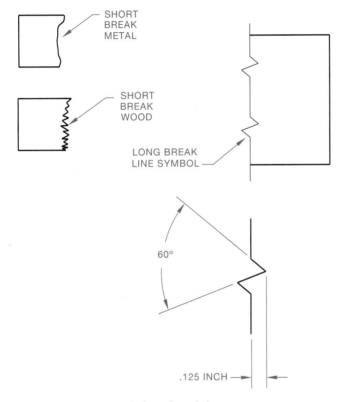

Figure 4–28 Long and short break lines.

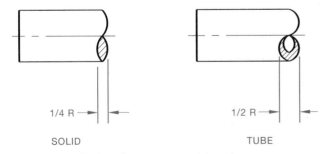

Figure 4–29 Cylindrical conventional breaks.

alternate positions of moving parts, adjacent positions of related parts, repetitive details, or the contour of filleted and rounded corners (see Figure 4–30).

## Chain Lines

*Chain lines* are thick lines of alternately spaced long and short dashes used to indicate that the portion of the surface next to

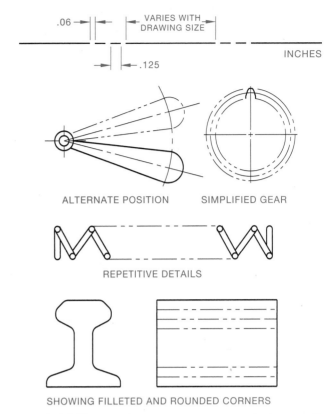

Figure 4–30 Phantom line representation and phantom line examples.

the chain line receives some specified treatment (see Figure 4–31).

## *PULLING IT ALL TOGETHER*

The previous discussion and examples explained and showed you the elements that make up the lines and lettering found on a drawing. While these elements are easily identified individually, putting them all together on a real print allows you to see how a drawing is created. Take some time to look at Figure 4–32 to see if you can find the lines and lettering that were introduced in this chapter. The line types are labeled for your reference.

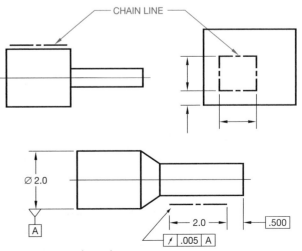

Figure 4–31 Chain lines.

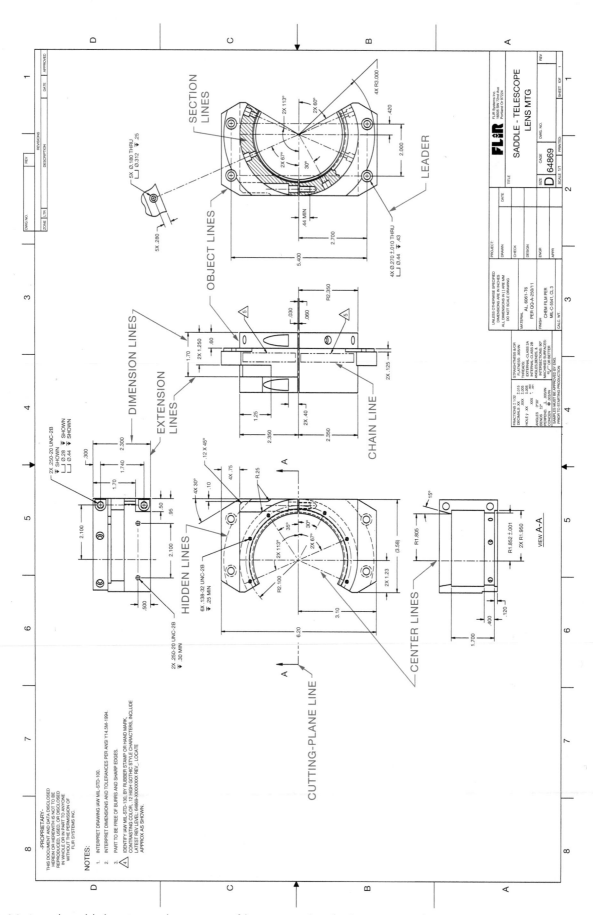

Figure 4–32 A real-world drawing with a variety of line types identified. *Courtesy Flir Systems, Inc.*

# CHAPTER 4 TEST

Multiple choice: Respond to the following by selecting a, b, c, or d to best answer the question or complete the statement.

1. Information on drawings that cannot be represented graphically by lines may be presented by:
   a. lettered dimensions
   b. notes
   c. titles
   d. all of the above

2. Lettering on drawings is generally:
   a. uppercase
   b. lowercase
   c. uppercase and lowercase
   d. none of the above

3. These types of notes are found anywhere on the drawing where they relate to information about specific features or processes:
   a. general notes
   b. specific notes
   c. common notes
   d. applied notes

4. These types of notes provide information that relates to the entire drawing.
   a. general notes
   b. specific notes
   c. common notes
   d. applied notes

5. Specific notes are also called:
   a. general notes
   c. common notes
   b. specific notes
   d. applied notes

6. General notes are usually found:
   a. grouped together in one location
   b. in the lower-left corner or near the title block
   c. in the upper-left corner
   d. any of the above

7. These lines describe the visible surface or edge of an object.
   a. object lines
   b. centerlines
   c. hidden lines
   d. extension lines

8. These lines represent invisible edges on an object.
   a. object lines
   b. centerlines
   c. hidden lines
   d. extension lines

9. These lines are used to show and locate the centers of circles and arcs.
   a. object lines
   b. centerlines
   c. hidden lines
   d. extension lines

10. These lines represent the axis of a circular or symmetrical form.
    a. object lines
    b. centerlines
    c. hidden lines
    d. phantom lines

11. These are thin lines used to establish the extent of a dimension.
    a. object lines
    b. leader lines
    c. dimension lines
    d. extension lines

12. These are thin lines capped on the ends with arrowheads and broken along their lengths for the dimension numeral placement.
    a. object lines
    b. leader lines
    c. dimension lines
    d. extension lines

13. These are thin lines used to connect a specific note to a feature.
    a. object lines
    b. leader lines
    c. dimension lines
    d. extension lines

14. These are thick lines used to identify where a sectional view is taken.
    a. cutting-plane lines
    b. section lines
    c. dimension lines
    d. extension lines

15. These are thin lines used in the view of a section to show where material has been cut.
    a. cutting-plane lines
    b. section lines
    c. dimension lines
    d. extension lines

16. These lines are used to provide a break in part of a feature that would otherwise be too long to show, or to remove a portion of an object to clarify a feature.

    a. cutting-plane lines
    b. section lines
    c. phantom lines
    d. break lines

17. These lines are used to identify alternate positions of moving parts, adjacent positions of related parts, or repetitive features.

    a. cutting-plane lines
    b. chain lines
    c. phantom lines
    d. break lines

18. These lines are used to indicate that the portion of the surface next to the line will receive some specific treatment.

    a. cutting-plane lines
    b. chain lines
    c. phantom lines
    d. break lines

19. These are thick lines used to identify where a view is taken, generally for view enlargements or placement of partial views.

    a. viewing-plane lines
    b. chain lines
    c. phantom lines
    d. cutting-plane lines

20. Section lines that are equally spaced are used to represent:

    a. general material
    b. cast iron
    c. malleable iron
    d. all of the above

## CHAPTER 4 PROBLEMS

**PROBLEM 4–1** Answer the following questions based on the print shown on the following page.

1. Are the items located at A specific or general notes?

2. Are the items located at B specific or general notes?

3. Name the type of line shown at C.

4. Name the type of line shown at D.

5. Name the type of line shown at E.

6. Name the type of line shown at F.

7. Name the type of line shown at G.

8. Name the type of line shown at H.

9. Name the type of line shown at I.

10. Name the type of line shown at J.

11. Name the type of line shown at K.

12. How many teeth are in the gear?

13. What is the drawing scale?

14. Give the drawing name.

15. What is the tolerance for three-place decimals?

16. What is the material used for this part?

17. What is the gear-tooth pressure angle?

18. What is the unspecified tolerance for angular dimensions?

19. What was the original sheet size of this drawing?

20. How many times has this drawing been revised?

**PROBLEM 4–2** Given the print found on the page 63, identify the lines labeled A through M.

A

B

C

D

E

F

G

H

I

J

K

L

M

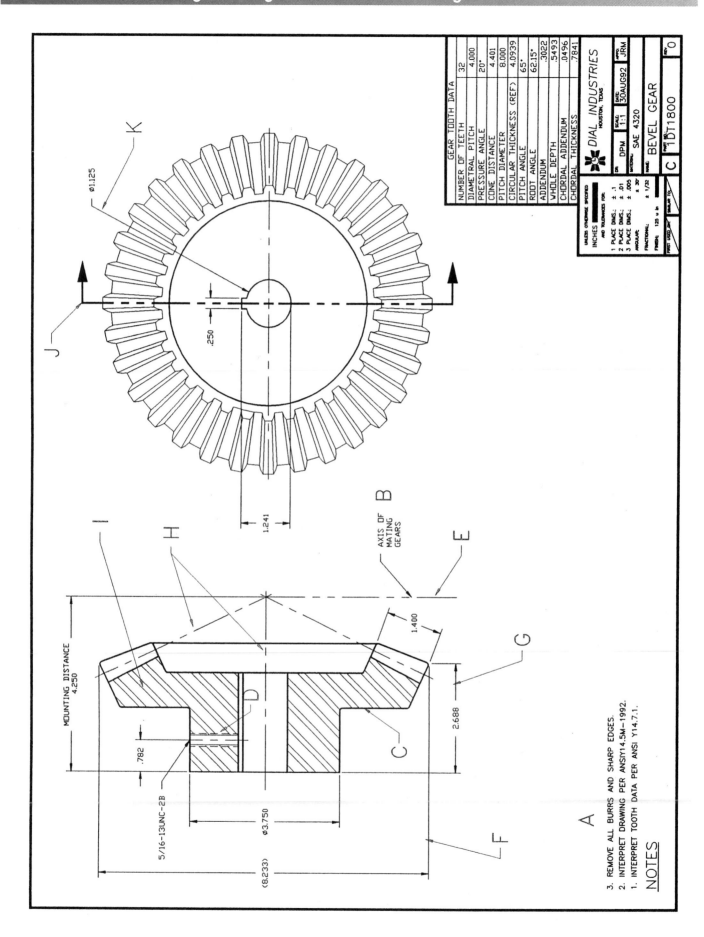

K

Ø1.125

J

.250

1.241

B

AXIS OF
MATING
GEARS

E

H

1.400

G

MOUNTING DISTANCE
4.250

2.688

.782

D

5/16-13UNC-2B

Ø3.750

(8.233)

F

A

NOTES

1. INTERPRET TOOTH DATA PER ANSI Y14.7.1.
2. INTERPRET DRAWING PER ANSI Y14.5M.—1992.
3. REMOVE ALL BURRS AND SHARP EDGES.

| GEAR TOOTH DATA | |
|---|---|
| NUMBER OF TEETH | 32 |
| DIAMETRAL PITCH | 4.000 |
| PRESSURE ANGLE | 20° |
| CONE DISTANCE | 4.401 |
| PITCH DIAMETER | 8.000 |
| CIRCULAR THICKNESS (REF) | 4.0939 |
| PITCH ANGLE | 65° |
| ROOT ANGLE | 62.15° |
| ADDENDUM | .3022 |
| WHOLE DEPTH | .5493 |
| CHORDAL ADDENDUM | .0496 |
| CHORDAL THICKNESS | .7841 |

DIAL INDUSTRIES
HOUSTON, TEXAS

UNLESS OTHERWISE SPECIFIED
AND TOLERANCES FOR:
INCHES
1 PLACE DIMS.: ± .1
2 PLACE DIMS.: ± .01
3 PLACE DIMS.: ± .005
ANGULAR: ± 30'
FRACTIONAL: ± 1/32
FINISH: 125 √ in

FIRST USED ON:

DR: DPM   DATE: 30AUG92   APPD: JRM
SCALE: 1:1
MATERIAL: SAE 4320
NAME: BEVEL GEAR
C   PART: 1DT1800   REV: 0
SIMILAR TO:

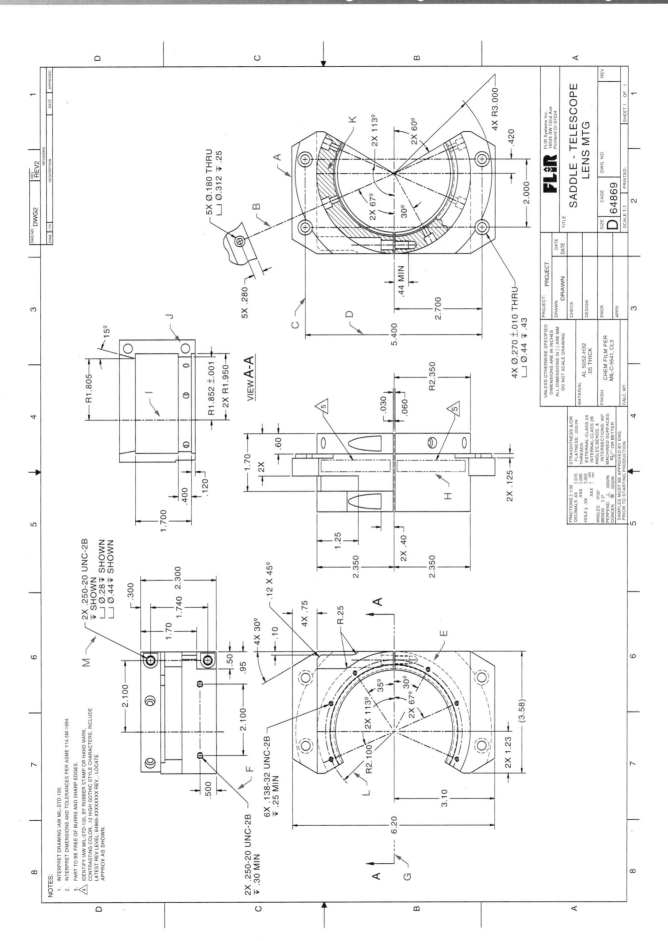

# Reading Multiview and Auxiliary Views

*ASME* This chapter is developed in accordance with the ANSI/ASME standard for multiview presentation, titled *Multi and Sectional View Drawings*, ANSI Y14.3. This standard is available from The American National Standards Institute, 1430 Broadway, New York, NY 10018, or the Society of Mechanical Engineers, 345 E. 47th Street, New York, NY 10017. The content of this discussion provides an in-depth analysis of the techniques and methods of multiview presentation.

*Orthographic projection* is any projection of the features of an object onto an imaginary plane called a plane of projection. The projection of the features of the object are made by lines of sight that are perpendicular to the plane of projection. When a surface of the object is parallel to the plane of projection, the surface appears in its true size and shape on the plane of projection. In Figure 5–1 the plane of projection is parallel to the surface of the object. The line of sight (projection from the object) is perpendicular to the plane of projection. Notice also that the object appears three-dimensional (width, height, and depth), while the view on the plane of projection has only two dimensions (width and height). In situations where the plane of projection is not parallel to the surface of the object, the resulting orthographic view is foreshortened, or shorter than true length (see Figure 5–2).

## MULTIVIEWS

Multiview projection establishes views of an object projected upon two or more planes of projection by using orthographic projection techniques. The result of multiview projection is a multiview drawing. A multiview drawing represents the shape of an object using two or more views. The drafter carefully considers the choice and

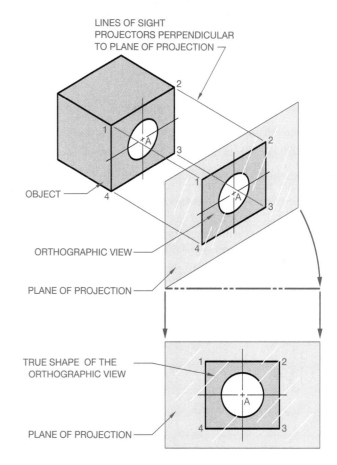

Figure 5–1 Orthographic projection to form orthographic view.

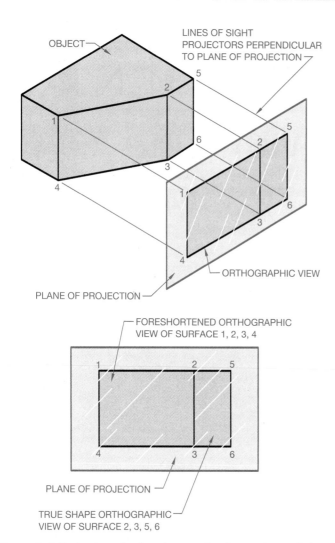

Figure 5–2 Projection of a foreshortened orthographic surface.

number of views used, so when possible, the surfaces of the object are shown in their true sizes and shapes.

It is often easier to visualize a three-dimensional picture of an object than it is to visualize a two-dimensional drawing. On manufacturing prints, however, the common practice is to show completely dimensioned and detailed drawings using two-dimensional views, known as *multiviews*. Figure 5–3 shows an object represented by a three-dimensional drawing, also called a *pictorial*, and three two-dimensional views, or multiviews. The multiviews represent the shape description of the object.

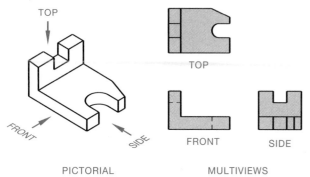

Figure 5–3 Pictorial view compared to multiview.

## Visualizing the Multiviews Using the Glass Box

If you place the object in Figure 5–3 in a glass box so the sides of the glass box are parallel to the major surfaces of the object, you can project those surfaces onto the sides of the glass box and create multiviews. Imagine that the sides of the glass box are the planes of projection that were previously discussed (see Figure 5–4). If you look at all sides of the glass box, you have six total views: front, top, right side, left side, bottom, and rear. Now, unfold the glass box as if the corners were hinged about the front (except the rear view), as demonstrated in Figure 5–5. These hinge lines are commonly called *fold lines*. Fold lines and projection lines are for drafting purposes only and are not displayed on the print, but knowing about them helps you learn how to correctly read a print.

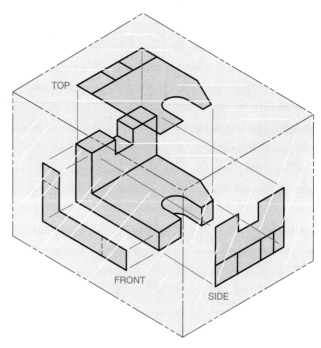

Figure 5–4 The glass-box principle.

Completely unfold the glass box onto a flat surface and you have the six views of an object represented in multiview. Figure 5–6 shows the glass box unfolded. Notice that the views are labeled front, top, right, left, rear, and bottom. You will always find the views in this arrangement when viewing multiviews on a print.

Now, analyze Figure 5–6 in more detail so you can see the items that are common between views. Knowing how to identify features of the object that are common between views will aid you in the visualization of multiviews. Notice in Figure 5–7 that the views are aligned. The top view is directly above and the bottom view is directly below the front view. The left-side view is directly to the left of the front view, while the right-side view is directly to the right of the front view. This format allows you to visually project points directly from one view to the next to help establish related features on each view.

The following discussion and related Figures 5–8 through 5–10 are intended to show you how a relationship between multiviews is created. Normally, a person reading prints does not physically draw or sketch lines between views to help find the desired information, but it can be helpful to use one of these techniques to coordinate information between views if necessary. Even if you do not physically draw fold lines and projection lines on a print, you can use a straightedge to help you correlate features between views. When you become familiar with reading prints, your eye will naturally look between views to coordinate information from one view to the next. This process of visualization is easier for some people than it is for others. You may want to rely on the use of one of these manual visualization methods if you have difficulty reading prints or until you become comfortable with it. Even if you can easily visualize the views found on prints, you may want to use one of the projection methods when you encounter a complex print.

Look closely at the relationship between the front, top, and right-side views. (A similar relationship exists with the left-side view.) Figure 5–8 shows a 45° line projected from the corner of the fold, or reference line (hinge), between the front, top, and side views.

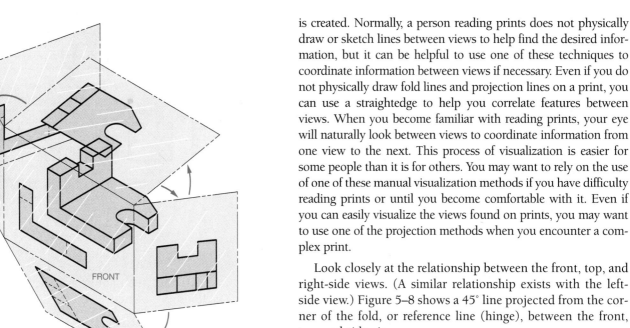

Figure 5–5 Unfolding the glass box at hinge lines, also called fold lines.

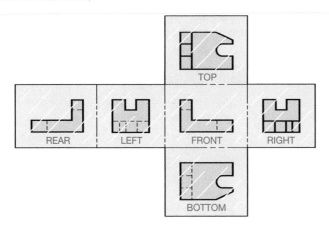

Figure 5–6 Glass box unfolded.

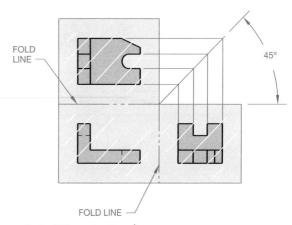

Figure 5–8 45° projection line.

This 45° line is used as an aid in projecting between views. All the features established on the top view can be projected to the 45° line and then down onto the side view. This projection works because the depth dimension is in the same relationship between the top and side views. The reverse is also true. Features from the side view may be projected to the 45° line and then over the top view.

The same concept of projection achieved in Figure 5–8 using the 45° line also works by using a compass at the intersection of the horizontal and vertical fold lines. The compass establishes the common relationship between the top and side views, as shown in Figure 5–9. Another method that is commonly used to transfer the size of features from one view to the next is to use dividers to transfer distances from the fold line of the top view to the fold line of the side view. The relationships between the fold line and these two views are the same, as shown in Figure 5–10.

The front view is usually the most important view and the one from which the other views are established. There is always one dimension common between views. For example, the width is

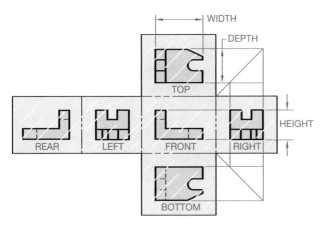

Figure 5–7 View alignment.

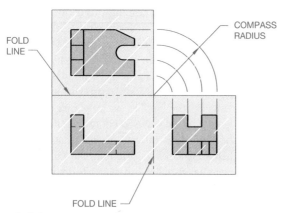

Figure 5–9 Projection with a compass.

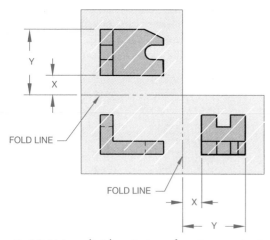

Figure 5–10 Using dividers to transfer view projections.

common between the front and top views and the height between the front and side views. Knowing this helps you to relate information from one view to another.

Take one more look at the relationship between the six views shown in Figure 5–11.

While this discussion tells you how the drafter transfers information from one view to the other, it is important for you to know how this is done to be able to accurately read prints. The drafter uses equipment to transfer features between views, but, as a print reader, you must be able to imagine the same procedure and transfer between views visually. Sometimes, especially with complex prints, you may even need to use a straightedge to physically assist you in coordinating information between views. Your ability to effectively read multiviews is a key to successful print reading.

## HOW VIEWS ARE SELECTED

While there are six primary views that can be selected to completely describe an object, it is seldom necessary to use all six views.

### The Front View

Usually, the front view is selected first. The front view is generally the most important view and, as you learned from the

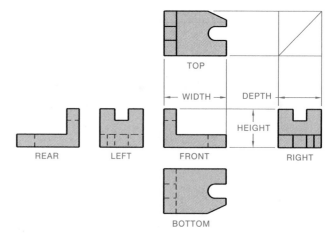

Figure 5–11 Multiview orientation.

glass-box description, the front view is the origin of all other views. The front view should:

■ Represent the most natural position of use

■ Provide the best shape description or most characteristic contours

■ Have the longest dimension

■ Have the fewest hidden features

■ Be the most stable position

Take a look at the pictorial drawing in Figure 5–12. Notice the front-view selection. This front-view selection violates the best shape-description and fewest hidden-features guidelines. However, the selection of any other view as the front would violate other rules; so, in this case, there is possibly no absolutely correct answer. Given the pictorial drawings in Figure 5–13, identify the view that you believe would be the best front view.

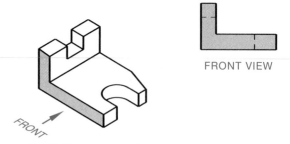

Figure 5–12 Front-view selection.

### Other View Selections

The same rules are used when selecting other needed views as when selecting the front view. Choose a view with the:

■ Most contours

■ Longest side

■ Least hidden features

■ Best balance or position

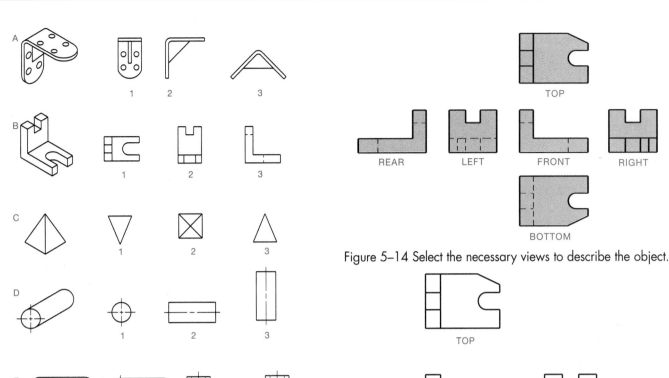

Figure 5–13 Select the best front views that correspond to the pictorial drawings at the left. You can make a first and second choice.

A. 2, 1. B. 3, 2. C. 3. D. 2, 1. E. 2, 1.

Given the six views of the object in Figure 5–14, which views would you select to completely describe the object? If your selection is the front, top, and right side, you are correct. A closer look at Figure 5–15 shows the three selected views. The front view shows the best position and the longest side, the top view clearly represents the angle and the arc, and the right-side view shows the notch. Any of the other views have many more hidden features.

Some objects can be completely described with two views or even one view (see Figure 5–16).

One-view drawings are also often practical. When an object has shape and a uniform thickness, more than one view is unnecessary. Figure 5–17 shows a gasket drawing where the thickness of the part is identified in the materials specifications of the title block. The types of parts that fit into this category include gaskets, washers, spacers, or similar features.

Although two-view drawings are generally considered the industry's minimum recommended views for a part, other objects that are clearly identified by dimensional shape can be drawn with one view, as in Figure 5–18.

In this example, the shape of the pin is clearly identified by the 25-mm diameter. In this case, the second view would be a circle and would not necessarily add any more valuable information to the drawing.

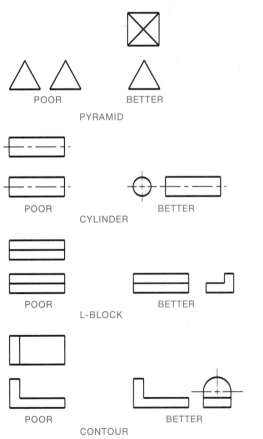

Figure 5–14 Select the necessary views to describe the object.

Figure 5–15 The selected views.

Figure 5–16 Selecting two views.

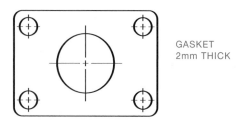

Figure 5–17 One-view drawing.

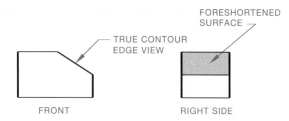

Figure 5–18 One-view drawing.

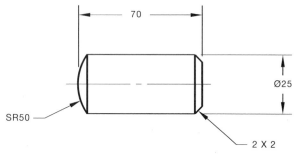

Figure 5–19 Contour representation.

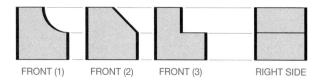

Figure 5–20 Select the front view that properly goes with the given right-side view.

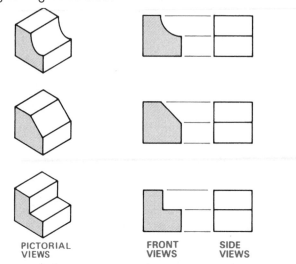

Figure 5–21 The importance of a view that clearly shows the contour of a surface.

## Contour Visualization

Some views do not clearly identify the shape of certain contours. You must then look to the next view to visualize the contour (see Figure 5–19).

The edge view of a surface shows that surface as a line. The true contour of the slanted surface is seen as an edge in the front view of Figure 5–19 and describes the surface as an angle, while the surface is foreshortened, slanting away from your line of sight, in the right-side view. In Figure 5–20, select the front view that properly describes the given right-side view. All three front views in Figure 5–20 could be correct. The side view does not help the shape description of the front-view contour (see Figure 5–21).

Cylindrical shapes appear round in one view and rectangular in another view, as seen in Figure 5–22. Both views in Figure 5–22 may be necessary, as one shows the diameter shape and the other shows the length. The ability to visualize from one view to the next is a skill that is critical for a print reader. You may have to train yourself to look at two-dimensional objects and picture three-dimensional shapes in your mind. You can also use some of the techniques discussed here to visualize features from one view to another.

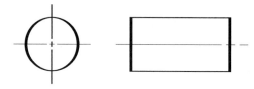

Figure 5–22 Cylindrical shape representation.

## Partial Views

When symmetrical objects are drawn in limited space or when there is a desire to save valuable drafting time, partial views can be used. The top view in Figure 5–23 is a partial view. Notice how the short break line is used to show that a portion of the view is omitted.

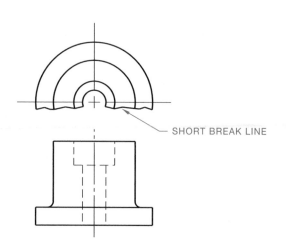

Figure 5–23 Partial view.

## View Enlargement

In some situations, when part of a view has detail that cannot be clearly dimensioned because of the drawing scale or complexity, a view enlargement can be used. When this is done, a thick phantom line circle is placed around the area to be enlarged. The circle is broken at a convenient location and an identification letter is centered in the break. Arrowheads are placed on the line next to the identification letter, as shown in Figure 5–24. An enlarged view of the detailed area is then shown in any convenient location in the field of the drawing. The view identification and scale is placed below the enlarged view. (see DETAIL A, Figure 5–24). The sample of view enlargement shown in Figure 5–24 conforms to the ASME standard, which is commonly used in industry. Many companies use a view-enlargement technique that is slightly different and more closely relates to military (MIL) standards. This method places a circle around the feature in the principal view with a leader note that says SEE DETAIL A, for example. The view enlargement is labeled with the title DETAIL A.

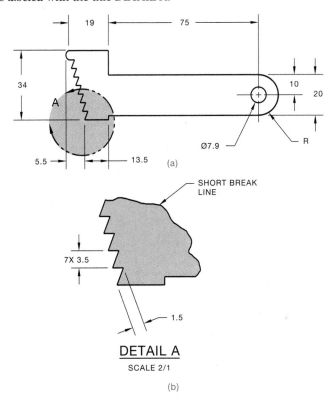

Figure 5–24 View enlargement example.

## PROJECTION OF CIRCLES AND ARCS

### Circles on Inclined Planes

When the line of sight in multiviews is perpendicular to a circular feature, such as a hole, the feature appears round, as shown in Figure 5–25. When a circle is projected onto an inclined surface, its view becomes elliptical in shape, as shown in Figure 5–26. The ellipse, shown in the top and right-side views of Figure 5–26, was established by projecting the major

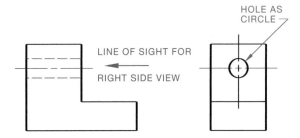

Figure 5–25 View of a hole projected as a circle and its hidden view through the part.

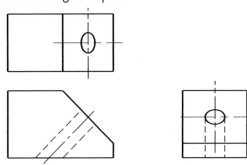

Figure 5–26 Hole represented on an inclined surface as an ellipse.

diameter from the top to the side view and the minor diameter to both views from the front view, as shown in Figure 5–27. The major diameter in this example is the hole diameter.

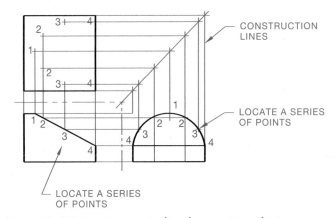

Figure 5–27 Locating an inclined curve in multiview.

### Arcs on Inclined Planes

When a curved surface from an inclined plane must be drawn in multiview, a series of points on the curve establishes the contour, as shown in Figure 5–27. See the completed curve in Figure 5–28.

### Fillets and Rounds

Fillets are slightly rounded, inside curves at corners, generally used to ease the machining of inside corners or to allow patterns to release more easily from casting and forgings. Fillets can also be designed into a part to allow additional material on inside corners for stress relief (see Figure 5–29).

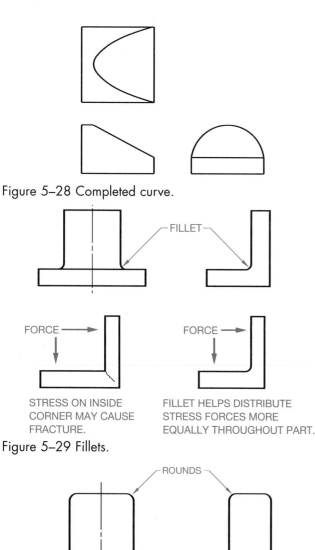

Figure 5–28 Completed curve.

Figure 5–29 Fillets.

STRESS ON INSIDE CORNER MAY CAUSE FRACTURE.

FILLET HELPS DISTRIBUTE STRESS FORCES MORE EQUALLY THROUGHOUT PART.

Figure 5–30 Rounds.

Figure 5–31 Break corner.

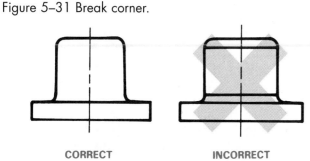

CORRECT         INCORRECT

Figure 5–32 Rounds and fillets in multiview.

Other than the concern of stress factors on parts, certain casting methods require that inside corners have fillets. The size of a fillet often depends on the precision of the casting method. For example, very precise casting methods may have smaller fillets than green-sand casting, where the exactness of the pattern requires large inside corners.

Rounds are rounded outside corners that are used to relieve sharp exterior edges. Rounds are also necessary in the casting and forging process for the same reasons as fillets. Figure 5–30 shows rounds represented in views.

A machined edge causes sharp corners, which may be desired in some situations. However, if these sharp corners are to be rounded, the extent of roundness depends on the function of the part. When a sharp corner has only a slight relief, it is referred to as a break corner, as shown in Figure 5–31.

## Rounded Corners in Multiview

An outside or inside rounded corner of an object is represented in multiview as a contour only. The extent of the round or fillet is not projected into the view, as shown in Figure 5–32. Cylindrical shapes can be represented with a front and top view, where the front identifies the height and the top shows the diameter. Figure 5–33 shows how these cylindrical shapes are represented in multiview. Figure 5–34 shows the representation of the contour of an object, as typically represented in multiview.

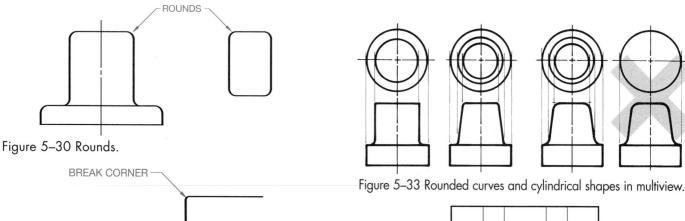

Figure 5–33 Rounded curves and cylindrical shapes in multiview.

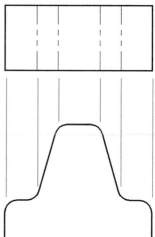

Figure 5–34 Contour in multiview.

## Runouts

The intersections of features with circular objects are projected in multiview to the extent where one shape runs into the other. The characteristics of the intersecting features are known as runouts. The runout of features intersecting cylindrical shapes is projected from the point of tangency of the intersecting feature, as shown in Figure 5–35. Notice also that the shape of the runout varies when drawn at the cylinder, depending upon the shape of the intersecting feature. Rectangular-shaped features have a fillet at the runout, while curved (elliptical or round) features contour toward the centerline at the runout. Runouts can also exist when a feature, such as a web, intersects another feature, as shown in Figure 5–36.

## THIRD-ANGLE PROJECTION

The method of multiview projection described in this chapter is also know as third-angle projection. This is the method of view arrangement that is commonly used in the United States.

In the previous discussion about multiview projection, the object was placed in a glass box so the sides of the glass box were parallel to the major surfaces of the object. Next, the object surfaces were projected onto the surfaces of the glass box. This achieved the same effect as if the viewer's line of sight was perpendicular to the surface of the box and looking directly at the object, as shown in Figure 5–37. With the multiview concept in mind, assume an area of space is divided into four quadrants, as shown in Figure 5–38.

If the object was placed in any of these quadrants, the surfaces of the object would be projected onto the adjacent planes. When placed in the first quadrant, the method of projection is known as first-angle projection. Projections in the other quadrants are termed second-, third-, and fourth-angle projections. Second- and fourth-angle projection is not used, while first- and third-angle projection is very common.

Third-angle projection is commonly used in the United States, and is achieved when the glass box from Figure 5–37 is placed in quadrant three from Figure 5–38. Figure 5–39 shows the relationship of the glass box to the projection planes in the

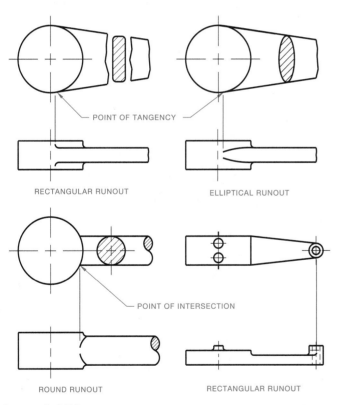

Figure 5–35 Runouts.

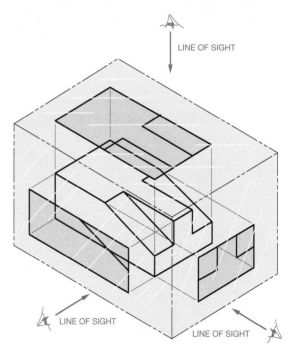

Figure 5–37 Glass box in third-angle projection.

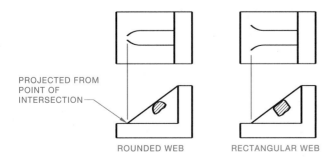

Figure 5–36 Other types of runouts.

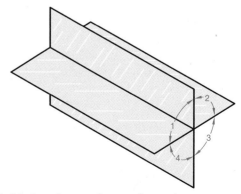

Figure 5–38 Quadrants of spatial visualization.

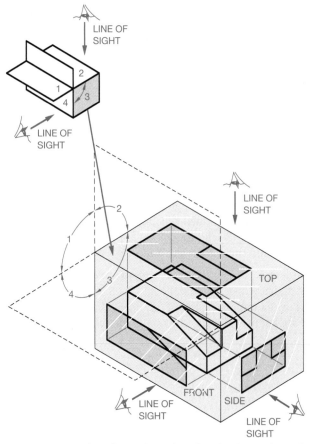

Figure 5–39 Glass box placed in the third quadrant for third-angle projection.

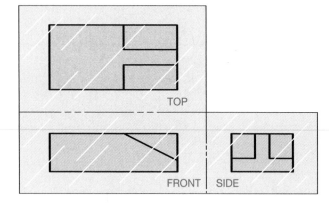

Figure 5–40 Third-angle projection.

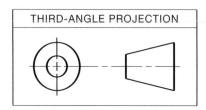

Figure 5–41 Third-angle projection symbol.

third-angle projection. In this quadrant, the projection plane is between the viewer's line of sight and the object. When the glass box in the third-angle projection quadrant is unfolded, the result is the multiview arrangement previously discussed and shown in Figure 5–40.

A third-angle projection drawing can be accompanied by a symbol on or next to the drawing title block. The standard third-angle projection symbol is shown in Figure 5–41.

## FIRST-ANGLE PROJECTION

*First-angle projection* is commonly used in Europe and other countries of the world. This method of projection places the glass box in the first quadrant. Views are established by projecting surfaces of the object onto the surface of the glass box. In this projection arrangement, however, the object is between the viewer's line of sight and the projection plane, as you can see in Figure 5–42. When the glass box in the first-angle projection quadrant is unfolded, the result is the multiview arrangement shown in Figure 5–43.

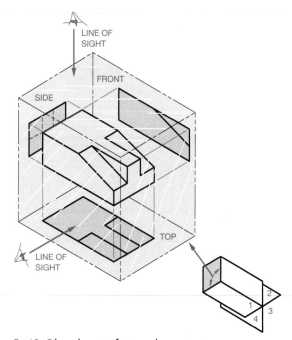

Figure 5–42 Glass box in first-angle projection.

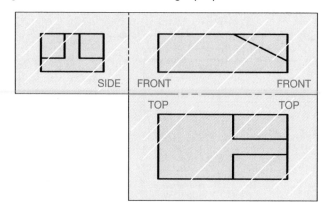

Figure 5–43 First-angle projection.

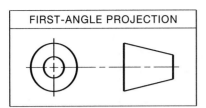

Figure 5–44 First-angle projection symbol.

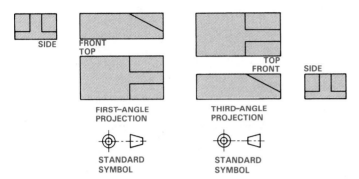

Figure 5–45 First-angle and third-angle projection.

A first-angle projection drawing can be accompanied by a symbol on or next to the drawing title block. The standard first-angle projection symbol is shown in Figure 5–44. Figure 5–45 shows a comparison of the same object in first- and third-angle projection.

## AUXILIARY VIEWS

*Auxiliary views* are used to show the true size and shape of a surface that is not parallel to any of the six principal views. When a surface feature is not perpendicular to the line of sight, the feature is said to be foreshortened, or shorter than true length. These foreshortened views do not give a clear or accurate representation of the feature. It is not proper to place dimensions on foreshortened views of objects. Figure 5–46 shows three views of an object with an inclined foreshortened surface.

An auxiliary view enables you to look directly at the inclined surface in Figure 5–46 so that you can view the surface and locate the hole in its true size and shape. An auxiliary view is projected from the inclined surface in the view where that surface appears as a line. The projection is at a 90° angle (see Figure 5–47). The height dimension, *H*, is taken from the view that shows the height in its true length.

Notice in Figure 5–47 that the auxiliary view shows only the true size and shape of the inclined surface. This is known as a partial auxiliary view. A full auxiliary view, Figure 5–48, would also show all the other features of the object projected onto the auxiliary plane. Normally the information needed from the auxiliary view is the inclined surface only, and the other areas do not usually add clarity to the view.

Look at the glass-box principle that was discussed earlier as it applies to auxiliary views, as shown in Figure 5–49. Figure 5–50 shows the glass box unfolded. Notice that the fold

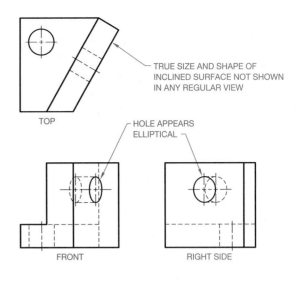

Figure 5–46 Foreshortened surface auxiliary view needed.

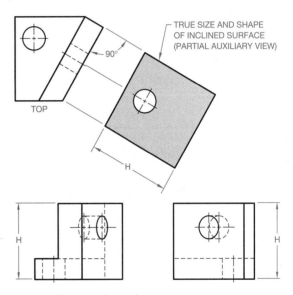

Figure 5–47 Partial auxiliary view.

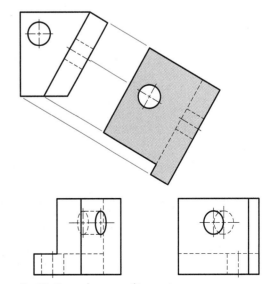

Figure 5–48 Complete auxiliary view.

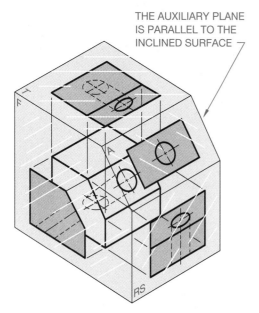

THE AUXILIARY PLANE IS PARALLEL TO THE INCLINED SURFACE

Figure 5–49 The object in a glass box.

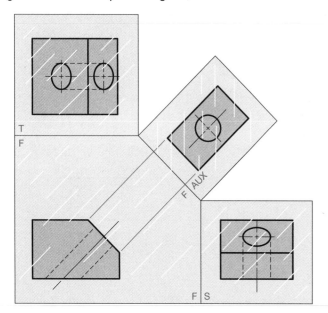

Figure 5–50 The glass box unfolded.

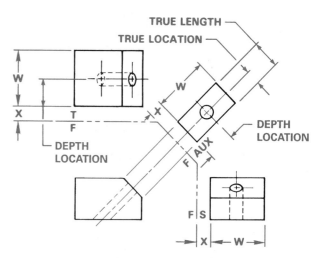

Figure 5–51 Establishing an auxiliary view with fold lines.

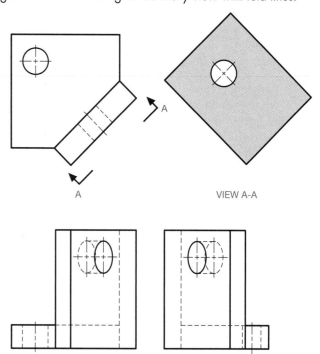

VIEW A-A

Figure 5–52 Establishing an auxiliary view with a viewing plane.

line between the front view and the auxiliary view is parallel to the edge view of the slanted surface.

The auxiliary view is projected from the edge view of the inclined surface, which establishes true length.

The true width of the inclined surface may be transferred from the fold lines of the top or side views, as shown in Figure 5–51. Hidden lines are generally not shown on auxiliary views, unless the use of hidden lines helps clarify certain features. The auxiliary viewcan be projected directly from the inclined surface as in Figure 5–47. When this is done, a centerline or projection line continues between the views to indicate alignment and view relationship. This centerline or projection line is often used as an extension line for dimensioning purposes. If view alignment is clearly obvious, the projection line can be omitted. When it is not possible to align the

auxiliary view directly from the inclined surface, viewing-plane lines can be used and the view placed in a convenient location on the drawing (Figure 5–52).

The viewing-plane lines are labeled with letters so the view can be clearly identified. This is especially necessary when several viewing-plane lines are used to label different auxiliary views. Views are placed in the same relationship as the viewing-plane lines indicate. Auxiliary views are not rotated; they are left in the same relationship as if they were projected from the slanted surface. Where multiple views are used, they are oriented from left to right and from top to bottom.

Figure 5–53 shows some other examples of partial auxiliary views in use. It is important to visualize the relationship of the slanted surfaces, edge view, and auxiliary view. If you have

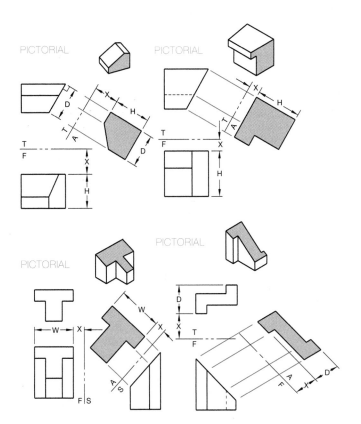

Figure 5–53 Partial auxiliary view examples.

trouble with visualization, it is possible to establish the auxiliary view through the mechanics of view projection. Use the following steps to do this:

**STEP 1** Number each corner of the inclined view so that the numbers coincide from one view to the other, as shown in Figure 5–53. Carefully project one point at a time from view to view. Some points have two numbers depending on the view.

**STEP 2** Establish the distance to each point from one view to the next and transfer the distance to the auxiliary view, as shown in Figure 5–54(a). Sometimes it is helpful to sketch a small pictorial to assist in visualization (see Figure 5–54(b)).

## CURVES REPRESENTED IN AUXILIARY VIEWS

Curves are shown in auxiliary views in the same manner as those shapes described previously. When corners exist, they are used to lay out the extent of the auxiliary surface. An auxiliary surface with irregular curved contours requires that the curve be divided into elements, so the element points can be transferred from one view to the next, as shown in Figure 5–55.

The contour of elliptical shapes may be displayed as shown in Figure 5–56.

## ENLARGEMENTS

In some situations the auxiliary view is enlarged to show a small detail more clearly. Figure 5–57 shows an object with a foreshort-

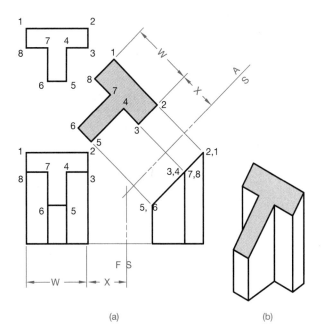

(a)    (b)

Figure 5–54 (a) Step 2: layout auxiliary view. (b) Pictorial to help visualize object.

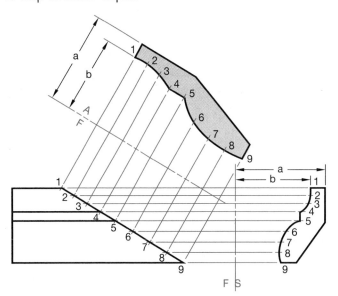

Figure 5–55 Plotting curves in auxiliary view.

ened surface. The two principal views are clearly dimensioned at a 1:1 scale. In this case, the view is too small to clarify the shape and size of the slot through the part. A viewing-plane line is placed to show the relationship of the auxiliary view. The auxiliary view is then shown in any convenient location at any desired scale; in this case, 2:1.

## SECONDARY AUXILIARY VIEWS

In some situations, a feature of an object is in an oblique position in relationship to the principal planes of projection. These inclined, or slanting, surfaces do not provide an edge view in any of the six possible multiviews. The inclined surface in Figure 5–58

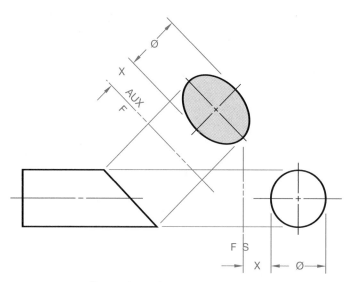

Figure 5–56 Elliptical auxiliary view.

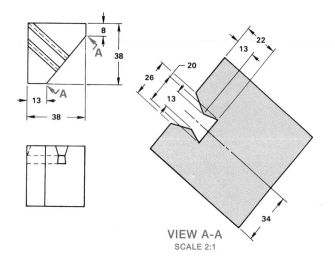

VIEW A-A
SCALE 2:1

Figure 5–57 Auxiliary view enlargement.

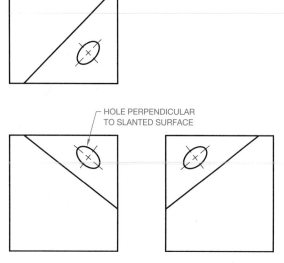

Figure 5–58 Oblique surface. There is no edge view in the principal views.

is foreshortened in each view and an edge view also does not exist. From the discussion on primary auxiliary views, you realize the projection must be from an edge view to establish the auxiliary.

A good way to learn how to read prints with secondary auxiliary views is to understand how they are created. This discussion takes you through the process of establishing a secondary auxiliary view.

To get the true size and shape of the inclined surface in Figure 5–58, a secondary auxiliary view is needed. The following steps outline how a secondary auxiliary view is found:

**STEP 1** Only two principal views are necessary to work from, as a third view does not add information. To establish the beginning projection for the primary auxiliary view, an element in one view is found in true length, as shown in Figure 5–59. The corners of the inclined surface are labeled and a fold line is placed perpendicular to the true-length element.

**STEP 2** The purpose of this step is to establish a primary auxiliary view which displays the slanted surface as an edge. The slanted surface is projected onto the primary auxiliary plane, as shown in Figure 5–60. This results in the inclined surface appearing as an edge view, or line.

**STEP 3** Now, with an edge view established, the next step is the same as the normal auxiliary view procedure. A fold line is placed parallel to the edge view. Points from the edge view are projected perpendicular to the secondary fold line to establish points for the secondary auxiliary view, as shown in Figure 5–61.

In Figure 5–61, the primary auxiliary view established all corners of the inclined surface in a line, or edge view. This edge view is nec-

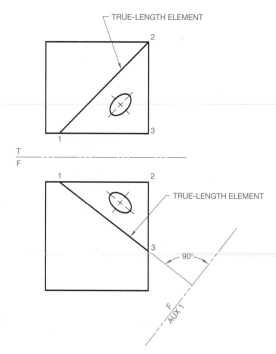

Figure 5–59 Step 1: draw a fold line perpendicular to a true-length element.

essary so the perpendicular line of projection (sight) for the secondary auxiliary assists in establishing the true size and shape of the surface. In many situations both the primary and secondary auxiliary views are used to establish the relationship between features of the object (see Figure 5–62). The primary auxiliary view shows the relationship of the inclined feature to the rest of the part, and the secondary auxiliary view shows true size and shape of the inclined features, as well as the true location of the holes.

## PULL IT ALL TOGETHER

If you follow the multiview and auxiliary view discussion found in this chapter, you should be able to read the views found on any print. Try to pull it all together by taking some time to look at the print shown in Figure 5–63. While this is a fairly complex print to read, your most important objective is to look at the print to see how the multiviews and auxiliary views are arranged, trying to visualize the part in your mind. There are three sectional views shown on this print. Sectional views are discussed in detail in Chapter 10. For now, just consider the sectional views as taking the place of a regular multiview or auxiliary view in the location shown. It is important for you to spend as much times as possible looking at prints, with the idea of visualizing the views and reading the information between views. Spend some more time looking at Figure 5–64, a part displayed using several carefully selected views, including front, top, right-side views; a view created from a viewing plane; a sectional view; and an auxiliary view.

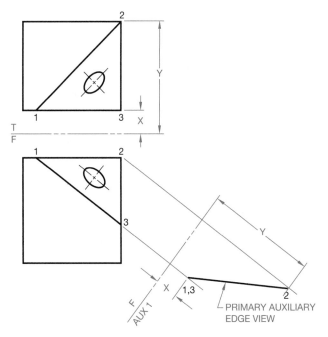

Figure 5–60 Step 2: primary auxiliary edge view of oblique surface.

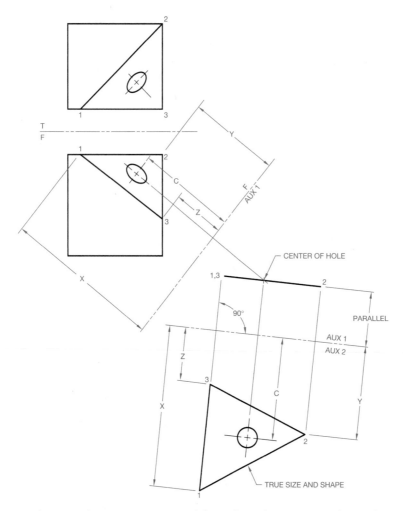

Figure 5–61 Step 3: the secondary auxiliary view projected from the edge view to obtain the true size and shape of the oblique surface.

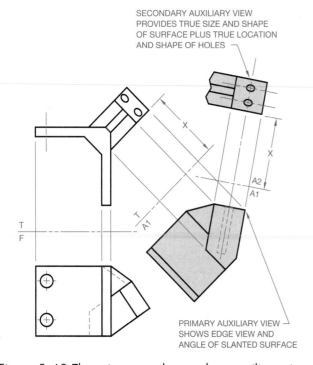

Figure 5–62 The primary and secondary auxiliary views.

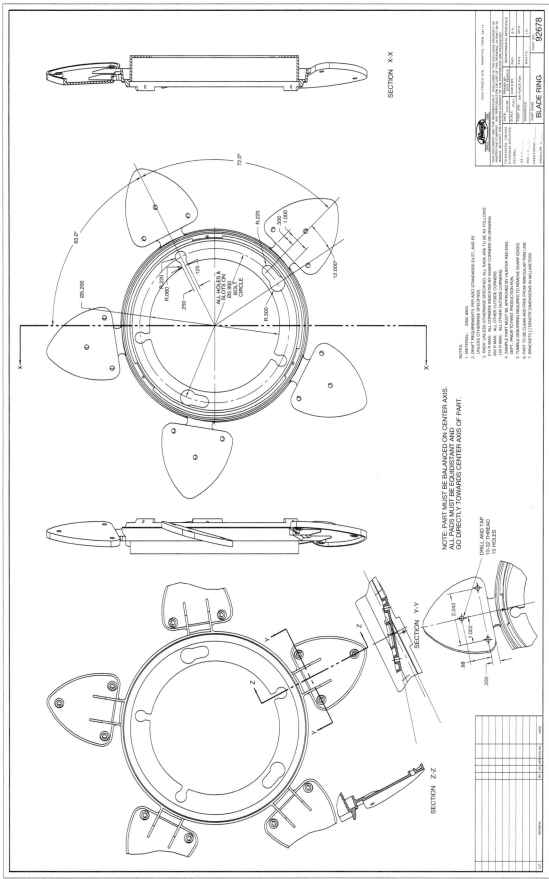

Figure 5–63 A complex part displayed on the drawing using several carefully selected views. *Courtesy of Hunter Fan Company.*

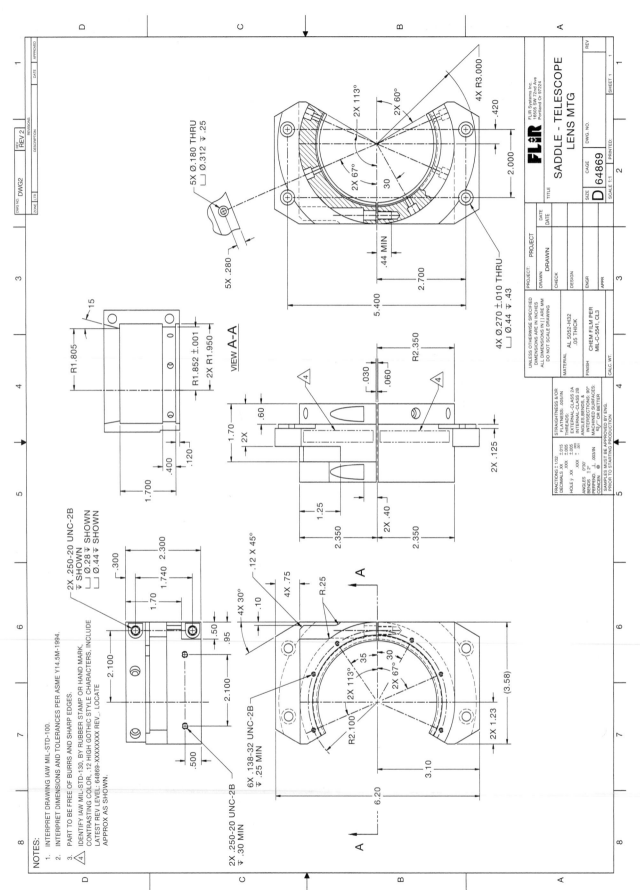

Figure 5–64 A complex part displayed on the drawing using several carefully selected views, including front, top, and right-side views; a view created from a viewing plane; a sectional view; and an auxiliary view. *Courtesy Flir Systems.*

# CHAPTER 5 TEST

Multiple choice: Respond to the following by selecting a, b, c, or d to best answer the question or complete the statement:

1. Any projection of a feature of an object onto an imaginary plane called a plane of projection is known as:
   a. graphic projection
   b. orthographic projection
   c. pictorial projection
   d. none of the above

2. This projection establishes views of an object projected upon two or more planes of projection.
   a. graphic projection
   b. orthographic projection
   c. pictorial projection
   d. multiview projection

3. A three-dimensional drawing is referred to as:
   a. a pictorial drawing
   b. an orthographic drawing
   c. an art drawing
   d. a multiview drawing

4. The top view is found directly above the:
   a. front view
   b. right-side view
   c. left-side view
   d. bottom view

5. The right-side view is found directly to the right of the:
   a. front view
   b. top view
   c. left-side view
   d. bottom view

6. The depth dimension has the same relationship between these two views.
   a. front and right side
   c. top and right side
   b. front and top
   d. none of the above

7. The width dimension has the same relationship between these two views.
   a. front and right side
   c. top and right side
   b. front and top
   d. none of the above

8. The height dimension has the same relationship between these two views.
   a. front and right side
   b. front and top
   c. top and right side
   d. none of the above

9. This view is usually the most important view, because it often represents the most natural position, has the least hidden features, shows the best shape, and is the most stable position.
   a. front
   b. top
   c. right side
   d. bottom

10. Your ability to effectively read multiview drawings is a key to successful:
    a. manufacturing
    b. job placement
    c. print reading
    d. graduation

11. This type of view may be used when symmetrical objects are drawn in limited space.
    a. small view
    c. top view
    b. partial view
    d. section view

12. This type of view may be used when part of a view has detail that cannot be clearly dimensioned because of the drawing scale or complexity.
    a. large view
    b. partial view
    c. view enlargement
    d. section view

13. If a hole is perpendicular to the edge view of a slanted surface in the right-side view, the hole appears _____ in shape in the front view.
    a. round
    c. as a line
    b. elliptical
    d. none of the above

14. If the line of sight is perpendicular to the edge view of a hole in the right-side view, the hole appears _____ in shape in the front view.
    a. round
    c. as a line
    b. elliptical
    d. none of the above

15. These are slightly rounded inside curves at corners, generally used to ease the machining, strengthen inside corners, or allow patterns to release more easily from casting and forgings.
    a. round
    b. break corner
    c. fillet
    d. runout

16. These are rounded outside corners that are used to relieve sharp exterior edges, or allow patterns to release more easily from casting and forgings.

    a. round
    c. fillet
    b. break corner
    d. runout

17. When a sharp corner has only a slight relief, it may be referred to as:

    a. round
    c. fillet
    b. break corner
    d. runout

18. The intersection of features with circular objects that are projected in multiview to the extent where one shape runs into the other is referred to as:

    a. round
    c. fillet
    b. break corner
    d. runout

19. The method of multiview projection used in the United States is called:

    a. first-angle projection
    b. second-angle projection
    c. third-angle projection
    d. fourth-angle projection

## Chapter 5 Multiview Problems

The following problems provide you with views that contain missing lines or missing views. Sketch the missing lines or missing views as appropriate. Some problems give you a pictorial view to aid in visualization.

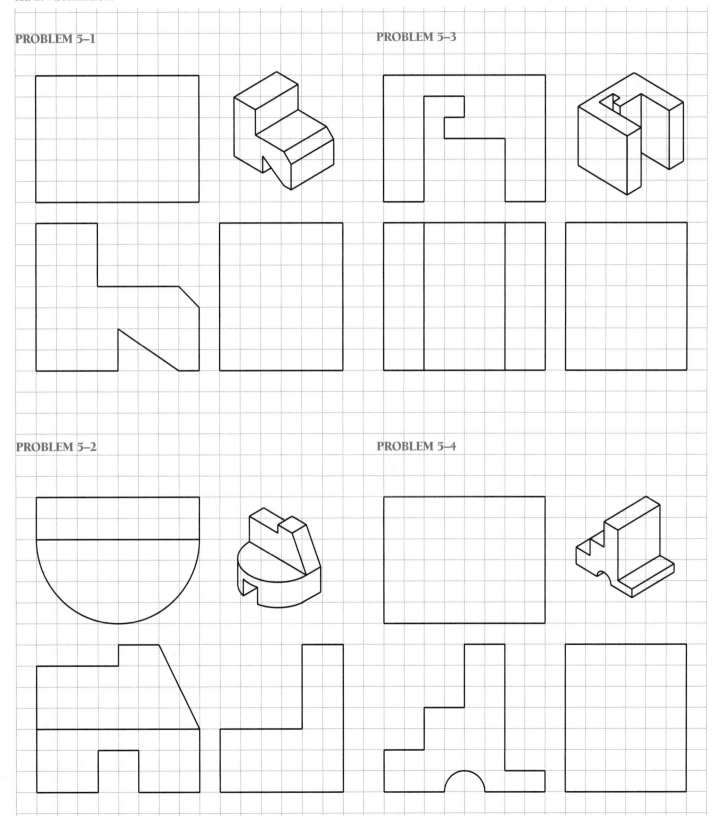

**PROBLEM 5–1**

**PROBLEM 5–3**

**PROBLEM 5–2**

**PROBLEM 5–4**

**PROBLEM 5–5**

**PROBLEM 5–7**

**PROBLEM 5–6**

**PROBLEM 5–8**

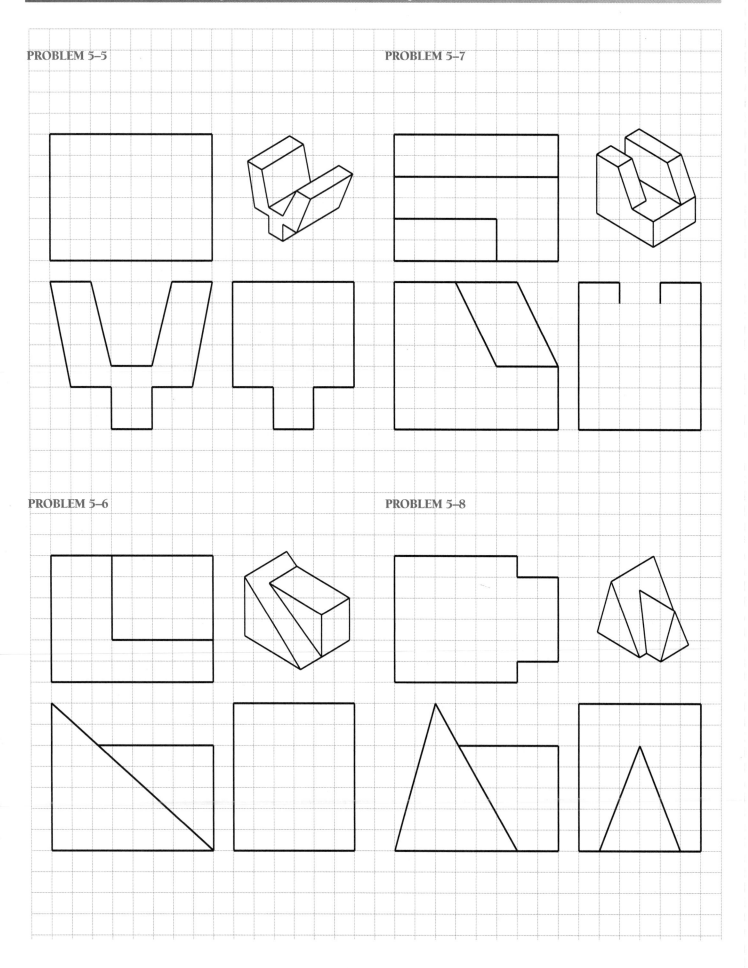

**PROBLEM 5–9**

**PROBLEM 5–11**

**PROBLEM 5–10**

**PROBLEM 5–12**

**PROBLEM 5–13**

**PROBLEM 5–15**

**PROBLEM 5–14**

**PROBLEM 5–16**

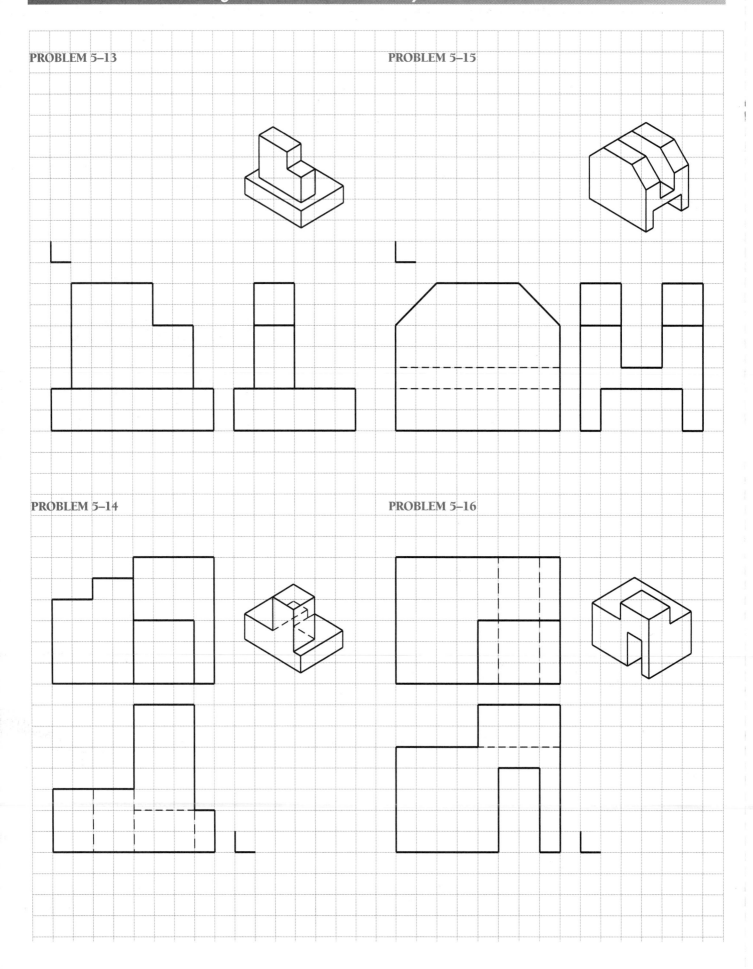

PROBLEM 5–17

PROBLEM 5–20

PROBLEM 5–18

PROBLEM 5–21

PROBLEM 5–19

PROBLEM 5–22

**PROBLEM 5–23**

**PROBLEM 5–26**

**PROBLEM 5–24**

**PROBLEM 5–27**

**PROBLEM 5–25**

**PROBLEM 5–28**

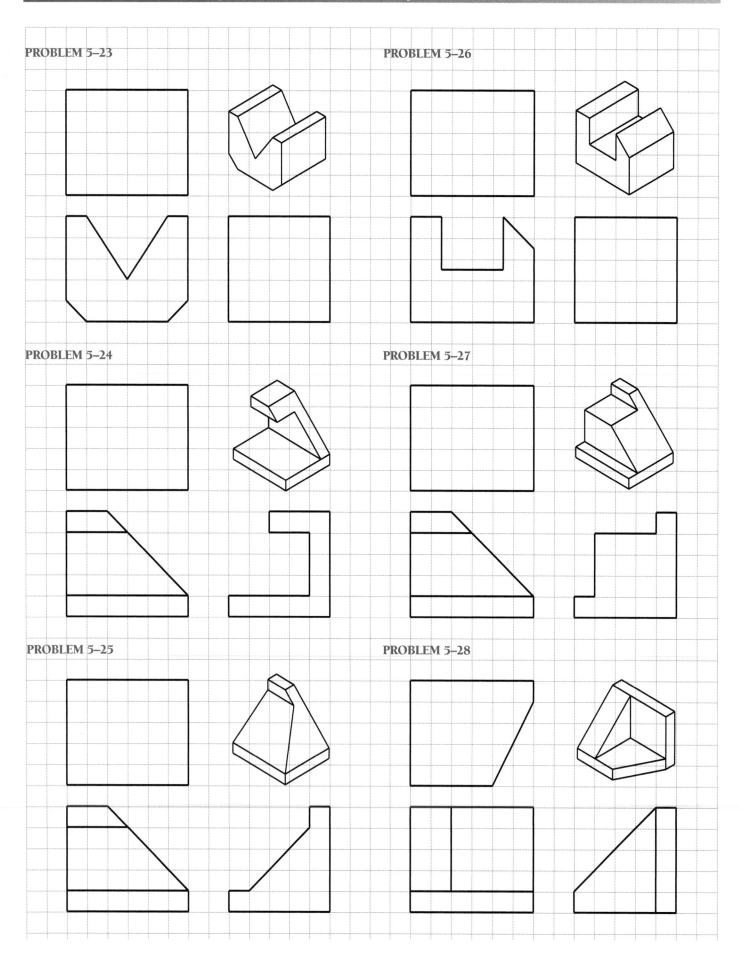

**PROBLEM 5–29**

**PROBLEM 5–32**

**PROBLEM 5–30**

**PROBLEM 5–33**

**PROBLEM 5–31**

**PROBLEM 5–34**

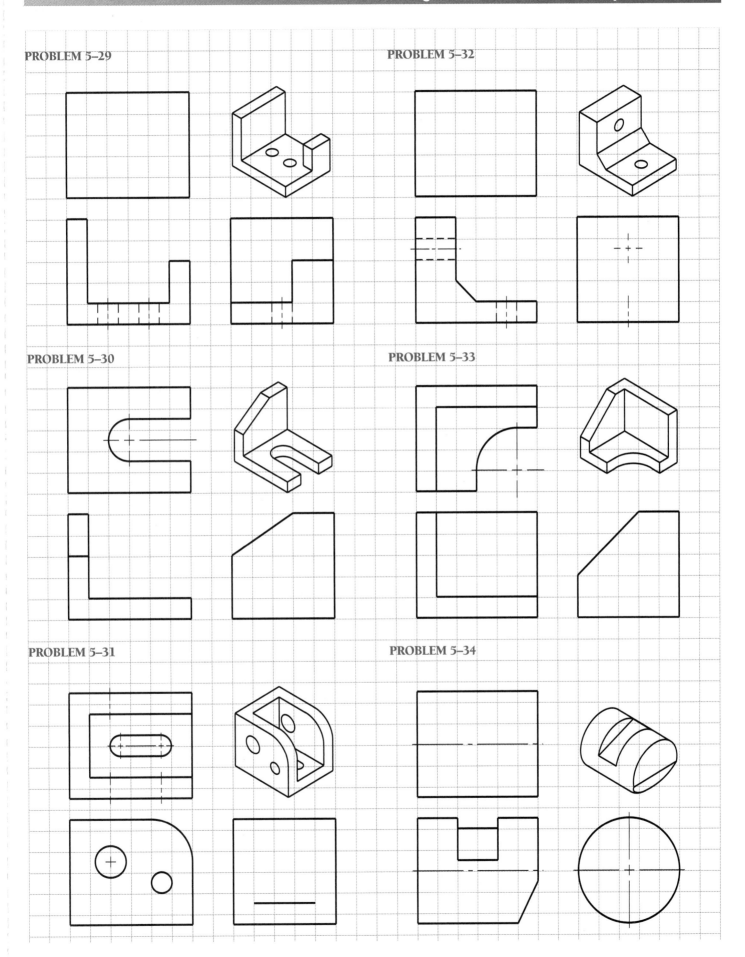

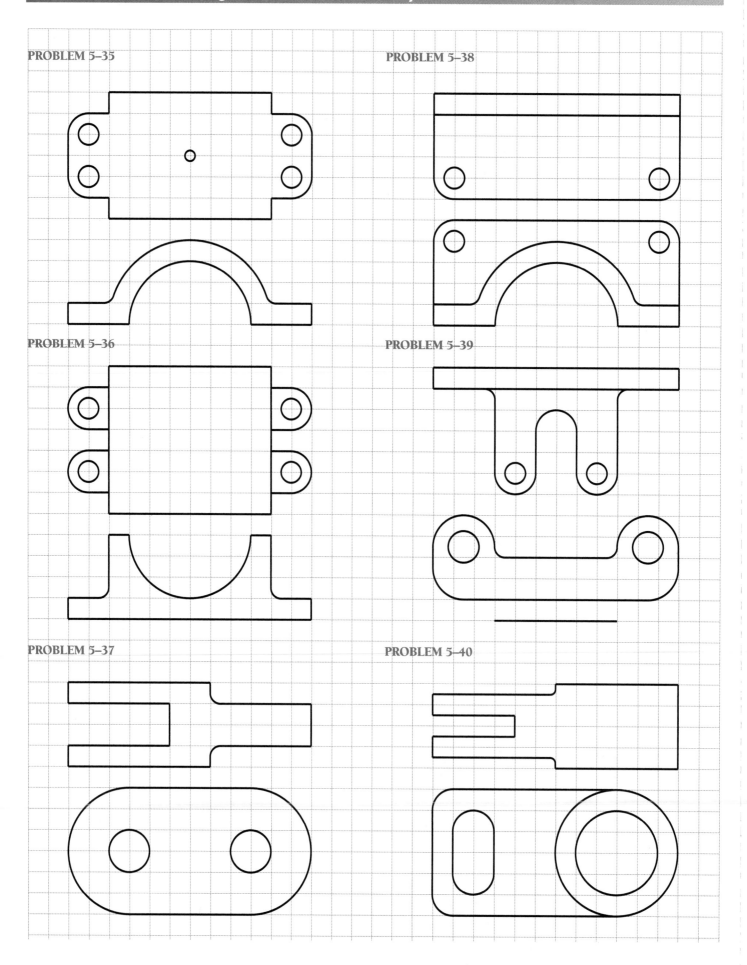

**PROBLEM 5–35**

**PROBLEM 5–36**

**PROBLEM 5–37**

**PROBLEM 5–38**

**PROBLEM 5–39**

**PROBLEM 5–40**

**PROBLEM 5–41** Pocket Block.

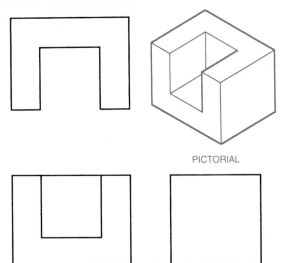

PICTORIAL

**PROBLEM 5–42** Angle Gage.

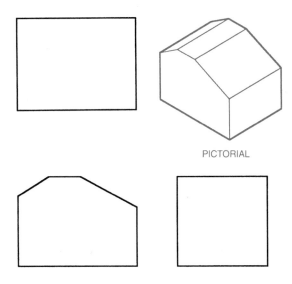

PICTORIAL

**PROBLEM 5–43** Base.

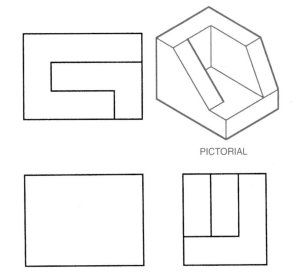

PICTORIAL

**PROBLEM 5–44** Corner Block.

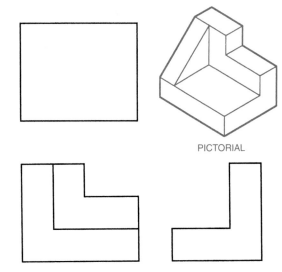

PICTORIAL

**PROBLEM 5–45** Cylinder Block.

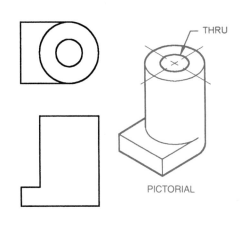

THRU

PICTORIAL

**PROBLEM 5–46** Shaft Block.

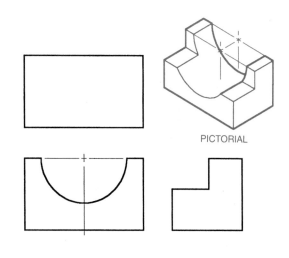

PICTORIAL

**PROBLEM 5–47** Gib.

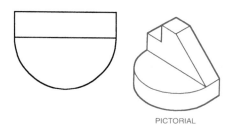

PICTORIAL

**PROBLEM 5–48** Eccentric.

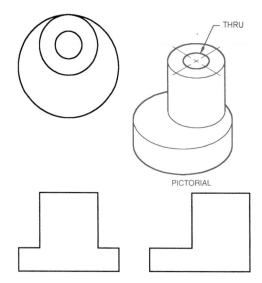

THRU

PICTORIAL

**PROBLEM 5–49** Guide Block.

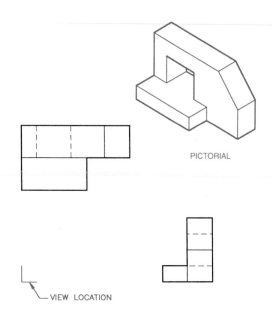

PICTORIAL

VIEW LOCATION

**PROBLEM 5–50** Key Slide.

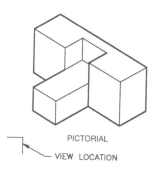

PICTORIAL

VIEW LOCATION

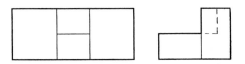

**PROBLEM 5–51** Angle Bracket.

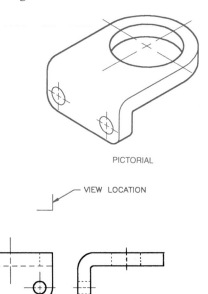

PICTORIAL

VIEW LOCATION

**PROBLEM 5–52** Clevis.

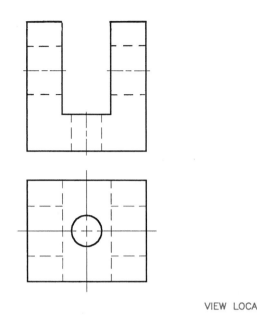

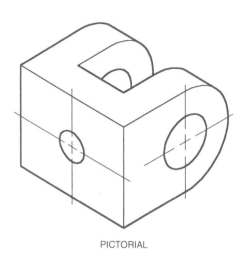

PICTORIAL

VIEW LOCATION

**PROBLEM 5–53** Given the following six views of an object, label each view on the line provided below the view.

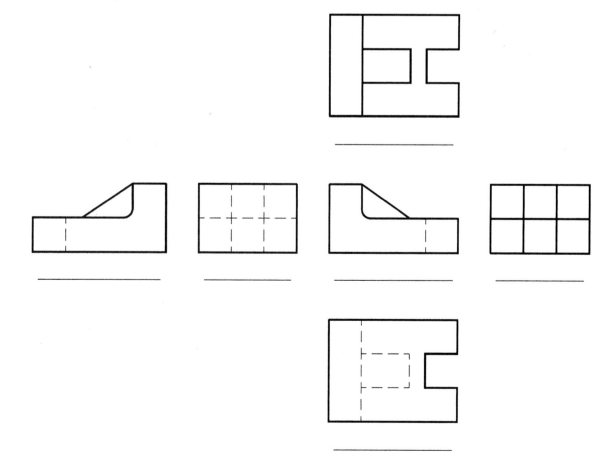

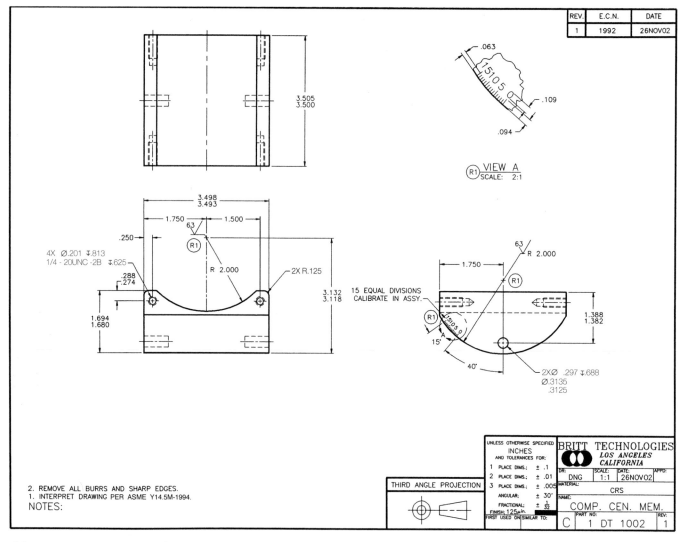

Problem 5–54 *Courtesy David George.*

**PROBLEM 5–54** Answer the following questions as you refer to the print shown on this page.

1.  Name the principal views shown on this print.
    _____

2.  Give the overall width, depth, and height dimensions.
    _____

3.  What is the scale of the principal views?
    _____

4.  What is the purpose of VIEW A?
    _____

5.  What is the scale of VIEW A?
    _____

6.  How many times has this drawing been revised?
    _____

7.  What is the current engineering change notice number?
    _____

8.  Give two indications that this drawing is prepared using third-angle projection.
    _____

9.  What is the tolerance of unspecified three-place dimensions?
    _____

10. Give an example of one general note found on this drawing.
    _____

11. Give at least one example of a specific note found on this drawing.
    _____

12. What does the hidden line that runs laterally from left to right across the right-side view represent?
    _____

13. Can you tell by looking at the front view alone that the bottom of the part is rounded?
    _____

.438

R.063

STOCK
SIZE

30°

2X 45° X .063

| UNLESS OTHERWISE SPECIFIED INCHES AND TOLERANCES FOR: | | DPM MARINE | JET SLEDS MINNEAPOLIS, MINNESOTA |
|---|---|---|---|

UNLESS OTHERWISE SPECIFIED
INCHES
AND TOLERANCES FOR:
1  PLACE DIMS.;    ± .1
2  PLACE DIMS.;    ± .01
3  PLACE DIMS.;    ± .005
   ANGULAR;        ± 30'
   FRACTIONAL;     ± 1/32
FINISH; 125μ in.
FIRST USED ON  SIMILAR TO:

DPM
MARINE

JET SLEDS
MINNEAPOLIS, MINNESOTA

DR: DNG | SCALE: 2:1 | DATE: 24OCT99 | APPD:

MATERIAL: Ø 5/16 BRASS ROD

NAME: CLAMP PLUG

A | PART NO: 1 DT 1010 | REV: 0

2. REMOVE ALL BURRS AND SHARP EDGES.
1. INTERPRET DRAWING PER ASME Y14.5M-1994.
NOTES:

Problem 5–55 *Courtesy David George.*

14. How do you know that the bottom of the part is rounded?

_____

15. If the top view had been omitted, in which view would the 3.505/3.500 dimension be placed?

_____

**PROBLEM 5–55** Answer the following questions as you refer to the print shown on this page.

1. Explain why only one view is used to fully describe this part.

_____

2. What is the material?

_____

3. How does the STOCK SIZE note on the print relate to the material specification in the title block?

_____

4. Is this drawing in inches or millimeters?

_____

5. What is the unspecified tolerance for angular dimensions?

_____

6. What is the sheet size?

_____

7. What does the centerline in this view represent?

_____

8. What is the scale of this drawing?

_____

9. How many times has this drawing been revised?

_____

10. What is the total length of the part?

_____

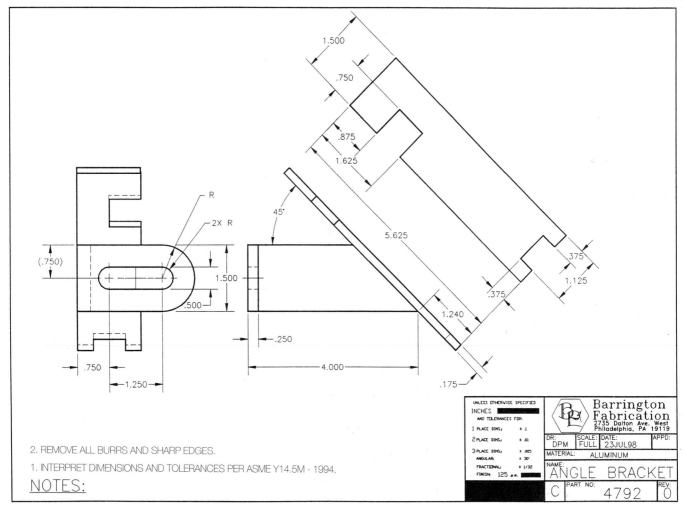

2. REMOVE ALL BURRS AND SHARP EDGES.

1. INTERPRET DIMENSIONS AND TOLERANCES PER ASME Y14.5M - 1994.

NOTES:

Problem 5–56 *Courtesy David George.*

## Chapter 5 Auxiliary View Problems

**PROBLEM 5–56** Answer the following questions as you refer to the print shown on this page.

1. Fully describe the views shown on this print.

2. What is the view called that is placed at an angle?

3. What is the purpose of the view described in Question 2?

4. Why is the view described in Question 2 necessary?

5. What is the angle from horizontal of the slanted surface?

6. Name the material used to make this part.

7. What is the part name?

8. Give the overall dimensions of the slanted surface.

9. What size sheet was used for this drawing?

10. What is the thickness of the slanted surface?

You are given an object with its surfaces projected to the planes of the glass box. This can help you visualize as you sketch the same views on the folded-out planes of the glass box at the left. Measure the given objects and transfer the measurements to your sketch.

**PROBLEM 5–57**

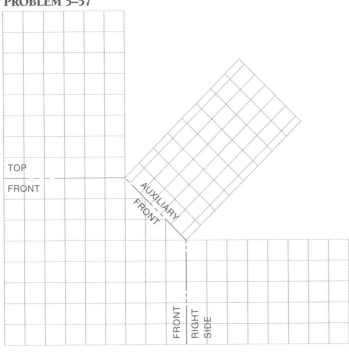

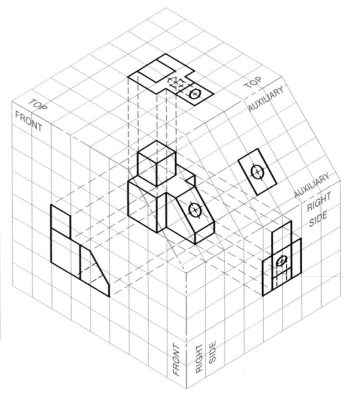

**PROBLEM 5–58**

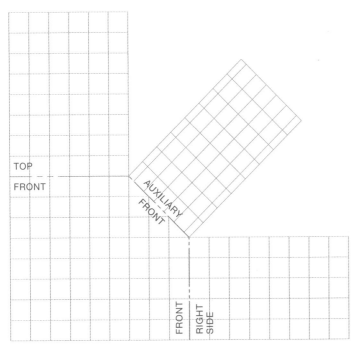

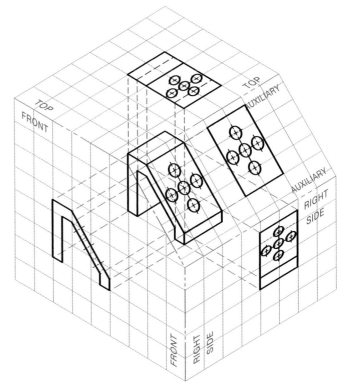

These problems provide you with views that contain missing auxiliary views. Sketch the missing auxiliary views as appropriate. These problems give you a pictorial view to aid in visualization.

**PROBLEM 5–59**

**PROBLEM 5–61**

**PROBLEM 5–60**

**PROBLEM 5–62**

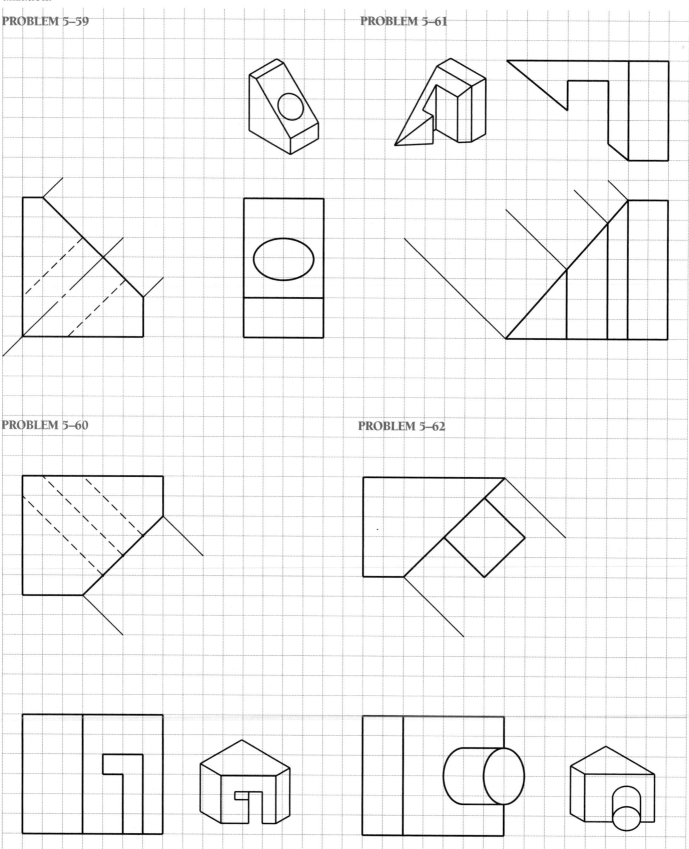

These problems provide you with views that contain missing auxiliary views. Sketch the missing auxiliary views as appropriate. The recommended auxiliary view projection and a started view or view location is given for your reference.

Measure the given views and transfer the measurements to your sketch.

**PROBLEM 5–63**

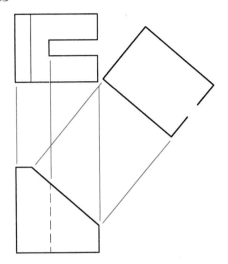

**PROBLEM 5–64**

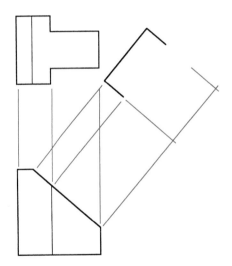

**PROBLEM 5–65**

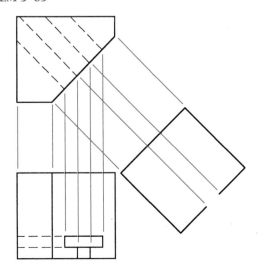

**PROBLEM 5–66**

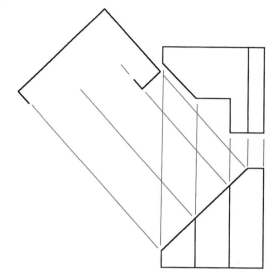

**PROBLEM 5–67**

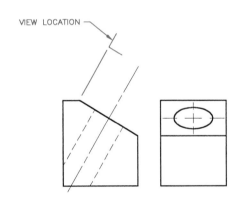

VIEW LOCATION

**PROBLEM 5–68**

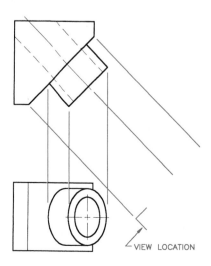

VIEW LOCATION

**PROBLEM 5–69**

**PROBLEM 5–70**

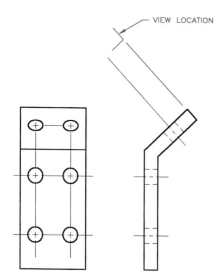

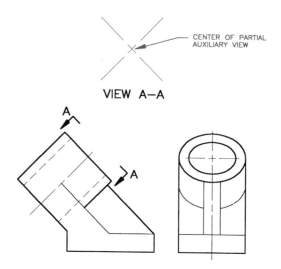

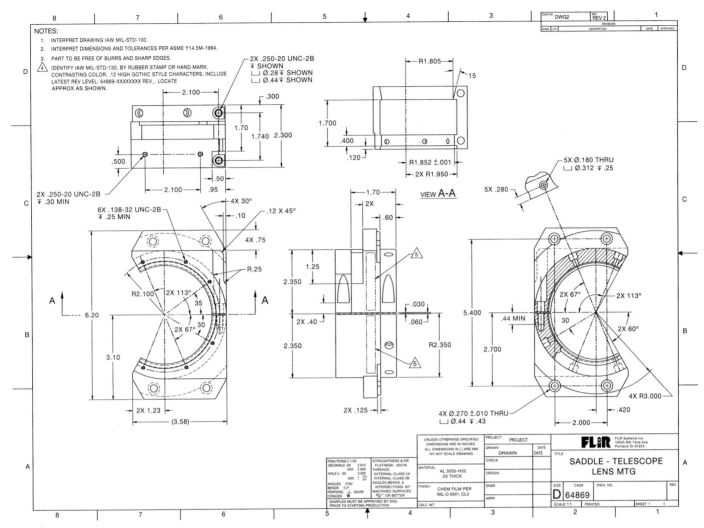

## Chapter 5 Problem: Reading a Complex Print

**PROBLEM 5–71** Answer the following questions as you refer to the print on the this page.

1. Give the overall width, height, and depth of the part.

2. Identify the radius to the center of the 6X .138-32UNC-2B features.

3. Name the view where the features identified in question number 2 are found.

4. Identify the material for this part.

5. Name the view where the 2X .250–20UNC–2B features are found.

6. How is the auxiliary view correlated to its adjacent view?

7. Give the location dimension to the center of the feature in the auxiliary view.

8. Identify the angle of the surface off the R1.805 dimension in VIEW A-A.

# Manufacturing Materials and Processes

6

## LEARNING OBJECTIVES

After completing this chapter you will be able to:

- Identify manufacturing materials from written descriptions.
- Identify casting, forging, and machining processes used in the manufacturing industry.
- Read prints to identify manufacturing materials, features, and processes.
- Identify surface finishes on a print.

- Answer questions related to tooling.
- Answer questions related to computer numerical control (CNC) and computer-integrated manufacturing (CIM).
- Answer questions regarding statistical process control (SPC).

## INTRODUCTION

A number of factors influence product manufacturing. Beginning with research and development (R & D), a product should be designed to meet a market demand, be of good quality, and be economically produced. The sequence of product development begins with an idea and results in a marketable commodity, as shown in Figure 6–1.

Drawings are created during the first three phases, and reading the prints is a critical part of the manufacturing and assembly phases of the process. It's important that you have a solid knowledge of manufacturing materials and processes to be able to effectively read engineering and manufacturing technology prints.

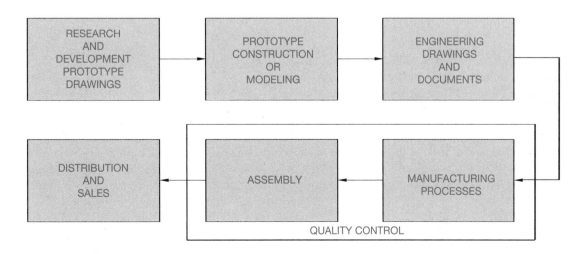

Figure 6–1 Sequence of product development.

# MANUFACTURING MATERIALS

A wide variety of materials available for product manufacturing falls into three general categories: metal, plastic, and inorganic materials. Metals are classified as *ferrous, nonferrous,* and *alloys.* Ferrous metals contain iron, such as cast iron and steel. Nonferrous metals do not have iron content; for example, copper and aluminum. Alloys are a mixture of two or more metals.

*Plastics* or polymers include two types of structure: thermoplastic and thermoset. *Thermoplastic* material can be heated and formed by pressure and, upon reheating, the shape can be changed. *Thermoset* plastics are formed into permanent shape by heat and pressure and cannot be altered by heating after curing. Plastics are molded into shape and require machining only for tight tolerance situations or when holes or other features are required that would be impractical to produce in a mold. It is common practice to machine some plastics for parts, such as gears and pinions.

Inorganic materials include carbon, ceramics, and composites. Carbon and graphite are classified together and have properties that allow molding by pressure. These materials have low *tensile strength* (ability to be stretched) and high compressive strength with increasing strength at increased temperatures. Ceramics are clay, glass, refractory, and inorganic cements. Ceramics are very hard, brittle materials that are resistant to heat, chemicals, and corrosion. Clay and glass materials have an amorphous structure, while *refractories* must be bonded together by applying temperatures. Because of their great heat resistance, refractories are used for high-temperature applications, such as furnace liners. *Composites* are two or more materials that are bonded together by adhesion. *Adhesion* is a force that holds together the molecules of unlike substances when the surfaces come in contact. These materials generally require carbide cutting tools or special methods for machining.

## Ferrous Metals

The two main types of ferrous metals are cast iron and steel. These are the metals that contain iron. Many classes of cast iron and steel are created for specific types of applications based on their composition.

## Cast Iron

There are several classes of cast iron, including gray, white, chilled, alloy, malleable, and nodular cast iron. Cast iron is primarily an alloy of iron and 1.7% to 4.5% of carbon, containing varying amounts of silicon, manganese, phosphorus, and sulfur.

### Gray Cast Iron

Gray iron is a popular casting material for automotive cylinder blocks, machine tools, agricultural implements, and cast iron pipe. Gray cast iron is easily cast and machined. It contains 1.7% to 4.5% carbon and 1% to 3% silicon.

*ANSI/ASTM* The American National Standards institute (ANSI) and *American Society for Testing Materials* (ASTM) specifications A48-76 group gray cast iron into two classes: easy to manufacture (20A, 20B, 20C, 25A, 25B, 25C, 30A, 30B, 30C, 35A, 35B, 35C);

and more difficult to manufacture (40B, 40C, 45B, 45C, 50B, 50C, 60B, 60C). The prefix denotes the minimum tensile strength in thousand pounds per square inch.

### White Cast Iron

White cast iron is extremely hard, brittle, and has almost no ductility. *Ductility* is the ability to be stretched, drawn, or hammered thin without breaking. Caution should be exercised when using this material, because thin sections and sharp corners may be weak, and the material is less resistant to impact uses. This cast iron is suited for products with more compressive strength requirements than gray cast iron (compare over 200,000 pounds per square inch (psi) with 65,000 to 160,000 psi). White cast iron is used where high wear resistance is required.

### Chilled Cast Iron

When gray iron castings are chilled rapidly, an outer surface of white cast iron results. This material has the internal characteristics of gray cast iron and the surface advantage of white cast iron.

### Alloy Cast Iron

Elements such as nickel, chromium, molybdenum, copper, or manganese may be alloyed with cast iron to increase the properties of strength, wear resistance, corrosion resistance, or heat resistance. Alloy iron castings are commonly used to make pistons, crankcases, brake drums, and crushing machinery.

### Malleable Cast Iron

The term *malleable* describes the ability to be hammered or pressed into shape without breaking. Malleable cast iron is produced by heat-treating white cast iron. The result is a stronger, more ductile, and shock-resistant cast iron that is easily machinable.

*ANSI/ASTM* The specifications for malleable cast iron are found in ANSI/ASTM A47-77.

### Nodular Cast Iron

Special processing procedures, along with the addition of magnesium- or cerium-bearing alloys, result in a cast iron with spherical-shaped graphite rather than flakes, as in gray cast iron. The results are iron castings with greater strength and ductility. Nodular cast iron can be chilled to form a wear-resistant surface ideal for use in crankshafts, anvils, wrenches, or heavy-use levers.

## Steel

Steel is an alloy of iron containing 0.8% to 1.5% carbon. Steel is a readily available material that may be worked in either a heated or cooled state. The properties of steel can be changed by altering the carbon content and heat treating. *Mild steel* (MS) is low in carbon (less than 0.3%), and is commonly used for forged and machined parts, but cannot be hardened. *Medium-carbon steel* (0.3% to 0.6% carbon) is harder than mild steel yet remains easy to forge and machine. *High-carbon steel* (0.6% to 1.50% carbon) can be hardened by heat treating, but is difficult to forge, machine, or weld.

*Hot-rolled steel* (HRS) characterizes steel that is formed into shape by pressure between rollers or by forging when in a red-hot state. When in this hot condition, the steel is easier to form than when it is cold. An added advantage of hot forming is a consistency in the grain structure of the steel, which results in a stronger, more ductile metal. The surface of hot-rolled steel is rough, with a blue-black oxide buildup. The term *cold-rolled steel* (CRS) implies the additional forming of steel after initial hot rolling. The cold-rolling process is used to clean up hot-formed steel; provide a smooth, clean surface; ensure dimensional accuracy; and increase the tensile strength of the finished product.

Steel alloys are used to increase the properties of hardness, strength, corrosion resistance, heat resistance, and wear resistance. Chromium steel is the basis for stainless steel and is used where corrosion and wear resistance is required. Manganese alloyed with steel is a purifying element that adds strength for parts that must be shock- and wear-resistant. Molybdenum is added to steel when the product must retain strength and wear resistance at high temperatures. When tungsten is added to steel, the result is a material that is very hard and ideal for use in cutting tools. Tool steels are high in carbon and/or alloy content so that the steel holds an edge when cutting other materials. When the cutting tool requires deep cutting at high speed, the alloy and hardness characteristics are improved for a classification known as *high-speed steel*. Vanadium alloy is used when a tough, strong, nonbrittle material is required.

Steel castings are used for machine parts where the use requires heavy loads and the ability to withstand shock. These castings are generally stronger and tougher than cast iron. Steel castings have uses as turbine wheels, forging presses, gears, machinery parts, and railroad car frames.

## Stainless Steel

Stainless steels are high-alloy chromium steels that have excellent corrosion resistance. In general, stainless steels contain at least 10.5% chromium, with some classifications of stainless steel having between 4% and 30% chromium. In addition to corrosion-resistance properties, stainless steel can resist oxidation and heat, and is very strong. Stainless steel is commonly used for restaurant and hospital equipment and for architectural and marine applications. The high strength-to-weight ratio also makes stainless steel a good material for some aircraft applications.

Stainless steel is known for its natural luster and shine that makes it look like chrome. You can also add color to stainless steel, which is used for architectural products—for example, roofing, hardware, furniture, and kitchen and bathroom fixtures.

The American Iron and Steel Institute (AISI) identifies stainless steels with a system of 200, 300, or 400 series numbers. The 200 series stainless steels contain chromium, nickel, and manganese; the 300 series steels have chromium and nickel; and the 400 series steels are straight-chromium stainless steels.

## Steel Numbering Systems

*AISI/SAE* The American Iron and Steel Institute (AISI) and the Society of Automotive Engineers (SAE) provide similar steel numbering systems. Steels are identified by four numbers, except for some chromium steels, which have five numbers. In steel identi-

fied as SAE 1020, the first two numbers (10) describe the type of steel and the last two numbers (20) specify the approximate amount of carbon in hundredths of a percent (0.20% carbon). The letter *L* or *B* may be placed between the first and second pair of numbers. When you encounter this, the *L* means that lead is added to improve machinability, and the *B* identifies a boron steel. The prefix *E* means that the steel is made using the electric furnace method. The prefix *H* indicates that the steel is produced to hardenability limits. Steel that is degassed and deoxidized before solidification is referred to as *killed* steel and is used for forging, heat treating, and difficult stampings. Steel that is cast with little or no degasification is known as *rimmed steel*, and has applications where sheets, strips, rods, and wires with excellent surface finish or drawing requirements are needed. General applications of SAE steels are shown in the online Appendix B, Table 25 (http://www.delmarlearning.com/companions/start.asp). For a more in-depth analysis of steel and other metals, refer to the *Machinery's Handbook*.* Additional information about numbering systems is provided later in this chapter.

## Hardening of Steel

The properties of steel can be altered by heat treating. Heat treating is a process of heating and cooling steel using specific controlled conditions and techniques. When steel is initially formed and allowed to cool naturally, it is somewhat soft. *Normalizing* is a process of heating the steel to a specified temperature and then allowing the material to cool slowly by air, which brings the steel to a normal state. To harden steel, the metal is first heated to a specified temperature that varies with different steels. Next the steel is quenched, or cooled suddenly by plunging into water, oil, or another liquid. Steel may also be *case hardened* using a process known as *carburization*. Case hardening refers to the hardening of the surface layer of the metal. Carburization is a process where carbon is introduced into the metal by heating to a specified temperature range while in contact with a solid, liquid, or gas material consisting of carbon. This process is often followed by quenching to enhance the hardening process. *Tempering* is a process of reheating a normalized or hardened steel through a controlled process of heating the metal to a specified temperature, followed by cooling at a predetermined rate to achieve certain hardening characteristics. For example, the tip of a tool can be hardened while the balance of the tool remains unchanged.

Under certain heating and cooling conditions and techniques, steel may also be softened using a process known as annealing.

## Hardness Testing

There are several methods of checking material hardness. The techniques have common characteristics based on the depth of penetration of a measuring device or other mechanical systems that evaluate hardness. The Brinell and Rockwell hardness tests are popular. The Brinell test is performed by placing a known load, using a ball of a specified diameter, in contact with the material surface. The diameter of the resulting impression in the material is measured and the Brinell Hardness Number (BHN) is then

---

*Erick Oberg, Franklin D. Jones, and Holbrook L. Horton, *Machinery's Handbook, 25th ed.* (New York: Industrial Press Inc., 1996).

calculated. The Rockwell hardness test is performed using a machine that measures hardness by determining the depth of penetration of a spherical-shaped device under controlled conditions. Several Rockwell hardness scales are used, depending on the type of material, the type of penetrator, and the load applied to the device. A general or specific note on a drawing that requires a hardness specification may read: CASE HARDEN 58 PER ROCKWELL "C" SCALE. For additional information, refer to the *Machinery's Handbook*.

## Nonferrous Metals

Metals that do not contain iron have properties that are better suited for certain applications where steel may not be appropriate.

### Aluminum

Aluminum is corrosion resistant, lightweight, easily cast, conductive of heat and electricity, can be easily *extruded*, and is very malleable. Extruding is shaping the metal by forcing it through a die. Pure aluminum is seldom used, but alloying it with other elements produces materials that have an extensive variety of applications. Some aluminum alloys lose strength at temperatures generally above 121°C (250°F), while they gain strength at cold temperatures. A variety of aluminum alloys are identified by numerical designations. A two- or three-digit number is used, with the first digit indicating the alloy type, as follows: 1 = 99% pure, 2 = copper, 3 = manganese, 4 = silicon, 5 = magnesium, 6 = magnesium and silicon, 7 = zinc, 8 = other. For designations above 99%, the last two digits of the code indicate the amount over 99%. For example, 1030 means 99.30% aluminum. The second digit is any number between 0 and 9, where 0 means no control of specific impurities and numbers 1 through 9 identify control of individual impurities.

### Copper Alloys

Copper is easily rolled and drawn into wire, has excellent corrosion resistance, is a great electrical conductor, and exhibits better ductility than any metal except for silver and gold. Copper is alloyed with many different metals for specific advantages, which can include improved hardness, casting ability, machinability, corrosion resistance, elastic properties, and lower cost.

### Brass

Brass is a widely used alloy of copper and zinc. Its properties include corrosion resistance, strength, and ductility. For most commercial applications, brass has about 90% copper and 10% zinc content. Brass is manufactured by any number of processes, including casting, forging, stamping, or drawing. Valves, plumbing pipe and fittings, and radiator cores are made from brass. Brass with greater zinc content can be used for applications requiring greater ductility, such as in producing cartridge cases, sheet metal, or tubing.

### Bronze

Bronze is an alloy of copper and tin. Tin, in small quantities, adds hardness and increases wear resistance. Tin content in coins and medallions, for example, ranges from 4% to 8%. Increasing amounts of tin also improves the hardness and wear resistance of the material, but causes brittleness. Phosphorus added to bronze (phosphor bronze) increases its casting ability and helps produce

more solid castings, which is important for thin shapes. Other materials, such as lead, aluminum, iron, and nickel, may be added to copper for specific applications.

## Precious and Other Specialty Metals

Precious metals include gold, silver, and platinum. These metals are valuable because they are rare, costly to produce, and have specific properties that influence use in certain applications.

### Gold

Gold for coins and jewelry is commonly hardened by adding copper. Gold coins, for example, are 90% gold and 10% copper. The term carat is used to refer to the purity of gold, where 1/24 gold is one carat. Therefore, 24 x 1/24, or 24 carats, is pure gold. Fourteen-carat gold, for example, is 14/24 gold and 10/24 copper. Gold is extremely malleable, corrosion resistant, and the best conductor of electricity. In addition to use in jewelry and coins, gold is used as a conductor in some electronic circuitry applications.

Gold is also used in applications where resistance to chemical corrosion is required.

### Silver

Silver is alloyed with 8% to 10% copper for use in jewelry and coins. Sterling silver that is used to make eating utensils and other household items is 925/1000 silver. Silver is easy to shape, cast, or form, and the finished product can be polished to a high-luster finish. Like gold, silver is useful because of its corrosion resistance and ability to conduct electricity.

### Platinum

Platinum is rarer and more expensive than gold. Industrial uses include applications where corrosion resistance and a high melting point are required. Platinum is used in catalytic converters because it has the unique ability to react with and reduce carbon monoxide and other harmful exhaust emissions in automobiles. The high melting point of platinum makes it desirable in certain aerospace applications.

### Columbium

Columbium is used in nuclear reactors because it has a very high melting point, 2403°C (4380°F), and is resistant to radiation.

### Titanium

Titanium has many uses in the aerospace and jet aircraft industries because it has the strength of steel, the approximate weight of aluminum, and is resistant to corrosion and temperatures up to 427°C (800°F).

### Tungsten

Tungsten has been used extensively as the filament in light bulbs because of its ability to be drawn into very fine wire and its high melting point. Tungsten, carbon, and cobalt are formed together under heat and pressure to create tungsten carbide, the hardest manmade material. Tungsten carbide is used to make cutting tools for any type of manufacturing application. Tungsten carbide saw-blade inserts are used in carpenters' saws so the cutting edge will last longer. Tungsten blades make a finer and faster cut than plain steel saw blades.

## A Unified Numbering System for Metals and Alloys

Many numbering systems have been developed for the identification of metals. The organizations that developed these numbering systems include the American Iron and Steel Institute (AISI), Society of Automotive Engineers (SAE), American Society for Testing Materials (ASTM), American National Standards Institute (ANSI), Steel Founders Society of America, American Society of Mechanical Engineers (ASME), American Welding Society (AWS), Aluminum Association, and Copper Development Association. Numbering systems for military specifications have been developed by the U.S. Department of Defense, and the General Accounting Office has created similar systems for federal specifications.

A combined numbering system created by the ASTM and the SAE was established in an effort to coordinate all the different numbering systems into one system. This system avoids the possibility that the same number might be used for two different metals. This combined system is the Unified Numbering System (UNS). The UNS is an identification numbering system for commercial metals and alloys; it does not provide metal and alloy specifications. The UNS system is divided into the following categories.

| UNS Series | Metal |
| --- | --- |
| **Nonferrous Metals and Alloys** | |
| A00001 to A99999 | Aluminum and aluminum alloys |
| C00001 to C99999 | Copper and copper alloys |
| E00001 to E99999 | Rare-earth and rare-earth-like metals and alloys |
| L00001 to L99999 | Low-melting metals and alloys |
| M00001 to M99999 | Miscellaneous metals and alloys |
| P00001 to P99999 | Precious metals and alloys |
| R00001 to R99999 | Reactive and refractory metals / alloys |
| Z00001 to Z99999 | Zinc and zinc alloys |
| | |
| **Ferrous Metals and Alloys** | |
| D00001 to D99999 | Specified mechanical properties steels |
| F00001 to F99999 | Cast irons |
| G00001 to G99999 | AISI and SAE carbon and alloy steels, excluding tool steels |
| H00001 to H99999 | AISI H-classification steels |
| J00001 to J99999 | Cast steels, excluding tool steels |
| K00001 to J99999 | Miscellaneous steels and ferrous alloys |
| S00001 to S99999 | Stainless steels |
| T00001 to T99999 | Tool steels |

The prefix letters of the UNS system often match the type of metal being identified. For example, A for aluminum, C for copper, and T for tool steels. Elements of the UNS numbers typically match numbers provided by other systems; for example, SAE1030 is G10300 in the UNS system.

## PLASTICS

A general definition of plastic is any complex, organic, polymerized compound capable of being formed into a desired shape by modeling, casting, or spinning. Plastic retains its shape under ordinary conditions of temperature. *Polymerization* is a process of joining two or more molecules to form a more complex molecule with physical properties that are different from the original molecules. The terms *plastic* and *polymer* are often used to mean the same thing. Plastics can appear in any state from liquid to solid. The main elements of plastic are generally common petroleum products, crude oil, and natural gas.

Many types of plastics are available for use in the design and manufacture of products. These plastics fall into two main categories, called thermoplastics and thermosets. *Thermoplastics* can be heated and formed by pressure, and the shape can change when reheated. Thermosets are formed into shape by heat and pressure and cannot be changed into a different shape after curing. Most plastic products are made with thermoplastics, because they are easy to make into shapes by heating, forming, and cooling. Thermoset plastics are the choice when the end product is used in an application where heat exists, such as the distributor cap and other plastic parts found on or near the engine of your car. Elastomers are polymer-based materials that have elastic qualities not found in thermoplastics and thermosets. You can stretch elastomers at least equal to their original length, after which they return to their original length

### Thermoplastics

Although there are thousands of different thermoplastic combinations, some more commonly used alternatives follow. You may recognize some of them by their acronyms, such as PVC.

*Acetal.* Acetal is a rigid thermoplastic that has good corrosion resistance and machinability. While it can burn, it works well in applications where friction, fatigue, toughness, and tensile strength are factors. Some applications for its use include manufacturing of gears, bushings, bearings, and products that come in contact with chemicals or petroleum.

*Acrylic.* Acrylics are used when a transparent plastic is needed clear or in a variety of colors. Acrylics are commonly used in products that you see through or through which light passes because, in addition to transparency, they are scratch- and abrasion-resistant and weatherproof. Acrylics make good windows, light fixtures, and lenses.

*Acrylic-styrene-acrylonitrile (ASA).* ASA has very good weatherability for use as siding, pools and spas, exterior car and marine parts, outdoor furniture, and garden equipment.

*Acrylonitrile-butadiene-styrene (ABS).* ABS is one of the most commonly used plastics because of its excellent impact strength, reasonable cost, and ease of processing. ABS also has good dimensional stability, temperature resistance above 212°F

(100°C), and chemical and electrical resistance. ABS is commonly used in products that you see daily, such as electronics enclosures, knobs, handles, and appliance parts.

*Cellulose.* Five versions of cellulose are nitrate, acetate, butyrate, propionate, and ethyl cellulose.

- Nitrate is tough, but very flammable, explosive, and difficult to process. Nitrate is commonly used to make products such as photo film, combs, brushes, and buttons.

- Acetate is not explosive (though flammable), is not solvent resistant, and becomes brittle with age. Acetate has the advantage of being transparent, can be made in bright colors, is tough, and is easy to process. It is often used for transparency film, magnetic tape, knobs, and sunglass frames.

- Butyrate is similar to acetate, but is used in applications where moisture resistance is needed. Butyrate is used for exterior light fixtures, handles, film, and outside products.

- Propionate has good resistance to weathering, is tough and impact-resistant, and becomes less brittle with age. Propionate is used for flashlights, automotive parts, small electronics cases, and pens.

- Ethyl cellulose has very high shock resistance and durability at low temperatures, but has poor weatherability.

*Fluoroplastics.* There are four types of fluoroplastics that have similar characteristics. These plastics are very resistant to chemicals, friction, and moisture. They have excellent dimensional stability for use as wire coating and insulation, nonstick surfaces, chemical containers, O-rings, and tubing.

*Ionomers.* Ionomers are very tough; resistant to abrasion, stress, cold, and electricity; and very transparent. They are commonly used for cold food containers, other packaging, and film.

*Liquid crystal polymers.* This plastic can be made very thin with excellent temperature, chemical, and electrical resistance. Uses include cookware and electrical products.

*Methyl pentenes.* Methyl pentenes have excellent heat and electrical resistance, and are very transparent. These plastics are used for items such as medical containers and products, cooking and cosmetic containers that require transparency, and pipe and tubing.

*Polyallomers.* This plastic is rigid with very high impact- and stress-fracture resistance at temperatures between −40° to 210°F (−40° to 99°C). Polyallomers are used when constant bending is required in the function of the material design.

*Polyamide (nylon).* Commonly called nylon, this plastic is tough; abrasion, heat, and friction-resistant; and strong. Nylon is corrosion-resistant to most chemicals, but is not as dimensionally stable as other plastics. Nylon is used for combs, brushes, tubing, gears, cams, and stocks.

*Polyarylate.* This material is impact-, weather-, and electricity-resistant, and extremely fire-resistant. Typical uses include electrical insulators, cookware, and other options where heat is an issue.

*Polycarbonates.* Excellent heat resistance, impact strength, dimensional stability, and transparency are positive characteristics of this plastic. Additionally, this material does not stain or corrode, but has moderate chemical resistance. Common uses include food containers, power tool housings, outside light fixtures, and appliance and cooking parts.

*Polyetheretherketone (PEEK).* PEEK has excellent heat-, fire-, abrasion-, and fatigue-resistance qualities. Typical uses include high electrical components, aircraft parts, engine parts, and medical products.

*Polyethylene.* This plastic has excellent chemical resistance and has properties that make it good to use for slippery or nonstick surfaces. Polyethylene is a common plastic that is used for chemical, petroleum, and food containers; plastic bags; pipe fittings; and wire insulation.

*Polyimides.* This plastic has excellent impact strength, wear resistance, and very high heat resistance, but it is difficult to produce. Products made from this plastic include bearings, bushings, gears, piston rings, and valves.

*Polyphenylene oxide (PPO).* PPO has an extensive range of temperature use from −275° to 375°F −170° to 191°C), and is fire-retardant and chemical-resistant. Applications include containers that require super-heated steam, pipe and fittings, and electrical insulators.

*Polyphenylene sulfide (PPS).* PPS has the same characteristics as PPO, but it is easier to manufacture.

*Polypropylene.* This is an inexpensive plastic to produce with many desirable properties, including heat, chemical, scratch, and moisture resistance. It is also resistant to continuous bending applications. Products include appliance parts, hinges, cabinets, and storage containers.

*Polystyrene.* This plastic is inexpensive and easy to manufacture, has excellent transparency, and is very rigid. However, it can be brittle and has poor impact, weather, and chemical resistance. Products include model kits, plastic glass, lenses, eating utensils, and containers.

*Polysulfones.* This material resists electricity and some chemicals, but can be damaged by certain hydrocarbons. While somewhat difficult to manufacture, it has good structural applications at high temperatures, and can be made in several colors. Common applications are hot water products, pump impellers, and engine parts.

*Polyvinylchloride (PVC).* PVC is one of the most common products found for use as plastic pipes and vinyl house siding because of its ability to resist chemicals and the weather.

*Thermoplastic polyesters.* There are two types of this plastic that exhibit strength and good electrical, stress, and chemical resistance. Common uses include electrical insulators, packaging, automobile parts, and cooking and chemical-use products.

*Thermoplastic rubbers (TPR).* This resilient material is used where tough, chemical-resistant plastic is needed. Uses include tires, toys, gaskets, and sports products.

## Thermosets

Thermoset plastics make up only about 15% of the plastics used because they are more expensive to produce; they are generally more brittle than thermoplastics; and once they are molded into shape, they cannot be remelted. However, their use is important in products that require a rigid and harder plastic than thermoplastic materials, and in applications where heat could melt thermoplastics. The following provides information about common thermoset plastics.

*Alkyds.* These plastics can be used in molding processes, but they are generally used as paint bases.

*Melamine formaldehyde.* This is a rigid thermoset plastic that is easily molded, economical, nontoxic, tough, and abrasion- and temperature-resistant. Common uses include electrical devices, surface laminates, plastic dishes, cookware, and containers.

*Phenolics.* The use of this material dates back to the late 1800s. This plastic is hard and rigid, has good compression strength, is tough, and does not absorb moisture, but it is brittle. Phenolic plastics are commonly used for the manufacture of electrical switches and insulators, electronic circuit boards, distributor caps, and binding material and adhesive.

*Unsaturated polyesters.* It is common to use this plastic for reinforced composites, also known as reinforced thermoset plastics (RTP). Typical uses are for boat and recreational vehicle construction, automobiles, fishing rods, tanks, and other structural products.

*Urea formaldehyde.* These plastics are used for many of the same applications as the previously described plastics, but they do not hold up to sunlight exposure. Common uses are for construction adhesives, as well as for limited internal plastic electrical devices.

## Elastomers

The two main types of plastics have been introduced as thermoplastics and thermosets. However, elastomers are also available types of plastics that are elastic, much like rubber. Elastomers can be referred to as synthetic rubbers. Synthetic rubbers are used to produce almost twice as many products as natural rubber. Natural rubber is a material that starts as the sap from some trees. Natural rubber and many synthetic rubbers are processed by combining with adhesives and using a process called vulcanization. Vulcanization is basically the heating of the material in a steel mold that forms the desired shape. The following information describes the most commonly used elastomers.

*Butyl rubber.* This material has a low air-penetration ability and very good resistance to ozone and aging, but has poor petroleum resistance. Common uses include tire tubes and puncture-proof tire liners.

*Chloroprene rubber (Neoprene).* The trade name Neoprene was the first commercial synthetic rubber. This material has better weather, sunlight, and petroleum resistance than natural rubber. It is also very flame-resistant, but does not resist electricity. Common uses include automotive hoses and other products where heat is found, like gaskets, seals, and conveyor belts.

*Chlorosulfonated polyethylene (CSM).* CSM has excellent chemical, weather, heat, electrical, and abrasion resistance. It is typically used in chemical tank liners and electrical resistors.

*Epichlorohydrin rubber (ECO).* This material has great petroleum resistance at very low temperatures. For this reason it is used in cold-weather applications such as snow-handling equipment and vehicles.

*Ethylene propylene rubber (EPM) and ethylene propylene diene monomer (EPDM).* This is a family of materials that has excellent weather, electrical, aging, and good heat resistance. These materials are used for weather stripping, wire insulation, conveyor belts, and many outdoor products.

*Fluoroelastomers (FPM).* FPM materials have excellent chemical and solvent resistance up to 400°F (204°C). FPM is expensive to produce, so it is used only when its positive characteristics are needed.

*Nitrile rubber.* This material resists swelling when immersed in petroleum. Nitrile rubber is used for any application involving fuels and hydraulic fluid such as hoses, gaskets, O-rings, and shoe soles.

*Polyacrylic rubber (ABR).* This material is able to resist hot oils and solvents. ABR is commonly used where the rubber is submerged in oil, such as with transmission seals.

*Polybutadiene.* This material has qualities that are similar to those of natural rubber. It is commonly mixed with other rubbers to improve tear resistance.

*Polyisoprene.* This material, developed during World War II to help with a shortage of natural rubber, has the same chemical structure as natural rubber. However, this synthetic is more expensive to produce.

*Polysulfide rubber.* The advantage of this rubber is that it is petroleum-, solvent-, gas-, moisture-, weather-, and age-resistant. The disadvantage is that it is low in tensile strength, resilience, and tear resistance. Applications include caulking and putty, sealants, and castings.

*Polyurethane.* This material has the ability to act like rubber or hard plastic. Because of its combined rubber and hard characteristics, products made from polyurethane include rollers, wear pads, furniture, and springs. In addition to these products, polyurethane is used to make foam insulation and floor covers.

*Silicones.* This material has a wide range of makeup from liquid to solid. Liquid and semiliquid forms are used for lubricants. Harder forms are used where nonstick surfaces are required.

*Styrene butadiene rubber (SBR).* SBR is very economical to produce and is heavily used for tires, hoses, belts, and mounts.

*Thermoplastic elastomers (TPE).* The other elastomers typically require the fairly expensive vulcanization process to produce. TPE, however, is able to be processed with injection molding just like other thermoplastics, and any scraps can be reused. There are different TPEs and, in general, they are less flexible than other types of rubber.

# MANUFACTURING PROCESSES

Casting, forging, and machining processes are used extensively in the manufacturing industry. It is a good idea for the entry-level drafter or pre-engineer to be generally familiar with types of casting, forging, and machining processes, and to know how to prepare related drawings. Any number of process methods may be used by industry. For this reason, it is best for the beginning drafting technician to remain flexible and adapt to the standards and techniques used by the specific company. As you, the drafting technician, gain knowledge of company products, processes, and design goals, you may begin to produce designs. It is common for a drafter to become a designer after 3 years of practical experience.

## Castings

Castings are the end result of a process sometimes called founding. *Founding*, or *casting*, as the process is commonly called, is the pouring of molten metal into a hollow or wax-filled mold. The mold is made in the shape of the desired casting. Several casting methods are used in industry. The results of some of the processes are castings that are made to very close tolerances and with smooth-finished surfaces. In the simplest terms, castings are made in three separate steps:

1. A pattern that is the same shape as the desired finished product is constructed.

2. Using the pattern as a guide, a mold is made by packing sand or other material around the pattern.

3. When the pattern is removed from the mold, molten metal is poured into the hollow cavity. After the molten metal solidifies, the surrounding material is removed and the casting is ready for cleanup or machining operations.

### Sand Casting

Sand casting is the most commonly used method of making castings. There are two general types of sand castings: green sand and dry sand molding. *Green sand* is a specially refined sand that is mixed with a specific moisture, clay, and resin, which work as binding agents during the molding and pouring procedures. New sand is light brown; the term "green sand" refers to the moisture content. In the *dry sand* molding process, the sand does not have any moisture content. The sand is bonded together with specially formulated resins. The end result of the green sand or the dry sand molds is the same.

Sand castings are made by pounding or pressing the sand around a split pattern. The first, or lower, half of the pattern is placed upside down on a molding board, then sand is pounded or compressed around the pattern in a box called a *drag*. The drag is then turned over, and the second, or upper, half of the pattern is formed when another box, called a *cope*, is packed with sand and joined to the drag. A fine powder is used as a parting agent between the cope and drag at the parting line. The parting line is the separating joint between the two parts of the pattern or mold. The entire box, made up of the cope and drag, is referred to as a flask (see Figure 6–2).

Before the molten metal can be poured into the cavity, a passageway for the metal must be made. The passageway is called a *runner* and *sprue*. The location and design of the sprue and runner are important to allow for a rapid and continuous flow of metal. Additionally, vent holes are established to allow gases, impurities, and metal to escape from the cavity. Finally, a riser (or group of risers) is used, depending on the size of the casting, to allow the excess metal to evacuate from the mold and, more important, to help reduce shrinking and incomplete filling of the casting. (See Figure 6–3.) After the casting has solidified and cooled, the filled risers, vent holes, and runners are removed.

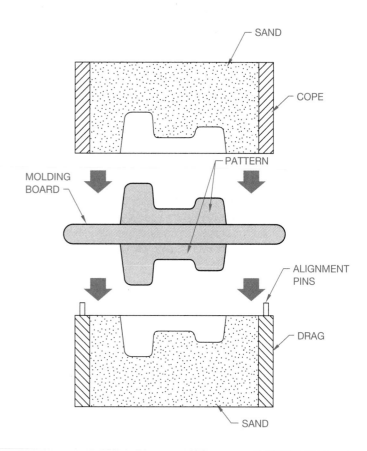

Figure 6–2 Components of the sand casting process.

### Cores

Often a hole or cavity is desired in a casting to help reduce the amount of material removal later or to establish a wall thickness. When this is necessary, a core is used. Cores are made from either clean sand mixed with binders such as resin and baked in an oven for hardening or ceramic products, when a more refined surface finish is required. When the pattern is made, a place for positioning the core in the mold is established; this is referred to as the core print. After the mold is made, the core is then placed in position in the mold at the core print. The molten metal, when poured into the mold, flows around the core. After the metal has cooled, the casting is removed from the flask and the core is cleaned, usually by shaking or tumbling. Figure 6–4 shows cores in place. Cored features help reduce casting weight and save on machining costs. Cores used in sand casting should generally be more than one inch

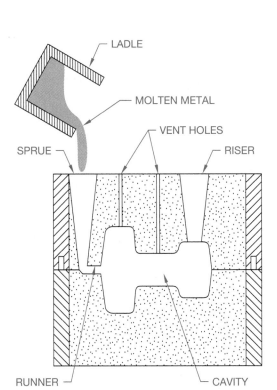

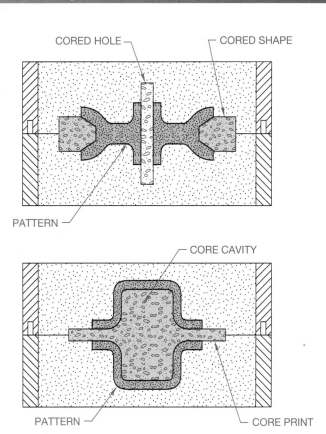

Figure 6–3 Pouring molten metal into a sand casting mold.

Figure 6–4 Cores in place.

(25.4 mm) in cross-section. Cores used in precision-casting methods may have much closer tolerances and fine detail. Certain considerations must be taken for supporting very large or long cores when placed in the mold. Usually in sand casting, cores require extra support when they are three times longer than the cross-sectional dimension. Depending on the casting method, the material, the machining required, and the quality, the core holes should be a specified dimension smaller than the desired end product if the hole is to be machined to its final dimension. Cores in sand castings should be between .125 to .5 inch (3.2 to 12.7 mm) smaller than the finished size.

### Centrifugal Casting

Objects with circular or cylindrical shapes lend themselves to centrifugal casting. In this casting process, a mold is revolved very rapidly while molten metal is poured into the cavity. The molten metal is forced outward into the mold cavity by centrifugal forces. No cores are needed because the fast revolution holds the metal against the surface of the mold. This casting method is especially useful for casting cylindrical shapes, such as tubing, pipes, or wheels (see Figure 6–5).

### Die Casting

Some nonferrous metal castings are made using the die casting process. Zinc alloy metals are the most common, although brass, bronze, aluminum, and other nonferrous products are also made using this process. Die casting is the injection of molten metal into a steel or cast iron die under high pressure. The advantage of die casting over other methods, such as sand casting, is that castings

can be produced quickly and economically on automated production equipment. When multiple dies are used, a number of parts can be cast in one operation. Another advantage of die casting is that high-quality precision parts can be cast with fine detail and a very smooth finish.

### Permanent Casting

Permanent casting refers to a process in which the mold can be used many times. This type of casting is similar to sand casting in that molten metal is poured into a mold. It is also similar to die casting because the mold is made of cast iron or steel. The result of permanent casting is a product that has better finished qualities than can be gained by sand casting.

### Investment Casting

*Investment casting* is one of the oldest casting methods. It was originally used in France for the production of ornamental figures. The process used today is a result of the *cire perdue*, or *lost-wax*, casting technique that was originally used. The reason that investment casting is called lost-wax casting is that the pattern is made of wax. This wax pattern enables the development of very close tolerances, fine detail, and precision castings. The wax pattern is coated with a ceramic paste. The shell is allowed to dry and is then baked in an oven so that the wax melts and flows out; thus "lost" as the name implies. The empty ceramic mold has a cavity that is the same shape as the precision wax pattern. Next, this cavity is filled with molten metal. When the metal solidifies, the shell is removed and the casting is complete. Generally very little cleanup or finishing is required on investment castings (see Figure 6–6).

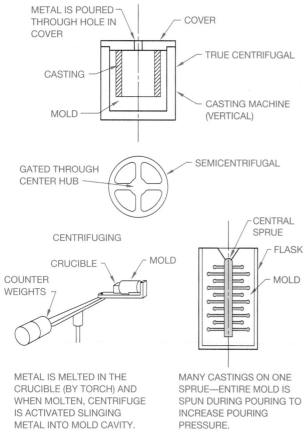

Figure 6–5 Centrifugal casting.

## Forgings

Forging is a process of shaping malleable metals by hammering or pressing between dies that duplicate the desired shape. The forging process is shown in Figure 6–7. Forging can be accomplished on hot or cold materials. Cold forging is possible on certain materials or material thicknesses where hole punching or bending is the required result. Some soft, nonferrous materials can be forged into shape while cold. Ferrous materials, such as iron and steel, must be heated to a temperature that results in an orange-red or yellow color. This color is usually achieved between 982° and 1066°C. Forging is used for a large variety of products and purposes. The advantage of forging over casting or machining operations is that the material is not only shaped into the desired form, but in the process it retains its original grain structure. Forged metal is generally stronger and more ductile than cast metal, and exhibits a greater resistance to fatigue and shock than machined parts. Notice in Figure 6–8 that the grain structure of the forged material remains parallel to the contour of the part, while the machined part cuts through the cross section of the material grain.

### Hand Forging

Hand forging is an ancient method of forming metals into desired shapes. The method of heating metal to a red color and then beating it into shape is called *smithing* or, more commonly, *blacksmithing*. Blacksmithing is used in industry only for finish work, but is still used for horseshoeing and the manufacture of specialty ornamental products.

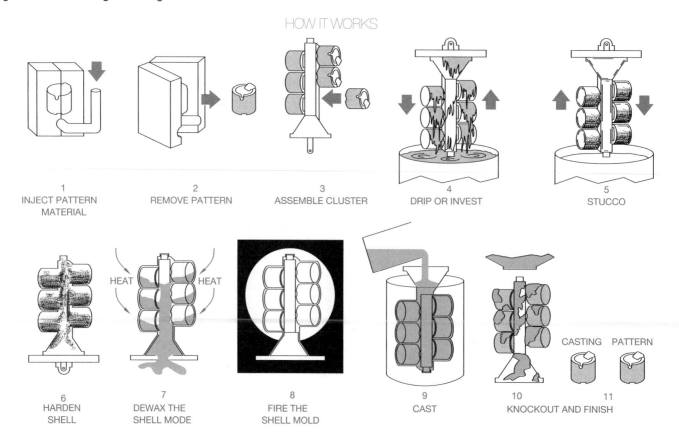

Figure 6–6 Investment Casting. *Courtesy Precision Castparts Corporation.*

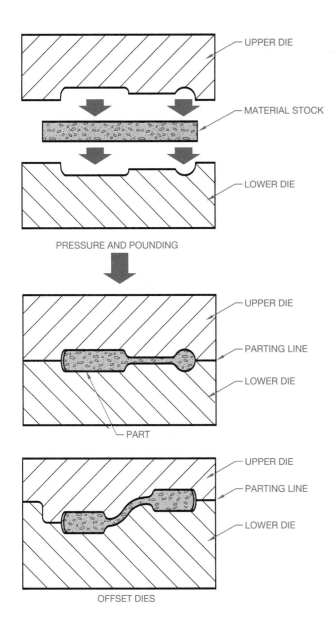

PRESSURE AND POUNDING

OFFSET DIES

Figure 6–7 The forging process.

### Machine Forging

Types of machine forging include upset, swaging, bending, punching, cutting, and welding. *Upset* forging is a process of forming metal by pressing along the longitudinal dimension to decrease length while increasing width. For example, bar stock is upset forged by pressing dies together from the ends of the stock to establish the desired shape. *Swaging* is the forming of metal by using concave tools or dies that result in a reduction in material thickness. *Bending* is accomplished by forming metal between dies, changing it from flat stock to a desired contour. Bending sheet metal is a cold forging process in which the metal is bent in a machine called a break. *Punching* and *cutting* are performed when the die penetrates the material to create a hole of any desired shape and depth or to remove material by cutting away. In forge *welding*, metals are joined together under extreme pressure. Material that is welded in this manner is very strong. The resulting weld takes on the same characteristics as the metal before joining.

Mass production forging methods allow for the rapid production of the high-quality products shown in Figure 6–9. In machine forging, the dies are arranged in sequence so that the finished forging is done in a series of steps. Complete shaping may take place after the material has been moved through several stages. Additional advantages of machine forging include:

1.  The part is formed uniformly throughout the length and width.

2.  The greater the pressure exerted on the material, the greater the improvement of the metallic properties.

3.  Fine grain structure is maintained to help increase the part's resistance to shock.

4.  A group of dies may be placed in the same press.

## Metal Stamping

Stamping is a process that produces sheet metal parts by the quick downward stroke of a ram die that is in the desired shape. The machine used is called a *punch press*. The punch press can "punch" holes of different sizes and shapes, cut metal, or form a

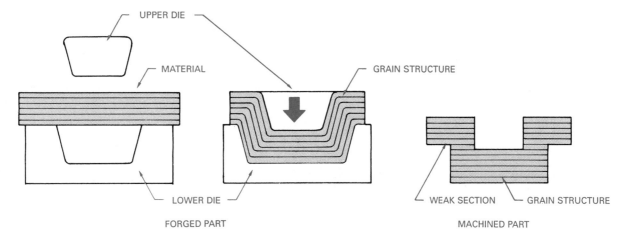

Figure 6–8 Forging compared to machining.

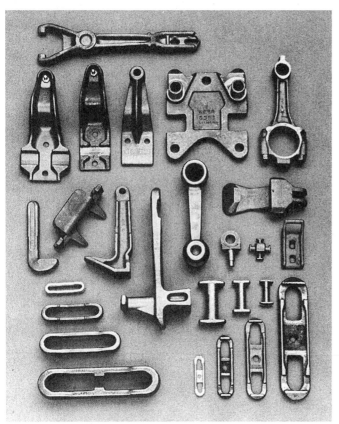

Figure 6–9 Forged products. *Courtesy Jarvis B. Webb Co.*

variety of shapes. Automobile parts, such as fenders and other body panels, are often produced using stamping. If the punch press is used to create holes, the ram contains a die in the shape of the desired hole or holes that pushes through the sheet metal. The punch press can also produce a detailed shape by pressing sheet metal between a die set, where the shape of the desired part is created between the die on the bed of the machine and the matching die on the ram. Stamping is generally performed on cold sheet metal, rather than, as with forging, performed on hot metal that is often much thicker. The stamping process is useful in producing a large number of parts. The process can often fabricate thousands of parts per hour. This is a very important consideration for mass production.

## Powder Metallurgy

The powder metallurgy process feeds metal-alloyed powders into a die where they are compacted under pressure to form the desired shape. The compacted metal is then removed from the die and heated at temperatures below the melting point of the metal. This heating process is referred to as *sintering*, which forms a bond between the metal powder particles. The process is the conventional powder metallurgy process that is called P/M.

Powder metallurgy processes can be a cost-effective alternative to casting, forging, and stamping. Powder metallurgy manufacturing can produce quality, precision parts at the rate of thousands per hour.

*Metal injection molding (MIM)* is another powder metallurgy process that can produce very complex parts. With this process, you inject a mixture of powder metal and a binder into a mold under pressure. The molded product is then sintered to create properties in the metal particles that are similar to those in a casting.

*Powder forging (P/F)* is another powder metallurgy process that places the formed metal particles in a closed die where pressure and heat are applied. Similar to the forging process, this process produces precision products that have good impact resistance and fatigue strength.

## MACHINE PROCESSES

The concepts discussed in this chapter serve as a basis for many of the dimensioning practices presented in Chapter 7. You need to understand machining processes and the drawing representations of these processes as a prerequisite to reading about the dimensions manufacturing process. In the following chapters, you'll find problem applications about dimensioning practice which is important to being able to effectively read prints.

### Machine Tools

#### Drilling Machine

The *drilling machine*, often referred to as a *drill press* (see Figure 6–10), is commonly used to machine-drill holes. Drilling machines are also used to perform other operations, such as reaming, boring, countersinking, counterboring, and tapping. During the drilling procedure the material is held on a table while the drill or other tool is held in a revolving spindle over the material. When drilling begins, a power or hand-feed mechanism is used to bring the rotating drill in contact with the material. Mass production drilling machines are designed with multiple spindles. Automatic drilling procedures are available on turret drills. These turret drills allow for automatic tool selections and spindle speed. Several operations can be performed in one setup—for example, drilling a hole to a given depth and tapping the hole with a specified thread.

#### Grinding Machine

A grinding machine uses a rotating abrasive wheel, rather than a cutting tool, for the purpose of removing material (see Figure 6–11). The grinding process is generally used when a smooth, accurate surface finish is required. Extremely smooth surface finishes can be achieved by honing or lapping. *Honing* is a fine abrasive process often used to establish a smooth finish inside cylinders. *Lapping* is the process of creating a very smooth surface finish, using a soft metal impregnated with fine abrasives or fine abrasives mixed in a coolant that floods over the part during the lapping process.

#### Lathe

One of the earliest machine tools, the lathe (Figure 6–12), is used to cut material by turning cylindrically shaped objects. The material to be turned is held between two rigid supports, called *centers*, or in a holding device called a *chuck* or *collet,* as shown in Figure 6–13. The material is rotated on a spindle while a cutting tool is brought

Figure 6–10 Drilling machine. *Courtesy Delta International Machining Corporation.*

Figure 6–11 Grinding machine. *Courtesy Litton Industrial Automation.*

Figure 6–12 Lathe. *Courtesy Hardinge Brothers, Inc.*

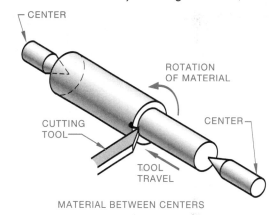

MATERIAL BETWEEN CENTERS

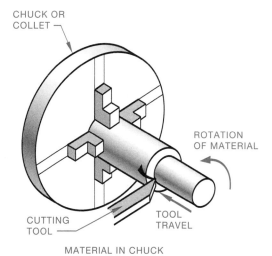

MATERIAL IN CHUCK

Figure 6–13 Holding material in a lathe.

into contact with the material. The cutting tool is supported by a tool holder on a carriage that slides along a bed as the lathe operation continues. The turret is used in mass production manufacturing where one machine setup must perform several operations. A turret lathe is designed to carry several cutting tools in place of the lathe tailstock or on the lathe carriage. The operation of the turret provides the operator with an automatic selection of cutting tools at preestablished fabrication stages. Figure 6–14 shows an example of eight turret stations and the tooling used.

## Milling Machine

The *milling machine* (see Figure 6–15) is one of the most versatile machine tools. The milling machine uses a rotary cutting tool to remove material from the work. The two general types of milling machines are *horizontal* and *vertical* mills. The difference is in the position of the cutting tool, which can be mounted on either a horizontal or vertical spindle. In the operation, the work is fastened to a table that is mechanically fed into the cutting tool, as shown in

Figure 6–16. A large variety of milling cutters are available to influence the flexibility of operations and shapes that can be performed using the milling machine. Figure 6–17 shows a few of the milling cutters available. Figure 6–18 shows a series of milling cutters grouped together to perform a milling operation. End milling cutters, as shown in Figure 6–19, are designed to cut on the end and the sides of the cutting tool. Milling machines that are commonly used in high-production manufacturing often have two or more cutting heads that are available to perform multiple operations. The machine tables of standard horizontal or vertical milling machines move from left to right (X-axis), forward and backward (Y-axis), and up and down (Z-axis), as shown in Figure 6–20.

### The Universal Milling Machine

Another type of milling machine, known as the universal milling machine, has table action that includes X-, Y-, and Z-axis movement, as well as angular rotation. The universal milling machine looks much the same as other milling machines, but has the advantage of additional angular table movement, as shown in Figure 6–21. This additional table movement allows the universal milling machine to produce machined features, such as spirals, not possible on conventional machines.

### Saw Machines

Saw machines can be used as cutoff tools to establish the length of material for further machining, or saw cutters can be

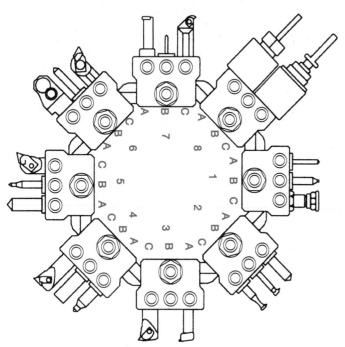

| STATION | | TOOLING |
|---|---|---|
| | Model | Description |
| 1A | | Drill with Bushing |
| 1B | | Center Drill with Bushing |
| 1C | T20 -5/8 | Adjustable Revolving Stock Stop |
| 2A | | Drill with Bushing |
| 2B | | Boring Bar with Bushing |
| 2C | | Threading Tool with Bushing |
| 3A | | Grooving Tool |
| 3C | | Insert Turning Tool |
| 4A | | Center Drill with Bushing |
| 4B | | Flat Bottom Drill with Bushing |
| 4C | | Insert Turning Tool |
| 5A | | Drill with Bushing |
| 5B | | Step Drill with Bushing |
| 5C | | Insert Turning Tool |
| 6A | T8 -5/8 | Knurling Tool |
| 6B | | Drill with Bushing |
| 6C | | Insert Turning Tool |
| 7A | | Grooving Tool |
| 7B | | Drill with Bushing |
| 7C | | Insert Threading Tool |
| 8A | TT -5/8 | "Collet Type" Releasing Tap Holder |
| | | Tap Collet |
| | | Tap |
| 8C | TE -5/8 | Tool Holder Extension |
| | T19 -5/8 | Floating Reamer Holder |
| | | Reamer with Bushing |

Figure 6–14 A turret with eight tooling stations and the tools used at each station. *Courtesy Toyoda Machinery USA, Inc.*

Figure 6–15 Close-up of a horizontal milling cutter.

Figure 6–16 Material removal with a vertical machine cutter. *Courtesy The Cleveland Twist Drill Company.*

SIDE CUTTER      HEAVY DUTY CUTTER

SQUARE END

BALL END

CORNER ROUNDING     ANGLE     CONVEX     CONCAVE

Figure 6–17 Milling cutters. *Courtesy The Cleveland Twist Drill Company.*

Figure 6–18 Grouping horizontal milling cutters for a specific machining operation. *Courtesy The Cleveland Twist Drill Company.*

used to perform certain machining operations, such as cutting a narrow slot (*kerf*).

Two types of machines that function as cutoff units only are the power *hacksaw* and the *band saw*. These saws are used to cut a wide variety of materials. The hacksaw, as shown in Figure 6–22, operates using a back-and-forth motion. The fixed blade in the

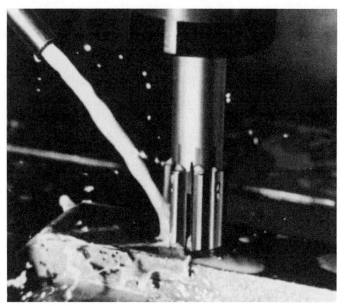

Figure 6–19 End mill operation. *Courtesy The Cleveland Twist Drill Company.*

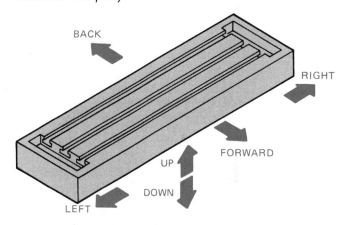

BACK

RIGHT

UP

FORWARD

DOWN

LEFT

TABLE MOVEMENTS OF THE PLAIN MILLING MACHINE

Figure 6–20 Table movements on a standard milling machine.

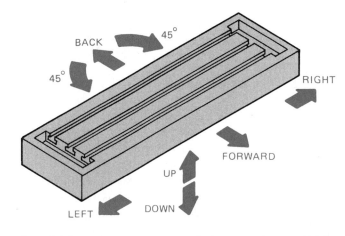

BACK    45°

45°

RIGHT

UP

FORWARD

LEFT    DOWN

TABLE MOVEMENTS OF THE UNIVERSAL MILLING MACHINE

Figure 6–21 Table movements on a universal milling machine.

power hacksaw cuts material on the forward motion. The metal-cutting band saw is available in a vertical or horizontal design, as shown in Figure 6–23. This type of cutoff saw has a continuous band that runs either vertically or horizontally around turning wheels. Vertical band saws may also be used to cut out irregular shapes.

Saw machines are also made with circular abrasive or metal-cutting wheels. The *abrasive saw* may be used for high-speed cutting where a narrow saw kerf is desirable or when very hard materials must be cut. One advantage of the abrasive saw is its ability to cut a variety of materials—from soft aluminum to case-hardened steels. (Cutting a variety of metals on the band or the power hacksaw requires blade and speed changes.) A disadvantage of the abrasive saw is the expense of abrasive discs. Many companies use this saw only when versatility is needed. The abrasive saw is usually found in the grinding room where abrasive particles can be contained, but may also be used in the shop for general-purpose cutting. Metal cutting saws with teeth, also known as *cold* saws, are used for precision cutoff operations, cutting saw kerfs, slitting metal, and other manufacturing uses. Figure 6–24 shows a circular saw blade.

Water jet cutting is used on composite materials and thin metal with a computer-controlled, 55,000-pounds-per-square-inch (psi) water jet. Cuts are made with holding tolerances of .0008 in. (0.020 mm) and without generating heat.

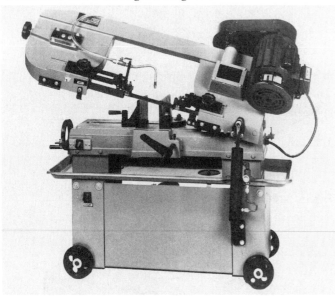

Figure 6–22 Power hacksaw. *Courtesy JET Equipment and Tool.*

### Shaper

The *shaper* is used primarily for production of horizontal, vertical, or angular flat surfaces. Shapers are generally becoming out of date and are rapidly being replaced by milling machines. A big problem with the shaper in mass-production industry is that it is very slow and cuts only in one direction. One of the main advantages of the shaper is its ability to cut irregular shapes that cannot be conveniently reproduced on a milling machine or other machine tools. However, other, more advanced multiaxis machine tools are now available that quickly and accurately cut irregular contours.

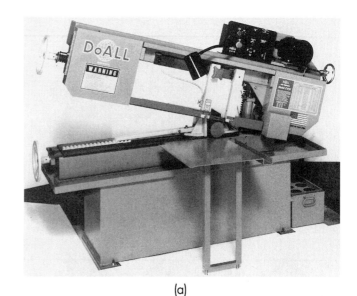

(a)

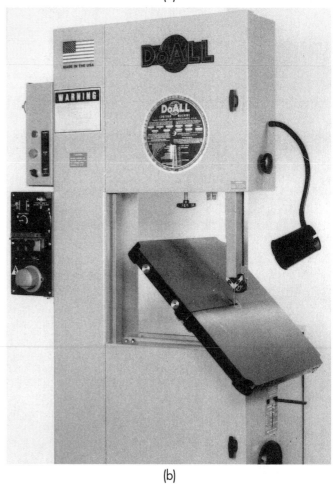

(b)

Figure 6–23 (a) Horizontal band saw; (b) vertical band saw. *Courtesy DoAll Company.*

### Chemical Machining

*Chemical machining* uses chemicals to remove material accurately. The chemicals are placed on the material to be removed while other areas are protected. The amount of time the chemical remains on the surface determines the extent of material removal.

Figure 6–24 Circular saw blade. *Courtesy The Cleveland Twist Drill Company.*

This process, also known as chemical milling, is generally used in situations where conventional machining operations are difficult. A similar method, referred to as chemical blanking, is used on thin material to remove unwanted thickness in certain areas while maintaining "foil" thin material at the machined area. Material may be machined to within .00008 in. (0.002 mm) using this technique.

### Electrochemical Machining

*Electrochemical machining (ECM)* is a process in which a direct current is passed through an electrolyte solution between an electrode and the work piece. A chemical reaction caused by the current in the electrolyte dissolves the metal, as shown in Figure 6–25.

### Electrodischarge Machining

*In electrodischarge machining (EDM)* the material to be machined and an electrode are submerged in a dielectric fluid that is a nonconductor, forming a barrier between the part and the electrode. A very small gap of about .001 inch is maintained between the electrode and the material. An arc occurs when the voltage across the gap causes the dielectric to break down. These arcs occur about 25,000 times per second, removing material with each arc. The compatibility of the material and the electrode is important for proper material removal. The advantages of EDM over conventional machining methods include its success in machining intricate parts and shapes that otherwise cannot be economically machined, and its use on materials that are difficult or impossible to work with, such as stainless steel, hardened steels, carbides, and titanium.

### Electron Beam (EB) Cutting and Machining

In this type of chemical machining, an electron beam generated by a heated tungsten filament is used to cut or "machine" very accurate features into a part. This process may be used to machine holes as small as .0002 in. (0.005 mm) or contour irregular shapes with tolerances of .0005 in. (0.013 mm). *Electron beam cutting* techniques are versatile and can be used to cut or machine any metal or nonmetal.

### Ultrasonic Machining

*Ultrasonic machining*, also known as impact grinding, is a process in which a high-frequency mechanical vibration is maintained in a tool designed to a specific shape. The tool and material to be machined are suspended in an abrasive fluid. The combination of the vibration and the abrasive causes the material removal.

### Laser Machining

The laser is a device that amplifies focused light waves and concentrates them in a narrow, very intense beam. The term LASER comes from the first letters of the words "*Light Amplification by Stimulated Emission of Radiation.*" Using this process, materials are cut or machined by instant temperatures up to 75,000°F (41,649°C). *Laser machining* can be used on any type of material and produces smooth surfaces without burrs or rough edges.

### CNC Post Processors

Computer-aided design (CAD) has a direct link to computer-aided manufacturing (CAM), and when combined, they are referred to as CAD/CAM. The software that converts CAD/CAM data to specific machine tool commands is called a *post processor*. Post-processor software is a key part of any CAD/CAM system used to produce machined parts. The CAD/CAM system produces numerical control (NC) output called a *cutter location file* (CL-file). This generic output represents the paths of the cutter while machining a part.

## Machined Features and Drawing Representations

The following discussion briefly defines common manufacturing-related terms. The figures that accompany each definition show an example of the tool, a pictorial of the feature, and the drawing representation. The terms are organized in categories of related features, rather than alphabetical order.

### Drill

A *drill* is used to machine new holes or enlarge existing holes in material. The drilled hole may go through the part, in which case the note THRU can be added to the diameter dimension. When the views of the hole clearly show that the hole goes through the part, the note THRU may be omitted. When the hole does not go through, the depth must be specified. This is referred to as a *blind hole.* The drill depth is the total usable depth to where the drill point begins to taper. A drill is a conical-shaped tool with cutting edges, normally used in a drill press. The drawing representation of a drill point is a 120° total angle (see Figure 6–26).

### Ream

The tool, called a *reamer*, is used to enlarge or finish a hole that has been drilled, bored, or cored. A cored hole is cast in place, as previously discussed. A reamer removes only a small amount of material; for example, .005 to 0.16 in. depending on the size of a hole. A reamed hole provides a smooth surface finish and a closer tolerance than is available with the existing hole. A reamer is a conical-shaped tool with cutting edges similar to a drill; however, you cannot use a reamer to create a hole as you can with a drill. Reamers may be used on a drill press, lathe, or mill (see Figure 6–27).

### Bore

*Boring* is the process of enlarging an existing hole to make a drilled or cored hole in a cylinder or part concentric with or perpendicu-

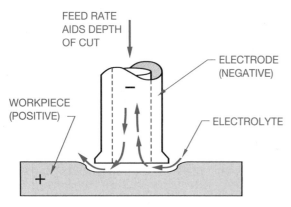

Figure 6–25 Electrochemical machining (ECM).

lar to other features of the part. A boring tool is used on machines such as a lathe, milling machine, or vertical bore mill for removing internal material (see Figure 6–28).

### Counterbore

The *counterbore* is used to enlarge the end or ends of a machined hole to a specified diameter and depth. You make the machined hole first, and then you align the counterbore during the machining process by means of a pilot shaft at the end of the tool. Counterbores are usually made to recess the head of a fastener below the surface of the object. You should be sure that the diameter and depth of the counterbore are adequate to accommodate the fastener head and fastening tools (see Figure 6–29).

### Countersink

A *countersink* is a conical feature in the end of a machined hole. Countersinks are used to recess the conically shaped head of a fastener, such as a flathead machine screw. As a drafter, you should specify the countersink note so the fastener head is recessed slightly below the surface. The total countersink angle should match the desired screw head. This angle is generally 80° to 82° or 99° to 101° (see Figure 6–30).

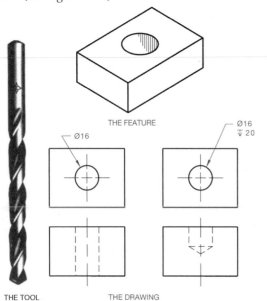

Figure 6–26 Drill. *Tool photo courtesy The Cleveland Twist Drill Company.*

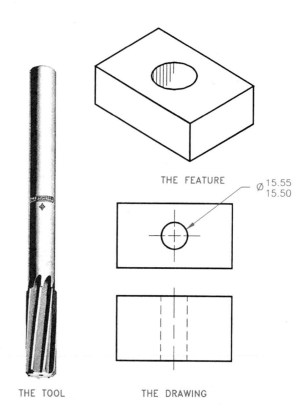

Figure 6–27 Reamer. *Tool photo courtesy The Cleveland Twist Drill Company.*

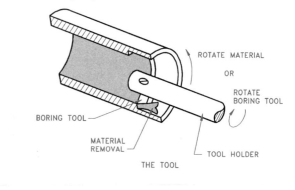

Figure 6–28 Bore.

### Counterdrill

A counterdrill is a combination of two drilled features. The first machined feature may go through the part, while the second feature is drilled, to a given depth, into one end of the first. The result is a machined hole that looks similar to a countersink-counterbore combination. The angle at the bottom of the counterdrill is a total of 120°, as shown in Figure 6–31.

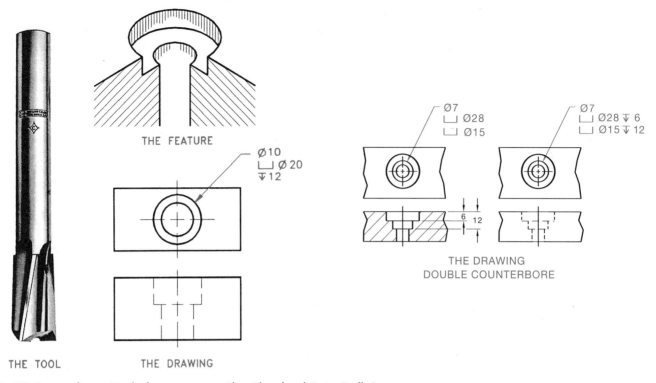

Figure 6–29 Counterbore. *Tool photo courtesy The Cleveland Twist Drill Company.*

### Spotface

A *spotface* is a machined, round surface on a casting, forging, or machined part on which a bolt head or washer can be seated. Spotfaces are similar in characteristics to counterbores, except that a spotface is generally only about 2 mm or less in depth. Rather than a depth specification, the dimension from the spotface surface to the opposite side of the part may be given. This is also true for counterbores; however, the depth dimension is commonly pro-

vided in a note. When no spotface depth is given, the machinist will spotface to a depth that establishes a smooth cylindrical surface (see Figure 6–32).

### Boss

A *boss* is a circular pad on forgings or castings that projects out from the body of the part. While more closely related to castings

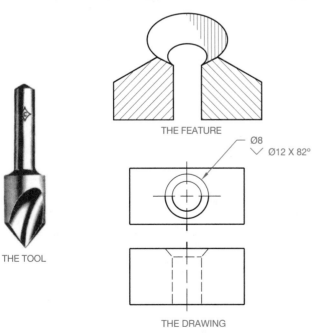

Figure 6–30 Countersink. *Tool photo courtesy The Cleveland Twist Drill Company.*

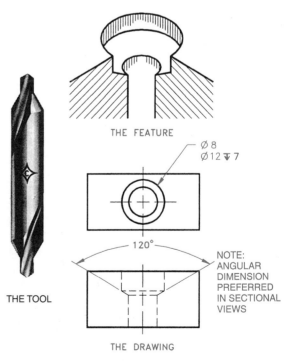

Figure 6–31 Counterdrill. *Tool photo courtesy The Cleveland Twist Drill Company.*

and forgings, the surface of the boss is often machined smooth for a bolt head or washer surface to sit on. Also, the boss commonly has a hole machined through it to accommodate the fastener's shank (see Figure 6–33).

### Lug

Generally cast or forged into place, a *lug* is a feature projecting out from the body of the part, usually rectangular in cross-section. Lugs are used as mounting brackets or function as holding devices for machining operations. Lugs are commonly machined with a drilled hole and a spotface to accommodate a bolt or other fastener (see Figure 6–34).

### Pad

A *pad* is a slightly raised surface projecting out from the body of a part. The pad surface can be any size or shape. The pad can be cast, forged, or machined into place. The surface is often machined to accommodate the mounting of an adjacent part. A boss is a type of pad, although the boss is always cylindrical in shape. (See Figure 6–35.)

### Chamfer

A *chamfer* is the cutting away of the sharp external or internal corner of an edge. Chamfers may be used as a slight angle to relieve a sharp edge or to assist the entry of a pin or thread into the mating feature (see Figure 6–36). Verify alternate methods of dimensioning chamfers in Chapter 7.

### Fillet

A fillet is a small radius formed between the inside angle of two surfaces. Fillets are often used to help reduce stress and strengthen an inside corner. Fillets are common on the inside corners of castings and forgings to strengthen corners. Fillets are also used to help a casting or forging release a mold or die. Fillets are arcs given as radius dimensions. The fillet size depends on the function of the part and the manufacturing process used to make the fillet (see Figure 6–37).

### Round

A *round* is a small-radius outside corner formed between two surfaces. Rounds are used to refine sharp corners, as shown in Figure 6–38. In some situations where a sharp corner must be relieved and a round is not required, a slight corner relief may be used, which is referred to as a break corner. The note BREAK CORNER may be used on the drawing. Another option is to provide a note that specifies REMOVE ALL BURRS AND SHARP EDGES. Burrs are machining fragments that are often left on a part after machining.

### Dovetail

A *dovetail* is a slot with angled sides that can be machined at any depth and width. Dovetails are commonly used as a sliding mechanism between two mating parts (see Figure 6–39).

### Kerf

A is a narrow slot formed by removing material while sawing or using some other machining operation (see Figure 6–40).

### Key, Keyseat, and Keyway

A *key* is a machine part that is used as a positive connection for transmitting torque between a shaft and a hub, pulley, or wheel. The key is placed in position in a *keyseat*, which is a groove or channel cut in a shaft. The shaft and key are then inserted into a hub, wheel, or pulley where the key mates with a groove, called a

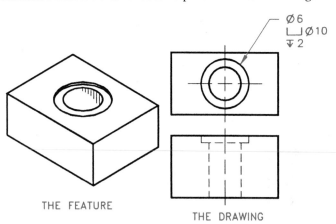

Figure 6–32 Spotface.

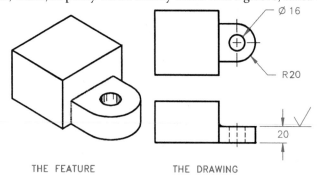

Figure 6–34 Lug.

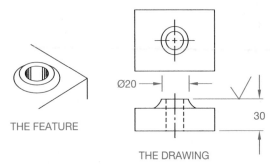

Figure 6–33 Boss.

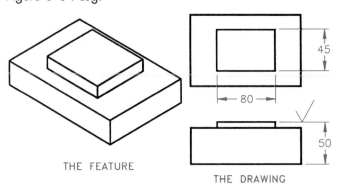

Figure 6–35 Pad.

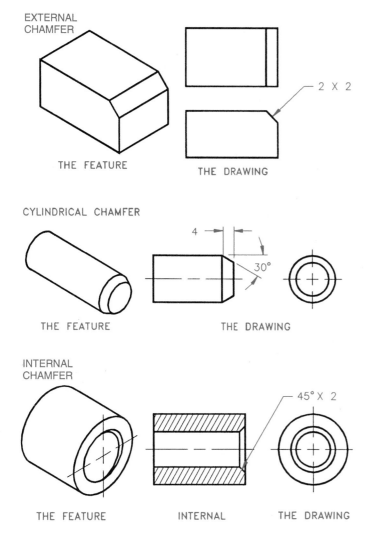

Figure 6–36 Chamfers.

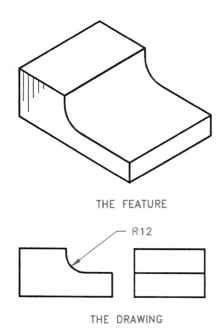

Figure 6–37 Fillet.

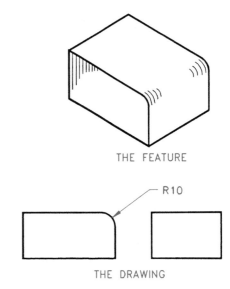

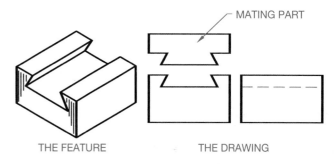

Figure 6–38 Round.

Figure 6–39 Dovetail.

*keyway*. There are several different types of keys. The key size is often determined by the shaft size (see Figure 6–41). Types of keys and key sizes are discussed in Chapter 8, Fasteners and Springs.

### Neck

A *neck* is the result of a machining operation that establishes a narrow groove on a cylindrical part or object. Figure 6–42 shows several different types of neck grooves. Dimensioning necks is explained in Chapter 7.

### Spline

A *spline* is a gearlike, serrated surface on a shaft and in a mating hub. Splines are used to transmit torque and allow for lateral sliding or movement between two shafts or mating parts. A spline can be used to take the place of a key when more torque strength is required or when the parts must have lateral movement (see Figure 6–43).

### Threads

Many different forms of threads are available for use as fasteners to hold parts together, to adjust parts in alignment with each other, or to transmit power. Threads that are used as fasteners are commonly referred to as screw *threads*. *External threads* are thread forms on an external feature, such as a bolt or shaft. The machine tool used to make external threads is commonly called a *die*. Threads can be machined on a lathe using a thread-cutting tool. (See Figure 6–44.) *Internal threads* are threaded features on the inside of a hole. The

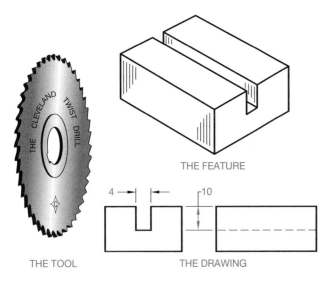

THE FEATURE

THE TOOL

4 | 10

THE DRAWING

Figure 6–40 Kerf. Tool photo courtesy The Cleveland Twist Drill Company.

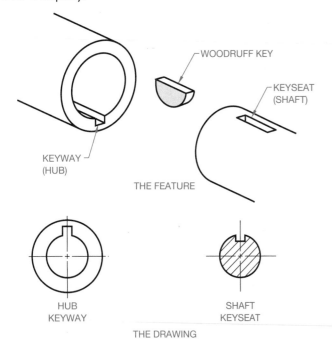

WOODRUFF KEY

KEYSEAT (SHAFT)

KEYWAY (HUB)

THE FEATURE

HUB KEYWAY

SHAFT KEYSEAT

THE DRAWING

Figure 6–41 Key, keyseat, and keyway.

machine tool that is commonly used to cut internal threads is called a *tap* (see Figure 6–45).

### T-slot

A *T-slot* is a slot of any dimension that is cut to resemble a "T." The T-slot can be used as a sliding mechanism between two mating parts (see Figure 6–46).

### Knurl

*Knurling* is a cold forming process used to uniformly roughen a cylindrical or flat surface with a diamond or straight pattern. Knurls are often used on handles or other gripping surfaces. Knurls may also be used to establish an interference (press) fit between two mating parts. The actual knurl texture is not displayed on the drawing (see Figure 6–47).

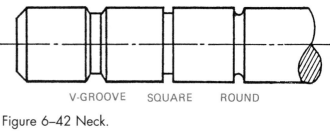

V-GROOVE    SQUARE    ROUND

Figure 6–42 Neck.

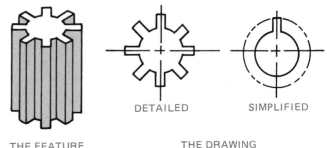

DETAILED    SIMPLIFIED

THE FEATURE    THE DRAWING

Figure 6–43 Spline.

### Surface Texture

*Surface texture*, or *surface finish*, is the intended condition of the material surface after manufacturing processes have been implemented. Surface texture includes such characteristics as roughness, waviness, lay, and flaws. Surface roughness is one of the most common characteristics of surface finish, which consists of the finer irregularities of the surface texture resulting, in part, from the manufacturing process. The surface roughness is measured in micrometers or microinches ($\mu$ – means "millionth"). The roughness averages for different manufacturing processes are shown in Appendix B, Table 26 (http://www.delmarlearning.com/companions/start.asp). Surface finish is discussed in detail in Chapter 7.

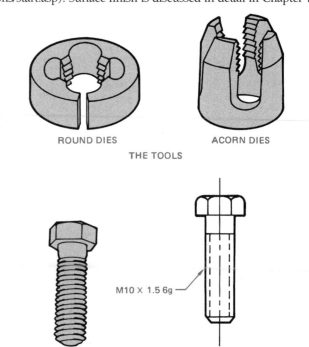

ROUND DIES    ACORN DIES

THE TOOLS

M10 × 1.5 6g

THE EXTERNAL THREAD    THE DRAWING (SIMPLIFIED)

Figure 6–44 External thread.

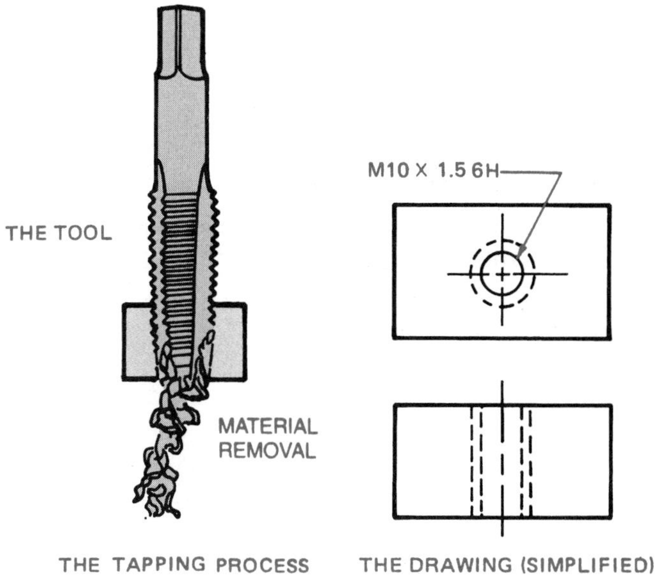

THE TOOL

M10 X 1.5 6H

MATERIAL
REMOVAL

THE TAPPING PROCESS          THE DRAWING (SIMPLIFIED)

Figure 6–45 Internal thread.

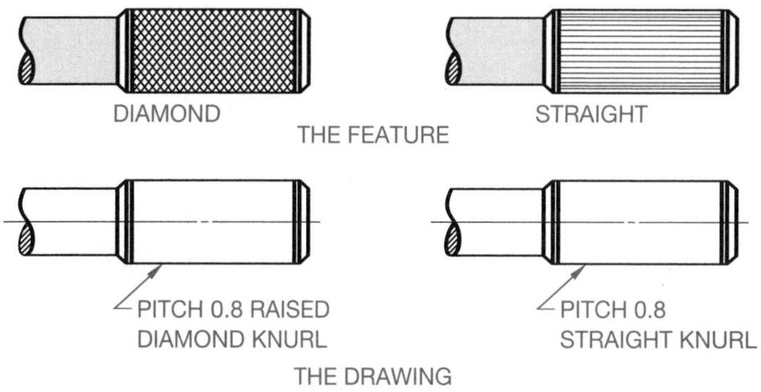

## Manufacturing Plastic Products

The previous discussions gave you some general information about the different types of plastics and synthetic rubbers, and many of the products that are commonly made from these materials. The traditional plastic product manufacturing processes include injection molding, extrusion, blow molding, compression molding, transfer molding, and thermoforming. These processes are explained in the following paragraphs.

### The Injection Molding Process

Injection molding is the most commonly used process for creating thermoplastic products. The process involves injecting molten plastic material into a mold that is in the form of the desired part product. The mold is in two parts that are pressed together during the molding process. The mold is then allowed to cool so the plastic can solidify. When the plastic has cooled and solidified, the press is opened and the part is removed from the mold. The injection molding machine has a hopper where either powder or granular material is placed. The material is then heated to a melting temperature. Next, the molten plastic is fed into the mold by an injection nozzle or a screw injection system. The injection system is similar to a large hypodermic needle and plunger that pushes the

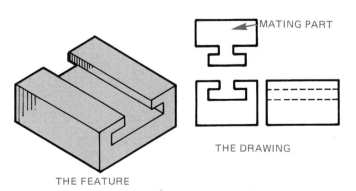

MATING PART

THE DRAWING

THE FEATURE

Figure 6–46 T-slot.

DIAMOND          THE FEATURE          STRAIGHT

PITCH 0.8 RAISED
DIAMOND KNURL

PITCH 0.8
STRAIGHT KNURL

THE DRAWING

Figure 6–47 Knurl.

molten plastic material into the mold. The most commonly used injection system is the screw machine. The machine has a screw design that transports the material to the injection nozzle. While the material moves toward the mold, the screw also mixes the plastic to a uniform consistency. This mixing process is an important advantage over the plunger system. Mixing is especially important when color is added or when using recycled material. The process of creating a product and the parts of the screw injection and injection nozzle systems are shown in Figure 6–48.

### The Extrusion Process

The extrusion process is used to make continuous shapes, such as moldings, tubing, bars, angles, hose, weather stripping, films, and any product that has a constant shape. This process creates the desired continuous shape by forcing molten plastic through a metal die. The extrusion process typically uses the same type of injection nozzle or a screw injection system that is used in the injection molding process. The contour of the die establishes the shape of the extruded plastic. Figure 6–49 shows the extrusion process in action.

### The Blow Molding Process

The blow molding process is commonly used to produce hollow products, such as bottles, containers, receptacles, and boxes. This process works by blowing hot polymer against the internal surfaces of a hollow mold. The molten plastic enters around a tube that also forces air inside the material, which forces it against the interior surface of the mold. The polymer expands to a uniform thickness against the mold. The mold is formed in two halves, so when the plastic cools, the mold is split to remove the product. The blow molding process is shown in Figure 6–50.

### The Calendering Process

The calendaring process is generally used to create products such as vinyl flooring, gaskets, and other sheet products. This process duplicates sheet or film thermoplastic or thermoset plastics by passing the material through a series of heated rollers. The space between the sets of rollers gets progressively smaller until the distance between the last set of rollers establishes the desired material thickness. Figure 6–51 illustrates the calendaring process used to produce sheet plastic products.

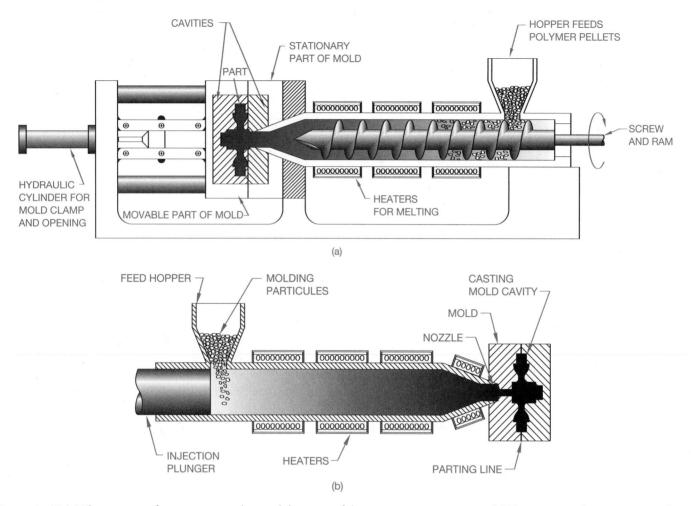

(a)

(b)

Figure 6–48 (a) The process of creating a product and the parts of the screw injection system. (b) The process of creating a product and the parts of the nozzle injection system.

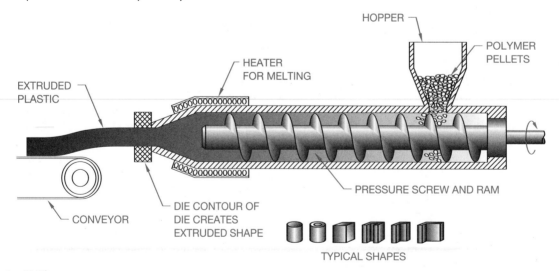

TYPICAL SHAPES

Figure 6–49 The extrusion process in action.

### The Rotational Molding Process

This process is typically used to produce large containers such as tanks, hollow objects such as floats, and other, similar types of large, hollow products. This process works by placing a specific amount of polymer pellets into a metal mold. The mold is then heated as it is rotated. This forces the molten material to form a thin coating against the sides of the mold. When the mold is cooled, the product is removed. The rotational molding process is shown in Figure 6–52.

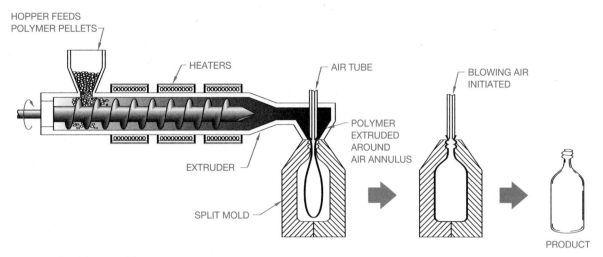

Figure 6–50 The blow molding process.

### The Solid Phase Forming Process

This process can be used to make a variety of objects including containers, electrical housings, automotive parts, and anything that has detailed shapes. Products can be stronger using this process because the polymer is heated but not melted. The solid phase forming process works by placing material into an initial hot die where it takes the preliminary shape. As the material cools, a die that matches the shape of the desired product forms the final shape. This process is similar to metal forging. Figure 6–53 shows the solid phase forming process.

### Thermoforming of Plastic

This process is similar to solid phase forming, but can be performed without a die. This process is used to make all types of thin-walled plastic shapes, such as containers, guards, fenders, and other, similar products. The process works by taking a sheet of material and heating it until it softens and sinks down by its own weight into a mold that conforms to the desired final shape. Vacuum pressure is commonly used to suck the hot material down against the mold, as shown in Figure 6–54.

### Free-form Fabrication of Plastic

The free-form fabrication process uses a computer model that is traced in thin cross-sections to control a laser that deposits layers of liquid resin or molten particles of plastic material to form the desired shape. Using the laser to fuse several thin coatings of powder polymer to form the desired shape is also an option for this process. This process is often referred to by the trade name Stereolithography. Figure 6–55 shows the free-form process.

## Thermoset Plastic Fabrication Processes

The manufacture of thermoset products can be more difficult than thermoplastic fabrication, because thermosets cannot be remelted once they have been melted and formed the first time. This characteristic makes it necessary to keep the manufacturing equipment operational for the longest period possible. When process equipment is shut down, it must be thoroughly cleaned before the thermoset material solidifies. The injection molding process that was discussed earlier can be used, but extreme care must be taken to

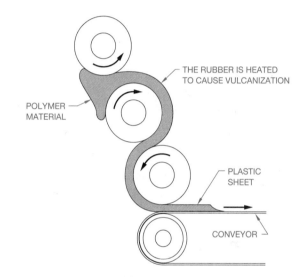

Figure 6–51 The calendaring process used to produce sheet plastic products.

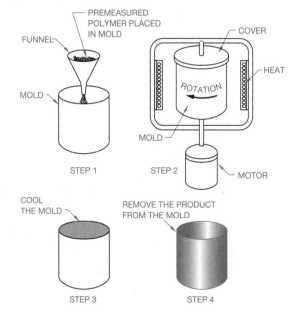

Figure 6–52 The rotational molding process.

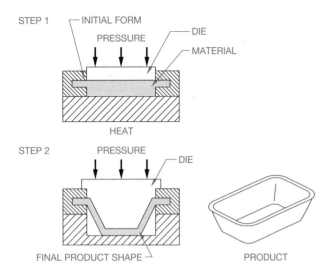

Figure 6–53 The solid phase forming process.

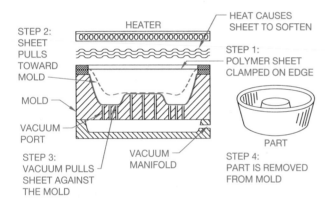

Figure 6–54 Thermoforming of plastic. Vacuum pressure is commonly used to suck the hot material down against the mold.

ensure that the equipment is kept clean. For this reason, other methods are commonly used for manufacturing thermoset products. Thermoset plastic production generally takes longer than thermoplastic production because thermoset plastics require a longer cure time. The most common production practices are casting, compression molding, foam molding, reaction injection molding, transfer molding, sintering, and vulcanization.

### Casting Thermoset Plastics

The method used to cast plastics is very similar to the permanent casting of metals that was explained earlier in this chapter. When casting plastic, the molten polymer or resin is poured into a metal flask with a mold that forms the desired shape of the product. The plastic product is removed when it has solidified. The casting process is shown in Figure 6–56.

### Compression Molding and Transfer Molding

Compression molding and transfer molding are common fabrication processes for thermosets. These processes use a specific amount of material that is heated and placed in a closed mold where additional heat and pressure are applied until the material takes the desired shape. The material is then cured and removed

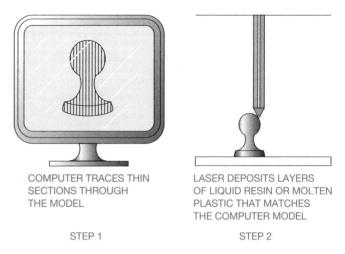

Figure 6–55 The free-form fabrication process.

from the mold. The compression molding process is shown in Figure 6–57.

Transfer molding is similar to compression molding for thermoset plastic products. In this process, the material is heated and then forced under pressure into the mold.

### Foam Molding and Reaction Injection Molding

Foam molding is similar to casting, but this process uses a foam material that expands during the cure to fill the desired mold. The foam molding process can be used to make products of any desired shape or sheets of foam products. The foam molding process is used to make a duck decoy in Figure 6–58.

The reaction injection molding process is similar to the foam process only because the material that is used expands to fill the mold. This process is often used to fabricate large parts, such as automobile dashboards and fenders. Polymer chemicals are mixed together under pressure and then poured into the mold, where they react and expand to fill the mold.

### The Sintering Process

Sintering is a process that takes powdered particles of material and produces detailed products under heat, pressure, and chemical reaction. The powder particles do not melt, but they join together into a dense and solid structure. This process is used when creating products for high-temperature applications.

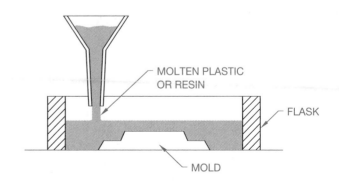

Figure 6–56 Casting thermoplastics.

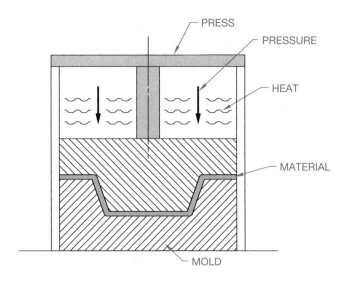

Figure 6–57 The compression molding process.

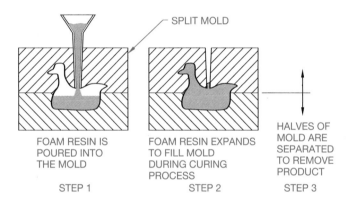

Figure 6–58 The foam molding process.

### The Vulcanization Process

Vulcanization is used to make rubber products, such as tires, and generally other circular and cylindrical shapes. This process works by wrapping measured layers of the polymer around a steel roll in the form of the desired product. The steel roll and material is placed inside an enclosure where steam is introduced to create heat and pressure. This process forces the molecules of the material to form a solid rubber coating on the roll. The vulcanization process is shown in Figure 6–59.

## Manufacturing Composites

Composites are also referred to as *reinforced plastics*. Earlier in this chapter you read about composite materials. These materials combine polymers with reinforcing material, such as glass, graphite, thermoplastic fibers, cotton, paper, and metal. The result is very strong material for use in products such as boats, planes, automobile parts, fishing rods, electrical devices, corrosion-resistant containers, and structural members. Composites are less expensive to produce and can be stronger and are lighter in weight than many metals, including aluminum, steel, and titanium. The basics of producing composite products are the layering of polymer and reinforcing material in alternate coats or in combination.

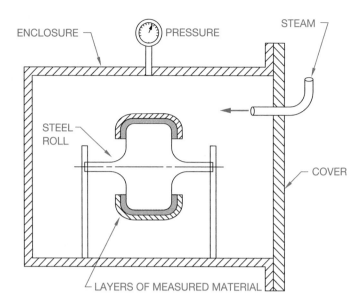

Figure 6–59 The vulcanization process.

The *layering process* combines alternating layers of polymer resin with reinforcing material such as glass. The number of layers determines the desired thickness. During the curing process, the resin saturates the reinforcing material, creating a unified composite. Rolling or spraying over the reinforcing material can apply layering in the resin. A similar method uses a spray of resin combined with pieces of reinforcing material. This method is called *chopped fiber spraying*.

Another process uses machines to wind resin-saturated reinforcement fibers around a shaft. That process is referred to as *filament winding*.

A process called *compression molding* is similar to the compression molding process that was explained earlier for thermoset plastic products. The difference is that mixing the polymer with reinforcing fibers creates composites. A similar process takes *continuous reinforcing* strands through a resin bath and then through a forming die. The die forms the desired shape. This method is often used to create continuous shapes of uniform cross-section.

The *resin transfer molding* process is used to make quality composite products with a smooth surface on both sides. This method places reinforcing material into a mold and then pumps resin into the mold.

A process referred to as *vacuum bag forming* uses vacuum pressure to force a thin layer of sheet-reinforced polymer around a mold. Figure 6–60 shows a variety of processes used to make composite products.

## *TOOL DESIGN*

In most production machining operations, special tools are required to either hold the workpiece or guide the machine tool. Tool design involves knowledge of kinematics (the study of mechanisms), machining operations, machine tool function, material handling, and material characteristics. *Tool design* is also known as *jig* and *fixture design*. In mass-production industries, jigs and fixtures are essential to ensure that each part is produced quickly and accurately within the dimensional specifications. These tools are used to hold the work piece so that machining operations are performed in the required positions. Application examples are shown in Figure 6–61. Jigs are either fixed or moving devices that are used to hold the work piece in position and guide the cutting tool.

Fixtures do not guide the cutting tool, but are used in a fixed position to hold the work piece. Fixtures are often used in the inspection of parts to ensure that the part is held in the same position each time a dimensional or other type of inspection is made.

Jig and fixture drawings are prepared as an assembly drawing where all of the components of the tool are shown as if they were

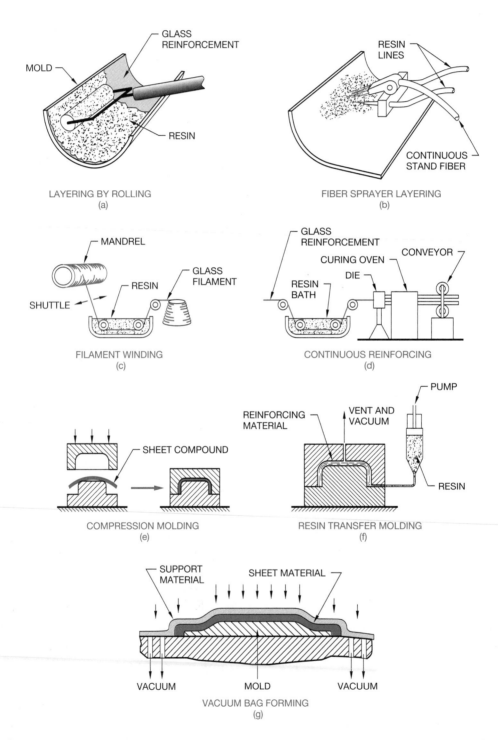

Figure 6–60 Common processes used to make composite products. (a) The layering process using rolling. (b) The layering process using spraying. (c) Filament winding. (d) The continuous reinforcing strands process. (e) Compression molding. (f) Resin transfer molding. (g) Vacuum bag forming.

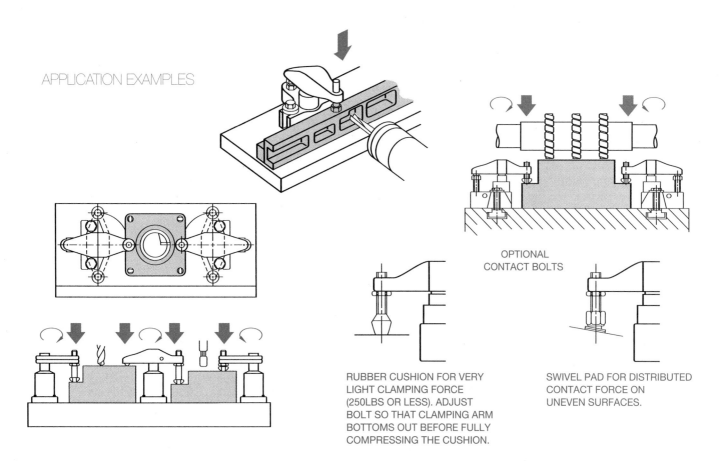

APPLICATION EXAMPLES

OPTIONAL
CONTACT BOLTS

RUBBER CUSHION FOR VERY
LIGHT CLAMPING FORCE
(250LBS OR LESS). ADJUST
BOLT SO THAT CLAMPING ARM
BOTTOMS OUT BEFORE FULLY
COMPRESSING THE CUSHION.

SWIVEL PAD FOR DISTRIBUTED
CONTACT FORCE ON
UNEVEN SURFACES.

Figure 6–61 Fixture application examples. *Courtesy Carr Lane Manufacturing Company*

assembled and ready for use (see Figure 6–62). Components of a jig or fixture often include such items as fast-acting clamps, spring-loaded positioners, clamp straps, quick-release locating pins, handles, knobs, and screw clamps (see Figure 6–63). Normally the part or workpiece is drawn in position, using phantom lines in a color, such as red, or a combination of phantom lines and color.

## The Tool Design Process*

Hammers, pliers, screwdrivers, wrenches, and other, similar items are tools, but they are not the kind of tools that are created by *tool designers*. *Tools*, as referred to in this discussion, are specially designed and built manufacturing aids, normally in a production environment, that are used to assist operators in the manufacture of specific parts. Many different kinds of tools fit this definition, including machining fixtures, welding fixtures, drill fixtures, drill jigs, inspection fixtures, progressive dies, injection molds, and others. Tools can range from very small and simple to very large and complicated. Some tools are mechanisms that resemble machinery, but there is a clear difference between tools and machines. Tools are dedicated to a specific part, family of parts, or product, while machines are for general usage across many parts and products. This differentiation is emphasized more for accounting and tax purposes than it is for technical reasons. Machines are capital investments, while tools are expense items related to a particular part or product. In many cases, the tool can be so complex that,

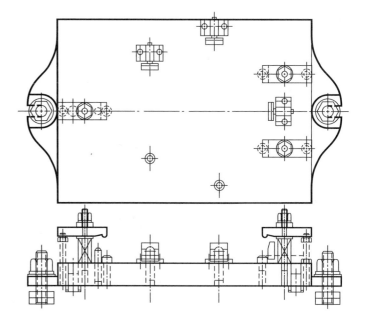

Figure 6–62 Fixture assembly drawing. *Courtesy Carr Lane Manufacturing Company.*

from a technical perspective, it is a machine, but from an accounting perspective, it is a tool. Some simplified definitions of the typical types of tools that are used in manufacturing follow.

■ *Drill jigs* are used in drilling operations using hand drills, drill presses, or radial drills, rather than milling machines. The drill jig registers the part relative to the critical datums. *Datums* are important points, axes, surfaces, or planes from which features are established and dimensioned. (Datums are discussed in detail in Chapters 7 and 11.)

■ The operator then drills through a drill bushing or bushings in the jig to locate the hole or holes in the part.

■ *Drill fixtures* are sometimes referred to as holding fixtures. Drill fixtures are used for drilling operations using milling machines, either manual or computer numerical controls (CNC). The drill fixture registers the part relative to the critical datums, but does not have drill bushings. The machine axes are used to locate the hole or holes. Drill fixtures must resist the force of the drilling operation. Typically, drilling forces are relatively low, and only in one direction.

■ *Machining fixtures* are used for machining operations using milling machines that are either manual or CNC. The machining fixture registers the part relative to the critical datums. The machine axes are used to locate the feature or features to be machined. Machining fixtures must resist the force of the machining operation; typically, machining forces are relatively high, and can be in any direction.

■ *Welding fixtures* are more accurately termed *welding jigs*, but the names have become synonymous in the industry. Welding fixtures are used to hold two or more pieces in the proper position and orientation so that the pieces can be welded together.

■ *Inspection fixtures* are very common in the casting industry. They are used to hold a part, registering it on the critical datums, while an inspector checks critical feature sizes and/or locations.

■ *Progressive dies* are the tooling used in punch press operations. A continuous stock of relatively thin material is fed into the punch press from coils. As the punch press cycles, the material is cut to appropriate shape, holes are punched, or other operations are performed to make a finished part. Cookie cutters, scissors, and paper hole punches are comparable to progressive dies.

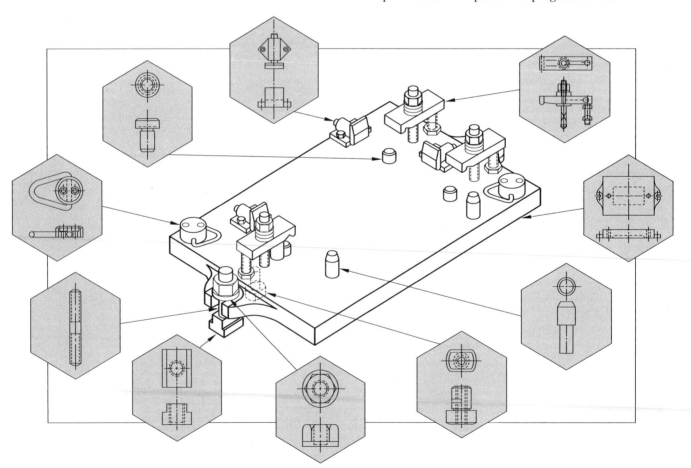

Figure 6–63 Fixture components. *Courtesy Carr Lane Manufacturing Company.*

Figure 6–64 shows a very simple drill jig tool. This drill jig is designed and manufactured to assist a shop employee in performing a specific task on a production part. The part is very simple, and an employee could measure and drill the hole without any tool, but for production work, the tool determines the location of the hole much faster, and the accuracy of the hole location is less dependent on employee expertise. This type of drill jig is referred to as a *pickoff jig*—pickoff because it sits on the part, rather than the part being put into the fixture; and jig because it actually locates a feature, which is the .25-in. diameter hole in this example.

Now that you have an idea of what makes up a tool, the following describes the three basic elements of tool design:

1.   Visualizing how shop personnel can accomplish a specific task.

2.   Conceptualizing hardware to assist in the accomplishment of the task.

3.   Creating drawings so the hardware can be manufactured.

*Visualizing* and *conceptualizing* are not to be confused with inventing. Tool designers are not inventors, but rather use existing products and items whenever possible. Many times, an existing design can be modified to satisfy a new requirement, and good tool designers often save their employer time and money by using existing designs. As a tool designer, you *must* be familiar with shop practices, and you *must* be able to visualize shop personnel accomplishing specific tasks within the shop environment. This visualization leads to a concept of tools that shop personnel can use to help them accomplish the task. The tool designer *must* also be a very good print reader.

The part print contains all the information the tool designer needs. The tool designer must be able to find the important information. Looking back to the example in Figure 6–64, you are shown only a small corner of the part. Even this small corner has much more information than is required for this drill jig. Print reading is the first step in the tool design process. The information necessary to design the drill jig must be found on the part print. Some dimensions are directly related to the drill jig, some are incidental to the tool designer, and some are totally unrelated to this drill jig. Looking at Figure 6–64, it is relatively easy to *sort out the applicable information*. Dimensions defining the feature size and location are directly related to the required tool. Dimensions that define the size and locations of other features that may impact the tool are of incidental interest. Dimensions that define the size and locations of features that do not impact the design of the tool are not relevant to the tool designer. More complicated part prints may require in-depth study to find the relevant information.

Tool designers often receive their assignments from a manufacturing engineer (ME). The ME defines the specific task and how it is to be accomplished. Consider the example in Figure 6–64 again. The tool designer receives a tooling design request (TDR) from the ME. The TDR has an engineering drawing of the part attached, which contains the following information:

| | |
|---|---|
| PART NAME: | PART |
| PART NO: | XXX-XX |
| OPERATION: | DRILL .25" DIA HOLE |
| MACHINE: | USE HAND DRILL |
| FIXTURE: | PICKOFF JIG LOCATING THE HOLE .50" AND 1.00" FROM THE EDGES OF THE PLATE |

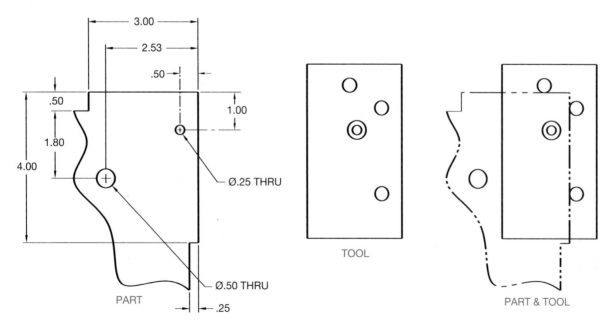

Figure 6–64 A simple example of a PART and a TOOL. The task of the employee is to drill a .25-in. diameter hole in a specific location. It is very efficient for an employee to use the TOOL to locate the hole. You simply place the TOOL on top of the PART, hold it with one hand, and drill the hole through the PART, using the hole in the TOOL to locate the hole in the PART.

The decision to use a pickoff jig and hand drill rather than a radial drill or even a CNC machine was made prior to the tool designer's involvement in the project. If the ME had determined that the best way to accomplish this task was to put this part on a CNC machine to drill the hole, the TDR would have been written accordingly. The tool designer designs a drill fixture that can be mounted on the CNC machine table. The fixture holds and clamps the part The part is located with accurate registration to the two critical edges that define the location of the .25-in diameter hole.

## The Tool Designer's Tools

A tool designer uses manual drafting or CAD practices to design a fixture. As a tool designer, you must capture the concept on paper or a computer screen so the fixture can be built. The tool designer must also be familiar with standard tooling components that are available from numerous manufacturers. Various components, such as rest pads, clamps, pins, and drill bushings, to name a few, are available in a wide variety of sizes and shapes. You want to remember that tool designers do not create something that already exists, but use standard components whenever possible.

Tools, in general, must possess certain qualities.

- Reliability
- Repeatability
- Ease of use
- Ease of manufacture
- Ease of maintenance and repair

Figure 6–65 shows the design that would go to the tool room for a machinist to use to build the pickoff drill jig previously discussed and shown in Figure 6–64. The drawing shown is a combination assembly drawing and detail drawing. More complex fixtures require the fixture assembly to be drawn on one sheet, and the component details to be drawn on separate sheets. Detail and assembly drawings are explained in Chapter 13, Reading Working Drawings.

Notice that the PART in Figure 6–64 is dimensioned using conventional dimension and extension lines, and the DRILL JIG drawing in Figure 6–65 is dimensioned using arrowless dimensioning. Another type of dimensioning practice can be used on both drawings. Chapter 7, Dimensioning and Tolerancing, completely explains each type of dimensioning practice.

The PLATE (Figure 6–65, item 1) is the only piece that must be made in the *tool room*. The tool room is the shop where tools are manufactured. The DOWEL PIN (Figure 6–65, item 2) and the DRILL BUSHING (Figure 6–65, item 3) are purchased components. Here is an example of how you can use a checklist to evaluate whether this jig meets quality requirements for tools:

- *Reliability.* Yes, this jig is simple, and very little could break or cause problems.
- *Repeatability.* Yes, again the simplicity makes repeatability inherent. The drill jig registers directly on the datum surfaces from which the hole location is defined.

- *Ease of use.* Yes, the jig offers the operator a simple means to locate the hole.
- *Ease of manufacture.* Yes, with only one manufactured part, and four purchased parts, it will be very easy to make.
- *Ease of maintenance and repair.* Yes, should something break or wear out, the component parts can be easily replaced.

In summary, you have learned about tools, tool design, and tool designers. Tools are shop aids, sometimes simple, and sometimes very complex. Tool design is the process of turning a concept of a tool into drawings so that the fixture can be manufactured. Tool designers are the people who imagine the concepts and turn them into drawings.

## COMPUTER-INTEGRATED MANUFACTURING (CIM)

Completely automated manufacturing systems combine computer-aided design and drafting (CADD), computer-aided engineering (CAE), and computer-aided manufacturing (CAM) into a controlled system known as *computer-integrated manufacturing* (CIM). In Figure 6–66, CIM brings together all the technologies in a management system, coordinating CADD, CAM, CNC, robotics, and material handling from the beginning of the design process through the packaging and shipment of the product. The computer system is used to control and monitor all the elements of the manufacturing system. Before a more complete discussion of the CIM systems, it is a good idea to define some of the individual elements.

## Computer-Aided Design and Drafting (CADD)

Engineers and designers use the computer as a flexible design tool. Designs can be created graphically using 3D models. The effects of changes can be quickly seen and analyzed. This allows the engineer to perform experimentation, perform stress analysis, and make calculations on the computer. In this manner, engineers can be more creative and improve the quality of the product at less cost. Computer-aided drafting is a partner to the design process.

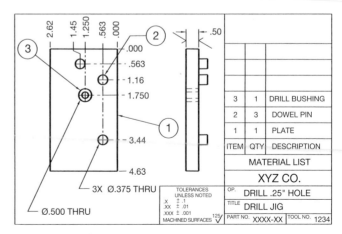

Figure 6–65 A simplified example of a tool design drawing.

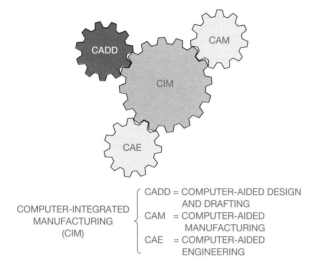

COMPUTER-INTEGRATED MANUFACTURING (CIM)
CADD = COMPUTER-AIDED DESIGN AND DRAFTING
CAM = COMPUTER-AIDED MANUFACTURING
CAE = COMPUTER-AIDED ENGINEERING

Figure 6–66 Computer-integrated manufacturing (CIM) is the bringing together of the technologies in a management system.

Accurate quality 2D drawings are created from the designs. Drafting with the computer has increased productivity from 2–20 times over manual techniques, depending on the project and skill of the drafter. The computer has also revolutionized the storage of drawings. There is no longer a need for rooms full of drawing file cabinets.

## Computer-Aided Engineering (CAE)

Three-dimensional models of the product are used for *finite element analysis*. This analysis can be accomplished with a program in which the computer breaks the model up into finite elements—small rectangular or triangular shapes. Then the computer is able to analyze each element and determine how it will act under given conditions. The program also evaluates how each element acts with the other elements and with the entire model. CAE also enables the engineer to simulate function and motion of the product without the need to build a real prototype. In this way, the product can be tested to see if it works as it should. This process has also been referred to as *predictive engineering,* in which a computer software prototype, rather than a physical prototype, is made to test the function and performance of the product. Design changes can be made directly on the computer screen. Key dimensions are placed on the computer model, and by changing a dimension, you can automatically change the design or you can change the values of variables in mathematical engineering equations to automatically alter the design.

## Computer Numerical Control (CNC)

Computer numerical control is the use of a computer to write, store, and edit numerical control programs to operate a machine tool. Numerical control (NC) is a method of controlling a machine tool using coded computer language.

## Computer-Aided Manufacturing (CAM)

CAM is a concept that surrounds any use of computers to aid in any manufacturing process. CAD and CAM work best when product programming is automatically performed from the model geometry created during the CAD process. Complex 3D geometry can be quickly and easily programmed for machining.

## Computer-Aided Quality Control (CAQC)

Information about the manufacturing process and quality control is collected automatically while parts are being manufactured. This information is fed back into the system and compared to the design specification for model tolerances. With this type of monitoring of the mass production process, maintaining the highest product quality is ensured.

## Robotics

According to the Society of Manufacturing Engineers (SME) Robot Institute of America, a *robot* is a reprogrammable multifunctional manipulator designed to move material, parts, tools, or specialized devices through variable programmed motions for the performance of a variety of tasks. *Reprogrammabale* means the robot's operating program can be changed to alter the motion of the arm or tooling. *Multifunctional* means the robot is able to perform a variety of operations based on the program and tooling it uses. A typical robot is shown in Figure 6–67.

To make a complete *manufacturing cell*, the robot is added to all of the other elements of the manufacturing process. The other elements of the cell include the computer controller, robotic program, tooling, associated machine tools, and material handling equipment.

*Closed-loop (servo)* and *open-loop (nonservo)* are the two types of control used to position the robot tooling. The robot is constantly monitored by position sensors in the closed-loop system. The movement of the robot arm must always conform to the desired path and speed. Open-loop robotic systems do not constantly monitor the position of the tool while the robot arm is moving. The control happens at the end of travel where limit controls position the accuracy of the tool at the desired place.

## The Human Factor

The properly operating elements of the CIM system provide the best manufacturing automation available. What has been described in this discussion might lead you to believe that the system can be set up, turned on, and the people can go home. Not so. An important part of CIM is the human element. People are needed to handle the many situations that happen in the manufacturing cell. The operators constantly observe the machine operation to ensure that material is feeding properly and quality is maintained. In addition to monitoring the system, people load raw material into the system, change machine tools when needed, maintain and repair the system as needed, and handle the completed work when the product is finished.

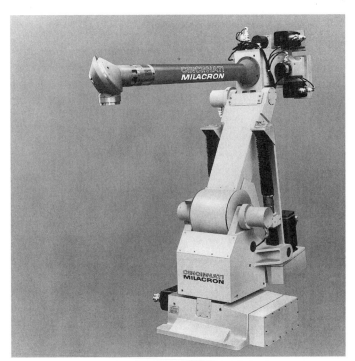

Figure 6–67 A typical robot. *Courtesy Cincinnati Milacron Marketing Company.*

## INTEGRATION OF COMPUTER-AIDED DESIGN AND COMPUTER-AIDED MANUFACTURING (CAD/CAM)

Computer-aided design (CAD) and computer-aided manufacturing (CAM) can be set up to create a direct link between the design and manufacture of a product. The CAD program is used to create the product geometry. This can be in the form of 2D multiview drawings as discussed in Chapter 5, Multiviews, of this text, or as 3D models as explained in Chapter 14, Solid Modeling, Animation, and Virtual Reality. The drawing geometry is then used in the CAM program to generate instructions for the computer numerical control (CNC) machine tools employing stamping, cutting, burning, bending, and other types of operations discussed throughout this chapter. This is commonly referred to as CAD/CAM *integration*. The CAD and CAM operations can be performed on the same computer, or the CAD work can be done at one location and the CAM program can be created at another location. If both the design and manufacturing are done within the company at one location, the computers can be linked through the local area network. If the design is done at one location and the manufacturing done at another, the computers can be linked through the Internet, or files can be transferred on disk or CD-ROM.

CAD/CAM is commonly used in modern manufacturing because it increases productivity over conventional manufacturing methods. The CAD geometry or model is created during the design and drafting process and is then used directly in the CAM process for the development of the CNC programming. The coordination can also continue into computerized quality control.

The CAD/CAM integration process allows the CAM program to import data from the CAD software. The CAM program then uses a series of commands to instruct CNC machine tools by setting up tool paths. The tool path includes the selection of specific tools to accomplish the desired operation. This can also include specifying tool feed rates and speeds, selecting tool paths and cutting methods, activating tool jigs and fixtures, and selecting coolants for material removal. Some CAM programs automatically calculate the tool offset based on the drawing geometry. An example of tool offset is shown in Figure 6–68. CAM programs, such as SurfCAM and MasterCAM, directly integrate the CAD drawing geometry from programs, such as AutoCAD, as a reference. The CAM programmer can then simply establish the desired tool and tool path. The final CNC program is generated when the postprocessor is run. A *postprocessor* is an integral piece of software that converts a generic CAM-system tool path into usable CNC machine code (G-code).

The *CCNC program* is a sequential list of machining operations in the form of a code that is used to machine the part as needed. Typically known as G & M code, these codes invoke preparatory functions and control tool machine movement, spindle speed and direction, and other operations such as clamping, part manipulation, and on-off switching. The machine programmer is trained to select the proper machine tools. A separate tool is used for each operation. These tools can include milling, drilling, turning, threading, and grinding. The CNC programmer orders the machining sequence, selects the tool for the specific operation, and determines the tool feed rate and cutting speed, depending on the material.

CAM software programs, such as SurfCAM and MasterCAM, increase productivity by helping the CNC programmer create the needed CNC code. CAD drawings from programs, such as AutoCAD, or solid models from software, such as Solid Works, can be transferred directly to the CAM program. A sequence of activities is commonly used to prepare the CAD/CAM integration.

1. Create the part drawing or model using programs, such as AutoCAD, Autodesk Inventor, or Solid Works. However, programmers commonly originate the model in the CAD/CAM software, such as MasterCAM or SurfCAM.

2. Open the CAD file in the CAM program.

3. Run the CAM software program, such as SurfCAM or MasterCAM, to establish the following:

   ■ Choose the machine or machines needed to manufacture the part.

   ■ Select the required tooling.

   ■ Determine the machining sequence.

   ■ Calculate the machine tool feed rates and speeds based on the type of material.

   ■ Verify the CNC program using the software's simulator.

   ■ Create the CNC code.

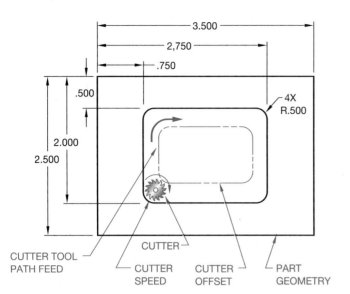

**Figure 6–68** The machine tool cutter, tool path, cutter speed, and cutter offset.

4. Prove-out the program on the CNC machine tool.

5. Run the program to manufacture the desired number of parts.

## STATISTICAL PROCESS CONTROL (SPC)

A system of quality improvement is helpful when you produce a product or engage in a service and want to improve the quality of work and, while increasing the output, do so with less labor and at reduced cost. Competition is here to stay regardless of the nature of the business. Improved quality means less waste and less rework, resulting in increased profits and an improved market position. Unfortunately, some managers see quality as a drag on profits. Quality is often placed after cost and delivery, because some managers believe that high quality can be achieved only through costly and slow inspection processes. Many managers are now seeing the influence of quality on sales because high quality has become an important criterion in their customers' purchase decisions. In addition, poor quality is expensive. It is estimated that between 15% and 40% of the American manufacturer's product cost is a result of unacceptable output. Regardless of the goods or service produced, it is always less costly to do it right the first time. Improved quality improves productivity, increases sales, reduces cost, and improves profitability. The net result is continued business success.

Traditionally, a type of quality control/detection system has been used in most organizations in the United States. This system comprises customer demand for a product, which is then manufactured in a process made up of a series of steps or procedures. Input to the process includes machines, materials, workforce, methods, and environment, as shown in Figure 6–69. Once the product or service is produced, it passes to an inspection operation where decisions are made to ship, scrap, rework, or otherwise correct any defects when discovered (if discovered). In actuality, if nonconforming products are being produced, some are being shipped.

Even the best inspection process screens out only a portion of the defective goods. Problems inherent in this system are that it does not work very well and it is costly. American businesses have become accustomed to accepting these limitations as the "costs of doing business."

The most effective way to improve quality is to alter the production process—the system—rather than the inspection process. This entails a major shift in the entire organization from the detection system to a prevention mode of operation. In this system, the elements (inputs, process, product or service, customer) remain the same, but the inspection method is significantly altered or eliminated. A primary difference between the two systems is that in the prevention system, statistical techniques and problem-solving tools are used to monitor, evaluate, and provide guidance for adjusting the process to improve quality. Statistical process control (SPC) is a method of monitoring a process quantitatively and using statistical signals to either leave the process alone, or change it (see Figure 6–70). SPC involves several fundamental elements:

1. The process, product, or service must be measured. It can be measured using either variables (a value that varies), or attributes (a property or characteristic) from the data collected. The data should be collected as close to the process as possible. If you are collecting data on a particular dimension of a manufactured part, data should be collected by the machinist responsible for holding that dimension.

2. The data can be analyzed using control charting techniques. Control charting techniques use the natural variation of a process, determining how much the process can be expected to vary if the process is operationally consistent. The control charts are used to evaluate whether the process is operating as designed, or if something has changed.

3. Action is taken based on signals from the control chart. If the chart indicates that the process is *in control* (operating consistently), the process is left alone at this point. On the other hand, if the process is found to be *out of control* (changing more than its normal variability allows), action is taken to bring it back into control. It is also important to determine how well the process meets specifications and how well it accomplishes the task. If a process is not in control, its ability to meet specifications is constantly changing. Process capability cannot be evaluated unless the process is in control. Process improvement generally involves changes to the process system that will improve quality or productivity or both. Unless the process is consistent over time, any actions to improve it may be ineffective.

Manufacturing quality control often uses computerized monitoring of dimensional inspections. When this is done, a chart is developed that shows feature dimensions obtained at inspection intervals. The chart shows the expected limits of sample averages as two dashed parallel horizontal lines, as shown in Figure 6–71. It is important not to confuse control limits with tolerances—they are not related to each other. The control limits come from the manufacturing process as it is operating. For example, if you set out to make a part 1.000±.005 inch, and you start the run and periodically take five samples and plot the averages (x) of the samples,

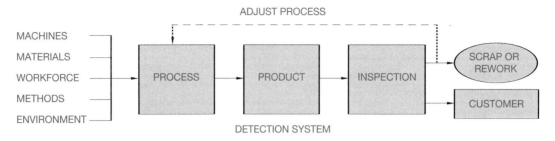

Figure 6–69 Quality control/detection system.

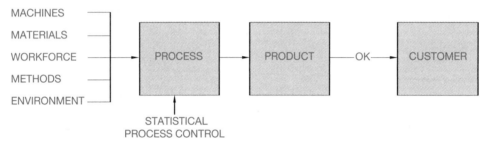

Figure 6–70 Quality control/prevention system.

the sample averages will vary less than the individual parts. The control limits represent the expected variation of the sample averages if the process is stable. If the process shifts or a problem occurs, the control limits signal that change. Notice in Figure 6–71 that the x values represent the average of each of five samples; x is the average of averages over a period of sample-taking. The *upper control limit (UCL)* and the *lower control limit (LCL)* represent the expected variation of the sample averages. A sample average may be "out of control," yet remain within tolerance. During this part of the monitoring process, the out-of-control point represents an individual situation that may not be a problem; however, if samples continue to be measured out of control limits, the process is out of control (no longer predictable) and therefore may be producing parts outside the specification. Action must then be taken to bring the process back into statistical control, or 100% inspection must be resumed. The SPC process only works when a minimum of 25 sample means are in control. When this process is used in manufacturing, part dimensions remain within tolerance limits and parts are guaranteed to have the quality designed.

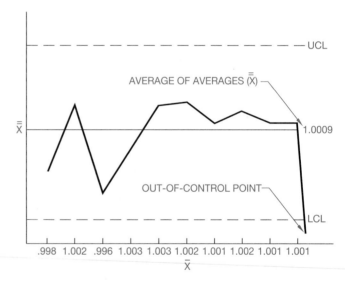

Figure 6–71 Quality control chart.

# CHAPTER 6 TEST

## Part I

Multiple choice: Respond to the following by selecting a, b, c, or d to best answer the question or complete the statement.

1.  These metals contain iron, such as iron and steel.
    a. ferrous
    b. nonferrous
    c. alloys
    d. inorganic

2.  These metals do not contain iron, such as copper and aluminum.
    a. ferrous
    b. nonferrous
    c. alloys
    d. inorganic

3.  This type of plastic material can be heated and formed by pressure, and the shape can be changed by reheating.
    a. plastic
    b. polymers
    c. thermoplastic
    d. thermoset

4.  This type of plastic material is formed into a permanent shape by heat and pressure, and cannot be altered by heating after curing.
    a. plastic
    b. polymers
    c. thermoplastic
    d. thermoset

5.  Carbon, ceramics, and composites are in a group of materials called:
    a. ferrous
    b. nonferrous
    c. alloys
    d. inorganic

6.  These materials are clay, glass, refractory, and inorganic cements that are very hard, brittle, and are resistant to heat, chemicals, and corrosion.
    a. carbon
    b. ceramic
    c. composites
    d. graphite

7.  Two or more materials that are bonded together by adhesion defined as:
    a. carbon
    c. composites
    b. ceramic
    d. graphite

8.  The ability to be stretched is characterized by:
    a. tensile strength
    b. ductility
    c. malleability
    d. normalizing

9.  The force that bonds together the molecules of unlike substances when the surfaces come in contact is referred to as:
    a. adhesion
    c. malleability
    b. ductility
    d. normalizing

10. This is a popular casting material for automotive cylinder blocks, machine tools, agricultural implements, and cast iron pipe.
    a. alloy cast iron
    b. chilled cast iron
    c. gray cast iron
    d. white cast iron

11. This is extremely hard, brittle, and has almost no ductility, and should be used where compressive strength and wear resistance is required.
    a. alloy cast iron
    b. chilled cast iron
    c. gray cast iron
    d. white cast iron

12. When gray cast iron is chilled rapidly, this material has the internal characteristics of gray cast iron and the surface advantage of white cast iron.
    a. alloy cast iron
    b. chilled cast iron
    c. malleable cast iron
    d. white cast iron

13. Elements such as nickel, chromium, molybdenum, copper, or manganese can be used with this type of cast iron to increase the properties of strength, wear resistance, corrosion resistance, or heat resistance.
    a. alloy cast iron
    b. nodular cast iron
    c. gray cast iron
    d. white cast iron

14. This term means the ability to be stretched, drawn, or hammered thin without breaking.

    a. adhesion
    b. ductility
    c. malleability
    d. normalizing

15. This term means the ability to be hammered or pressed into shape without breaking.

    a. adhesion
    b. ductility
    c. malleability
    d. extruded

16. Special processing procedures, along with the addition of magnesium- or cerium-bearing alloys, result in a cast iron with spherical-shaped graphite rather than flakes, as in gray cast iron, resulting in a material with greater strength and ductility.

    a. alloy cast iron
    b. nodular cast iron
    c. malleable cast iron
    d. white cast iron

17. This is an alloy of iron containing 0.8–1.5% carbon.

    a. alloy cast iron
    b. nodular cast iron
    c. malleable cast iron
    d. steel

18. The properties of this material can be altered by changing the carbon content.

    a. alloy cast iron
    c. malleable cast iron
    b. nodular cast iron
    d. steel

19. This material has 0.3 – .6% carbon, has some hardness, but remains easy to forge and machine.

    a. mild steel
    b. medium-carbon steel
    c. high-carbon steel
    d. hot-rolled steel

20. This material is low in carbon and is commonly used for forged and machined parts, but cannot be hardened.

    a. mild steel
    b. medium-carbon steel
    c. high-carbon steel
    d. hot-rolled steel

21. This carbon steel can be hardened by heat treating, but is difficult to forge, machine, or weld.

    a. mild steel
    c. high carbon steel
    b. medium carbon steel
    d. hot-rolled steel

22. This steel is formed into shape by pressure between rollers or by forging when in a red-hot state.

    a. MS
    b. HRS
    c. CRS
    d. SAE

23. This steel has additional forming after initial hot rolling and is used to clean up hot-forged steel, and provides a smooth, clean surface, ensures dimensional accuracy, and increases tensile strength.

    a. MS
    b. HRS
    c. CRS
    d. SAE

24. This steel alloy is the basis for stainless steel and is used where corrosion and wear resistance is required.

    a. tungsten
    b. chromium
    c. manganese
    d. molybdenum

25. This steel alloy results in a material that is very hard and ideal for use in cutting tools.

    a. tungsten
    b. chromium
    c. manganese
    d. molybdenum

26. This steel alloy is a purifying element that adds strength for parts that must be shock- and wear-resistant.

    a. tungsten
    b. chromium
    c. manganese
    d. molybdenum

27. This alloy is added to steel when the product must retain strength and wear resistance at high temperatures.

    a. tungsten
    b. chromium
    c. manganese
    d. molybdenum

28. This steel alloy is used when a tough, strong, nonbrittle material is required.

    a. tungsten
    c. manganese
    b. vanadium
    d. molybdenum

29. This is a good agricultural steel.

    a. SAE 4320
    b. SAE 1040
    c. SAE 1070
    d. SAE 2330

30. This steel is good for truck and bus gears.
    a. SAE 4320
    b. SAE 1040
    c. SAE 1070
    d. SAE 2330

31. This steel is good for carbon steel forgings.
    a. SAE 4320
    c. SAE 1070
    b. SAE 1040
    d. SAE 2330

32. This is a process of heating and cooling steel using specific, controlled conditions and techniques.
    a. killed steel
    b. heat treating
    c. rimmed steel
    d. carburization

33. This steel is degassed and deoxidized and is used for forging, heat treating, and difficult stampings.
    a. killed steel
    b. normalized steel
    c. rimmed steel
    d. carburized steel

34. This steel is cast with little or no degasification and is used for sheets, strips, rods, and wires with excellent surface finish.
    a. killed steel
    b. normalized steel
    c. rimmed steel
    d. carburized

35. This is a process of heating steel to a specified temperature and then allowing it to cool slowly by air, which brings the steel to a normal state.
    a. killed
    b. normalization
    c. rimmed
    d. carburization

36. During the heat treating process the metal is heated to a specified temperature and then cooled suddenly by plunging in water, oil, or other liquid. This sudden cooling process is referred to as:
    a. case hardening
    b. normalization
    c. quenching
    d. carburization

37. This is a process that case hardens steel by introducing carbon into the metal by heating to a specified temperature range while in contact with a solid, liquid, or gas material containing carbon.
    a. annealing
    b. normalization
    c. tempering
    d. carburization

38. This process achieves certain hardening characteristics by reheating a normalized steel through a controlled process to a specified temperature followed by cooling at a predetermined rate.
    a. annealing
    c. tempering
    b. normalization
    d. carburization

39. This process is used to soften steel under certain heating and cooling conditions and techniques.
    a. annealing
    b. normalization
    c. tempering
    d. carburization

40. This hardness testing system is performed by placing a known load, using a ball of specified diameter, in contact with the material surface. The diameter of the resulting impression in the material is measured.
    a. Brinell
    b. Rockwell

41. This hardness testing system uses a machine that measures hardness by determining the depth of penetration of a spherical-shaped device under controlled conditions.
    a. Brinell
    b. Rockwell

42. The ability of a metal to be shaped by forcing it through a die is called:
    a. adhesion
    b. ductility
    c. malleability
    d. extrudation

43. This material is a widely used alloy of copper and zinc; it is corrosion-resistant, strong, and ductile.
    a. aluminum
    b. columbium
    c. brass
    d. bronze

44. This material is lightweight, easily cast, conductive of heat and electricity, can be easily extruded, and is very malleable.
    a. aluminum
    b. copper
    c. brass
    d. bronze

45. This metal is an alloy of copper and tin, where the tin adds hardness and increases wear resistance.
    a. aluminum
    c. brass
    b. columbium
    d. bronze

46. This material is easily rolled and drawn into wire, has excellent corrosion resistance, is a great electrical conductor, and has better ductility than any other metal except silver or gold.
    a. aluminum
    c. brass
    b. copper
    d. bronze

47. This material is used for coins and jewelry and is commonly hardened by adding copper.
    a. gold
    c. platinum
    b. silver
    d. titanium

48. This metal is rare and expensive and has industrial uses, including applications where corrosion resistance and a high melting point are required.
    a. gold
    b. silver
    c. platinum
    d. titanium

49. This metal has the unique ability to react with and reduce carbon monoxide and other harmful exhaust emissions in automobiles.
    a. columbium
    b. tungsten
    c. platinum
    d. titanium

50. This metal, known as sterling, is alloyed with 8–10% copper for use in jewelry and coins and household items, and is 925/1000 pure.
    a. gold
    b. silver
    c. platinum
    d. titanium

51. This material is used in nuclear reactors because it has a very high melting point of 2403°C and is resistant to radiation.
    a. columbium
    b. tungsten
    c. platinum
    d. titanium

52. This metal has been used extensively as a filament in light bulbs because it can be drawn into very fine wire and because of its high melting point.
    a. columbium
    b. tungsten
    c. platinum
    d. titanium

53. This metal has many uses in the aerospace and jet aircraft industries because it has the strength of steel, the approximate weight of aluminum, and is resistant to corrosion and temperatures up to 427°C.
    a. columbium
    b. tungsten
    c. platinum
    d. titanium

54. This is the hardest human-made material and is used to make cutting tools for any type of manufacturing application.
    a. columbium carbide
    b. tungsten carbide
    c. platinum carbide
    d. titanium carbide

55. Fourteen-carat gold is:
    a. 14/24 copper and 10/24 gold
    b. 14% gold
    c. 14/24 gold and 10/24 copper
    d. 14% copper

56. Casting is the end result of a method called:
    a. forging
    b. founding
    c. machining
    d. stamping

57. This is the most commonly used method of making castings.
    a. centrifugal casting
    b. die casting
    c. permanent casting
    d. investment casting

58. Objects with circular or cylindrical shapes lend themselves to this type of casting, where a mold is revolved rapidly while molten metal is poured into the cavity.
    a. centrifugal casting
    b. die casting
    c. permanent casting
    d. sand casting

59. This type of casting refers to a process in which a steel or iron mold can be used many times and produces a better finished quality than with sand casting.
    a. centrifugal casting
    b. die casting
    c. permanent casting
    d. sand casting

60. This method of casting injects molten nonferrous metal into a steel or iron die under high pressure, where castings are produced quickly and economically.
    a. centrifugal casting
    b. die casting
    c. permanent casting
    d. sand casting

61. Also known as the lost-wax method, this casting technique enables the development of very close tolerances. A wax mold is coated with ceramic, and after the ceramic is hardened, the wax is melted and flows out, leaving a cavity to be filled with molten metal.

    a. centrifugal casting
    b. die casting
    c. permanent casting
    d. investment casting

62. This is a manufacturing process of shaping malleable metals by hammering or pressing between dies that duplicate the desired shape.

    a. founding
    b. forging
    c. casting
    d. stamping

63. Forged metal is generally stronger and more ductile than cast metal, and exhibits a greater resistance to fatigue and shock than machined parts.

    a. true
    b. false

64. This machine tool uses a rotating abrasive wheel for removing material and is generally used when a smooth accurate surface finish is required.

    a. drilling machine
    c. lathe
    b. grinding machine
    d. milling machine

65. This is one of the most versatile machine tools, and uses a vertical or horizontal rotary cutting tool to remove material from the work:

    a. drilling machine
    b. grinding machine
    c. lathe
    d. milling machine

66. This machine tool is commonly used to machine drilled holes, and is also used for reaming, boring, countersinking, counterboring, and tapping:

    a. drilling machine
    b. grinding machine
    c. lathe
    d. milling machine

67. This machine tool is used to cut material by turning cylindrically shaped objects between centers or in a holding device called a chuck:

    a. drilling machine
    b. grinding machine
    c. lathe
    d. milling machine

68. This machine can be used to cut material to length or for cutting a narrow slot called a kerf:

    a. saw machine
    b. grinding machine
    c. lathe
    d. milling machine

69. This machining operation uses a direct current that is passed through an electrolyte solution between an electrode and the workpiece. Chemical reaction, caused by the current in the electrolyte, dissolves the metal.

    a. electrochemical machining (ECM)
    b. electrodischarge machining (EDM)
    c. electron beam (EB)
    d. ultrasonic machining

70. This process, also known as impact grinding, is a process in which a high-frequency mechanical vibration is maintained in a tool designed to a specific shape:

    a. electrochemical machining (ECM)
    b. electrodischarge machining (EDM)
    c. electron beam (EB)
    d. ultrasonic machining

71. In this type of machining, an electron beam generated by a heated tungsten filament is used to cut very accurate features into a part.

    a. electrochemical machining (ECM)
    b. electrodischarge machining (EDM)
    c. electron beam (EB)
    c. ultrasonic machining

72. In this process, the material and an electrode are submerged in a dielectric fluid, which is a nonconductor, forming a barrier between the part and the electrode. A very small gap of about .001 inch is maintained between the electrode and the material. A high-speed arc causes the material to be removed.

    a. electrochemical machining (ECM)
    b. electrodischarge machining (EDM)
    c. electron beam (EB)
    d. ultrasonic machining

73. This process uses light amplification by stimulated emission of radiation, which cuts material with instant temperatures of up to 75,000°F and produces smooth surfaces without burrs or rough edges:

    a. laser machining
    b. electrodischarge machining (EDM)
    c. electron beam (EB)
    d. ultrasonic machining

74. This tool is used to enlarge or finish a hole by removing a small amount of material between .005 and .016 inch:

    a. drill
    b. ream
    c. bore
    d. counterbore

75. This tool is used to enlarge the end of a machined hole to a specified diameter and depth usually for recessing the head of a bolt or screw:
    a. drill
    b. ream
    c. bore
    d. counterbore

76. This tool is used to machine new holes or enlarge existing holes in material:
    a. drill
    b. ream
    c. bore
    d. counterbore

77. This tool is used on a lathe for removing internal material:
    a. drill
    b. ream
    c. bore
    d. counterbore

78. This is a conical-shaped tool used to cut the recess for the conically shaped head of a fastener:
    a. counterdrill
    b. countersink
    c. spotface
    d. counterbore

79. This is a machined round surface that is not very deep and is usually used to clean up a surface for a washer or bolt head to rest on:
    a. counterdrill
    b. countersink
    c. spotface
    d. counterbore

80. This is a circular pad on forgings or castings that project out from the body of the part:
    a. boss
    b. lug
    c. pad
    d. chamfer

81. This is a slightly raised surface of any size and shape:
    a. boss
    b. lug
    c. pad
    d. chamfer

82. This is the cutting away of the sharp external or internal corner of an edge:
    a. round
    b. fillet
    c. pad
    d. chamfer

83. This is a small radius formed between the inside angle of two surfaces:
    a. round
    b. fillet
    c. pad
    d. chamfer

84. This is a small-radius outside corner formed between two surfaces:
    a. round
    b. fillet
    c. kerf
    d. chamfer

85. This is a narrow slot formed by removing material while sawing:
    a. round
    b. fillet
    c. kerf
    d. chamfer

86. Threads that are used as fasteners are commonly referred to as:
    a. fastener threads
    b. screw threads
    c. bolt threads
    d. taps

87. The machine tool used to make external threads is called:
    a. a die
    b. a tap
    c. an external threader
    d. a threader

88. The machine tool used to cut internal threads is called:
    a. a die
    b. a tap
    c. an internal threader
    d. a threader

89. This is a cold forming process used to uniformly roughen a cylindrical or flat surface with a diamond or straight pattern:
    a. roughening
    b. pressing
    c. forming
    d. knurling

90. This is the intended condition of the material surface after manufacturing processes have been implemented, which includes roughness, waviness, lay, and flaws:
    a. surface texture
    b. surface finish
    c. both a and b
    d. only b

91. These tooling devices are either fixed or moving devices that are used to hold the work piece in position and guide the cutting tool:

    a. jig
    b. fixture
    c. mass production
    d. assembly

92. These tooling devices do not guide the cutting tool, but are used in a fixed position to hold the workpiece:

    a. jig
    b. fixture
    c. mass production
    d. assembly

93. Completely automated manufacturing combines computer-aided design and drafting, computer-aided engineering, and computer-aided manufacturing into a controlled system known as:

    a. CADD
    b. CAM
    c. CAE
    d. CIM

94. This manufacturing system uses statistical techniques and problem-solving tools to monitor, evaluate, and provide guidance for adjusting the process to improve quality:

    a. quality control detection system
    b. quality control prevention system
    c. quality control monitoring system
    d. quality control statistical system

95. This is a method of monitoring a manufacturing process quantitatively and using statistical signals to either leave the process alone, or change the process:

    a. quality control detection system
    b. quality control statistical system
    c. quality control monitoring system
    d. statistical process control

96. Surface roughness is one of the most common characteristics of surface finish, which consists of the finer irregularities of the surface texture resulting, in part, from the manufacturing process. When the drawing is in inches, the surface roughness is measured in:

    a. thousandths of an inch
    b. micrometers
    c. microinches
    d. ten-thousandths of an inch

97. When the drawing is in metric, the surface roughness is measured in:

    a. thousandths of an inch
    b. micrometers
    c. microinches
    d. ten-thousandths of an inch

98. Micro means:

    a. thousandths
    b. ten-thousandths
    c. millionths
    d. microminiature

99. The reaming process produces a surface roughness microinch average of:

    a. 500–1000
    c. 32–125
    b. 63–250
    d. 4–16

100. The polishing process produces a surface roughness microinch average of:

    a. 500–1000
    b. 63–250
    c. 32–125
    d. 4–16

## Part II

Short answers: Answer the questions with short, complete statements or sketches as needed.

**Manufacturing Materials**

1. Define ferrous metals.

    _____
    _____

2. Identify the two main types of ferrous metals.

    _____
    _____

3. Define nonferrous metals.

    _____
    _____

4. What is another name for plastics?

    _____
    _____

5. Define thermoplastic.

    _____
    _____

6. Define thermoset.

    _____
    _____

7. Define tensile strength.

    _____
    _____

8. Why are refractories used for ≈applications such as furnace liners?

    _____
    _____

9. What are composites?

    _____
    _____

10. Describe the characteristics of gray cast iron.
_____
_____

11. Given the gray cast iron material specification 30A, what does the prefix 30 denote?
_____
_____

12. Define ductility.
_____
_____

13. What are the properties of white cast iron?
_____
_____

14. Which cast iron has the internal characteristics of gray cast iron and the exterior properties of white cast iron?
_____
_____

15. Define malleable.
_____
_____

16. Identify at least two uses for nodular cast iron.
_____
_____

17. How is it possible to alter the properties of steel?
_____
_____

18. Name a steel that is low in carbon and is commonly used for forged and machined parts.
_____
_____

19. Describe high-carbon steel.
_____
_____

20. Describe the difference between hot- and cold-rolled steel.
_____
_____

21. Describe the properties of the following steel alloying elements: manganese, molybdenum, and tungsten.
_____
_____
_____

22. Describe stainless steel.
_____
_____

23. Give the typical contents of stainless steel.
_____
_____

24. Identify the contents of the 200, 300, and 400 series stainless steels as identified by the American Iron and Steel Institute.
_____
_____

25. Identify at least four common uses for stainless steel.
_____
_____

26. Given the steel identification number SAE 1020, describe the components SAE, 10, and 20.
_____
_____

27. Identify the steel recommended for the following general applications: agricultural steel, bolts, and screws; car and truck gears; and transmission shafts.
_____
_____

28. Define heat treating.
_____
_____

29. Define normalizing.
_____
_____

30. Define case hardening.
_____
_____

31. Define carburization.
_____
_____

32. Define tempering.
_____
_____

33. How is the Rockwell hardness test performed?
_____
_____

34. Describe the properties of aluminum.
_____
_____

35. Define extruded.
_____
_____

36. Identify the alloying elements in brass.
_____
_____

37. Identify the alloying elements in bronze.
_____
_____

38. What is the advantage of adding phosphorus to bronze?
_____
_____

39. Describe at least one industrial use for gold.
_____
_____

40. Identify the metal that has the weight advantage of aluminum and the strength of steel.

41. What are the elements of and process for making tungsten carbide?

In Questions 42 through 57, identify the type of metal or metals that are part of the given Unified Numbering System (UNS) series:

42. A00001 to A99999

43. C00001 to C99999

44. E00001 to E99999

45. L00001 to L99999

46. M00001 to M99999

47. P00001 to P99999

48. R00001 to R99999

49. Z00001 to Z99999

50. D00001 to D99999

51. F00001 to F99999

52. G00001 to G99999

53. H00001 to H99999

54. J00001 to J99999

55. K00001 to K99999

56. S00001 to S99999

57. T00001 to T99999

58. Give the general definition of plastic.

59. Name the process of joining two or more molecules to form a more complex molecule with physical properties that are different from the original molecules.

60. Identify the term that refers to polymer-based materials that have elastic qualities not found in thermoplastics and thermosets.

61. Give the name and basic characteristics of ABS.

62. Give the name and basic characteristics of ASA.

63. Give the name of the thermoplastic that has the versions nitrate, acetate, butyrate, propionate, and ethyl cellulose.

64. Name the type of thermoplastics that is very resistant to chemicals, friction, and moisture. These materials have excellent dimensional stability for use as wire coating and insulation, nonstick surfaces, chemical containers, O-rings, and tubing.

65. Identify the common name for the polyamide thermoplastic.

66. Name a thermoplastic that is rigid with very high impact and stress fracture resistance at temperatures between −40° and 210°F (−40° to 99°C), and is used when constant bending is required in the function of the material design.

67. Give the name of an inexpensive plastic to produce that has many desirable properties, including heat, chemical, scratch, and moisture resistance. It is also resistant to continuous bending applications. Products include appliance parts, hinges, cabinets, and storage containers.

68. Name the plastic that is inexpensive and easy to manufacture, has excellent transparency, and is very rigid. However, it can be brittle and has poor impact, weather, and chemical resistance. Products include model kits, plastic glass, lenses, eating utensils, and containers.

69. Name the plastic that is commonly called PVC.

70. Identify as least two applications for PVC and its characteristics for these applications.

71. Briefly discuss the major differences between thermoplastics and thermosets.

72. Name the plastic material that dates back to the late 1800s. This plastic is hard and rigid, has good compression strength, is tough and does not absorb moisture, but it is brittle. These plastics are commonly used for the manufacture of electrical switches and insulators, electronic circuit boards, distributor caps, and binding material and adhesive.

73. Name the material that is commonly used as a plastic for reinforced composites, also known as reinforced thermoset plastics (RTP). Typical uses are for boat and recreational vehicle construction, automobiles, fishing rods, tanks, and other structural products.

74. Identify the type of plastics that are also referred to as synthetic rubber.

75. Give the term that refers to the heating of the material in a steel mold that forms the desired shape.

76. Name the material that has the trade name Neoprene and was the first commercial synthetic rubber. This material has better weather, sunlight, and petroleum resistance than natural rubber. It is also very flame-resistant, but does not resist electricity. Common uses include automotive hoses and other products where heat is found, such as gaskets, seals, and conveyor belts.

77. Name the material that has a wide range of makeup from liquid to solid. Liquid and semiliquid forms are used for lubricants. Harder forms are used where non-stick surfaces are required.

**Manufacturing Processes**

78. What is another name for casting?

79. Define casting.

80. Name the most commonly used method of making castings.

81. Discuss the function of cores in the casting process.

82. List at least two advantages of using cores.

83. Describe centrifugal casting.

84. List at least two advantages of the die casting process.

85. Describe the permanent casting process.

86. Name the casting technique that is referred to as "lost wax."

87. Why is shrinkage allowance required in casting design?

88. What is the estimated shrinkage for most irons?

89. Define draft.

90. In casting design, is draft added to the minimum or maximum design sizes of the part?

91. Identify at least two reasons why fillets are used in casting design.

92. List at least three considerations that influence the amount of extra material that must be left on a casting for machining allowance.

_____

_____

93. What is the recommended standard finish allowance for iron or steel?

_____

_____

94. Describe hot spots in casting.

_____

95. Define forging.

_____

96. What is the grain structure advantage of using the forging process to manufacture a part rather than making a machined part?

_____

_____

97. Describe upset forging.

_____

_____

98. Describe swaging.

_____

_____

99. Describe bending.

_____

_____

100. Describe punching.

_____

_____

101. List at least two advantages of machine forging.

_____

_____

102. Briefly describe the stamping process.

_____

_____

103. Give the name of the machine that performs the stamping operation.

_____

_____

104. Name the process that takes metal-alloyed powders and feeds them into a die where they are compacted under pressure to form the desired shape. The compacted metal is then removed from the die and heated at temperatures below the melting point of the metal.

_____

_____

105. What is the process called that takes the compacted metal that has been removed from the die in the previous question and then heats it at temperatures below the melting point of the metal?

_____

_____

106. Briefly describe the metal injection molding (MIM) process.

_____

_____

107. Name the manufacturing process that is the most commonly used process for creating thermoplastic products. The process involves injecting molten plastic material into a mold that is in the form of the desired part or product. The mold is in two parts that are pressed together during the molding process. The mold is then allowed to cool so the plastic can solidify. When the plastic has cooled and solidified, the press is opened and the part is removed from the mold.

_____

_____

108. Name the manufacturing process that is used to make continuous shapes, such as moldings, tubing, bars, angles, hoses, weather stripping, films, and any product that has a constant shape. This process creates the desired continuous shape by forcing molten plastic through a metal die.

_____

_____

109. Name the manufacturing process that is commonly used to produce hollow products, such as bottles, containers, receptacles, and boxes. This process works by blowing hot polymer against the internal surfaces of a hollow mold.

_____

_____

110. Give the name of the manufacturing process that is generally used to create products such as vinyl flooring, gaskets, and other sheet products. This process fabricates sheet or film thermoplastic or thermoset plastics by passing the material through a series of heated rollers.

_____

_____

111. Name the process that works by placing a specific amount of polymer pellets into a metal mold. The mold is then heated as it is rotated. This forces the molten material to form a thin coating against the sides of the mold. When the mold is cooled, the product is removed.

_____

_____

112. Give the name of the process that works by placing material into an initial hot die, where it takes the preliminary shape. As the material cools, a die that matches to the shape of the desired product forms the final shape.

_____

_____

113. Name the process that works by taking a sheet of material and heating it until it softens and sinks down by its own weight into a mold that conforms to the desired final shape. Vacuum pressure is commonly used to suck the hot material down against the mold.

114. Give the name of the process that uses a computer model that is traced in thin cross-sections to control a laser that deposits layers of liquid resin or molten particles of plastic material to form the desired shape.

115. Name the common fabrication process for thermosets that uses a specific amount of material that is heated and placed in a closed mold, where additional heat and pressure are applied until the material takes the desired shape. The material is then cured and removed from the mold.

116. Give the name of the process that is similar to casting, but uses a foam material that expands during the cure to fill the desired mold.

117. Give the name of the material that is also commonly referred to as reinforced plastic.

118. Briefly describe the layering process that is used for making reinforced plastic.

119. Give the name of process where resin is used to make quality composite products with a smooth surface on both sides. This method places reinforcing material into a mold and then pumps resin into the mold.

120. Name the process that uses vacuum pressure to force a thin layer of sheet-reinforced polymer around a mold.

121. Give the name of the process that uses machines to wind resin-saturated reinforcement fibers around a shaft.

## Machine Processes

122. List at least two reasons why the mechanical drafter should be familiar with machining processes.

123. List four types of machining operations that can be performed using a drilling machine.

124. Identify one of the primary functions of the grinding machine.

125. Describe the main function of a lathe.

126. Describe a feature of milling machines that influences the ability of operations and shapes that can be performed.

127. Identify two types of saw tools that can be used for cut-off and machining operations.

128. Describe chemical machining.

129. Describe electrochemical machining (ECM).

130. Describe electrodischarge machining (EDM).

131. Describe electron beam (EB) machining.

132. What is another name for ultrasonic machining?

133. Describe the basic function of a laser device.

134. Laser machining may be used on what materials?

135. What does the abbreviation CNC mean?

136. Briefly explain the CNC process. Describe computer-integrated manufacturing.

137. What is the purpose of boring?

138. Identify at least one function of a counterbore.

139. Describe drill depth.

140. Describe and give one application of a knurl.

141. Describe the primary function of a key.

142. Define keyseat.

143. What is the function of a reamer?

144. What action will the machinist take when no spotface depth is given?

145. Define surface texture.

146. In what units is surface roughness height measured?

147. Discuss the results of designing a part with specifications that require overmachining.

148. Describe the difference between jigs and fixtures.

149. Normally a jig or fixture is drawn as an assembly of the unit ready for use, and the workpiece, or part to be held, is drawn in position. How is the workpiece drawn in relationship to the jig or fixture?

150. Describe the use of drill jigs.

151. What are drill fixtures sometimes referred to as?

152. Describe the use of drill fixtures.

153. Explain how machining fixtures work.

154. Describe the function of a welding fixture.

155. Briefly explain the use of inspection fixtures.

156. Define and describe the use of progressive dies.

157. What is a pickoff jig?

158. Why is it important for a tool designer to be a good print reader?

159. Identify at least four qualities that tools must possess.

160. Name the quality control/detection system that uses statistical techniques and problem solving to monitor, evaluate, and provide guidance for adjusting the process to improve quality.

161. Describe the basic function of control charting in quality control.

## CHAPTER 6 PROBLEMS

Select one or more of the topics listed below (or as assigned by your instructor) and write a report of approximately 500 words about each. The completion of these reports requires research that can include reviews of articles from current professional magazines, visits to local industries, interviews with professionals, or searching the Internet.

**PROBLEM 6–1** Casting

**PROBLEM 6–2** Forging

**PROBLEM 6–3** Conventional machine shop

**PROBLEM 6–4** Computer numerical control (CNC) machining

**PROBLEM 6–5** Surface roughness

**PROBLEM 6–6** Tool design

**PROBLEM 6–7** Chemical machining

**PROBLEM 6–8** Electrochemical machining (ECM)

**PROBLEM 6–9** Electrodischarge machining (EDM)

**PROBLEM 6–10** Electron beam (EB) machining

**PROBLEM 6–11** Ultrasonic machining

**PROBLEM 6–12** Laser machining

**PROBLEM 6–13** Statistical process control (SPC)

**PROBLEM 6–14** Thermoplastics

**PROBLEM 6–15** Thermosets

**PROBLEM 6–16** Elastomers

**PROBLEM 6–17** Metal stamping

**PROBLEM 6–18** Powder metallurgy

**PROBLEM 6–19** Manufacturing thermoplastic products

**PROBLEM 6–20** Manufacturing thermoset plastic products

**PROBLEM 6–21** Manufacturing composites

**PROBLEM 6–22** Cast iron

**PROBLEM 6–23** Steel

**PROBLEM 6–24** Aluminum

**PROBLEM 6–25** Copper alloys

**PROBLEM 6–26** Precious and other specialty metals

**PROBLEM 6–27** Computer-aided manufacturing (CAD)

**PROBLEM 6–28** Computer-aided engineering (CAE)

**PROBLEM 6–29** Computer-integrated manufacturing (CIM)

**PROBLEM 6–30** Robotics

**PROBLEM 6–31** Computer-aided design

**PROBLEM 6–32** Computer-aided drafting

**PROBLEM 6–33** Manufacturing cell

**PROBLEM 6–34** CAD/CAM integration

**PROBLEM 6–35** Look at the PART in Figure 6–64 and answer the following questions:

a.  Which dimensions are critical to this drill jig?
_____
_____

b.  Which dimensions are of incidental interest to the tool designer?
_____
_____

c.  Which dimensions are not relevant to this drill jig?
_____
_____

**PROBLEM 6–36** Sketch a drill jig on the drawing below to satisfy the listed TDR. You can also redraw the drawing using manual drafting or CADD and then draw the drill jig, if preferred by your course objectives.

**PROBLEM 6–37** Given the following drawing of the sand casting components, place the proper term on the blank line provided for the identification of each feature.

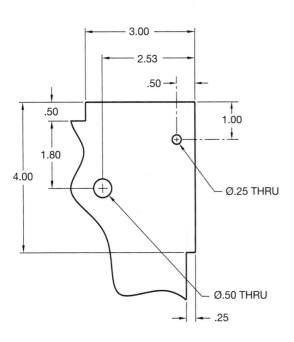

Problem 6–36

PART NAME: ------------------------PART
PART NO: ------------------------------XXXX-XX
OPERATION: ------------------------DRILL .50" DIA HOLE
MACHINE: ----------------------------USE HAND DRILL
FIXTURE: -----------------------------PICKOFF JIG LOCATING
                                      THE HOLE 1.80" AND
                                      2.53" FROM THE EDGES
                                      OF THE PLATE

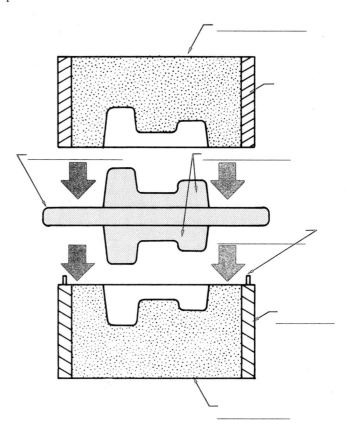

Problem 6–37

**PROBLEM 6–38** Given the following drawing of pouring molten metal into a sand casting mold, place the proper term on the blank line provided for the identification of each feature.

**PROBLEM 6–39** Given the following drawing of the sand casting cores, place the proper term on the blank line provided for the identification of each feature.

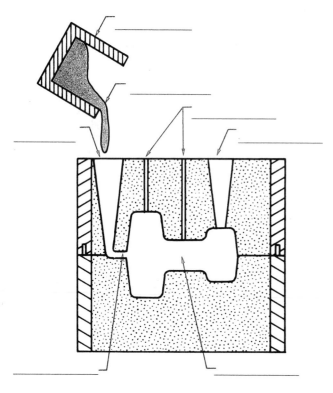

Problem 6–38

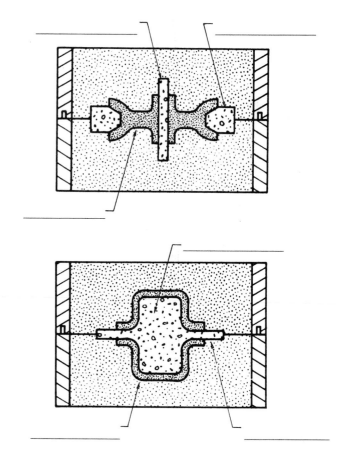

Problem 6–39

**PROBLEM 6–40** Given the following drawing of the forging process, place the proper term on the blank line provided for the identification of each feature.

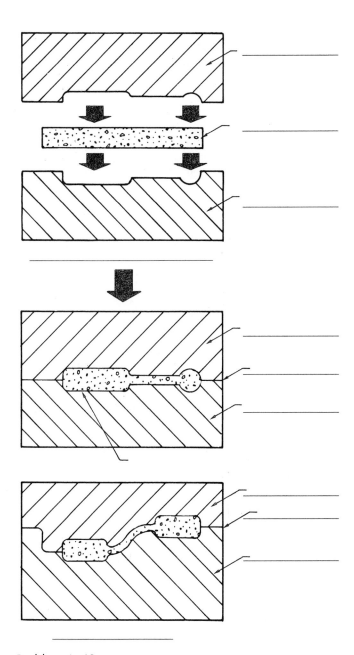

Problem 6–40

**PROBLEM 6–41** Completely identify the following machine process drawings (shown on this and the following two pages) using short, complete statements on the blank lines provided with each drawing.

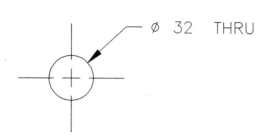

a. _____

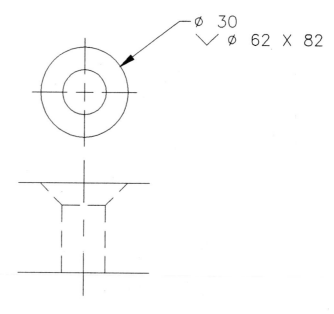

c. _____

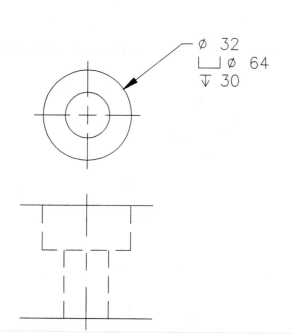

b. _____

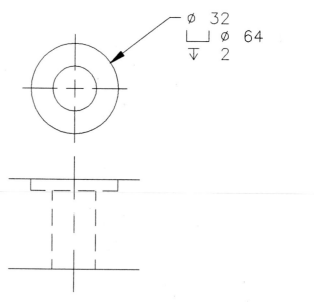

d. _____

Problem 6–41(a)

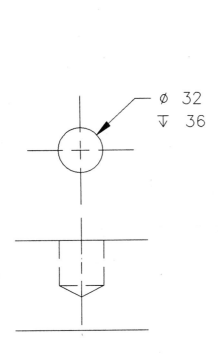

Ø 32
▼ 36

e. _____

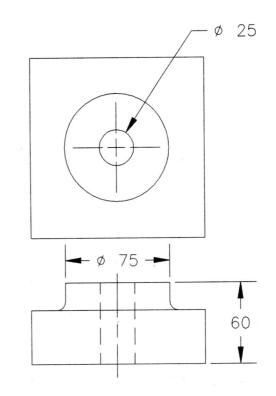

Ø 25

Ø 75

60

g. _____

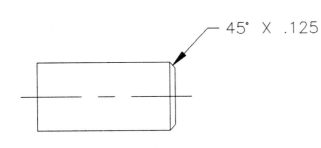

45° X .125

f. _____

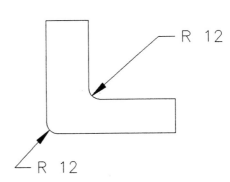

R 12

R 12

h. _____

Problem 6–41(b)

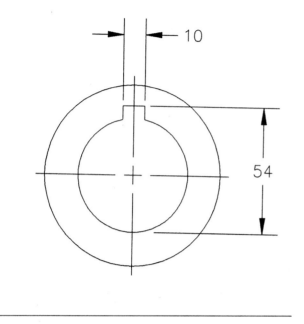

i. _____

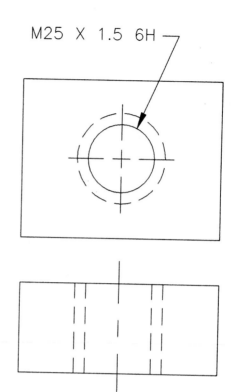

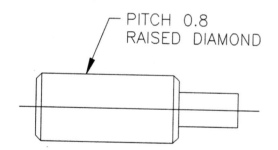

PITCH 0.8
RAISED DIAMOND

k. _____

j. _____

Problem 6–41(c)

**PROBLEM 6–42** Given the chart on this page, fill in the blanks with the proper elements of the quality control prevention system.

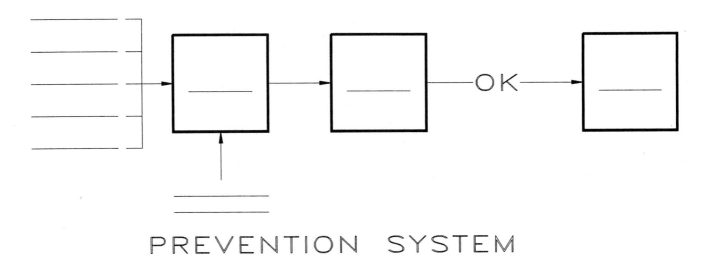

Problem 6–42

# CHAPTER 7

## *Reading Dimensions*

## LEARNING OBJECTIVES

After completing this chapter, you will be able to:

- Identify the ASME standard for dimensioning and tolerancing.
- Read prints displaying a variety of dimensioning applications.
- Read and calculate dimension tolerances on a print.
- Determine the maximum material condition of given features.
- Calculate clearance and allowance between mating features.
- Answer questions about basic fits.
- Read casting and forging dimensions and related information.
- Read the information given in surface finish symbols.

## ANSI STANDARD

The standard adopted by the American National Standards Institute (ANSI) and published by The American Society of Mechanical Engineers (ASME) is titled *Dimensioning and Tolerancing,* ASME Y14.5M_1994. This standard is available through the ASME at 345 East 47th Street, New York, NY 10017. The standard that controls general dimensional tolerances found in the title block or in general notes is ASME Y14.1M–1992, Metric Drawing Sheet Size and Format (ANSI Y14.1 for inches).

A complete detail drawing includes multiviews and dimensions, which provide both shape and size descriptions. A dimension is defined as a measurement given on a drawing. There are two classifications of dimensions: size and location. Size dimensions are placed directly on a feature to identify a specific size, or may be connected to a feature in the form of a note. The relationship of features of an object is defined with location dimensions.

Notes are a type of dimension that generally identifies the size of a feature or features with more than a numerical specification. For example, a note for a counterbore gives size and identification of the machine process used in manufacturing. There are basically two types of notes: local (or specific) notes and general notes. Local notes are connected to specific features on the views of the drawing. General notes are placed separate from the views and relate to the entire drawing.

## DIMENSIONING SYSTEMS

Dimensioning systems vary according to the nature of the work and approach a company adopts in creating its drawings. Several types of dimensioning systems are used in industry, including: Unidirectional, Aligned, Tabular, and Arrowless.

### Unidirectional

Unidirectional dimensioning is commonly used in mechanical drafting. It requires all numerals, figures, and notes to be lettered horizontally and read from the bottom of the drawing sheet. Figure 7–1 shows unidirectional dimensioning in use.

### Aligned

*Aligned* dimensioning requires all numerals, figures, and notes to be aligned with the dimension lines so they may be read from the bottom (for horizontal dimensions) and from the right side (for vertical dimensions). This method of dimensioning is commonly used in architectural and structural drafting (see Figure 7–2).

### Tabular

*Tabular* dimensioning is a system in which size and location dimensions from datums, or coordinates (X, Y, Z axis), are given in a table identifying features on the drawing. Figure 7–3 shows a method of tabular dimensioning.

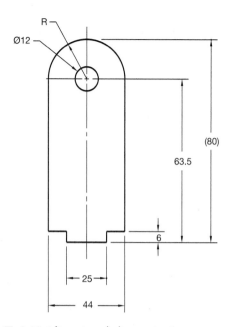

Figure 7–1 Unidirectional dimensioning.

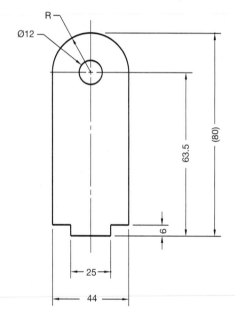

Figure 7–2 Aligned dimensioning.

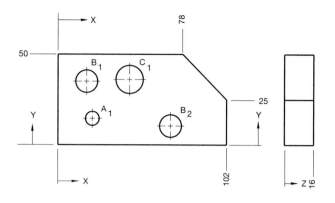

Figure 7–3 Tabular dimensioning.

| HOLE SYMBOL | HOLE DIA | LOCATION | | DEPTH Z |
|---|---|---|---|---|
| | | X | Y | |
| $A_1$ | 6 | 15 | 14 | THRU |
| $B_1$ | 9 | 12 | 38 | 9 |
| $B_2$ | 9 | 57 | 7 | 12 |
| $C_1$ | 12 | 43 | 38 | THRU |

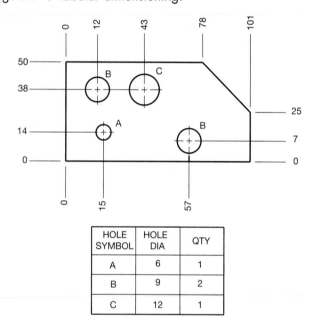

| HOLE SYMBOL | HOLE DIA | QTY |
|---|---|---|
| A | 6 | 1 |
| B | 9 | 2 |
| C | 12 | 1 |

Figure 7–4 Arrowless dimensioning.

## Arrowless

Also known as dimensioning without dimension lines, or ordinate dimensioning, *arrowless* dimensioning is similar to tabular dimensioning in that features are identified with letters and keyed to a table. Location dimensions are established with extension lines as coordinates of a specific feature known as a datum (see Figure 7–4).

## Chart Drawing

*Chart drawings* are used when a particular part or assembly has one or more dimensions that change depending on the specific application. For example, the diameter of a part may remain constant, with several alternate lengths required for different purposes.

The variable dimension is usually labeled on the drawing with a letter in the place of the dimension. The letter is then placed in a chart where the changing values are identified. Figure 7–5 is a chart drawing that shows two dimensions that have alternate sizes. The view drawn represents a typical part and the dimensions are labeled A and B. The correlated chart identifies the various lengths (A) available at given diameters (B). The chart, in this example, also shows purchase part numbers for each specific item. This method of dimensioning is commonly used in vendor or specification catalogs for alternate part identification.

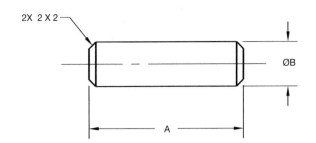

| LENGTH A | B=20.3 | B=38.1 | B=50.8 | B=57.2 |
|----------|---------|---------|---------|---------|
|  | PART NO. | PART NO. | PART NO. | PART NO. |
| 76 | DP20.3-76.2 | DP38.1-76.2 | DP50.8-76.2 | DP57.2-76.2 |
| 101 | DP20.3-101.6 | DP38.1-101.6 | DP50.8-101.6 | DP57.2-101.6 |
| 127 | DP20.3-127 | DP38.1-127 | DP50.8-127 | DP57.2-127 |
| 152 | DP20.3-152.4 | DP38.1-152.4 | DP50.8-152.4 | DP57.2-152.4 |

Figure 7–5 Chart drawing.

## DIMENSION RULES AND DEFINITIONS

*ANSI/ASME* The document that is the American National Standards Institute (ANSI) (published by The American Society of Mechanical Engineers, or ASME) standard for dimensioning is titled *Dimensioning and Tolerancing* ASME Y14.5M. This standard is available from the American National Standards Institute, 1430 Broadway, New York, NY 10018. This standard is all metric, indicated by the M following the number. The following fundamental rules for dimensioning are adapted from ASME Y14.5M.

1. Each dimension shall have a tolerance, except for those dimensions specifically identified as reference, maximum, minimum, or stock. The tolerance can be applied directly to the dimension or indicated by a general note or located in the title block of the drawing.

2. Dimensioning and tolerancing shall be complete to the extent that there is full understanding of the characteristics of each feature. Neither measuring the drawing nor assumption of a distance or size is permitted. Except drawings such as loft, printed wiring, templates, and master layouts prepared on stable material, provided the necessary control dimensions are given.

3. Each necessary dimension of an end product shall be shown. No more dimensions than those necessary for complete definition shall be given. The use of reference dimensions on a drawing should be minimized.

4. Dimensions shall be selected and arranged to suit the function and mating relationship of a part and shall not be subject to more than one interpretation.

5. The drawing should define a part without specifying manufacturing methods. However, in those cases in which manufacturing, processing, quality assurance, or environmental information is essential to the definition of engineering requirements, it shall be specified on the drawing or in document references on the drawing.

6. It is permissible to identify as nonmandatory certain processing dimensions that provide for finish allowance, shrink allowance, and other requirements, provided the final dimensions are given on the drawing. Nonmandatory processing dimensions shall be identified by an appropriate note, such as NONMANDATORY (MFG DATA).

7. Dimensions should be arranged to provide required information for optimum readability. Dimensions should be shown in true profile views and refer to visible outlines.

8. Wires, cables, sheets, rods, and other materials manufactured to gage or code numbers shall be specified by linear dimensions indicating the diameter or thickness. Gage or code numbers may be shown in parentheses following the dimension.

9. A 90° angle is implied where centerlines and lines depicting features are shown on a drawing at right angles and no angle is specified. The tolerance for these 90° angles is the same as the general angular tolerance specified in the title block or general notes.

10. A 90° basic angle applies where centerlines of features in a pattern or surfaces shown at right angles on the drawing are located or defined by basic dimensions and no angle is specified.

11. Unless otherwise specified, all dimensions are applicable at 20°C (68°F). Compensation may be made for measurements made at other temperatures.

12. All dimensions and tolerances apply in a *free-state condition* except for nonrigid parts. Free-state condition describes distortion of a part after removal of forces applied during manufacturing. *Nonrigid* parts are those that may have dimensional fluctuation due to thin wall characteristics.

13. Unless otherwise specified all geometric tolerances apply for full depth, length, and width of the feature.

14. Dimensions apply on the drawing where specified.

## Dimensioning Definitions

*Actual Size.* The part size as measured after production is known as actual size, also known as produced size.

*Allowance.* The tightest possible fit between two mating parts. Maximum material condition (MMC) internal feature – MMC external feature = allowance.

*Basic Dimension.* A basic dimension is considered to be a theoretically exact size, location, profile, or orientation of a feature or point. The basic dimension provides a basis for the application of tolerance from other dimensions or notes. Basic dimensions are drawn with a rectangle around the numerical value. For example: $\boxed{.625}$ or $\boxed{30°}$

*Bilateral Tolerance.* A bilateral tolerance is allowed to vary in two directions from the specified dimension.

Inch examples are $.250 \, ^{+\,.002}_{-\,.005}$ and $.500 \pm .005$.

Metric examples are $12 \, ^{+\,0.1}_{-\,0.2}$ and $12 \pm 0.2$.

*Datum.* A datum is considered to be a theoretically exact surface, plane, axis, center plane, or point from which dimensions for related features are established.

*Datum Feature.* The datum feature is the actual feature of the part that is used to establish a datum.

*Dimension.* A dimension is a numerical value used on a drawing to describe size, shape, location, geometric characteristic, or surface texture.

*Feature.* A feature is any physical portion of an object, such as a surface or hole.

*Least Material Condition (LMC).* Least material condition is the opposite of maximum material condition. The LMC is the lower limit for an external feature and the upper limit for an internal feature.

*Limits of Dimension.* The limits of a dimension are the largest and smallest possible boundary or location to which a feature may be made as related to the tolerance of the dimension. Consider the following inch dimension and tolerance: .750 ± .005. The limits of this dimension are calculated as follows: .750 + .005 = .755 upper limit and .750 − .005 = .745 lower limit. For the metric dimension 19 ± 0.1, 19 + 0.1 = 19.1 is the upper limit and 19 − 0.1 = 18.9 is the lower limit.

*Maximum Material Condition (MMC).* The maximum material condition, given the limits of the dimension, is the situation where a feature contains the most material possible. MMC is the largest limit for an internal feature.

*Geometric Tolerance.* The general term applied to the category of tolerances used to control form, profile, orientation, location, and runout.

*Nominal Size.* A dimension used for general identification such as stock size or thread diameter.

*Reference Dimension.* A dimension usually without tolerance, used for information purposes only. A reference is a repeat of a given dimension or established from other values shown on the drawing. It does not govern production or inspection. See Figures 7–1 and 7–2.

*Specified Dimension.* A specified dimension is a part of the dimension from which the limits are calculated. For example, the specified dimension of .625 ± .010 inch is .625. In the following metric example, 15 ± 0. mm, 15 is the specified dimension.

*Tolerance.* The tolerance of a dimension is the total permissible variation in size or location. Tolerance is the difference of the lower limit from the upper limit. For example, the limits of .500 ± .005 inch are .505 and .495, making the tolerance equal to .010 inch. The tolerance for the following metric example, 12 ± 0.1 mm, is 0.2 mm.

*Unilateral Tolerance.* A unilateral tolerance is a tolerance that has a variation in only one direction from the specified dimension, as in:

$.875 \, ^{+\,.000}_{-\,.002}$ in.

or

$22 \, 0 - 0.2; \; 22 \, ^{+\,0.2}_{-\,0}$

## Dimensioning Units

The metric *International System of Units (SI)* is featured predominately in this workbook because SI units (millimeters) supersede United States (U.S.) customary units specified on engineering drawings.

Metric units expressed in millimeters or U.S. customary units expressed in decimal inches are considered the standard units of linear measurement on engineering documents and drawings. The selection of millimeters or inches depends on the needs of the individual company. When all dimensions are either in millimeters or inches, the general note, UNLESS OTHERWISE SPECIFIED, ALL DIMENSIONS ARE IN MILLIMETERS (or INCHES), is usually found on the drawing. Inch dimensions should be followed by IN. on predominantly millimeter drawings and mm should follow millimeters on predominantly inch-dimensioned drawings.

## Decimal Points

A specified dimension in inches is expressed to the same number of decimal places as its tolerance, and zeros are added to the right of the decimal point if needed. For example, .500 ± .002.

Neither the decimal point or a zero is shown when the metric dimension is a whole number; 24 is an example. Also, when the metric dimension is greater than a whole number by a fraction of a millimeter, the last digit to the right of the decimal point is not followed by a zero, as in 24.5. This is true unless tolerance numerals are involved. Look at Figure 7–9. Both the plus and minus values of a metric or inch tolerance have the same number of decimal places, and zeros are added to fill in where needed.

A way to recognize metric or inch drawings is: Where millimeter dimensions are less than one, a zero precedes the decimal, as in 0.8. Decimal inch dimensions do not have a zero before the decimal, as in .75.

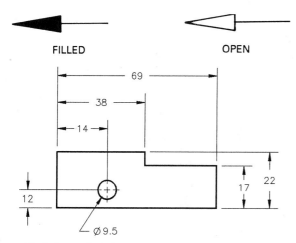

Figure 7–6 Arrowheads.

## Fractions

Fractions are used on engineering drawings, but they are not as common as decimal inches or millimeters. Fraction dimensions generally represent a larger tolerance than decimal numerals.

## Arrowheads

Arrowheads are used to terminate dimension lines and leaders.

Individual company preference dictates if arrowheads are filled in solid or left open, as shown (see Figure 7–6).

## *BASIC DIMENSIONING CONCEPTS*

### Dimension Placement

A properly dimensioned print places the smallest dimensions closest to the object and progressively larger dimensions outward from the object. Dimensions are grouped and placed between views when possible. This makes the drawing easier for you to read as shown in Figure 7–7.

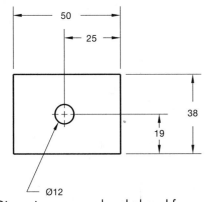

Figure 7–7 Dimensions grouped and placed for easy reading.

### Relationship of Dimension Lines to Numerals

Dimension numerals are centered on the dimension line. The numeral, dimension line, and arrowheads are placed between

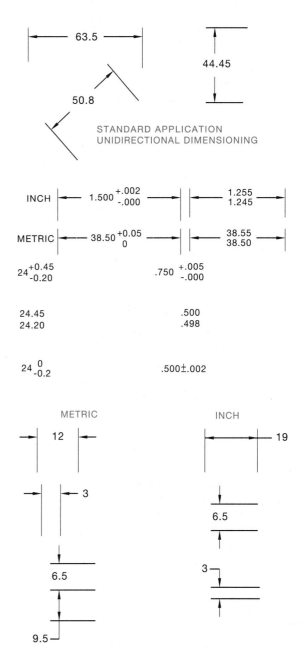

Figure 7–8 Dimensioning applications.

extension lines when space allows. When space is limited, other options can be used. Figure 7–8 shows several dimensioning options.

As you look at the dimensioning examples, you need to become familiar with the manner in which dimensions are shown. This helps you choose the correct dimension for the feature you are reading a print about. A key to reading dimensions correctly is to look for the relationship between the arrowheads and dimension line to the extension lines. The dimension numeral correlates in a logical manner with these dimension components, as shown in Figure 7–9.

### Chain Dimensioning

*Chain* dimensioning, also known as point-to-point dimensioning, is a method of dimensioning from one feature to the next. Each

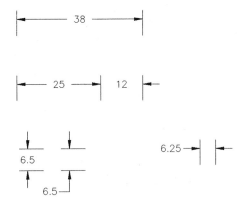

Figure 7–9 Look for the relationship between the placement of the arrowheads, dimension line, and dimension numeral.

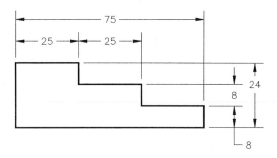

Figure 7–10 Chain dimensioning.

dimension is dependent on the previous dimension or dimensions. This is a common practice, although the tolerance of each dimension builds on the next, which is known as tolerance *buildup*, or *stacking*. See Figure 7–10, which also shows the common mechanical drafting practice of providing an overall dimension while leaving one of the intermediate dimensions blank. The overall dimension is often a critical dimension that is independent in its relationship to the other dimensions. Also, if all dimensions are given, the actual size may not equal the given overall dimension because of tolerance buildup. An example of tolerance buildup is when three chain dimensions have individual tolerances of ± 0.1 and each feature is manufactured at or toward the + l0.1 limit—the potential tolerance buildup is three times + 0.1, or a total of 0.3. The overall dimension would have to carry a tolerance of ± 0.3 to accommodate this buildup. If the overall dimension is critical, such an amount may not be possible. A dimension can be given only as reference. A *reference dimension* is enclosed in parentheses, as seen in Figure 7–11(a). Figure 7–11(b) shows the overall dimension of a symmetrical object as a reference.

## Datum Dimensioning

*Datum* dimensioning is a common method of dimensioning machine parts whereby each feature dimension originates from a common surface, axis, or center plane (see Figure 7–12). Each dimension in datum dimensioning is independent so there is no tolerance buildup. Figure 7–13 shows how dimensions can be symmetrical about a center plane used as a datum. Also notice in Figure 7–13 how extension lines break when they cross dimension

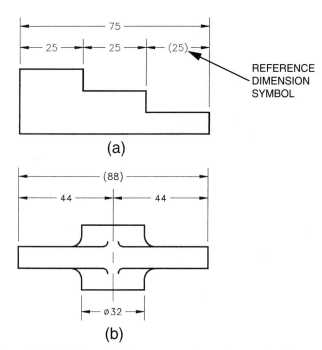

REFERENCE DIMENSION SYMBOL

Figure 7–11 Reference dimension examples, and reference dimension symbol.

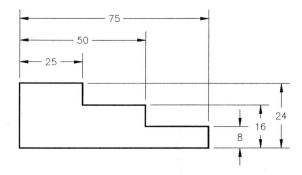

Figure 7–12 Datum dimensioning from a common surface.

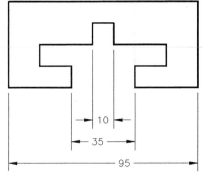

Figure 7–13 Symmetrical features dimensioned about a datum center plane.

lines, and how dimension numerals are staggered, rather than being stacked directly above one another, for easier reading. Figure 7–14 shows another method of dimensioning from a center plane datum.

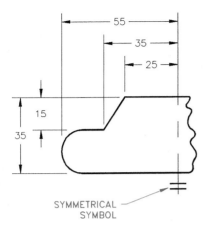

Figure 7–14 Datum dimensioning from a center plane datum. Notice the symmetrical symbol used to define a symmetrical object. In this view, the right part is removed and a short break line is used. The entire view could have been shown.

## Dimensions on Cylinders and Squares

Cylindrical shapes are dimensioned in the view where the cylinders appear as rectangles. The diameters are identified by the diameter symbol, and the circular view may be omitted (see Figure 7–15). Square features can be dimensioned in a similar manner using the square symbol, as shown in Figure 7–16.

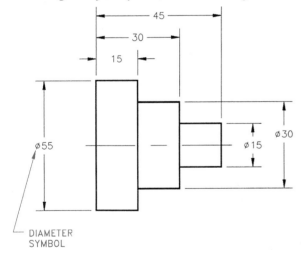

Figure 7–15 Dimensioning cylindrical shapes.

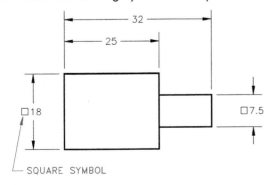

Figure 7–16 Dimensioning square feature.

## Dimensioning Angles

Angular surfaces can be dimensioned as coordinates, as angles in degrees, or as a flat tape. (see Figure 7–17). Angles are calibrated in degrees, symbol °. (There are 360° in a circle. Each degree contains 60 minutes, symbol '. Each minute has 60 seconds, symbol ". 1° = 60'; 1' = 60".

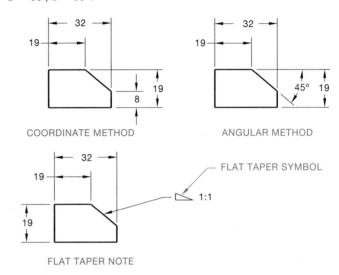

Figure 7–17 Dimensioning angular surfaces.

## Dimensioning Chamfers

A *chamfer* is a slight surface angle used to relieve a sharp corner. Chamfers of 45° can be dimensioned with a note, while other chamfers require an angle and size dimension, as seen in Figure 7–18. A note is used on 45° chamfers because both sides of a 45° angle are equal.

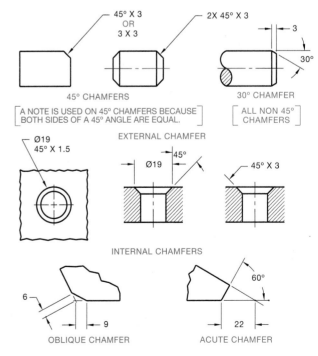

Figure 7–18 Dimensioning chamfers.

## Dimensioning Conical Shapes

Conical shapes are dimensioned where possible in the view where the cone appears as a triangle, as in Figure 7–19. A conical taper may be treated in one of three possible ways, as shown in Figure 7–20. Other geometric dimensioning methods are possible.

## Dimensioning Hexagons and Other Polygons

Dimension hexagons and other polygons are dimensioned across the flats in the views where the true shape is shown. A length dimension is provided in the side view, as shown in Figure 7–21.

## Dimensioning Arcs

Arcs are dimensioned with leaders and radius dimensions in the views where they are shown as arcs. The leader may extend from the center to the arc or may point to the arc. The letter *R* precedes all radius dimensions where reversals in the contour of the radius are permitted. The symbol CR refers to "controlled radius." *Controlled radius* means that the limits of the radius tolerance zone must be tangent to the adjacent surfaces, and there can be no reversals in the contour. Use of the CR control is more restrictive

than use of the R radius symbol. In a situation where an arc lies on an inclined plane and the true representation is not shown, the note TRUE R can be used to specify the actual radius. However, dimensioning the arc in an auxiliary view of the inclined surface is better, if possible (see Figure 7–22). Depending on the situation, arcs can be dimensioned with or without their centers located. Figure 7–23 shows a very large arc with the center moved closer to the object. To save space, a break line is used in the leader and the shortened locating dimension. The length of an arc may also be dimensioned one of three ways, as shown in Figure 7–24.

A spherical radius can be dimensioned with the abbreviation SR preceding the numerical value, as shown in Figure 7–25.

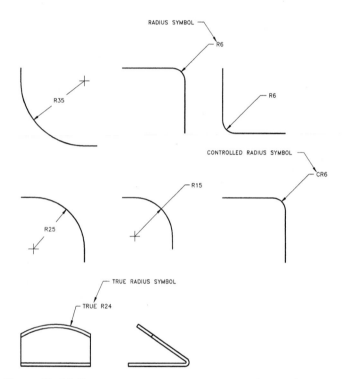

Figure 7–22 Dimensioning arcs—no centers located.

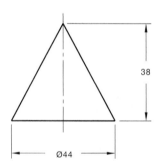

Figure 7–19 Dimensioning conical shapes.

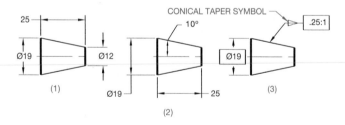

Figure 7–20 Dimensioning conical tapers.

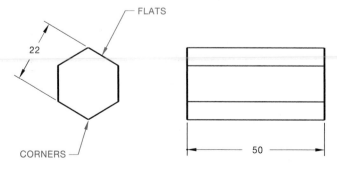

Figure 7–21 Dimensioning hexagons.

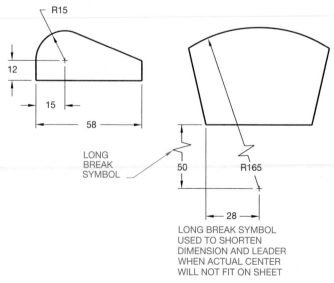

Figure 7–23 Dimensioning arcs—centers located.

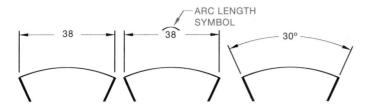

Figure 7–24 Dimensioning arc length.

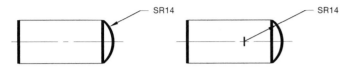

Figure 7–25 Dimensioning a spherical radius.

## Dimensioning Contours Not Defined as Arcs

Coordinates, or points along the contour, are located from common surfaces or data, as shown in Figure 7–26. Figure 7–27 shows a curved contour dimensioned using oblique extension lines.

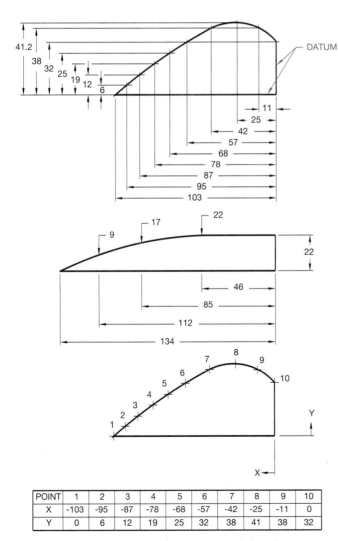

| POINT | 1 | 2 | 3 | 4 | 5 | 6 | 7 | 8 | 9 | 10 |
|-------|-----|-----|-----|-----|-----|-----|-----|-----|-----|-----|
| X | -103 | -95 | -87 | -78 | -68 | -57 | -42 | -25 | -11 | 0 |
| Y | 0 | 6 | 12 | 19 | 25 | 32 | 38 | 41 | 38 | 32 |

Figure 7–26 Dimensioning contours not defined as arcs.

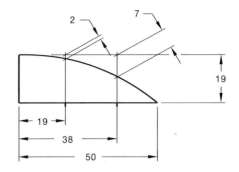

Figure 7–27 Dimensioning a curved contour using oblique extension lines.

## NOTES FOR SIZE FEATURES

Notes in the form of words and abbreviations are used, rather than symbols, in previous ASME standards. You can read old prints or prints from companies that have not adapted new standards. The ASME standard symbols are also shown in an effort to reinforce the taper dimensioning symbols that are used on engineering and manufacturing drawings.

### Holes

Hole sizes are dimensioned with leaders to the view where they appear as circles, or dimensioned in a sectional view (see Figure 7–28).

A hole through a part can be noted if it is not obvious. The diameter symbol precedes the diameter numeral, as dimensioned in Figure 7–29. If a hole does not go through the part, the depth is noted in the circular view or in section, as shown in Figure 7–30.

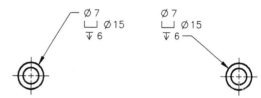

Figure 7–28 Hole features are commonly dimensioned with a note and leader pointing to the circular view.

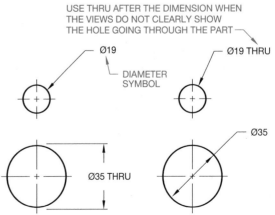

Figure 7–29 Dimensioning hole diameters, and the diameter symbol.

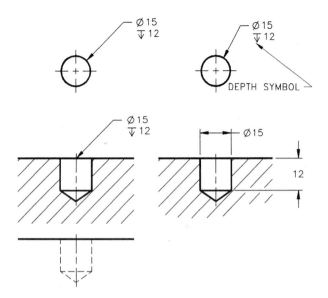

Figure 7–30 Dimensioning hole diameters and depths.

## Counterbore

A counterbore is often used to machine a diameter below the surface of a part so a bolt head or other fastener can be recessed. Counterbore and other similar notes are given in order of machine operations with a leader in the view where they appear as circles. Multiple counterbores are dimensioned in a similar manner, and the counterbore depth may be given in a sectional view (see Figure 7–31).

## Countersink or Counterdrill

A countersink, or counterdrill, is often used to recess the head of a fastener below the surface of a part (see Figure 7–32).

## Spotface

A spotface is used to provide a flat bearing surface for a washer face or bolt head (see Figure 7–33). The spotface and counterbore symbols look the same.

## Multiple Features

When a part has more than one feature of the same size, the features can be dimensioned with a note that specifies the number of like features, as shown in Figure 7–34. If a part contains several features all of the same size, the method shown in Figure 7–35 can be used.

## Slots

Slotted holes can be dimensioned in one of three ways, as shown in Figure 7–36. These methods are used only when the ends are fully rounded and tangent to the sides. When the ends of a slot or external feature have a radius greater than the width of the feature, the size of the radius is given, as shown in Figure 7–37.

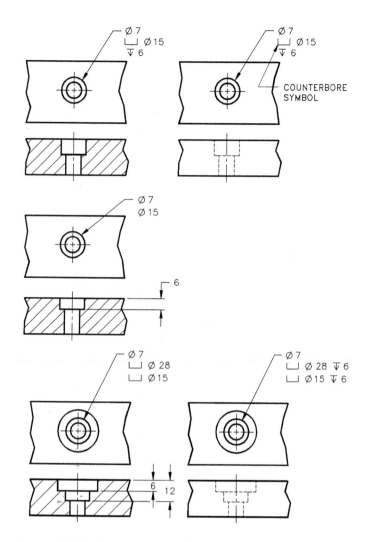

Figure 7–31 Common counterbore notes.

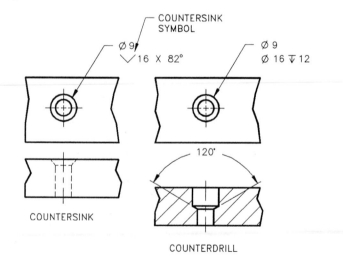

Figure 7–32 Countersink, and counterdrill notes.

## Keyseats

Keyseats are dimensioned in the view that clearly shows their shapes by width, depth, length, and location, as in Figure 7–38.

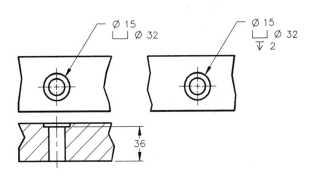

Figure 7–33 Spotface note.

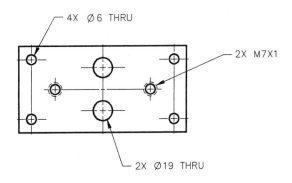

Figure 7–34 Dimension notes for multiple features.

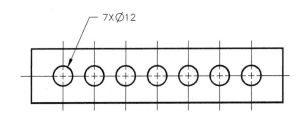

Figure 7–35 Dimension notes for multiple features all of common size.

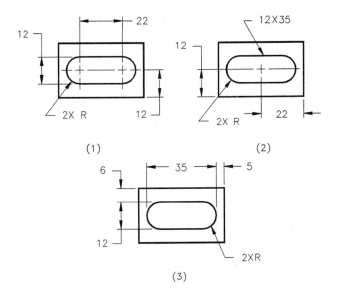

Figure 7–36 Dimensioning slotted holes with full radius ends.

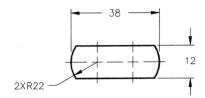

Figure 7–37 Dimensioning slot or external feature with end radius larger than feature width.

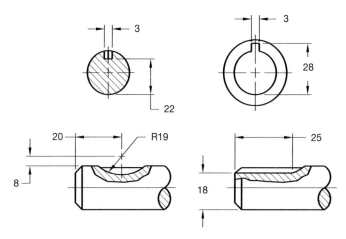

Figure 7–38 Dimensioning keyseats.

## Knurls

Knurls are dimensioned with notes and leaders that point to the knurl in the rectangular view, as shown in Figure 7–39(a). Although not a recommended ASME standard, some companies prefer showing a knurl representation on the view, as seen in Figure 7–39(b).

## Necks and Grooves

Necks and grooves can be dimensioned, as shown in Figure 7–40.

# LOCATION DIMENSIONS

In general, dimensions either identify a size or a location. The previous dimensioning discussions focused on size dimensions. Location dimensions to cylindrical features, such as holes, are given to the center of the feature in the view where they appear as circles, or in a sectional view. Rectangular shapes are located to their centerlines or center planes. Some location dimensions also control size.

## Locating Holes

A hole is located to its center in the view where the hole appears as a circle, as shown in Figure 7–41.

## Rectangular Coordinates

Linear dimensions are used to locate features from planes or centerlines, as shown in Figure 7–42.

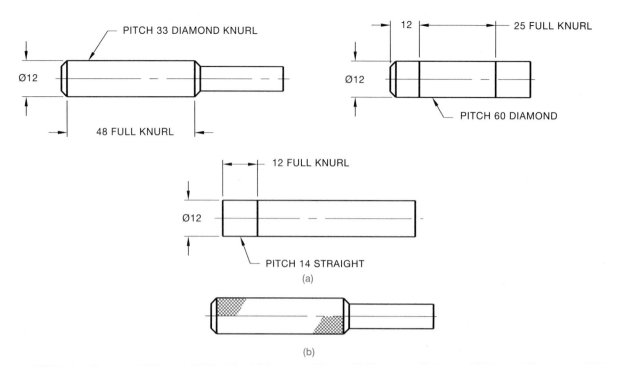

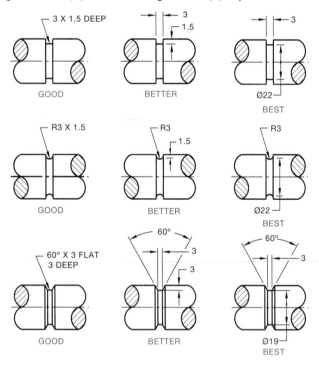

Figure 7–39 (a) Dimensioning knurls. (b) Optional knurl representation.

Figure 7–40 Dimensions for necks and grooves.

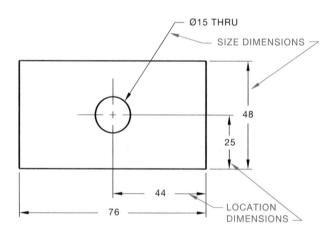

Figure 7–41 Size and location dimensions.

total length as reference. This method is acceptable for chain dimensioning (see Figure 7–44(a). A similar method is used when locating multiple tabs and slots, as in Figure 7–44(b). When repetitive features are nearly the same size, they may be shown with an identification letter, such as Y, as shown in Figure 7–44(c).

## DIMENSION ORIGIN

When the dimension between two features must clearly show from which feature the dimension originates, the dimension origin symbol may be used. This method of dimensioning means that the origin feature must be established first and the related feature can then be dimensioned from the origin (see Figure 7–45).

## Polar Coordinates

Angular dimensions locate features from planes or centerlines, as shown in Figure 7–43.

## Repetitive Features

Repetitive features are located by noting the number of times a dimension is repeated, and giving one typical dimension and the

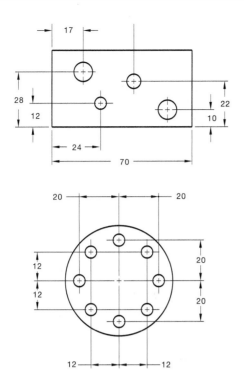

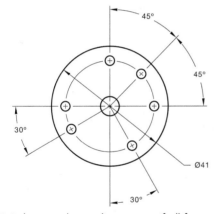

Figure 7–42 Rectangular coordinate location dimensions.

Figure 7–43 Polar coordinate dimensions. If all features are equally spaced, a note such as 6 x 60° is used in one location.

## DIMENSIONING ON AUXILIARY VIEWS

Dimensions are normally placed on views that provide the best size and shape description of an object. In many instances the surfaces of a part are foreshortened and require auxiliary views to completely describe true size, shape, and the location of features. When foreshortened views occur, dimensions are placed on the auxiliary view for clarity (see Figure 7–46).

## GENERAL NOTES

Notes, discussed previously, are classified as specific notes, because they refer to specific features of an object, while general notes relate to the entire drawing. Each drawing contains a certain amount of general notes either in or near the drawing title block, at the lower-left or upper-left corner of the drawing (see Figure 7–47). General notes include many items, such as:

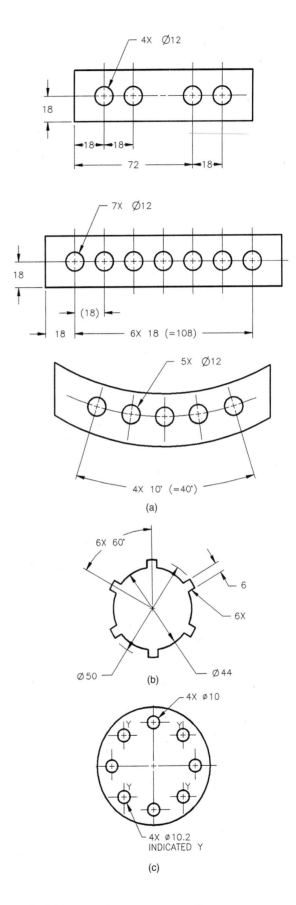

Figure 7–44 Dimensioning repetitive features.

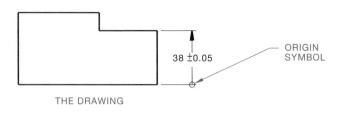

THE DRAWING

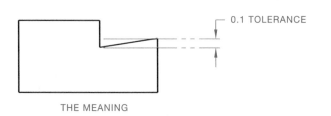

THE MEANING

Figure 7–45 Dimension origin, and the dimension origin symbol.

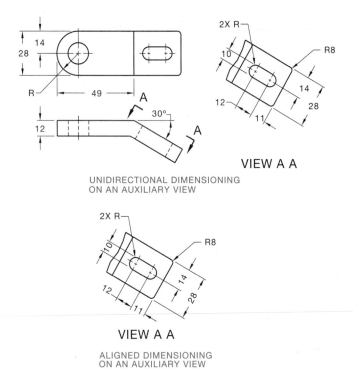

UNIDIRECTIONAL DIMENSIONING
ON AN AUXILIARY VIEW

VIEW A A

ALIGNED DIMENSIONING
ON AN AUXILIARY VIEW

Figure 7–46 Dimensioning auxiliary views.

- Material specifications
- Dimensions: inches or millimeters
- General tolerances
- Confidentiality note, copyrights, or patents
- Name of person that made the drawing
- Scale
- Date
- Part name
- Drawing size
- Part number
- Number of revisions
- First-angle or third angle projection symbol

*ASME* Other general notes are located next to the title block for ASME drawings. The exact location of the general notes depends on specific company standards. Military standards (MIL) specify that general notes must be located near the upper-left corner of the drawing. A common location for general notes is the lower-left corner of the drawing. Figure 7–48(a) shows some general notes that could commonly be included on ASME standard format drawings. Other common locations for general notes are the lower-right corner of the drawing, directly above the title block, and just to the left of the title block.

## DELTA NOTES

The term delta refers to a triangle placed on the drawing for reference. The triangle is commonly placed next to a dimension, such as 2.65△5, or other location where it applies to a feature or item. When you see a triangle, refer to a general note that relates to this item—the method is often used when it applies to a specific feature. Placing the note directly on the drawing is very difficult, however, because the note is too long, or the note applies to several dimensions or features. Look at the note △5 in Figure 7–48(b). This note relates to the same delta item located somewhere on the drawing. Some companies use symbols other than a triangle. Hexagons and circles can also be used, but the triangle is common.

## TOLERANCING

Each dimension has a tolerance, except for those dimensions specifically identified as reference, maximum, minimum, or stock.

| THIS DRAWING IS THE PROPERTY OF PEERLESS-WINSMITH. IT SHALL BE USED ONLY IN CONNECTION WITH CONTRACTS AND PROPOSALS OF THE COMPANY. RECIPIENT AGREES NOT TO USE THIS DRAWING OR THE INFORMATION IT CONTAINS IN ANY WAY DETRIMENTAL TO PEERLESS-WINSMITH. | | | | | MATERIAL | | PART NO. | | | WINSMITH PEERLESS-WINSMITH SPRINGVILLE, NEW YORK 14141 | | | |
|---|---|---|---|---|---|---|---|---|---|---|---|---|---|
| | | | | | SPEC. | | DWN. | DATE | | | | |
| REV. | CHANGE | BY | CHK. | DATE | PATTERN NO. | | CHKD. | DATE | NAME | | | |
| | | | | | TOLERANCES UNLESS OTHERWISE SPECIFIED | | APP. | DATE | UNIT | | | |
| | | | | | X.X = ±.050" X.XX = ±.030" X.XXX = ±.010" | | SCALE | | SIZE | DWG NO. | | REV |
| | | | | | ANGLES = ±.5° FINISH 125 ✓ | | HEAT TREAT | | | | | |

Figure 7–47 General title block information. Courtesy Peerless-Winsmith, Inc.

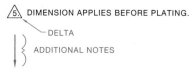

ADDITIONAL NOTES

3. ALL FILLETS AND ROUNDS R.125 UNLESS OTHERWISE SPECIFIED.

2. REMOVE ALL BURRS AND SHARP EDGES.

1. INTERPRET DIMENSIONS AND TOLERANCES PER ASME Y14.5M-1994.

## NOTES:

(a)

NOTES:

1. INTERPRET DRAWING IAW MIL-STD-100. CLASSIFICATION PER MIL-T-31000, PARA 3.6.4.

2. INTERPRET DIMENSIONS AND TOLERANCES PER ASME Y14.5M-1994.

3. PART TO BE FREE OF BURRS AND SHARP EDGES.

4. BAG ITEM AND IDENTIFY IAW MIL-STD-130, INCLUDE CURRENT REV LEVEL: 64869-0956356 REV 42375.

⑤ DIMENSION APPLIES BEFORE PLATING.

—— DELTA

ADDITIONAL NOTES

(b)

Figure 7–48(a) General notes located in the lower-left corner of the sheet. (b) Notes can be conveniently placed to read from the first note downward with CADD. This makes it easy to continue from one note to the next.

The tolerance can be applied directly to the dimension or indicated by a general note or located in the title block of the drawing.

## Definitions

*Tolerance* is the total permissible variation in a size or location dimension.

A *specified dimension*, also known as nominal size, is that part of the dimension from which the limits are calculated. For example, 15.8 is the specified dimension of 15.8 ± 0.2.

A *bilateral tolerance* is allowed to vary in two directions from the specified dimension.

Bilateral tolerances can be equal or unequal and are represented differently in metric when compared to inch units.

| Equal bilateral | Unequal bilateral |
|---|---|
| *Metric* | |
| 26.50 ± 0.25 | $26.50 \, {}^{+\,0.35}_{-\,0.25}$ |
| *Inch* | |
| 1.625 ± .005 | $1.625 \, {}^{+\,0.10}_{-\,.005}$ |

A *unilateral tolerance* varies in only one direction from the specified dimension and is shown differently in metric and inches.

**Unilateral**

*Metric*

$26.50 \, {}^{+\,0}_{-\,0.50}$         $26.50 \, {}^{+\,0.50}_{-\,0}$

*Inch*

$1.625 \, {}^{+\,.000}_{-\,0}$         $1.625 \, {}^{+\,.005}_{-\,.000}$

Limits are the largest and smallest possible sizes at which a feature can be produced, in relation to the tolerance of the dimension.

| **Example 1:** | 19.0 ± 0.1 |
|---|---|
| Upper limit: | 19.0 + 0.1 = 19.1 |
| Lower limit: | 19.0 − 0.1 = 18.9 |
| **Example 2:** | $9.5 \, {}^{0}_{-\,0.5}$ |
| Upper limit: | 9.5 + 0 = 9.5 |
| Lower limit: | 9.5 − 0.5 = 9.0 |

The tolerance, being the total permissible variation in the dimension, is easily calculated by subtracting the lower limit from the upper limit.

| **Example 3:** | 22.0 + 0.1 |
|---|---|
| Upper limit: | 22.1 |
| Lower limit: | −21.9 |
| Tolerance | 0.2 |
| **Example 4:** | $31.75 \, {}^{+\,0.10}_{-\,0}$ |
| Upper limit: | 31.85 |
| Lower limit: | −31.75 |
| Tolerance | 0.10 |

As previously explained, all dimensions on a drawing have a tolerance, except reference, maximum, minimum, or stock size dimensions. Dimensions on a drawing can read, as in Figure 7–49, with general tolerances specified in the title block of the drawing.

General tolerance specifications, as given in a typical industry title block, are shown in Figure 7–50. Using this title block as an example, the tolerances of the dimensions in Figure 7–50 follow (given in inches).

2.500 is a three-place decimal and would have .xxx ± .005 applied; 2.500 ± .005, tolerance equals 0.010.

2.50 is a two-place decimal and would have .xx ± .010 applied; 2.50 ± .010, tolerance equals .020.

2.5 is a one-place decimal and would have .x ± .020 applied;

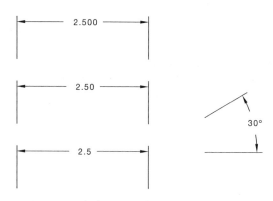

Figure 7–49 Typical drawing dimensions.

2.5 ± .020, tolerance equals .040.

30° would have angles ± 0.5° applied; 30° ± 0.5, tolerance equals 1°.

Dimensions that require tolerances different from the general tolerances given in the title block are specified in the dimension, as in Figure 7–51, and are referred to as dimensions with specified tolerances.

## Maximum and Least Material Conditions

*Maximum material condition*, abbreviated MMC, is the condition of a part or feature when it contains the most amount of material. The key is *most material*. The MMC of an external feature is the upper limit, or largest size (see Figure 7–52). The MMC of an internal feature is the lower limit or smallest size (see Figure 7–53).

The *least material condition*, LMC, is the opposite of MMC. LMC is the least amount of material possible in the size of a part. The LMC of an external feature is its lower limit. The LMC of an internal feature is its upper limit.

## Clearance Fit

A *clearance fit* is a condition when, because of the limits of dimensions, there is always a clearance between mating parts. The features in Figure 7–54 have a clearance fit. Notice the largest size of the shaft is smaller than the smallest hole size.

## Allowance

The *allowance* of a clearance fit between mating parts is the tightest possible fit between the parts. The allowance is calculated with the formula:

$$\begin{array}{ll} & \text{MMC Internal Feature} \\ - & \underline{\text{MMC External Feature}} \\ & \text{Allowance} \end{array}$$

The allowance of the parts in Figure 7–56 is:

| | |
|---|---|
| MMC Internal Feature | 16.15 |
| – MMC External Feature | – 16.00 |
| Allowance | 0.15 |

## Interference Fit

An *interference fit*, also known as a force or shrink fit, is the condition that exists when, because of the limits of the dimensions, mating parts must be pressed together. Interference fits are used, for example, when a bushing must be pressed onto a housing or when a pin is pressed into a hole (see Figure 7–55).

## Types of Fits

A fit is the relationship between two mating parts. There is a variety of fit classifications available for use when mating parts are designed. The following briefly discusses the options.

### Selection of Fits

In selecting the limits of size for any application, the type of fit is determined, first, based on the use or service required from the equipment being designed. Then, the limits of size of the mating parts are established to ensure that the desired fit will be produced. Theoretically, an infinite number of fits could be chosen, but the number of standard fits described here should cover most applications.

### Designation of Standard ANSI Fits

Standard fits are designated by letter symbols to make it easy for you to reference classes of fit as you are learning about them. The symbols are not intended to be shown on manufacturing drawings; instead, sizes are specified on drawings.

| | |
|---|---|
| RC | Running, or sliding, clearance fit |
| LC | Locational clearance fit |
| LT | Transition clearance fit |
| LN | Locational interference fit |
| FN | Force, or shrink, fit |

Figure 7–50 General tolerances from a company title block. *Courtesy Althin Medical, Inc.*

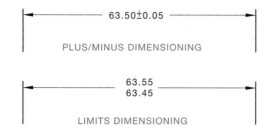

Figure 7–51 Specific tolerance dimensions.

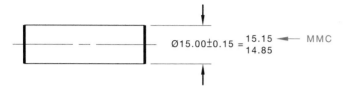

Figure 7–52 Maximum material condition (MMC) of an external feature. Plus/minus dimension shown.

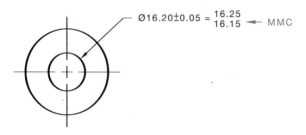

Figure 7–53 Maximum material condition (MMC) of an internal feature. Plus/minus dimension shown.

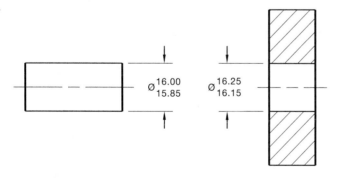

Figure 7–54 A clearance fit between two mating parts. Limits dimensions shown.

These letter symbols are used in conjunction with a number representing the class of fit: thus, FN 4 represents a Class 4 force fit.

**Description of Standard ANSI Fits**

The classes of fits are arranged in three general groups known as running and sliding fits, locational fits, and force fits.

*Running and Sliding Fits (RC).* Running and sliding fits are intended to provide a similar running performance with suitable lubrication allowance, throughout the range of their sizes. The clearances for the first two classes, used chiefly as sliding fits, increase more slowly with the diameter than do the clearances for the other classes. In this way, accurate location is maintained even

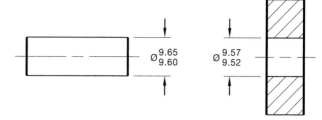

Figure 7–55 An interference fit between two mating parts. Limits dimensions shown.

at the expense of free relative motion. These fits are described as follows:

■ RC1. Close sliding fits are intended for the accurate location of parts that must assemble without perceptible play.

■ RC2. Sliding fits are intended for accurate location, but with greater maximum clearance than class RC1. Parts made to this fit move and turn easily, but are not intended to run freely, and in the larger sizes may seize with small temperature changes.

■ RC3. Precision running fits are about the closest fits that can be expected to run freely, and are intended for precision work at slow speeds and light journal pressures, but are not suitable where appreciable temperature differences are likely to be encountered.

■ RC4. Close running fits are intended chiefly for running fits on accurate machinery with moderate surface speeds and journal pressures, where accurate location and minimum play is desired.

■ RC5 and RC6. Medium running fits are intended for high running speeds or heavy journal pressures, or both.

■ RC7. Free running fits are intended for use where accuracy is not essential, or where large temperature variations are likely to be encountered, or under both these conditions.

■ RC8 and RC9. Loose running fits are intended for use where wide commercial tolerances may be necessary; together with an allowance on the external member.

*Locational Fits (LC, LT, and LN).* Locational fits are fits intended to determine only the location of the mating parts. They can provide rigid or accurate location, as with interference fits, or provide some freedom of location, as with clearance fits. Accordingly, locational fits are divided into three groups: clearance fits (LC), transition fits (LT), and interference fits (LN). These fits are described as follows:

■ LC. Locational clearance fits are intended for parts which are normally stationary, but which can be freely assembled or disassembled. They range from snug fits for parts requiring accuracy of location, through medium clearance fits for parts, such as spigots, to looser fastener fits where freedom of assembly is of prime importance.

■ LT. Locational transition fits are a compromise between clearance and interference fits. They are used in applications

| NOMINAL SIZE RANGE IN INCHES | RC4 STANDARD TOLERANCE LIMITS | |
|---|---|---|
| | Hole | Shaft |
| 0–.12 | +.0006 | −.0003 |
| | 0 | −.0007 |
| .12–.24 | +.0007 | −.0004 |
| | 0 | −.0009 |
| .24–.40 | +.0009 | −.0005 |
| | 0 | −.0011 |
| .40–.71 | +.0010 | −.0006 |
| | 0 | −.0013 |
| .71–1.19 | +.0012 | −.0008 |
| | 0 | −.0016 |
| 1.19–1.97 | +.0016 | −.0010 |
| | 0 | −.0020 |

Figure 7–56 Standard RC4 fits for nominal sizes ranging from 0 to 1.97 inches.

where accuracy of location is important, but either a small amount of clearance or interference is permissible.

■ LN. Locational interference fits are used where accuracy of location is of prime importance, and for parts requiring rigidity and alignment with no special requirements for bore pressure. Such fits are not intended for parts designed to transmit frictional loads from one part to another by virtue of the tightness of fit. Such conditions are covered by force fits.

*Force Fits (FN).* Force, or shrink, fits constitute a special type of interference fit normally characterized by maintenance of constant bore pressures throughout its range of sizes. The interference, therefore, varies almost directly with diameter, and the difference between its minimum and maximum value is small so as to maintain the resulting pressures within reasonable limits. These fits are described as follows:

■ FN1. Light drive fits are those requiring light assembly pressures and produce more or less permanent assemblies. They are suitable for thin sections or long fits, or in external cast-iron members.

■ FN2. Medium drive fits are suitable for ordinary steel parts or for shrink fits on light sections. They are about the tightest fits that can be used with high-grade, cast-iron, external members.

■ FN3. Heavy drive fits are suitable for heavy steel parts or for shrink fits in medium sections.

■ FN4 and FN5. Force fits are suitable for parts which can be highly stressed, or for shrink fits where the heavy pressing forces required are impractical.

### Establishing Dimensions for Standard ANSI Fits.

The fit used in a specific situation is determined by the operating requirements of the machine. While this information is already provided for you in the form of dimensions on the drawing, it is a good idea for you to understand the basis for establishing these dimensions. As you become familiar with reading prints, you should also refer to the *Machinery's Handbook* for additional in-depth information about fits. Tolerances are based on the type of fit and nominal size ranges, such as 0–.12, .12–.24, .24–.40, .40–.71, .71–1.19, and 1.19–1.97 inches. So, if you have a 1-inch shaft diameter and an RC4 fit, refer to Figure 7–56 to determine the shaft and hole limits. The hole and shaft limits for a 1-inch nominal diameter are shown here.

Upper hole limit = 1.000 + .0012 = 1.0012

Lower hole limit = 1.000 + 0 = 1.000

Upper shaft limit = 1.000 − .0008 = .9992

Lower shaft limit = 1.000 − .0016 = .9984

The hole is then dimensioned as Ø 1.0012–1.000, and the shaft as Ø .9992–.9984.

### Standard ANSI/ISO Metric Limits and Fits

The standard for the control of metric limits and fits is governed by the document ANSI B4.2–1978 (reaffirmed in 1984), *Preferred Metric Limits and Fits*. The system is based on symbols and numbers that relate to the internal or external application and the type of fit. The specifications and terminology for fits are slightly different from the ANSI standard fits previously described. The metric limits and fits are divided into three general categories: clearance fits, transition fits, and interference fits. *Clearance fits* are generally the same as the running and sliding fits explained earlier. With clearance fits, a clearance always occurs between the mating parts under all tolerance conditions. With *transition fits*, a clearance or interference may result because of the range of limits of the mating parts. When interference fits are specified, a press or force situation exists under all tolerance conditions. Refer to Figure 7–57 for the ISO symbol and descriptions of the different types of metric fits.

The metric limits and fits may be designated in a dimension in one of three ways. The method used depends on individual company standards and the extent of use of the ISO systems. When most companies begin using this system, the tolerance limits are calculated and shown on the drawing followed by the tolerance symbol in parentheses, for example, 25.000–24.979 (25 h7). The symbol in parentheses represents the basic size, 25, and the shaft tolerance code, h7. The term *basic size* means the dimension from which the limits are calculated. When companies become accustomed to using the system, they represent dimensions with the code followed by the limits in parentheses, as follows: 25 h7 (25.000–24.979). Finally, when a company has used the system long enough for interpreters to understand the designations, the code is placed alone on the drawing, like this: 25 h7. When it is necessary to determine the dimension limits from code dimensions, use the charts in ANSI B4.2–1978 or the *Machinery's Handbook*. For example, if you want to determine the limits of the mating parts with a basic size of 30 and a close running fit, refer to the chart shown in Figure 7–58. The hole limits for the 30-mm basic size in Figure 7–58 are Ø30.033–30.000, and the shaft dimension limits are Ø29.980–19.959.

| TYPE OF FIT | ISO SYMBOL | | DESCRIPTION OF FIT |
|---|---|---|---|
| | Hole | Shaft | |
| **Clearance Fit** | H11/c11 | C11/h11 | *Loose running* |
| | H9/d9 | D9/h9 | *Free running* |
| | H8/f7 | F8/h7 | *Close running* |
| | H7/g6 | G7/h6 | *Sliding* |
| | H7/h6 | H7/h6 | *Locational clearance* |
| **Transition Fit** | H7/k6 | K7/h6 | *Locational transition* |
| | H7/n6 | N7/h6 | *Locational transition* |
| **Interference Fit** | H7/p6[1] | P7/h6 | *Locational interference* |
| | H7/s6 | S7/h6 | *Medium drive* |
| | H7/u6 | U7/h6 | *Force* |

Figure 7–57 ISO symbols and description of metric fits.

| BASIC SIZE | CLOSE RUNNING FIT | |
|---|---|---|
| | Hole (H8) | Shaft (f7) |
| 20 | 20.033 | 19.993 |
| | 20.000 | 19.959 |
| 25 | 25.033 | 24.980 |
| | 25.000 | 24.959 |
| 30 | 30.033 | 29.980 |
| | 30.000 | 29.959 |
| 40 | 40.039 | 39.975 |
| | 40.000 | 39.950 |
| 50 | 50.039 | 49.975 |
| | 50.000 | 49.950 |

Figure 7–58 Tolerances of close-running fits for basic sizes ranging from 20 to 50 mm.

## Statistical Tolerancing

*Statistical tolerancing* is the assigning of tolerances to related dimensions in an assembly based on the requirements of statistical process control (SPC). SPC was discussed in Chapter 6. Statistical tolerancing is displayed in dimensioning, as shown in Figure 7–59(a). When the feature can be manufactured using SPC or conventional means, it is necessary to show both the statistical tolerance and the conventional tolerance, as in Figure 7–59(b). The appropriate general note should also accompany the drawing, as shown in Figure 7–59.

## DIMENSIONS APPLIED TO PLATING AND COATINGS

When platings (such as chromium, copper, and brass) or coatings (such as galvanizing, polyurethane, and silicone) are applied to a part or feature, the specified dimensions should be defined in relationship to the coating or plating process. A general note that indicates that the dimensions apply before or after plating or coating is commonly used and specifies the desired variables; for example,

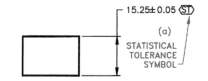

FEATURES IDENTIFIED AS STATISTICALLY TOLERANCED SHALL BE PRODUCED WITH STATISTICAL PROCESS CONTROLS.

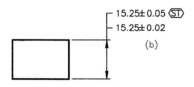

FEATURES IDENTIFIED AS STATISTICALLY TOLERANCED SHALL BE PRODUCED WITH STATISTICAL PROCESS CONTROLS, OR TO THE MORE RESTRICTIVE ARITHMETIC LIMITS.

Figure 7–59 Statistical tolerancing application and notes, and the statistical tolerancing symbol.

DIMENSIONAL LIMITS APPLY BEFORE (AFTER) PLATING (COATING). A leader connecting a specific note to the surface can also be used, as shown in Figure 7–60. Notice that a dot replaces the arrowhead on the leader.

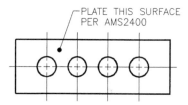

Figure 7–60 A dot replaces the arrowhead on a leader connecting a specific note to a surface.

## MAXIMUM AND MINIMUM DIMENSIONS

In some situations, a dimension with an unspecified tolerance requires that the general tolerance is applied in one direction only from the specified dimension. For example, when a 12 mm radius cannot exceed 12mm, the dimension can read R12 MAX. Therefore, when it is desirable to establish a maximum or minimum dimension, the abbreviations MAX or MIN are applied to the dimension. A dimension with a specified tolerance reads as previously discussed; for example:

$$R12 \, {}^{\;\;0}_{-0.05} \text{ or } R12 \, {}^{+0.05}_{\;\;0}$$

### Casting Drawing and Design

The end result of a casting drawing is the fabrication of a pattern. The preparation of casting drawings depends on the casting process used, the material to be cast, and the design or shape of the part.

## Shrinkage Allowance

When metals are heated and then cooled, the material shrinks until the final temperature is reached. The amount of shrinkage depends on the material used. The shrinkage for most iron is about .125 inch per foot, .250 inch per foot for steel, .125–.156 inch per foot for aluminum, .22 inch per foot for brass, and .156 inch per foot for bronze. Values for shrinkage allowance are approximate, because the exact allowance depends on the size and shape of the casting, and the contraction of the casting during cooling. The print normally does not show shrinkage considerations, because the pattern maker uses shrink rules that use expanded scales to take into account the shrinkage of various materials.

## Draft

*Draft* is the taper allowance on all vertical surfaces of a pattern, which is necessary to facilitate the removal of the pattern from the mold. Draft is not necessary on horizontal surfaces, because the pattern easily separates from these surfaces without sticking.

Draft angles begin at the parting line and taper away from the molding material (see Figure 7–61). The draft is added to the minimum design sizes of the product. Draft varies with different materials, size and shape of the part, and casting methods. Little, if any, draft is necessary in investment casting. The factors that influence the amount of draft are the height of vertical surfaces, the quality of the pattern, and the ease with which the pattern must be drawn from the mold. A typical draft angle for cast iron and steel is .125 inch per foot. Whether draft is taken into consideration on a drawing depends on company standards. Some companies leave draft angles to the pattern maker, while others place draft angles on the drawing.

## Fillets and Rounds

One of the purposes of fillets and rounds on a pattern is the same as that of draft angles: to allow the pattern to eject freely from the mold. Also, the use of fillets on inside corners helps reduce the tendency of cracks to develop during shrinkage (see Figure 7–62).

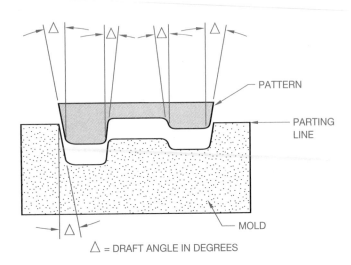

Figure 7–61 Draft angles for castings.

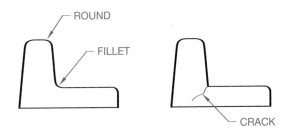

Figure 7–62 Fillets and rounds for casting.

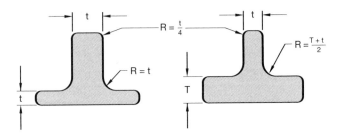

Figure 7–63 Recommended fillet and round radii for sand castings.

The radius for fillets and rounds depends on the material to be cast, the casting method, and the thickness of the part. The recommended radii for fillets and rounds used in sand casting is determined by part thickness, as shown in Figure 7–63.

## Machining Allowance

Extra material must be left on the casting for any face that will be machined. As with other casting design characteristics, the machining allowance depends on the casting process, material, size and shape of the casting, and the machining process to be used for finishing. The standard finish allowance for iron and steel is .125 inch, and for nonferrous metals, such as brass, bronze, and aluminum, .062 inch. In some situations the finish allowance may be as much as .5 or .75 inch for castings that are very large or have a tendency to warp.

Other machining allowances can be the addition of lugs, hubs, or bosses on castings that are otherwise hard to hold. The drafter can add these items to the drawing or the pattern maker may add them to the pattern. These features may not be added for product function, but can serve as aids for chucking or clamping the casting in a machine (see Figure 7–64).

There are several methods that can be used on drawings for casting and machining operations. The method used depends on company standards. A commonly used technique is to provide two drawings, one a casting drawing and the other a machining drawing. The casting drawing, as shown in Figure 7–65, shows the part as a casting. Only dimensions necessary to make the casting are shown on this drawing. The other drawing is of the same part, but this time only machining information and dimensions are given. The actual casting goes to the machine shop along with the machining drawing so the features can be machined as specified. The machining drawing is shown in Figure 7–66.

Another method of preparing casting and machining drawings is to show both casting and machining information together on

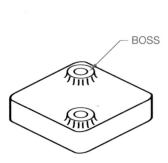

MOUNTING BOSSES FOR
BASE PLATE. ONLY REQUIRES
MACHINING BOSS SURFACES.

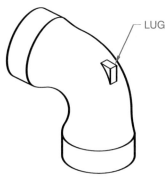

HOLDING LUG FOR
SURFACE MACHINING ELBOW

MACHINING HUB
EXTENSION FOR PULLEY

Figure 7–64 Cast features added for machining.

one drawing. This technique requires the pattern maker to add machining allowances. The drawing may have draft angles specified in the form of a note. The pattern maker must add the draft angles to the finished sizes given. A combination casting/machining drawing is shown in Figure 7–67.

Another technique is to show the part as a machining drawing and with phantom lines used to show the extra material for machining allowance and draft angles, as in Figure 7–68.

## Forging, Design, and Drawing

Draft for forgings serve much the same purpose as draft for castings. The draft associated with forging is found in the dies. The sides of the dies must be angled to facilitate the release of the metal during the forging process. If the vertical sides of the dies did not have draft angle, the metal would become stuck in the die. Internal and external draft angles may be specified differently because the internal drafts in some materials may have to be greater to help reduce the tendency of the part to stick in the die. While draft angles may change slightly with different materials, the common exterior draft angle recommended is 7°. The internal draft angles for most soft materials is also 7°, but the recommended interior draft for iron and steel is 10°.

The application of fillets and rounds to forging dies is to improve the ejection of the metal from the die. Another reason, similar to that for casting, is increased inside corner strength. One factor that applies to forgings, unlike castings, is that fillets and rounds that are produced too small in forging dies may substantially reduce the life of the dies. Recommended fillet and round radius dimensions are shown in Figure 7–69.

## Forging Drawings

Any of several methods can be used to prepare forging drawings. One technique used to create forging drawings that is clearly different from the preparation of casting drawings is the addition of draft angles. Casting drawings usually do not show draft angles, while draft angles are usually shown in forging detail drawings

Before a forging can be made, the dimensions of the stock material to be used for the forging must be determined. Some companies leave this information to the forging shop to determine, while other companies expect their engineering departments to make these calculations. After the stock size is determined, a drawing showing size and shape of the stock material is prepared. The blank material is dimensioned, and the outline of the end product is drawn inside the stock view using phantom lines (see Figure 7–70).

Forgings are made with extra material added to surfaces that must be machined. Forging detail drawings may show the desired end product with the outline of the forging shown in phantom lines at areas that require machining (see Figure 7–71). Notice the double line around the perimeter representing draft angle. Another option that is used by some companies is to make two separate drawings, one a forging drawing and the other a machine drawing. The forging drawing shows all of the views, dimensions, and specifications that relate only to the production of the forging as shown in Figure 7–72. The machining drawing does not duplicate the information that was provided for the forging. The only information on the machining drawing is views, dimensions, and specifications related to the machining processes as shown in Figure 7–73.

## Drawings for Plastic Part Manufacturing

As discussed in Chapter 6, a variety of plastic materials and manufacturing processes is available for creating plastic parts. Much like castings and forgings of metal parts, plastic manufacturing often requires draft angles to be applied to the design. This depends on the type of plastic and the manufacturing method used. Draft angles allow the finished plastic part to be ejected or removed from the mold without difficulty. Also associated with plastic parts, as with metal castings and forgings, is a parting line. The parting line is the location on the part where it has been separated from the mold. When labeled on the drawing, the abbreviation PL is used. Figure 7–74 shows some examples of parting lines.

Draft angles can be specified in a note, such as .010 MAX DRAFT ANGLE. In this case, the patternmaker uses this amount of draft as a guide and produces a pattern that has draft angles that are less than .010 on each side where needed for the manufacturing

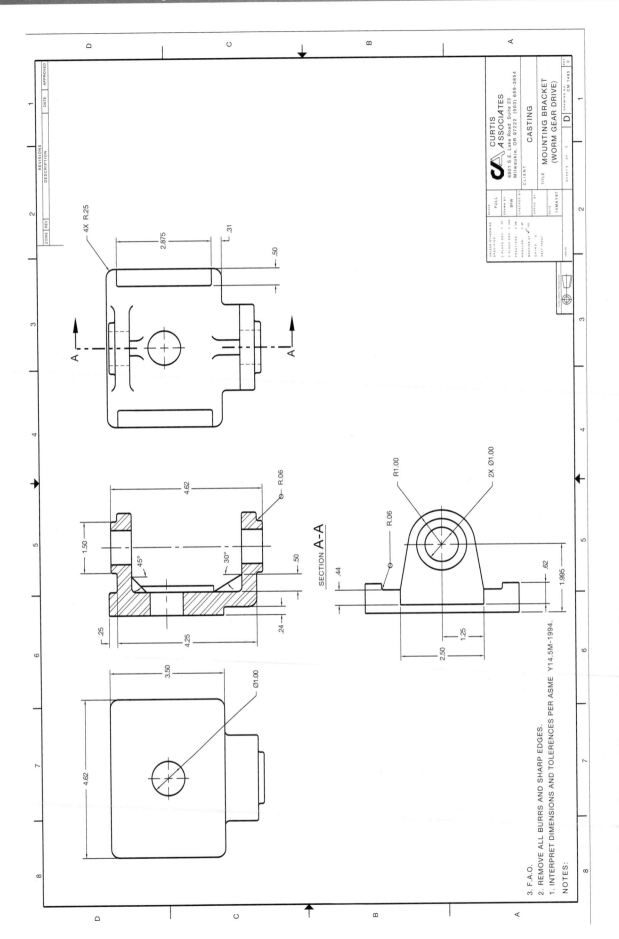

Figure 7–65 Casting drawing. *Courtesy Curtis Associates.*

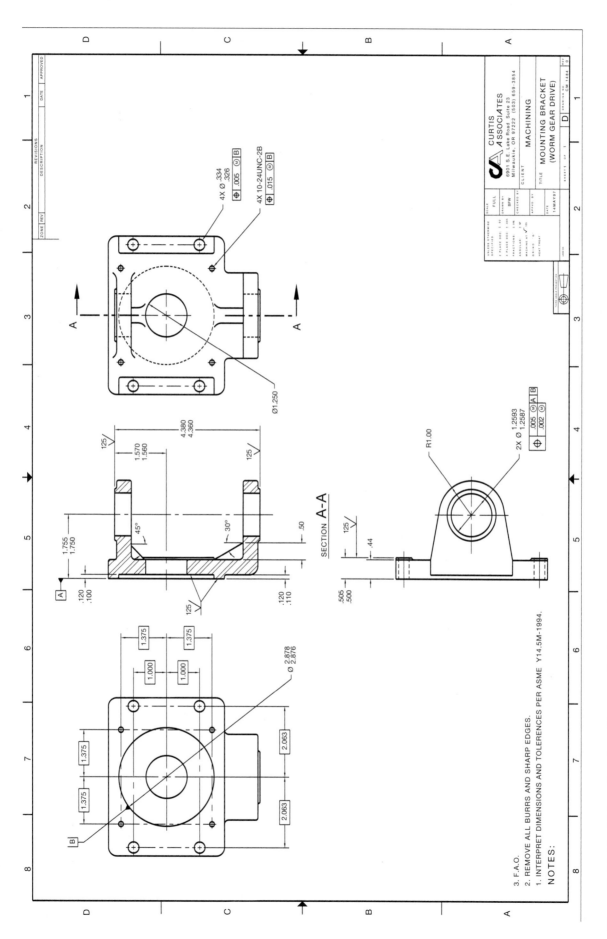

Figure 7–66 Machining drawing. *Courtesy Curtis Associates.*

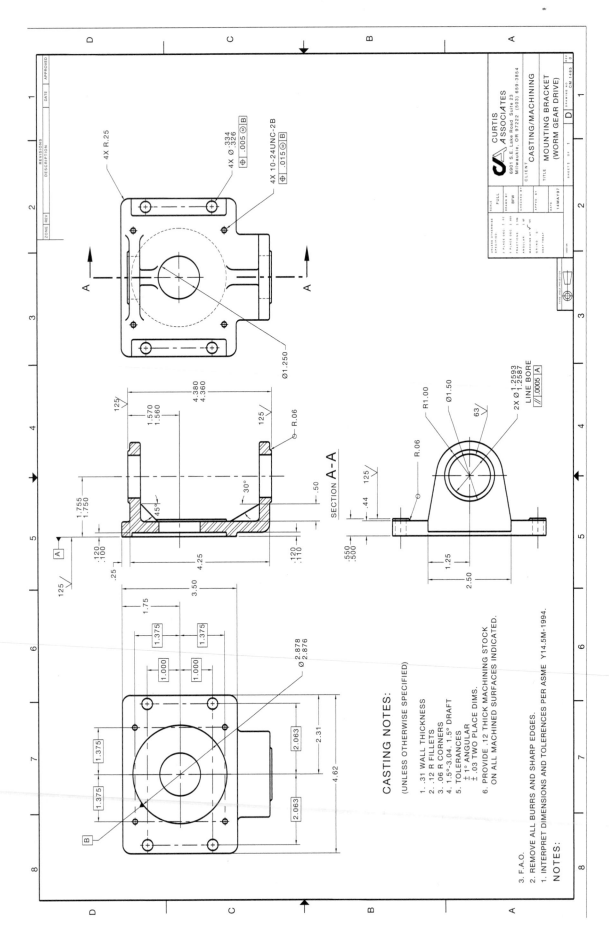

Figure 7-67 Drawing with casting and machining information combined. *Courtesy Curtis Associates.*

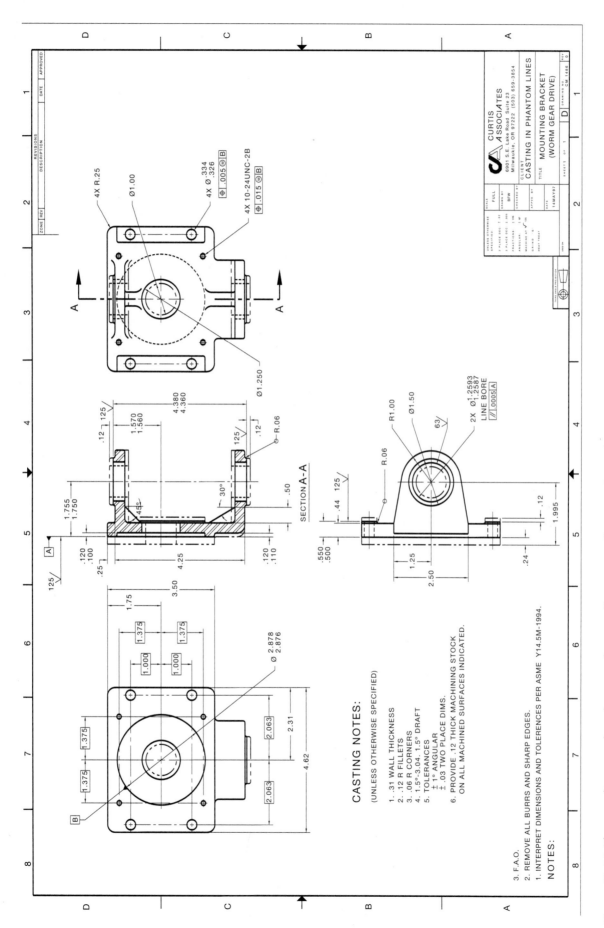

Figure 7–68 Phantom lines used to show machining allowances. *Courtesy Curtis Associates.*

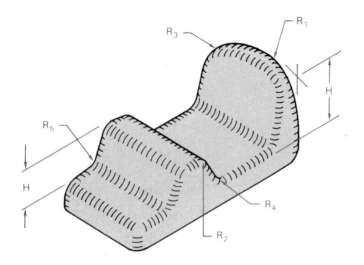

| DIMENSIONS IN MILLIMETERS | | | | | |
|---|---|---|---|---|---|
| H | R₁ | R₂ | R₃ | R₄ | R₅ |
| 6 | 1.5 | 1.5 | 4.5 | 3 | 3 |
| 13 | 1.5 | 1.5 | 4.5 | 3 | 3 |
| 25 | 3 | 3 | 9 | 6 | 9 |
| 50 | 4.5 | 6 | 13 | 13 | 15 |
| 75 | 6 | 7.5 | 16 | 16 | 25 |
| 100 | 7.5 | 10.5 | 25 | 25 | 35 |
| 125 | 9 | 13 | 23 | 32 | 44 |
| 150 | 10.5 | 15 | 32 | 38 | 50 |

Figure 7–69 Recommended fillets and rounds for forgings.

process. These draft angles are generally established within the specified part dimensions rather than added on to the part dimensions. A draft angle tolerance can also be specified as a zone. This can be shown on the drawing, as in Figure 7–75, or it can be specified as a tolerance in a general note or in the title block. A general note might read: ALL DRAFT ANGLES .010, or ALL DRAFT ANGLES 6°. The engineer or the mold maker determines the amount of draft angle.

Another method of specifying draft is the plus draft and minus draft methods. This is abbreviated as +DFT or –DFT and is placed with the feature dimensions on the part. In the +DFT application, the draft is added to the dimension for external dimensions and removed from internal dimensions, as shown in Figure 7–76. In the –DFT method, the draft is removed from the external dimension and added to the internal dimension as shown in Figure 7–76(b). Both +DFT and –DFT can be combined on a drawing, as shown in Figure 7–76(c).

## MACHINED SURFACES

The machine tools used in the manufacturing industry were discussed in Chapter 6. The following topics explain how machined surfaces are treated on a drawing and what you need to look for when reading prints.

### Surface Finish Definitions

There are several terms related to surface finishes that you should know as a print reader. The following defines and discusses this terminology.

**Surface Finish**

*Surface finish* refers to the roughness, waviness, lay, and flaws of a surface. Surface finish is the specified smoothness required on the finished surface of a part that is obtained by machining, grinding, honing, or lapping. The drawing symbol associated with surface finish is shown in Figure 7–77.

**Surface Roughness**

*Surface roughness* refers to fine irregularities in the surface finish and is a result of the manufacturing process used. Roughness height is measured in micrometers, mm, (millionths of a meter) or in microinches, min, (millionths of an inch).

**Surface Waviness**

*Surface waviness* is the often widely spaced condition of surface texture usually caused by machine chatter, vibrations, work deflection, warpage, or heat treatment. Waviness is rated in millimeters or inches.

**Lay**

*Lay* is the term used to describe the direction or configuration of the predominant surface pattern. The lay symbol is used if considered essential to a particular surface finish. The characteristic lay symbol may be attached to the surface finish symbol, as shown in Figure 7–78.

### Surface finish symbol

Some of the surfaces of an object can be machined to certain specifications. When this is done, a surface finish symbol is placed on

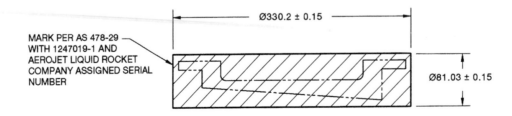

Figure 7–70 Blank material for forging process. *Courtesy Aerojet Propulsion Division.*

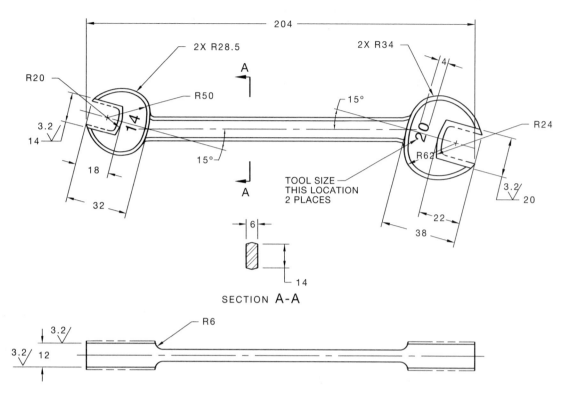

Figure 7–71 Phantom lines used to show machining allowance on a forging drawing.

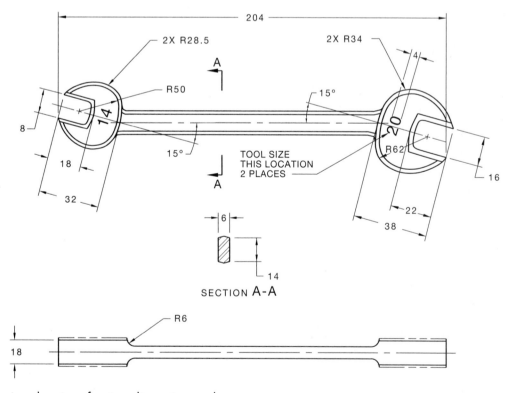

Figure 7–72 Forging drawing, forging dimensions only.

the view where the surface or surfaces appear as lines (edge view) (see Figure 7–79). The finish symbol on a machine drawing alerts the machinist that the surface must be machined to the given specification. The finish symbol also tells the pattern or die maker that extra material is required in a casting or forging.

The surface finish symbol is properly displayed on a drawing, as shown in Figure 7–80.

Often only the surface roughness height is used with the surface finish symbol, as, for example, the number 3.2 above the finish

Figure 7–73 Machining drawing with machining dimensions only.

Figure 7–74 Parting lines labeled on a drawing.

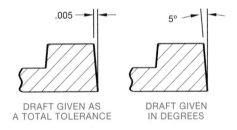

Figure 7–75 Draft angle tolerance shown on a drawing.

symbol in Figure 7–81, which means 3.2 micrometers. When other characteristics of a surface texture are specified, they are shown in the format represented in Figure 7–82. For example, roughness-width cutoff is a numerical value that establishes the maximum width of surface irregularities to be included in the roughness-height measurement. Standard roughness-width cutoff values for inch specifications are .003, .010, .030, .100, .300, and 1.000; .030 is implied when no specification is given.

Figure 7–83 is a magnified pictorial representation of the characteristics of a surface finish symbol. Figure 7–84 shows some common roughness height values in micrometers and microinches, a description of the resulting surface, and the process by which the surface can be produced. When a maximum and minimum limit is specified, the average roughness height must lie within the two limits.

When a standard or general surface finish is specified in the drawing title block or in a general note, a surface finish symbol, without roughness height specified, is used on all surfaces that are the same as the general specification. When a part is completely finished to a given specification, one of these general notes is used: FINISH ALL OVER, or abbreviation FAO, or FAO 125 mIN. The placement of surface finish symbols on a drawing can be accom-

plished in a number of ways, as shown in Figure 7–85. Additional elements may be applied to the surface finish symbol, as shown in Figure 7–86.

## DESIGN AND DRAFTING OF MACHINED FEATURES

As a print reader, you may not be involved in the design of machine features. However, experienced machinists are often called upon to help in the design or redesign of a part or product. Whether you are directly involved in design or not, it is a good idea for you to be familiar with the design process of the parts and products that you manufacture.

## DIMENSIONING FOR CADD/CAM

The implementation of computer-aided design and drafting (CADD) and computer-aided manufacturing (CAM) in industry is best accomplished when common control of the computer exists between engineering and manufacturing. The success of this automation is, in part, related to the standardization of operating and documentation procedures. CADD can be accomplished through the same coordinate dimensioning systems already described in this chapter. Standard dimensioning systems are used to establish a geometric model of the part that, in turn, is displayed at a graphics workstation. The data retrieved from this model is the mathematical description of the part to be produced. The drafter must dimension the part completely and accurately so that each contour or geometric shape of the part is continuous. The dimensioning systems that locate features or points on a feature in relation to X, Y, and Z axes derived from a common origin are most effective, such as datum, tabular, arrowless, and polar coordinate dimensioning. The X, Y, and Z axes originate from three mutually perpendicular planes, which are generally the geometric counterpart of the sides of the part when the surfaces are at right angles, as seen in Figure 7–87. If the part is cylindrical, two of the planes intersect at right angles to establish the axis of the cylinder, and the third is perpendicular to the intersecting planes, as shown in Figure 7–88. The X, Y, and Z coordinates that are used to establish

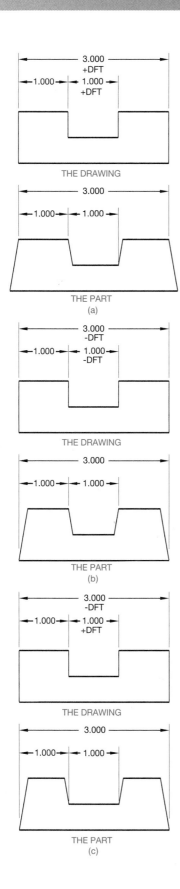

Figure 7–76 Draft can be specified on a drawing with (a) the plus draft method, (b) the minus draft method, or (c) a combination of both.

Figure 7–77 Surface finish symbol.

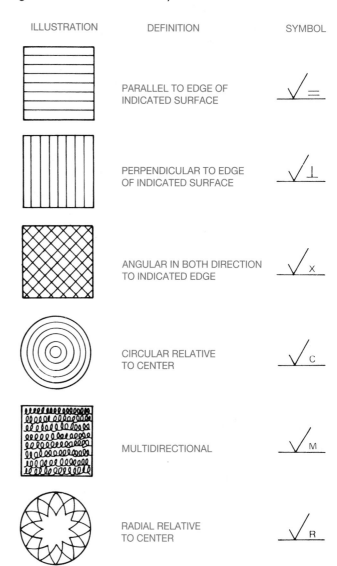

Figure 7–78 Characteristic lay added to the finish symbol.

features on a drawing are converted to X, Y, and Z axes that correspond to the linear and rotary motions that occur in CAM.

The position of the part in relation to mathematical quadrants determines whether the X, Y, and Z values are positive or negative. The preferred position is the mathematical quadrant that enables programming positive commands for the machine tool. Notice, in Figure 7–89, that the positive x and y values occur in quadrant 1.

In CAD/CAM and computer-integrated manufacturing (CIM) programs, the drawing is made on the computer screen and sent directly to the computer numerical control machine tool without generating a hard copy of the drawing. Figure 7–90 shows a computer drawing created for CAD/CAM. The CAD/CAM program also

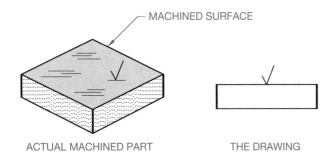

Figure 7–79 Standard surface finish symbol placed on the edge view. The symbol should always be placed horizontally when unidirectional dimensioning is used.

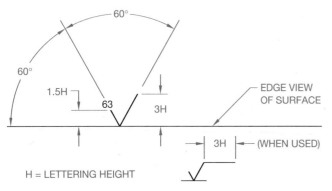

Figure 7–80 Properly drawn surface finish symbol.

Figure 7–81 Surface finish symbol with surface roughness height given in micrometers.

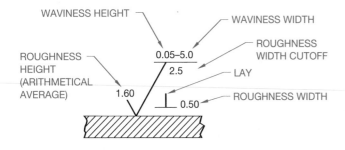

Figure 7–82 Elements of a complete surface finish symbol.

allows the operator to show the tool path, as shown in Figure 7–91. The tool path display allows the operator to determine if the machine tool correctly performs the assigned machining operation.

## AN INTRODUCTION TO ISO 9000

The ISO 9000 Quality Systems Standard was established to encourage the development of standards, testing, quality control, and the certification of companies, organizations, and institutions where these practices are implemented. *ISO* stands for the International Organization for Standardization. ISO is an international organiza-

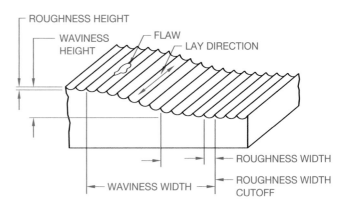

Figure 7–83 Surface finish characteristics magnified.

tion that is made up of nearly 100 countries. The United States is a member country that is represented by the American National Standards Institute (ANSI). ISO 9000 can certify companies, organizations, and institutions when their engineering, drafting standards, manufacturing, and quality control meet the requirements of a model quality management system established by the ISO 9000 organization. A certification is obtained by passing an inspection by an independent representative of the ISO. ISO 9000 certification is also referred to as registration. There are a number of reasons why a company may want to become ISO 9000 certified.

- ■ To conduct business with customers requiring ISO 9000 certification. This includes agencies of the U.S. government.
- ■ To compete in a competitive market that requires strict attention to quality control.
- ■ To reduce cost and improve profits.
- ■ To improve and maintain customer confidence.
- ■ To achieve improved product quality and strengthen customer relations.
- ■ To manufacture and sell products in the European Union markets where this certification is required.
- ■ To require that subcontractors meet the same ISO expectations for quality control.
- ■ To obtain ISO 9000 certification that also satisfies requirements established by other local and national organizations.
- ■ To improve company standards.

The ISO 9000 Quality Systems Standard is made up of a series of five international standards that provide leadership in the development and completion of successful quality-management systems. These five standards are briefly described here.

*ISO 9000-1.* This standard provides direction and definitions that describe what each standard contains and assists companies in the selection and use of the appropriate ISO standard for the desired results.

*ISO 90001.* This is the model that can be used by any organization for designing, documenting, and implementing ISO standards. The model takes a product through the process of design, drafting, manufacturing, quality control, installation, and service.

| MICROMETERS | ROUGHNESS HEIGHT RATING MICRO INCHES | SURFACE DESCRIPTION | PROCESS |
|---|---|---|---|
| 25 | 1000 | VERY ROUGH | SAW AND TORCH CUTTING, FORGING, OR SAND CASTING. |
| 12.5 | 500 | ROUGH MACHINING | HEAVY CUTS AND COARSE FEEDS IN TURNING, MILLING, AND BORING. |
| 6.3 | 250 | COARSE | VERY COARSE SURFACE GRIND, RAPID FEEDS IN TURNING, PLANING, MILLING, BORING, AND FILING. |
| 3.2 | 125 | MEDIUM | MACHINING OPERATIONS WITH SHARP TOOLS, HIGH SPEEDS, FINE FEEDS, AND LIGHT CUTS. |
| 1.6 | 63 | GOOD MACHINE FINISH | SHARP TOOLS, HIGH SPEEDS, EXTRA-FINE FEEDS AND CUTS. |
| 0.80 | 32 | HIGH-GRADE MACHINE FINISH | EXTREMELY FINE FEEDS AND CUTS ON LATHE, MILL, AND SHAPERS REQUIRED. EASILY PRODUCED BY CENTERLESS, CYLINDRICAL, AND SURFACE GRINDING. |
| 0.40 | 16 | | |
| 0.20 | 8 | VERY FINE MACHINE FINISH | FINE HONING AND LAPPING OF SURFACE. |
| 0.050 0.100 | 2 – 4 | EXTREMELY SMOOTH MACHINE FINISH | EXTRA-FINE HONING AND LAPPING OF SURFACE. MIRROR FINISH. |
| 0.025 | 1 | SUPER FINISH | DIAMOND ABRASIVES. |

Figure 7–84 Common roughness height values with a surface description and associated machining process.

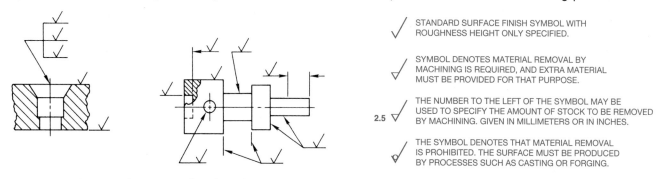

Figure 7–85 Proper placement of surface finish symbols.

STANDARD SURFACE FINISH SYMBOL WITH ROUGHNESS HEIGHT ONLY SPECIFIED.

SYMBOL DENOTES MATERIAL REMOVAL BY MACHINING IS REQUIRED, AND EXTRA MATERIAL MUST BE PROVIDED FOR THAT PURPOSE.

2.5 THE NUMBER TO THE LEFT OF THE SYMBOL MAY BE USED TO SPECIFY THE AMOUNT OF STOCK TO BE REMOVED BY MACHINING. GIVEN IN MILLIMETERS OR IN INCHES.

THE SYMBOL DENOTES THAT MATERIAL REMOVAL IS PROHIBITED. THE SURFACE MUST BE PRODUCED BY PROCESSES SUCH AS CASTING OR FORGING.

Figure 7–86 Material removal elements added to the surface finish symbol.

*ISO 9002.* This standard is the same as ISO 9001, except that it does not contain the requirement of documenting the design and development process.

*ISO 9003.* This standard is for companies or organizations that only need to demonstrate through inspection and testing methods that they are providing the desired product or service.

*ISO 9004-1.* This is a set of guidelines that can be used to assist organizations in their development and implementation of quality management systems.

In addition to the ISO 9000 series are the QS 9000 and AS 9000 standards. The QS 9000 is Automotive Requirements, and AS 9000 is the Aerospace Standard. These standards contain all of ISO 9001, plus requirements beyond ISO 9001. These standards were developed by the U.S. automotive and aerospace industries for specific applications and needs that customize the standard for their industries.

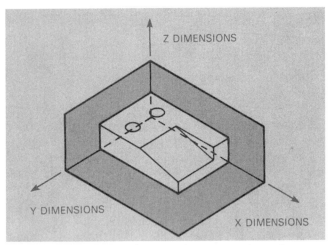

Figure 7–87 Features of a part dimensioned in relation to X, Y, and Z axes.

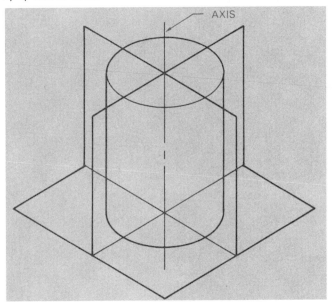

Figure 7–88 Features of a part dimensioned in relation to X, Y, and Z axes.

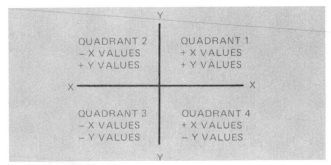

Figure 7–89 Features of a part dimensioned in relation to X, Y, and Z axes.

Now that you have learned about all the different dimensioning practices, take some time to look at the print found in Figure 7–92. A complex drawing with dimensions placed in the proper locations helps you read the drawing. Begin by reading the title block information, followed by the general notes, and then read the dimensions and notes on each view as you correlate between views.

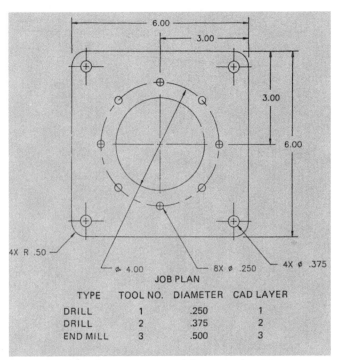

JOB PLAN

| TYPE | TOOL NO. | DIAMETER | CAD LAYER |
|------|----------|----------|-----------|
| DRILL | 1 | .250 | 1 |
| DRILL | 2 | .375 | 2 |
| END MILL | 3 | .500 | 3 |

Figure 7–90 A drawing created for CAD/CAM.

Figure 7–91 The tool path display, shown in color, allows the operator to determine if the machine tool will perform the assigned machining operation.

Some companies place a pictorial view of the part, as shown in the lower-left corner of Figure 7–92. This gives you more help to visualize the part.

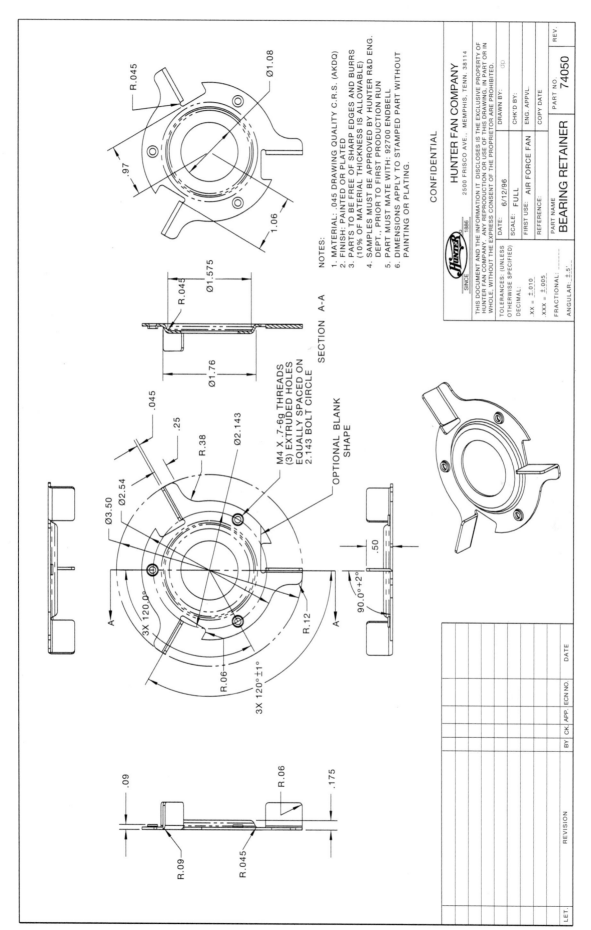

Figure 7–92 A complex drawing with dimensions placed in the proper locations helps in reading the drawing. Notice the pictorial drawing in the bottom center, which helps you visualize the part. *Courtesy Hunter Fan Company, Memphis, TN.*

# CHAPTER 7 TEST

Multiple choice: Respond to the following by selecting a, b, c, or d to best answer the question or complete the statement.

1. This dimensioning system has all numerals, figures, and notes to be read along with the dimension lines, where horizontal dimensions read from the bottom and vertical dimensions read from the right side of the sheet.
   - a. unidirectional
   - b. aligned
   - c. tabular
   - d. arrowless

2. This dimensioning system requires that all numerals, figures, and notes be placed horizontally and read from the bottom of the drawing sheet.
   - a. unidirectional
   - b. aligned
   - c. tabular
   - d. arrowless

3. This is a dimensioning system that works without dimension lines, where features are identified with letters and keyed to a table, and location dimensions are established with extension lines as coordinates from datums.
   - a. unidirectional
   - b. aligned
   - c. tabular
   - d. arrowless

4. This is a dimensioning system in which size and location dimensions from datums or coordinates are given in a table.
   - a. unidirectional
   - b. aligned
   - c. tabular
   - d. arrowless

5. These drawings are used when a particular part or assembly has one or more dimensions that change depending on the specific application.
   - a. basic dimension
   - b. chart drawing
   - c. tabular drawing
   - d. arrowless drawing

6. Each dimension shall have a tolerance, except for those dimensions specifically identified as reference, maximum, minimum, or stock.
   - a. true
   - b. false

7. The tolerance may be applied directly to the dimension or indicated by a general note or located in the drawing title block.
   - a. true
   - b. false

8. It is normal to assume or scale a distance or size on a drawing.
   - a. true
   - b. false

9. Wires, cables, sheets, rods, and other materials manufactured by gage or code numbers shall be specified by linear dimensions indicating the diameter or thickness.
   - a. true
   - b. false

10. A 90° angle is assumed where centerlines and lines depicting features are shown on a drawing at right angles and no angle is specified.
    - a. true
    - b. false

11. Unless otherwise specified, all dimensions are applicable at 98.8° F.
    - a. true
    - b. false

12. This dimension is considered to be a theoretically exact size, location, profile, or orientation of a feature or point.
    - a. actual size
    - b. allowance
    - c. basic dimension
    - d. bilateral dimension

13. This is the part size as measured after production.
    - a. actual size
    - b. allowance
    - c. basic dimension
    - d. feature

14. This is the physical portion of a part.
    - a. feature
    - b. tolerance
    - c. actual size
    - d. produced size

15. This is the tightest possible fit between two mating parts.
    - a. specified dimension
    - b. allowance
    - c. basic dimension
    - d. limits of dimension

16. This is the largest and smallest possible size or location to which a part may be made as related to the tolerance of the dimension.
    - a. specified dimension
    - b. allowance
    - c. basic dimension
    - d. limits of dimension

17. This type of tolerance is allowed to vary in two directions from the specified dimension.

   a. bilateral
   b. unilateral

18. This type of tolerance is allowed to vary in one direction from the specified dimension.

   a. bilateral
   b. unilateral

19. This is a physical portion of an object, such as a surface or hole.

   a. actual size
   b. allowance
   c. basic dimension
   d. feature

20. This is the part of the dimension from which the limits are calculated.

   a. actual size
   b. specified dimension
   c. basic dimension
   d. tolerance

21. This is the total amount of permissible variation in size or location.

   a. actual size
   b. specified dimension
   c. basic dimension
   d. tolerance

22. The maximum material condition of the shaft dimension ∅.625/.620 is:

   a. .625
   b. .620

23. The maximum material condition of the shaft dimension ∅.625/.620 is:

   a. .625
   b. .620

24. The least material condition of the hole dimension ∅.625/.620 is:

   a. .625
   b. .620

25. The least material condition of the shaft dimension ∅.625/.620 is:

   a. .625
   b. .620

26. Metric units expressed in millimeters or U.S. customary units expressed in decimal inches are considered the standard units of linear measurement on engineering documents and drawings.

   a. true
   b. false

27. A situation where both inch and metric dimensions are shown on a drawing is known as:

   a. multidimensions
   b. both dimensions
   c. dual dimensions
   d. inch/millimeter

28. Where millimeter dimensions are less than one, a zero precedes the decimal.

   a. true
   b. false

29. Where inch dimensions are less than one, a zero precedes the decimal.

   a. true
   b. false

30. Fraction dimensions generally represent a larger tolerance than decimal dimensions.

   a. true
   b. false

31. A properly dimensioned print places the smallest dimensions closest to the object and the progressively larger dimensions outward from the object.

   a. true
   b. false

32. This is also known as point-to-point dimensioning:

   a. chain dimensioning
   b. datum dimensioning

33. This is a method of dimensioning where each feature dimension originates from a common surface, axis, or center plane:

   a. chain dimensioning
   b. datum dimensioning

34. A circle contains _____degrees.

   a. 180°
   b. 270°
   c. 360°
   d. 380°

35. Arcs are dimensioned with:

   a. diameter ∅
   b. radius (R)
   c. length (L)
   d. circumference (C)

36. Diameters are dimensioned with:

   a. diameter ∅
   b. radius (R)
   c. length (L)
   d. circumference (C)

37. Hole sizes are dimensioned with leaders to the view where they appear as _____, or dimensioned in a sectional view.

    a. rectangles
    b. hidden features
    c. circles
    d. lines

38. Location dimensions to cylindrical features are given to the _____ of the feature in the view where they appear as a circle, or in a sectional view.

    a. edge
    b. center
    c. surface
    d. coordinates

39. This type of coordinate dimensioning is where linear dimensions are used to locate features from planes or centerlines.

    a. polar coordinate
    b. angular coordinate
    c. linear coordinate
    d. rectangular coordinate

40. When foreshortened views occur, dimensions are placed on this view for clarity.

    a. front view
    b. side view
    c. auxiliary view
    d. view enlargement

41. These notes relate to the entire drawing.

    a. related notes
    b. general notes
    c. specific notes
    d. common notes

42. The limits of this dimension, .750 +/− .005 inch, are:

    a. .750/.755
    b. .745/.750
    c. .700/.800
    d. .745/.755

43. The limits of this dimension, 24.50 +/− 0.24 mm, are:

    a. 24.25/24.50
    b. 24.25/24.75
    c. 24.50/24.75
    d. 24.00/25.00

44. This is a condition when, due to the limits of dimensions, there is always clearance between mating parts.

    a. clearance fit
    b. allowance
    c. interference fit
    d. maximum material condition

45. Also known as a force or shrink fit, this is a condition that exists when, because of the limits of the dimensions, mating parts must be pressed together

    a. clearance fit
    b. allowance
    c. interference fit
    d. maximum material condition

46. What would the shaft and hole size be for a Ø .50 nominal size using an RC4 fit?

    a. shaft Ø .4870/.4094, hole Ø .5000/.5100
    b. shaft Ø .4987/.4994, hole Ø .5000/.50001
    c. shaft Ø .4987/.4994, hole Ø .5000/.5100
    d. shaft Ø .4987/.4994, hole Ø .5000/.5010

47. The shrinkage for most iron in castings is about:

    a. .125 inch per foot
    b. .250 inch per foot
    c. .156 inch per foot
    d. .22 inch per foot

48. This is the taper allowance on all vertical surfaces of a casting or forging.

    a. chamfer
    b. allowance
    c. draft
    d. clearance angle

49. The standard machining allowance for iron and steel is:

    a. .125
    b. .166
    c. .250
    d. .188

50. The general term that refers to the roughness, waviness, lay, and flaws of a surface is:

    a. surface roughness
    b. surface finish
    c. surface texture
    d. surface condition

51. The term used to describe the direction of configuration of the predominant surface pattern is:

    a. surface texture
    b. surface roughness
    c. surface waviness
    d. lay

52. The fine irregularities in the surface finish and a result of the manufacturing process used is known as:

    a. surface texture
    b. surface roughness
    c. surface waviness
    d. lay

53. This is the often widely spaced condition of surface texture usually caused by such factors as machine chatter, vibrations, work deflection, warpage, or heat treatment.
    a. surface texture
    b. surface roughness
    c. surface waviness
    d. lay

54. This surface finish roughness height, in microinches, is a good machine finish made with sharp tools, high speeds, and extrafine feeds and cuts.
    a. 16
    b. 32
    c. 63
    d. 125

55. This surface finish roughness height, in microinches, is a medium machine finish made with sharp tools, high speeds, and fine feeds and light cuts.
    a. 16
    b. 32
    c. 63
    d. 125

56. This surface finish roughness height, in microinches, is a very fine machine finish made by honing and lapping of the surface.
    a. 8
    b. 16
    c. 32
    d. 63

57. It is possible in CAD/CAM/CIM to make a drawing on the computer screen and send the information directly to the computer numerical control machine tool without generating a hard copy or print of the drawing.
    a. true
    b. false

58. The triangle on a drawing with a number inside that is used to refer you to a general note is a:
    a. triangle note
    b. specific note
    c. delta note
    d. reference note

59. The assigning of tolerances to related dimensions in an assembly based on the requirements of statistical process control.
    a. tolerance statistics
    b. statistical tolerancing
    c. statistical dimensioning
    d. statistical process control

60. In plastics manufacturing, these angles allow the finished plastic part to be ejected or removed from the mold without difficulty.
    a. draft angles
    b. mold angles
    c. parting line angles
    d. ejection angles

61. The location on the part where it has been separated from the mold in plastics manufacturing.
    a. separation line
    b. reference line
    c. parting line
    d. draft line

62. In this application of specifying draft on plastic parts, draft is added to the dimension for external dimensions and subtracted from the dimension for internal dimensions.
    a. +DFT and –DFT
    b. +DFT
    c. –DFT
    d. +DFT external and –DFT internal

63. This standard was established to encourage the development of standards, testing, quality control, and the certification of companies, organizations, and institutions where these practices are implemented.
    a. ISO Quality Systems Standard
    b. ISO Quality Standard
    c. ISO International Organization for Standardization
    d. ISO 9000 Quality Systems Standard

64. This is the model that can be used by any organization for designing, documenting, and implementing ISO standards. The model takes a product through the process of design, drafting, manufacturing, quality control, installation, and service.
    a. ISO 9000–1
    b. ISO 9002
    c. ISO 9001
    d. ISO 9003

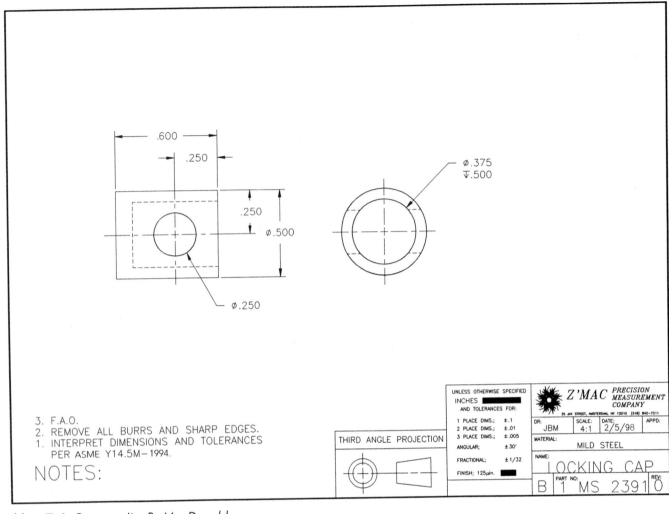

Notes:

3. F.A.O.
2. REMOVE ALL BURRS AND SHARP EDGES.
1. INTERPRET DIMENSIONS AND TOLERANCES
   PER ASME Y14.5M–1994.

NOTES:

THIRD ANGLE PROJECTION

UNLESS OTHERWISE SPECIFIED
INCHES ▇▇▇
AND TOLERANCES FOR:

| 1 PLACE DIMS.; | ±.1 |
| 2 PLACE DIMS.; | ±.01 |
| 3 PLACE DIMS.; | ±.005 |
| ANGULAR; | ±30' |
| FRACTIONAL; | ±1/32 |
| FINISH; 125μin. | ▇ |

*Z'MAC* PRECISION MEASUREMENT COMPANY
35 JAY STREET, AMSTERDAM, NY 12010  (518) 842-7211

| DR: JBM | SCALE: 4:1 | DATE; 2/5/98 | APPD: |
| MATERIAL: | | MILD STEEL | |
| NAME: | | LOCKING CAP | |
| B | 1 | PART NO: MS 2391 | REV: 0 |

Problem 7–1 *Courtesy Jim B. MacDonald.*

## CHAPTER 7 PROBLEMS

**PROBLEM 7–1** Answer the following questions as you refer to the print shown on this page.

1. Describe the views shown on this print.

2. What is the overall shape of the part?

3. Give the overall dimensions.

4. Give the specifications of the internal feature that runs along the axis of the part.

5. What is the dimension of the feature that runs perpendicular to the axis of the part?

6. Give the location dimensions to the Ø.250 feature.

7. Name the material used to make this part.

8. Are the dimensions in inches or millimeters?

9. What is the specified surface finish?

10. Give the tolerance of the Ø.375 dimension.

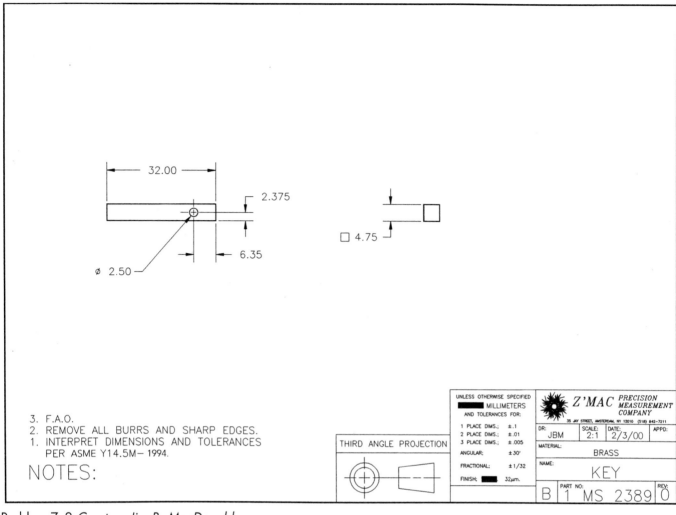

3. F.A.O.
2. REMOVE ALL BURRS AND SHARP EDGES.
1. INTERPRET DIMENSIONS AND TOLERANCES
   PER ASME Y14.5M– 1994.

NOTES:

THIRD ANGLE PROJECTION

UNLESS OTHERWISE SPECIFIED
■ MILLIMETERS
AND TOLERANCES FOR:

| 1 PLACE DIMS.; | ±.1 |
| 2 PLACE DIMS.; | ±.01 |
| 3 PLACE DIMS.; | ±.005 |
| ANGULAR; | ±30' |
| FRACTIONAL; | ±1/32 |
| FINISH; ■ | 32μm. |

*Z'MAC* PRECISION
MEASUREMENT
COMPANY
35 JAY STREET, AMSTERDAM, NY 12010  (518) 842–7211

| DR: JBM | SCALE: 2:1 | DATE: 2/3/00 | APPD: |

MATERIAL: BRASS

NAME: KEY

| B | PART NO: 1 | MS 2389 | REV: 0 |

Problem *7–2 Courtesy Jim B. MacDonald.*

**PROBLEM 7–2** Answer the following questions as you refer to the print shown on this page.

1. Describe the views shown on this print.
   _____
   _____

2. What is the overall shape of the part?
   _____

3. Give the overall dimensions.
   _____
   _____

4. What does the symbol □ mean?
   _____
   _____

5. What is the dimension of the feature that runs perpendicular to the length of the part?
   _____
   _____

6. Give the location dimensions to the Ø2.50 feature.
   _____
   _____

7. Name the material used to make this part.
   _____

8. Are the dimensions in inches or millimeters?
   _____

9. What is the specified surface finish?
   _____

10. Give the tolerance of the 6.35 dimension.
    _____

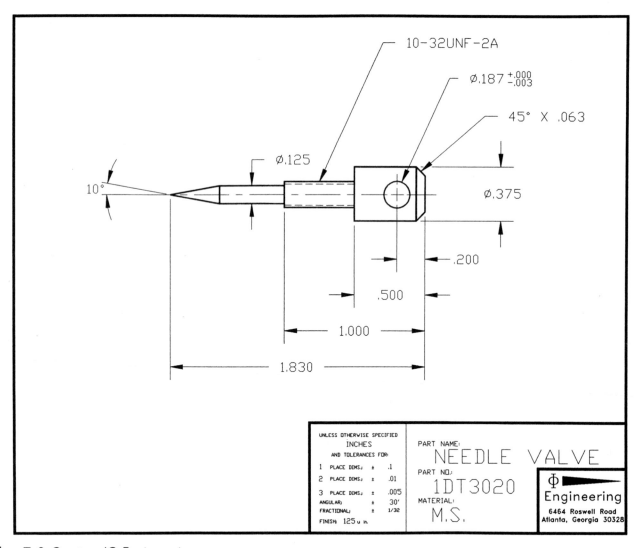

Problem 7–3 *Courtesy IO Engineeering.*

**PROBLEM 7–3** Answer the following questions as you refer to the print shown on this page.

1. How is it possible for this print to display only one view of the part?

_____
_____
_____
_____

2. What are the overall dimensions of the part?

_____
_____

3. Give the total angle of the tapered end.

_____

4. What is the diameter of the feature next to the tapered end?

_____

5. Give the limits of the dimension for the hole through the part.

_____

6. What type of dimensioning is used to dimension the features along the length of this part?

_____
_____

7. Give the tolerance of the Ø.187 hole.

_____

8. What is the tolerance of the 10° angle?

_____

9. Give the specification of the chamfer.

_____

10. What is the full name of the material for this part?

_____

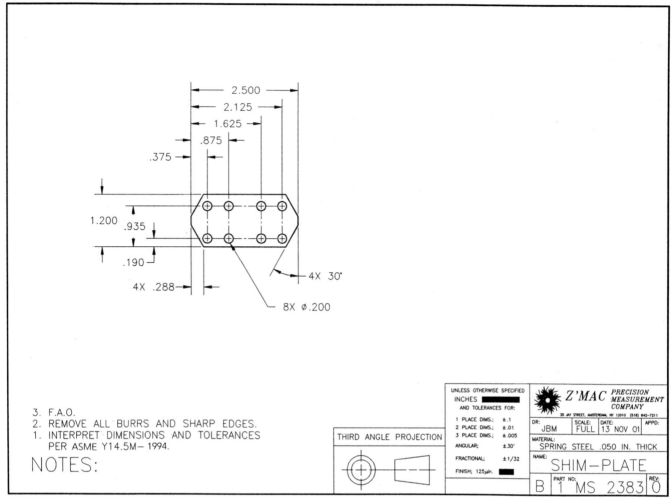

*Problem 7–4 Courtesy Jim B. MacDonald.*

**PROBLEM 7–4** Answer the following questions as you refer to the print shown on this page.

1. Describe the views shown on this print.
   _____
   _____
   _____
   _____

2. How is it possible to show only one view?
   _____
   _____
   _____

3. Give the overall dimensions.
   _____
   _____

4. How many holes are there in the part?
   _____

5. What is the size dimension of the holes?
   _____

6. What type of dimensioning is used to locate the holes?
   _____

7. Name the material used to make this part.
   _____

8. Are the dimensions in inches or millimeters?
   _____

9. What is the specified surface finish?
   _____

10. Give the tolerance of the .375 dimension.
    _____

11. How many angled corners are there?
    _____

12. Give the coordinate and angular dimension of the angled corners.
    _____
    _____

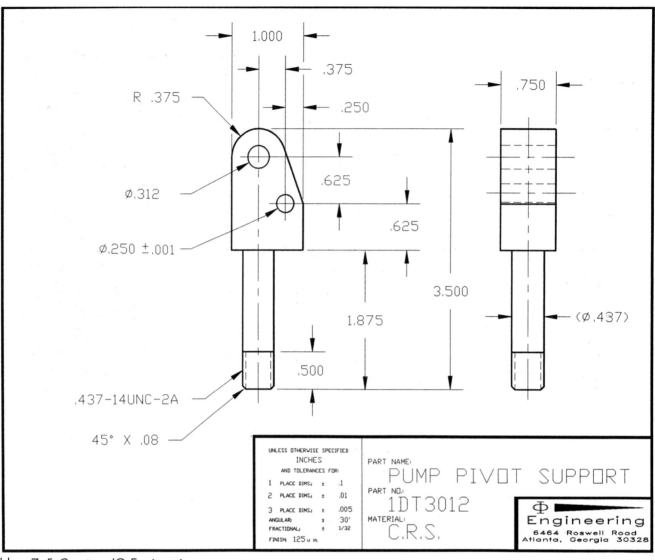

R .375

Ø.312

Ø.250 ±.001

.437–14UNC–2A

45° X .08

1.000

.375

.250

.625

.625

3.500

1.875

.500

.750

(Ø.437)

UNLESS OTHERWISE SPECIFIED
INCHES
AND TOLERANCES FOR:

1  PLACE DIMS.｝  ±   .1
2  PLACE DIMS.｝  ±   .01
3  PLACE DIMS.｝  ±   .005
ANGULAR｝        ±   30'
FRACTIONAL｝     ±   1/32
FINISH: 125 u in.

PART NAME:
PUMP PIVOT SUPPORT
PART NO.:
1DT3012
MATERIAL:
C.R.S.

Φ▬
Engineering
6464 Roswell Road
Atlanta, Georgia 30328

Problem 7–5 *Courtesy IO Engineering.*

**PROBLEM 7–5** Answer the following questions as you refer to the print shown on this page.

1. Describe the views shown on this print. _____
   _____
   _____

2. Give the overall dimensions of the part.
   _____

3. Give the size dimensions of the holes through this part.
   _____
   _____

4. Give the diameter dimension at the shaft body of the part.
   _____
   _____

5. Name the type of dimensioning practice used to locate the holes.
   _____
   _____

6. What is the tolerance of the Ø .250 hole?
   _____

7. What is the tolerance of the Ø .312 hole?
   _____

8. What is the tolerance of the .08 dimension?
   _____

9. What do the parentheses around the Ø .437 dimension mean?
   _____

10. Give the full name of the material for this part.
    _____

11. The dimension numerals and notes on this print are placed so they read from the bottom of the sheet. What is this system called?
    _____

12. Give the chamfer specification.
    _____

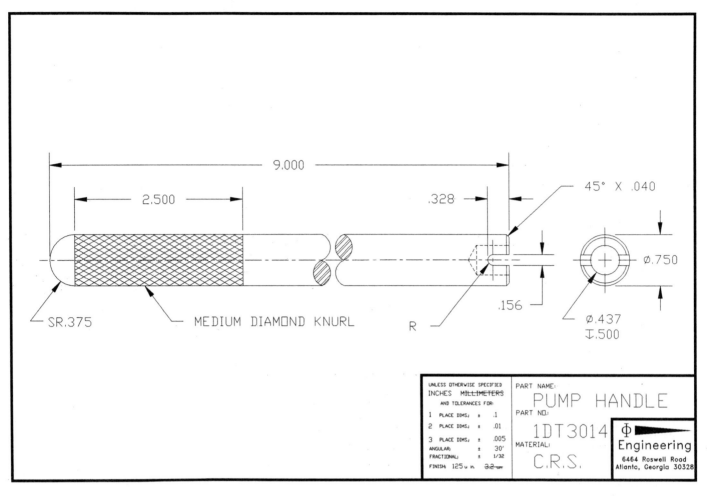

Problem 7–6 *Courtesy IO Engineering.*

**PROBLEM 7–6** Answer the following questions as you refer to the print shown on this page.

1. Describe the views shown on this print.
   _____
   _____
   _____

2. Give the overall dimensions of the part.
   _____
   _____

3. Give the length of the knurl.
   _____

4. Give the length of the part to the center of the spherical radius.
   _____

5. Give the specifications for the hole.
   _____
   _____

6. What are the dimensions of the slot?
   _____
   _____

7. Give the specifications of the chamfer.
   _____

8. Give the full name of the material used for this part.
   _____
   _____

9. What does SR.375 mean?
   _____

10. What does the single R with a leader mean?
    _____
    _____
    _____

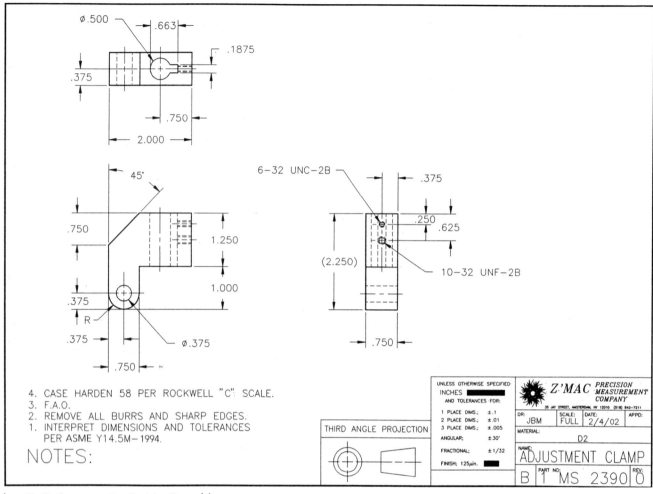

Problem 7–7 Courtesy Jim B. MacDonald.

**PROBLEM 7–7** Answer the following questions as you refer to the print shown on this page.

1.  Describe the views shown on this print.
    _____
    _____
    _____

2.  Give the overall dimensions.
    _____
    _____

3.  What type of dimensioning is used to dimension the height in the front view?
    _____

4.  Name the material used to make this part.
    _____

5.  Are the dimensions in inches or millimeters?
    _____

6.  What is the specified surface finish?
    _____

7.  Give the tolerance of the .375 dimension.
    _____

8.  How many angled corners are there?
    _____

9.  Give the coordinate and angular dimension of the angled corner or corners.
    _____
    _____

10. What is the tolerance of the 45° dimension?
    _____

11. Give the case hardening specification.
    _____

12. What are the keyway dimensions?
    _____

13. Give the radius of the rounded feature represented in the front view.
    _____

14. Why do you think the radius of the rounded feature, described in Question 13, is dimensioned with only an R?
    _____

15. What do the parentheses on the 2.250 dimension mean?
    _____
    _____

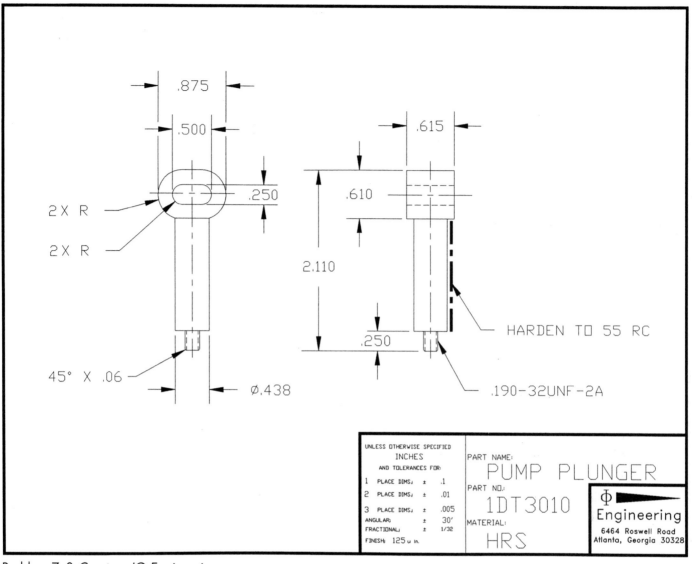

Problem 7–8 *Courtesy IO Engineering.*

**PROBLEM 7–8** Answer the following questions as you refer to the print shown on this page.

1. Describe the views shown on this print.
   _____
   _____
   _____

2. Give the overall dimensions of the part.
   _____

3. Give the dimensions of the slot through this part.
   _____

4. Explain the 2 **X** R note.
   _____
   _____
   _____

5. What is the radius of the slot ends?
   _____

6. Explain the purpose of the thick line located at the HARDEN note leader.
   _____
   _____

7. Give the chamfer specification.
   _____

8. What is the diameter of the main body?
   _____

9. Give the tolerance of the .06 dimension.
   _____

10. Give the full name of the material used to make this part.
    _____

| *Enoch Mfg. Co.* | Curr ECN: None | Drawing 1 of 5 | Enoch #: 1443 |
| | Last ECN: None | Wk/Order#: | |
| Rev Updated: N/A | Display Scale: 5 : 1 | Rel.Date: | Op.#s: 10 |
| Title: ANVIL | Matl: 5/16" HEX 41L40 | Start Date: | Dept: PRIMARY |
| | Date: 2/92 Engr: | Cust Rev: G | Nxt.Dept: ABRADING |

.490 / .484

.098 / .086

.079 / .071

35° / 25°

Ø .284 / .276

Ø .200 / .193

5/16" HEX

CUTOFF BURR O.K. THIS END

.043 / .035

.059 / .051

.303 / .287

.303 / .287

Problem 7–9 Anvil.

**PROBLEM 7–9** Answer the following questions as you refer to the prints shown on this and the following page.

1. You are given two sheets from the set of drawings. How many total sheets are there in the set?

2. Give the dimensions of the grooves.

3. Give the dimension across the flats of the hexagon.

4. One of the optional parts has a rounded end. What is the dimension of this end?

5. What type of dimensioning system is used to dimension the features along the length of the part?

6. Give the height of the letter to be stamped in the part.

7. What is the depth of the letter to be stamped in the part?

8. One of the optional parts has an angled recessed feature. What is the angle?

9. What is the surface finish specified for the feature described in Question 8?

10. What is the tolerance of the .484/.468 dimension?

11. What is the tolerance of the .200/.193 dimension?

12. What is the tolerance of the .1085/.0985 dimension?

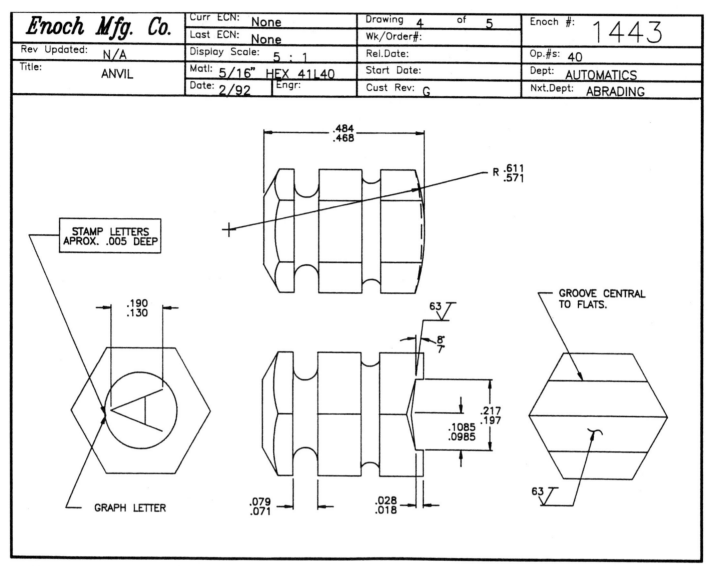

| *Enoch Mfg. Co.* | Curr ECN: None | Drawing 4 of 5 | Enoch #: 1443 |
|---|---|---|---|
| | Last ECN: None | Wk/Order#: | |
| Rev Updated: N/A | Display Scale: 5 : 1 | Rel.Date: | Op.#s: 40 |
| Title: ANVIL | Matl: 5/16" HEX 41L40 | Start Date: | Dept: AUTOMATICS |
| | Date: 2/92 | Engr: | Cust Rev: G | Nxt.Dept: ABRADING |

Problem 7–9 Anvil (continued). *Courtesy Enoch Mafg. Co.*

# OFFSET RIVET SETS—CUPPED

FUCTION OF TOOL:   TO SQUEEZE UNIVERSAL HEAD RIVETS.  THIS RIVET SET
ASSEMBLY MOUNTS IN PORTABLE COMPRESSION RIVETERS.

TO ORDER:   SPECIFY THE 6 DIGIT MTS NUMBER ADJACENT TO THE DESIRED
RIVET SET DIMENSION.

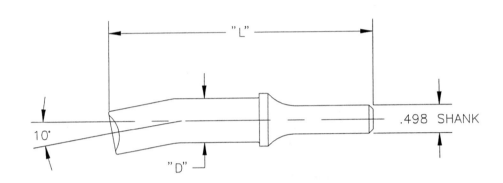

| MTS # | | |
|---|---|---|
| L | 3/16 RIV | 1/4 RIV |
| 3 1/2 | 011201 | 011206 |
| 5 1/2 | 011202 | 011207 |
| 7 1/2 | 011203 | 011208 |
| 10 1/2 | 011204 | 011209 |
| 13 1/2 | 011205 | 011210 |

Problem 7–10 *Courtesy Manufacturer's Tool Service.*

**PROBLEM 7–10** Answer the following questions as you refer to the print shown on this page.

1.  What type of dimensioning and drawing system is displayed on this print with the changing "L" values?

_____

2.  Give the L value for the 1/4 RIVET number 011209.

_____

3.  Give the L value for the 3/16 RIVET number 011202.

_____

4.  Give the L value for the 1/4 RIVET number 011206.

_____

5.  Give the optional "D" values.

_____

6.  What is the shank diameter?

_____

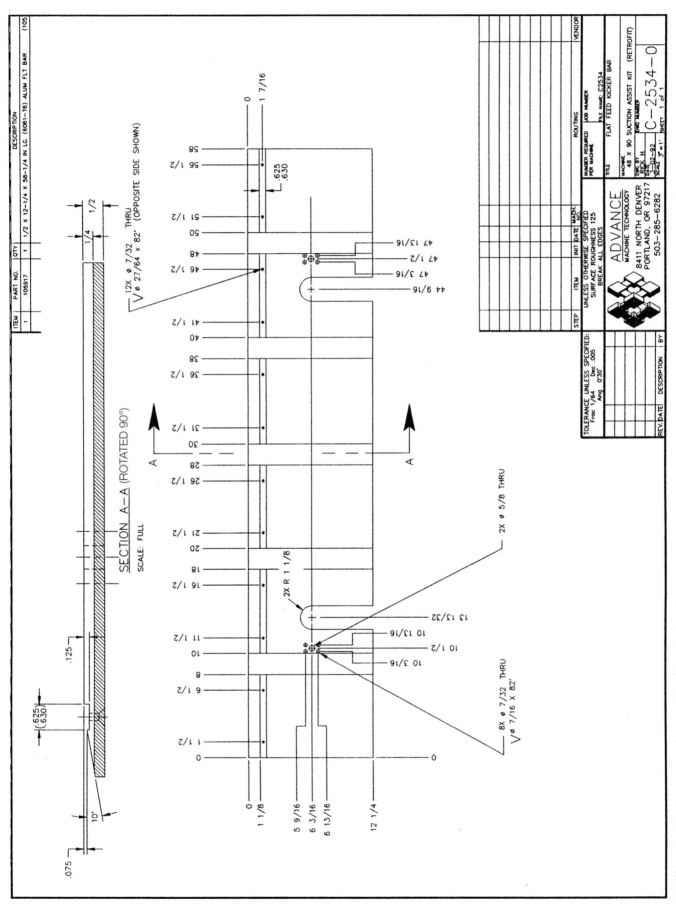

Problem 7–11 Courtesy Advance Machine Technology.

**PROBLEM 7–11** Answer the following questions as you refer to the print shown on the previous page.

1.  What is the type of dimensioning system that uses no dimension lines as displayed on this print?
    _____
    _____

2.  The dimension numerals on this print are placed so they read from the bottom and right side of the sheet. What is this system called?
    _____

3.  Give the overall length, width, and thickness of the part.
    _____

4.  What is the width and depth of the slot that runs the entire length of the part?
    _____

5.  How many slots run the width of the part?
    _____

6.  Give the width and depth of the slots described in Question 5.
    _____

7.  There are two features with rounded ends. Give the location dimensions to each of these features along the X axis.
    _____

8.  Give the location dimension along the Y axis to the features described in Question 7.
    _____

9.  Give the width of the features described in question 7.
    _____

10. List the number and specification of each different-sized hole in this part.
    _____
    _____
    _____
    _____

11. Give the location dimensions to the Ø5/8 features along the X axis.
    _____
    _____

12. Give the location dimensions to the Ø5/8 features along the Y axis.
    _____
    _____

**PROBLEM 7–12** Answer the following questions as you refer to the print shown on the following page.

1.  What is the type of dimensioning system that uses no dimension lines and keys features to a table displayed on this print?
    _____
    _____

2.  How many different holes are in this part?
    _____

3.  Give the overall length, width, and thickness.
    _____
    _____

4.  Give the quantity and specifications of the slots.
    _____
    _____
    _____

5.  What is the tolerance of the 5.875 dimension?
    _____

6.  Explain the meaning of the parentheses on the (4.250) dimension.
    _____
    _____

7.  How many "B" holes are displayed?
    _____

8.  What is the diameter of the "B" holes?
    _____

9.  Give the full name of the material used to make this part.
    _____
    _____

10. What is the bend radius?
    _____

11. Give the finish specification.
    _____

12. Give the paint specification.
    _____
    _____

13. Give the X coordinate location of the two "D" holes.
    _____

14. Give the Y coordinate location of the two "D" holes.
    _____

15. Give the X coordinate location of the upper right "A" feature.
    _____

16. Give the Y coordinate location of the upper right "A" feature.
    _____

17. Give the dimensions to the bend lines.
    _____

18. What is to be done with sharp edges?
    _____

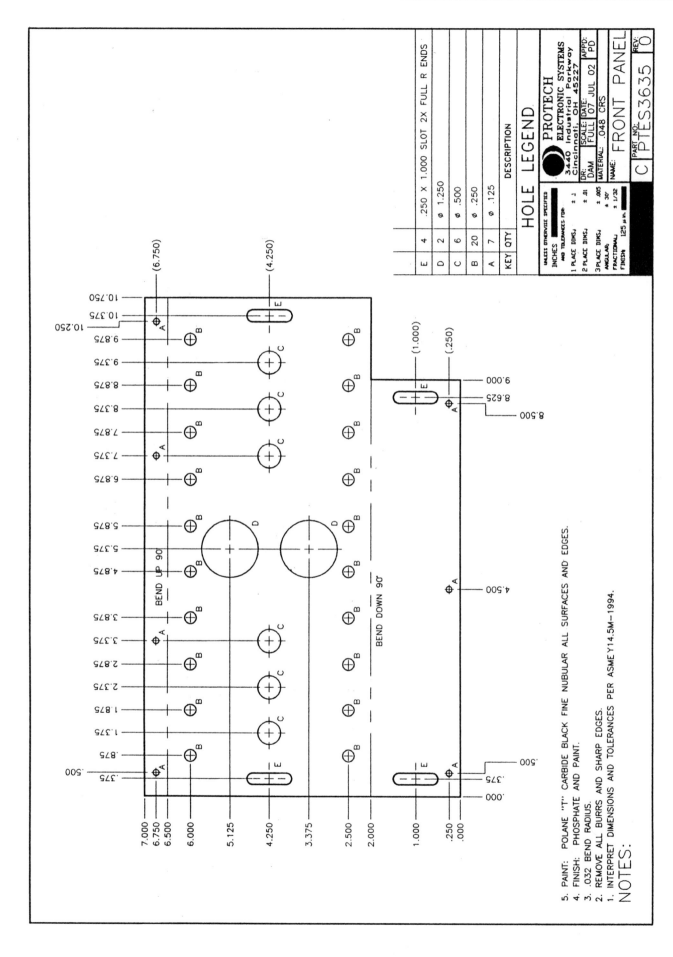

Problem 7-12 Courtesy Protech Electronic Systems.

# *Reading Drawings with Fasteners and Springs*

## LEARNING OBJECTIVES

After completing this chapter you will be able to:

- Identify the ANSI standard for screw thread representations.
- Define screw thread terminology.
- Read prints with screw thread representations and thread notes.
- Read and identify the parts of metric and Unified and American National threads.

- Identify fastener head types.
- Read prints and answer questions with regards to pins, rivets, washers, retaining rings, keys, keyways, and keyseats.
- Define spring terminology and read prints with spring representations.

*ASME* This chapter introduces you to the methods of presenting drawings and specifications for fasteners and springs. Fasteners include screw threads, keys, pins, rivets, and weldments. There are two types of springs, helical and flat. The American National Standards Institute documents that govern the standards for fasteners and springs are *Screw Thread Representation* ANSI Y14.6, *Screw Thread Representation (Metric Supplement)* ANSI Y14.6aM, *Symbols for Welding and Nondestructive Testing Including Brazing* ANSI/AWS A2.4, and *Mechanical Spring Representation* ANSI Y14.13M.

## SCREW-THREAD FASTENERS

The standardization of screw threads was achieved among the United States, the United Kingdom, and Canada in 1949. A need for interchangeability of screw-thread fasteners was the purpose of this standardization and resulted in the Unified Thread Series. The Unified Thread Series is now the American Standard for screw threads. Prior to 1949 the United States standard was the American National screw threads. The unification standard occurred as a result of combining some of the characteristics of the American National screw threads with the United Kingdom's long-accepted Whitworth screw threads. Screw thread systems were revised again in 1974 for metric application. The modifications were minor and primarily based on metric translation. To emphasize that unified screw threads evolved from inch calibration, the term unified inch screw threads is used, while the term unified screw threads metric translation is used for the metric conversion.

Screw threads are a helix or conical spiral form on the external surface of a shaft or internal surface of a cylindrical hole, as shown in Figures 8–1 and 8–2. Screw threads are used for a myriad of services, such as for holding parts together as fasteners, for leveling and adjusting objects, and for transmitting power from one object or feature to another.

## SCREW-THREAD TECHNOLOGY

Refer to Figure 8–1 and Figure 8–2 as a reference for the following definitions related to external and internal threads.

*Axis.* The thread axis is the centerline of the cylindrical thread shape.

*Body.* That portion of a screw shaft that is left unthreaded.

*Chamfer.* An angular relief at the last thread to help allow the thread to more easily engage with a mating part.

*Classes of Threads.* A designation of the amount of tolerance and allowance specified for a thread.

*Crest.* The top of external and the bottom of internal threads.

*Depth of Thread.* Depth is the distance between the crest and the root of a thread, measured perpendicular to the axis.

*Die.* A machine tool used for cutting external threads.

*Fit.* Identifies a range of thread tightness or looseness.

*Included Angle.* The angle between the flanks (sides) of the thread.

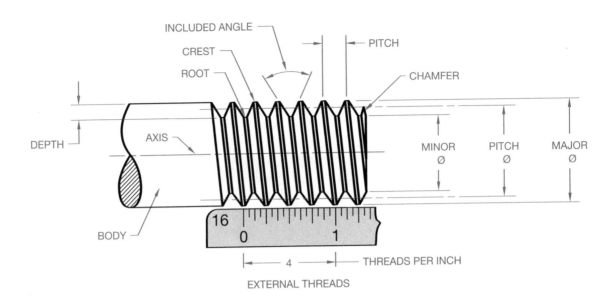

Figure 8–1 External screw thread components.

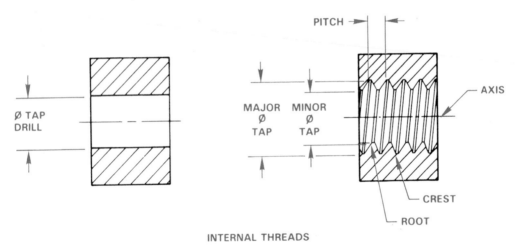

Figure 8–2 Internal screw thread components.

*Lead.* The lateral distance a thread travels during one complete rotation.

*Left-hand Thread.* A thread that engages with a mating thread by rotating counterclockwise, or with a turn to the left when viewed toward the mating thread.

*Major Diameter.* The distance on an external thread from crest to crest through the axis. For an internal thread the major diameter is measured from root to root across the axis.

*Minor Diameter.* The dimension from root to root through the axis on an external thread and measured across the crests through the center for an internal thread.

*Pitch.* The distance measured parallel to the axis from a point on one thread to the corresponding point on the next thread.

*Pitch Diameter.* A diameter measured from a point halfway between the major and minor diameters through the axis to a corresponding point on the opposite side.

*Right-hand Thread.* A thread that engages with a mating thread by rotating clockwise, or with a turn to the right when viewed toward the mating thread.

*Root.* The bottom of external and the top of internal threads.

*Tap.* A tap is the machine tool used to form an interior thread. Tapping is the process of making an internal thread.

*Tap Drill.* A tap drill is used to make a hole in metal before tapping.

*Thread.* The part of a screw thread represented by one pitch.

*Thread Form.* The design of a thread determined by its profile.

*Thread Series.* Groups of common major diameter and pitch characteristics determined by the number of threads per inch.

*Threads per Inch.* The number of threads measured in one inch. The reciprocal of the pitch in inches.

## THREAD-CUTTING TOOLS

The tap is a machine tool used to form an internal thread, as shown in Figure 8–3. A die is a machine tool used to form external threads (see Figure 8–4).

Figure 8–3 Tap. *Courtesy Greenfield Tap & Die, Division of TRW, Inc.*

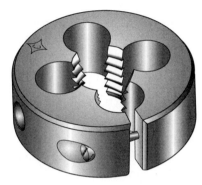

Figure 8–4 Die. *Courtesy The Cleveland Twist Drill, an Acme-Cleveland Company.*

A *tap set* is made up of a taper tap, a plug tap, and a bottoming tap, as shown in Figure 8–5. The taper tap is generally used for starting a thread. The threads are tapered to within 10 threads from the end. The tap is tapered so the tool more evenly distributes the cutting edges through the depth of the hole. The plug tap has the threads tapered to within five threads from the end. The plug tap can be used to completely thread through material or thread a blind hole (a hole that does not go through the material) if full threads are not required all the way to the bottom. The bottoming tap is used when threads are needed to the bottom of a blind hole.

The die is a machine tool used to cut external threads. Thread cutting dies are available for standard thread sizes and designations.

External and internal threads can also be cut on a lathe. A lathe is a machine which holds a piece of material between two centers or in a chucking device. The material is rotated as a cutting tool removes material while traversing along a carriage that slides along a bed. Figure 8–6 shows how a cutting tool can make an external thread.

## THREAD FORMS

*Unified threads* are the most common threads used on threaded fasteners. Figure 8–7 shows the profile of a Unified thread.

*American National threads*, shown in profile in Figure 8–8, are similar to the Unified thread, but have a flat root. Still in use today,

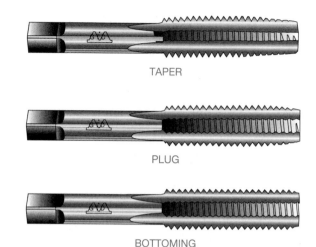

TAPER

PLUG

BOTTOMING

Figure 8–5 A tap set includes taper, plug, and bottoming taps. *Courtesy Greenfield Tap & Die, Division of TRW, Inc.*

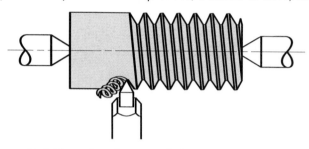

Figure 8–6 Thread cutting on a lathe.

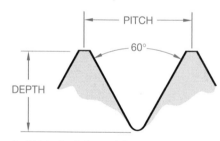

Figure 8–7 Unified thread form.

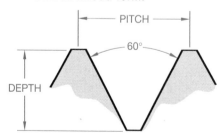

Figure 8–8 American National thread form.

the American National thread generally replaced the sharp-V thread form.

The *sharp-V thread*, although not commonly used, is a thread that will fit and seal tightly. It is difficult to manufacture because the sharp crests and roots of the threads are easily damaged (see Figure 8–9). The sharp-V thread was the original U.S. standard thread form.

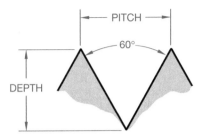

Figure 8–9 Sharp-V thread form.

*Metric thread* forms vary slightly from one European country to the next. The International Organization for Standardization (ISO) was established to standardize metric screw threads. The ISO thread specifications are similar to the Unified thread form (see Figure 8–10).

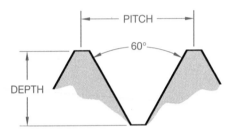

Figure 8–10 Metric thread form.

*Whitworth threads* are the original British standard thread forms developed in 1841. These threads have been referred to as parallel screw threads. The Whitworth thread forms are primarily being used for replacement parts (see Figure 8–11)

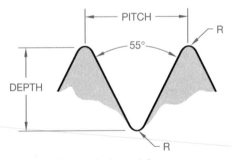

Figure 8–11 Whitworth thread form.

*Square thread* forms, shown in Figure 8–12, have a longer pitch than unified threads. Square threads were developed as threads that would effectively transmit power; however, they are difficult to manufacture because of their perpendicular sides. There are modified square threads with ten-degree sides. The square thread is generally replaced by Acme threads.

*Acme thread* forms are commonly used when rapid traversing movement is a design requirement. Acme threads are popular on such designs as screw jacks, vice screws, and other equipment and machinery that requires rapid screw action. A profile of the Acme thread form is shown in Figure 8–13.

*Buttress threads* are designed for applications where high stress occurs in one direction along the thread axis. The thread flank or

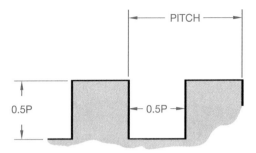

Figure 8–12 Square thread form.

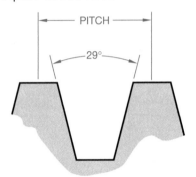

Figure 8–13 Acme thread form.

side which distributes the thrust or force is within 7{insert degree symbol} of perpendicularity to the axis. This helps reduce the radial component of the thrust. The buttress thread is commonly used in situations where tubular features are screwed together and lateral forces are exerted in one direction (see Figure 8–14).

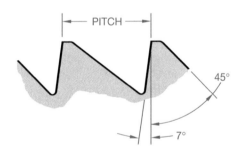

Figure 8–14 Buttress thread form.

*Dardelet thread* forms are primarily used in situations where a self-locking thread is required. These threads resist vibrations and remain tight without auxiliary locking devices (see Figure 8–15).

*Rolled thread* forms are used for screw shells of electric sockets and lamp bases (see Figure 8–16).

*American National Standard Taper pipe threads* are standard threads used on pipes and pipe fittings. These threads are designed to provide pressure-tight joints or not, depending on the intended function and materials used. American pipe threads are measured by the nominal pipe size, which is the inside pipe diameter. For example, a 1/2-inch pipe size has an outside pipe diameter of .840 inch (see Figure 8–17).

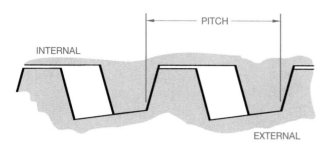

Figure 8–15 Dardelet self-locking thread form.

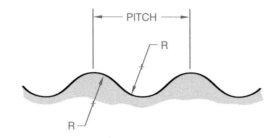

Figure 8–16 Rolled thread form.

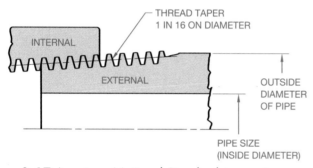

Figure 8–17 American National Standard taper pipe thread form.

## THREAD REPRESENTATIONS

There are three methods of thread representation in use: detailed, schematic, and simplified, as shown in Figure 8–18. The detailed representation can be used in special situations that require a pictorial display of threads, such as in a sales catalog or a display drawing. Detailed threads are not common on most manufacturing drawings since they are much too time consuming to draw. *Schematic* representations are also not commonly used in industry. Although they do not take the time of detailed symbols, they do

require extra time to draw. Some companies, however, use the schematic thread representation.

The simplified representation is the most common method of drawing thread symbols. *Simplified* representations clearly describe threads, and they are quick and easy to draw. Figure 8–19 shows simplified threads in different applications.

Also, notice how the use of a thread chamfer slightly changes the appearance of the thread. Chamfers are commonly applied to the first thread to help start a thread in its mating part.

When an internal screw thread does not go through the part, it is common to drill deeper than the depth of the required thread when possible. This process saves time and reduces the chance of breaking a tap. The thread may go to the bottom of a hole, but to produce it requires an extra process using a bottoming tap. Figure 8–20 shows a simplified representation of a thread that does not go through. The bolt should be shorter than the depth of

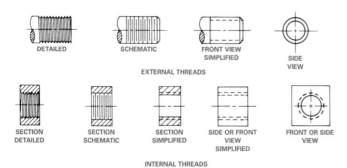

Figure 8–18 Thread representations.

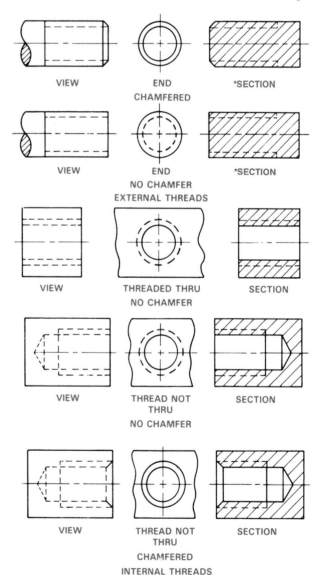

Figure 8–19 Simplified thread representations. *Threaded shafts are not sectioned unless there is a need to expose an internal feature.

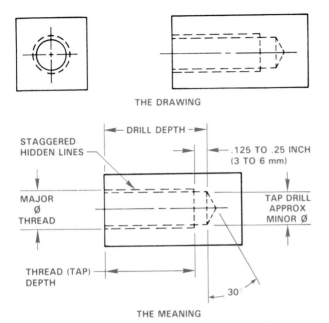

THE DRAWING

THE MEANING

Figure 8–20 The simplified internal thread that does not go through the part.

thread so the bolt does not hit bottom. Figure 8–21 shows a bolt fastener as it appears in assembly with two parts using simplified thread representation. The detailed thread drawings for American National Standard Taper pipe threads are the same as unified threads, except that the major and minor diameters taper. Figure 8–22 shows that pipe threads can be drawn tapered or straight depending on company preference. The thread note clearly defines the type of thread.

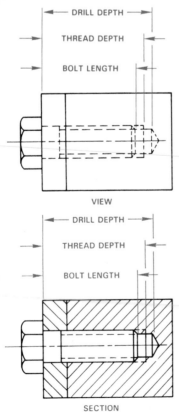

VIEW

SECTION

Figure 8–21 The simplified external thread in assembly.

## THREAD NOTES

Simplified, schematic, and detailed thread representations clearly show where threads are displayed on a drawing. However, the representations alone do not give the full information about the thread. As the term "representation" implies, the symbols are not meant to be exact, but they are meant to describe the location of a thread when used. The information that does clearly and completely identify the thread being used is the *thread note*. The thread note must be accurate; otherwise, the thread will be manufactured incorrectly.

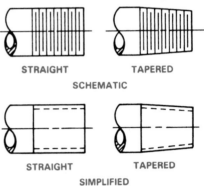

STRAIGHT          TAPERED
SCHEMATIC

STRAIGHT          TAPERED
SIMPLIFIED

Figure 8–22 Straight and tapered pipe-thread representations.

### Metric Threads

*ISO* The metric thread notes shown in Figure 8–23 are the recommended standard as specified by the ISO. You'll find descriptions of the note components in the following paragraphs.

A   M is the symbol for ISO metric threads.

B   The nominal major diameter in millimeters, followed by the symbol **X**, meaning *by*.

C   The thread pitch in millimeters, followed by a dash (–).

D   The number may be a 3, 4, 5, 6, 7, 8, or 9, which identifies the grade of tolerance from fine to coarse. The larger the number, the larger the tolerance. Grades 3 through 5 are fine, and 7 through 9 are coarse. Grade 3 is very fine and grade 9 is very coarse. Grade 6 is most commonly used and is the medium-tolerance metric thread. The grade 6 metric thread is comparable to the class 2 Unified Screw Thread. A letter placed after the number means the thread tolerance class of the internal or external thread. Internal threads are designated by uppercase letters, such as G or H, where G denotes a tight allowance and H identifies an internal thread with no allowance. The term "allowance" refers to the tightness of fit between the mating parts. External threads are defined with lowercase letters, such as e, g, or h. For external threads, *e* denotes a large allowance, *g* is a tight allowance, and *h* establishes no allowance. Grades and tolerances below 5 are intended for tight fits with mating parts, above 7 are a free class of fit intended for quick and easy assembly. When the grade and allowance are the same for both the major diameter and the pitch diameter of the metric thread, the designation is given as shown, 6H. In some situations where precise tolerances and allowances are critical between the major and pitch diameters, separate specifications could be used, for example, 4H 5H, or 4g 5g, where the first group (4g) refers to the grade and allowance of the pitch diameter, and the second group (5g) refers to the grade and tolerance of the major diameter. A fit between a pair of threads is indicated in the same thread note by specifying the internal thread followed by the external thread specification separated by a slash. For example, 6H/6g.

E   A blank space at (E) means a right-hand thread, a thread that engages when turned to the right. A right-hand thread is assumed unless an LH is lettered in this space. LH, which describes a left-hand thread, must be specified for a thread that engages when rotated to the left.

F   The depth of internal threads or the length of external threads in millimeters is provided at the end of the note. When the thread goes through the part, this space is left blank, although some companies prefer to letter the description, THRU.

Remember, the thread note is always given in the order shown in Figure 8–23.

M 10X 1.5–6H
(A) (B) (C) (D)(E)(F)

Figure 8–23 Metric thread note.

## Unified and American National Threads

The thread note is always shown in the order given in Figure 8–24. The components of the note are described below:

1/2- 13 UNC–2   A
(A) (B)   (C)  (D)(E)(F)(G)(H)

Figure 8–24 Unified and American National thread note.

A   The major diameter of the thread in inches followed by a dash (–).

B   Number of threads per inch.

C   Series of threads are classified by the number of threads per inch as applied to specific diameters and thread forms, such as coarse or fine threads. UNC (in the example) means Unified National Coarse. Others include UNF for Unified National Fine, UNEF for Unified National Extra Fine, or UNS for the Unified National Special. The UNEF and UNS thread designations are for special combinations of diameter, pitch, and length of engagement. American National screw threads are identified with UN for external and internal

threads, or UNR, a thread designed to improve fatigue strength of external threads only. The series designation is followed by a dash (–).

D   Class of fit is the amount of tolerance. 1 means a large tolerance, 2 is a general-purpose moderate tolerance, and 3 is for applications requiring a close tolerance.

E   A means an external thread (shown in the example) while B means an internal thread. (B replaces A in this location.) The A or B may be omitted if the thread is clearly external or internal, as shown on the drawing.

F   A blank space at F means a right-hand thread or a right-hand thread is assumed. LH in this space identifies a left-hand thread.

G   A blank space at G means a thread with a single lead, that is, a thread that engages one pitch when rotated 360°. If a double or triple lead is required, the word DOUBLE or TRIPLE must be lettered here.

H   This location is for internal thread depth or external thread length in inches. When the drawing clearly shows that the thread goes through, this space is left blank. If clarification is needed, the word THRU can be lettered here.

## Other Thread Forms

Other thread forms, such as Acme, are noted on a drawing using the same format. For example, 5/8-8 ACME-2G describes an Acme thread with a 5/8-inch major diameter, eight threads per inch and a general purpose (G) class 2 thread fit.

For a complete analysis of threads and thread forms, refer to the *Machinery's Handbook* published by Industrial Press, Inc. American National Standard Taper pipe threads are noted in the same manner with the Letters NPT (National Pipe Thread) used to designate the thread form. A typical note may read 3/4–14 NPT.

## Thread Notes on a Drawing

The thread note is usually applied to a drawing with a leader in the view where the thread appears as a circle for internal threads, as shown in Figure 8–25. External threads may be dimensioned with a leader, as shown in Figure 8–26, with the thread length given as a dimension or at the end of the note. An internal thread that does not go through the part may be dimensioned, as in Figure 8–27. Some companies may require the drafter to indicate the complete process required to machine a thread. This includes noting the tap drill size, tap drill depth if not through, the thread note, and thread not through (see Figure 8–28). A thread chamfer can also be specified in the note, shown in Figure 8–29.

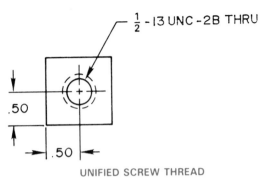

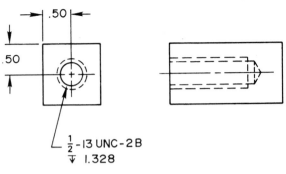

UNIFIED SCREW THREAD

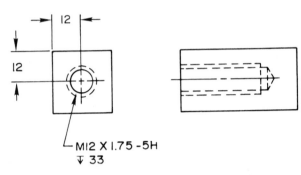

METRIC SCREW THREAD

**Figure 8–25** Drawing and noting internal screw threads (simplified representation).

**Figure 8–27** Drawing and noting internal screw threads with a given depth.

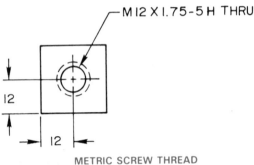

UNIFIED SCREW THREAD

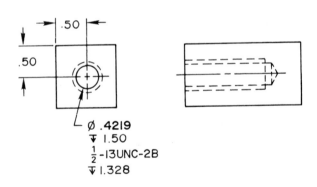

**Figure 8–28** Showing tap drill depth and thread depth.

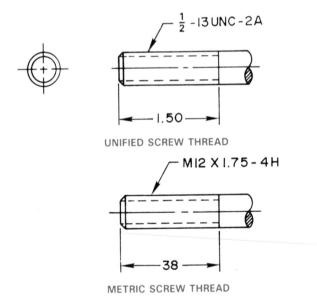

METRIC SCREW THREAD

**Figure 8–26** Drawing and noting external screw threads.

## MEASURING SCREW THREADS

When measuring features from prototypes or existing parts, the screw thread size can be determined on a fastener or threaded part by measurement. Measure the major diameter with a vernier caliper or micrometer. Determine the number of threads per inch when a rule or scale is the only available tool by counting the number of threads between inch graduations, as shown in Figure 8–30. The quickest and easiest way to determine the thread specification is with a screw pitch gage,

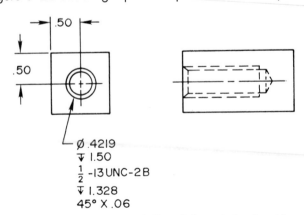

**Figure 8–29** Showing tap drill and thread depth with a chamfer.

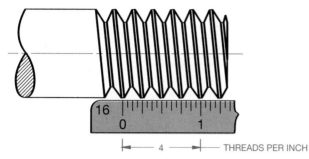

Figure 8–30 Determining the number of threads per inch by scale.

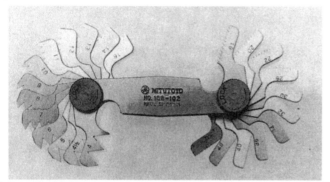

Figure 8–31 Screw pitch gage. Courtesy Mitutoyo/MI Corporation.

Figure 8–31, which is a set of thin leaves with teeth on the edge of each leaf that correspond to standard thread sections. Each leaf is stamped to show the number of threads per inch. Therefore, if the major diameter measures 5/8 inch and the number of teeth per inch is 18, by looking at a thread variation chart you find that you have a 5/8–18 UNF thread.

## THREADED FASTENERS

### Bolts and Nuts

A *bolt* is a threaded fastener with a head on one end, designed to hold two or more parts together with a nut or threaded feature. The nut is tightened upon the bolt or the bolt head can be tightened into a threaded feature. Bolts can be tightened or released by torque applied to the head or to the nut. Bolts are identified by a thread note, length, and head type. For example, 5/8–11 UNC-2 X 1-1/2 LONG HEXAGON HEAD BOLT.

Figure 8–32 shows various types of bolt heads. Figure 8–33 shows common types of nuts. Nuts are classified by thread specifications and type. Nuts are available with a flat base or a washer face.

### Machine Screws

*Machine screws* are popular screw-thread fasteners used for general assembly of machine parts. Machine screws are available in coarse (UNC) and fine (UNF) threads, in diameters ranging from .060 inch to .5 inch, and in lengths from 1/8 to 3 inch.es Machine

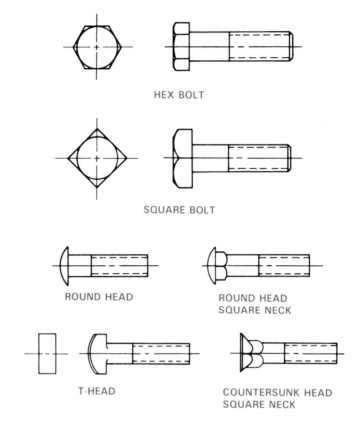

Figure 8–32 Bolt head types.

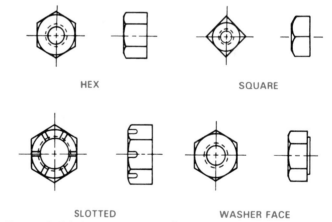

Figure 8–33 Common types of nuts.

screws are specified by thread, length, and head type. There are several types of heads available for use on prints (see Figure 8–34).

### Cap Screws

*Cap screws* are fine-finished machine screws, which are generally used without a nut. Mating parts are fastened where one feature is threaded. Cap screws have a variety of head types and range in diameter from .060 inch to 4 inches, with a large range of lengths. Lengths vary with diameter. For example, lengths increase in 1/16-inch increments for diameters up to 1 inch. For diameters larger than 1 inch, lengths increase in increments of 1/8 or 1/4 inch; the other extreme is a 2-inch increment for lengths over

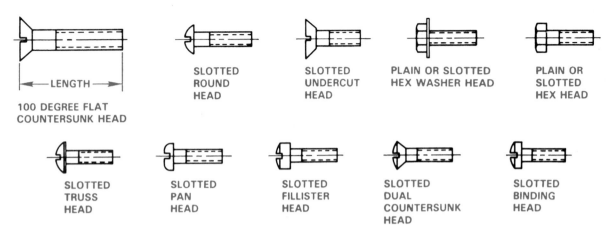

Figure 8–34 Types of machine screw heads.

10 inches. Cap screws have a chamfer to the depth of the first thread. Standard cap screw head types are shown in Figure 8–35.

## Set Screws

*Set screws* are used to help prevent rotary motion and to transmit power between two parts, such as a pulley and shaft. Purchased with or without a head, set screws are ordered by specifying thread, length, head or headless, and type of point. Headless set screws are available in slotted sockets and with hex or spline sockets. The shape of a set screw head is usually square. Standard square-head set screws have cup points, although other points are available. Figure 8–36 shows optional set types of screws.

## Lag Screws and Wood Screws

*Lag screws* are designed to attach metal to wood or wood to wood. Before assembly with a lag screw, a pilot hole is cut into the wood.

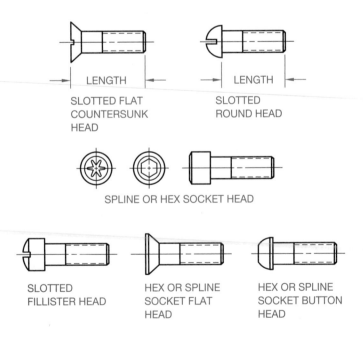

Figure 8–35 Cap screw head styles.

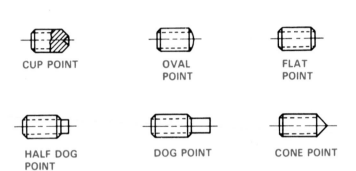

Figure 8–36 Set screw point styles.

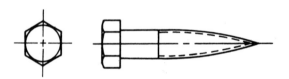

Figure 8–37 Hex head lag screw.

The threads of the lag screw then form their own mating thread in the wood. Lag screws are sized by diameter and length. A lag screw is shown in Figure 8–37. Wood screws are similar in function to lag screws and are available in a wide variety of sizes, head styles, and materials.

## Self-tapping Screws

*Self-tapping screws* are designed for use in situations where the mating thread is created by the fastener. These screws are used to hold two or more mating parts when one of the parts becomes a fastening device. A clearance fit is required through the first series of features or parts, while the last feature receives a pilot hole similar to a tap drill for unified threads. The self-tapping screw then forms its own threads by cutting or displacing material as it enters the pilot hole. There are several different types of self-tapping screws with head variations similar to cap screws. The specific function of the screw is important, as these screws may be designed for applications ranging from sheet metal to hard-metal fastening.

## Thread Inserts

*Screw thread inserts* are helically formed coils of diamond-shaped wire made of stainless steel or phosphor bronze. The inserts are used by being screwed into a threaded hole to form a mating internal thread for a threaded fastener. Inserts are used to repair worn or damaged internal threads and to provide a strong thread surface in soft materials. Some screw thread inserts are designed to provide a secure mating of fasteners in situations where vibration or movement could cause parts to loosen. Figure 8–38 shows the relationship between the fastener, thread insert, and tapped hole.

## Nuts

A nut is used as a fastening device in combination with a bolt to hold two or more pieces of material together. The nut thread must match the bolt thread for acceptable mating. Figure 8–39 shows the nut–and–bolt relationship. The hole in the parts must be drilled larger than the bolt for clearance.

There are a variety of nuts in hexagon or square shapes. Nuts are also designed slotted to allow them to be secured with a pin or key. Acorn nuts are capped for appearance. Self-locking nuts are available with neoprene gaskets that help keep the nut tight when movement or vibration is a factor. Figure 8–40 shows some common nuts.

## WASHERS

*Washers* are flat, disk-shaped objects with a center hole to allow a fastener to pass through. Washers are made of metal, plastic, or other materials for use under a nut or bolt head, or at other machinery wear points, to serve as a cushion, or a bearing surface to prevent leakage or to relieve friction. There are several different types of washers, any of which can serve as a cushion, bearing sur-

face, or locking device (see Figure 8–41). Washer thickness varies from .016 inch to .633 inch.

## SELF-CLINCHING FASTENERS

A *self-clinching fastener* is any device, usually threaded, that displaces the material around a mounting hole when pressed into a properly sized drilled or punched hole. This pressing or squeezing process causes the displaced sheet material to cold flow into a specially designed annular recess in the shank or pilot of the fastener. A serrated clinching ring, knurl, ribs, or hex head prevents the fastener from rotating in the metal when tightening torque is applied to the mating screw or nut. When properly installed, self-clinching fasteners become a permanent and integral part of the panel, chassis bracket, printed circuit board, or other item in which they are installed. They meet high-performance standards and enable easier disassembly of components for repair or service.

Self-clinching fasteners generally take less space and require fewer assembly operations than caged and other types of locking nuts. They also have greater reusability and more holding power than sheet metal screws. They are used mainly where good pull-out and torque loads are required in sheet metal that is too thin to provide secure fastening by any other methods.

Self-clinching fasteners traditionally fall into the categories of nuts, spacers and standoffs, and studs.

## Self-clinching Nuts

These types of nuts feature thread strengths greater than those of mild screws and are commonly used wherever strong internal threads are needed for component attachment or fabrication assembly. During installation, a clinching ring locks the displaced

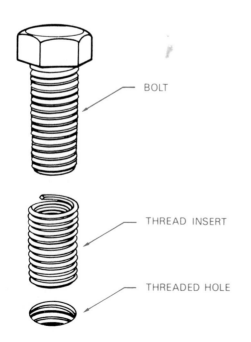

Figure 8–38  Thread insert.

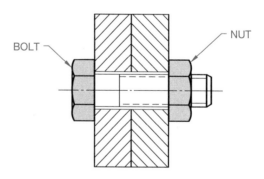

Figure 8–39 Nut-and-bolt relationship known as a floating fastener.

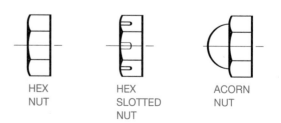

Figure 8–40 Common nuts.

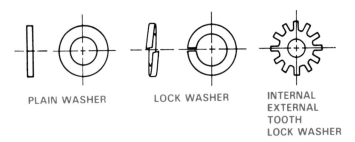

**Figure 8–41** Types of washers.

metal behind the fastener's tapered shank, resulting in a high push-out resistance. High torque-out resistance is achieved when the knurled platform is embedded in the sheet metal. The clinching action of self-clinching nuts occurs on the fastener side of the thin sheet, with the reverse side remaining flush. A self-clinching nut is shown in Figure 8–42(a).

## Self-clinching Spacers and Standoffs

Self-clinching spacers and standoffs are used where it is necessary to space or stack components away from a panel. Thru-threaded or blind types are generally standard, but variations have been developed to meet emerging applications, primarily in the electronics industry. These types include standoffs with concealed heads, others that allow boards to snap into place for easier assembly and removal, and those designed specifically for use in printed circuit boards. Figure 8–42(b) shows an example of a self-clinching standoff.

## Self-clinching Studs

Self-clinching studs are externally threaded self-clinching fasteners that are used where the attachment must be positioned before being fastened. Flush-head studs are normally specified, but variations are available for desired high torque, thin sheet metal, or electrical applications. Figure 8–42(c) shows a self-clinching stud.

## DOWEL PINS

*Dowel pins* used in machine fabrication are metal cylindrical fasteners which retain parts in a fixed position or keep parts aligned. Generally, depending on the function of the parts, one or two dowel pins are sufficient for holding. Dowel pins must generally be pressed into a hole with an interference tolerance of between .0002 and .001 inch depending on the material and the function of the parts. Figure 8–43 shows the section of two parts and a dowel pin.

## TAPER AND OTHER PINS

For applications that require perfect alignment of accurately constructed parts, *tapered dowel pins* may be better than straight dowel pins. Taper pins are also used for parts that frequently have to be taken apart or where removal of straight dowel pins may cause excess hole wear. Figure 8–44 shows an example of a taper pin

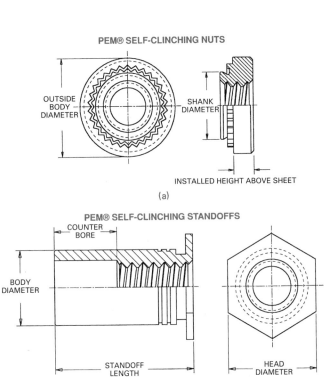

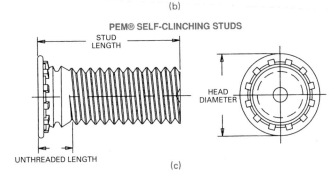

**Figure 8–42** Self-clinching fasteners: (a) a self-clinching nut; (b) a self-clinching standoff; (c) a self-clinching stud. *Courtesy of Penn Engineering & Manufacturing Corp. and Hammer Inc. Advertising & Public Relations.*

assembly. Taper pins, shown in Figure 8–45, range in diameter, D, from 7/0, which is .0625–.875 inch, and lengths, L, vary from .375 inch to 8 inches.

Other types of pins serve functions similar to taper pins, such as holding parts together, aligning parts, locking parts, and transmitting power from one feature to another. Common pins are shown in Figure 8–46.

## RETAINING RINGS

*Internal* and *external retaining rings* are available as fasteners to provide a stop or shoulder for holding bearings or other parts on a shaft. They are also used internally to hold a cylindrical feature in a housing. Common retaining rings require a groove in the shaft or housing for mounting with a special plier tool. Also available are self-locking retaining rings for certain applications (see Figure 8–47).

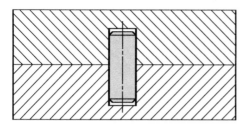

Figure 8–43 Dowel pin in place, sectional view.

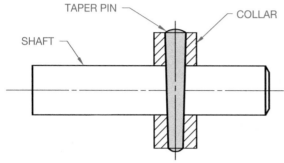

Figure 8–44 Taper pin assembly, sectional view.

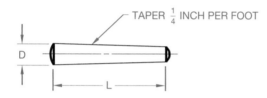

Figure 8–45 Taper pin.

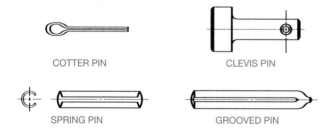

Figure 8–46 Other common pins.

## KEYS, KEYWAYS, AND KEYSEATS

Standards for keys were established to control the relationship between key sizes, shaft sizes, and tolerances for key applications. A key is an important machine element which is used to provide a positive connection for transmitting torque between a shaft and hub, pulley, or wheels. The key is placed in position in a keyseat, which is a groove or channel cut in a shaft. The shaft and key are then inserted into the hub, wheel, or pulley, where the key mates with a groove called a keyway. Figure 8–48 shows the relationship between the key, keyseat, shaft, and hub.

Standard key sizes are determined by shaft diameter. Keyseat depth dimensions are established in relationship to the shaft diameter. Figure 8–49 shows the standard dimensions for the related features. Types of keys are shown in Figure 8–50.

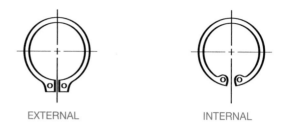

EXTERNAL    INTERNAL

Figure 8–47 Retaining rings.

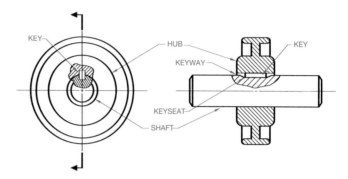

Figure 8–48 Relationship between the key, keyseat, keyway, shaft, and hub.

## RIVETS

A rivet is a metal pin with a head used to fasten two or more materials together. The rivet is placed through holes in mating parts and the end without a head extends through the parts to be headed-over (formed into a head) by hammering, pressing, or forging. The end with the head is held in place with a solid steel bar known as a dolly, while a head is formed on the other end. Rivets are classified by body diameter, length, and head type (see Figure 8–51).

### Springs

A *spring* is a mechanical device, often in the form of a helical coil, that yields by expansion or contraction because of pressure, force, or stress applied. Springs are made to return to normal form when the force or stress is removed. Springs are designed to store energy for the purpose of pushing or pulling machine parts by reflex action into certain desired positions. Improved spring technology provides springs with the ability to function a long time under high stresses. The effective use of springs in machine design includes five basic dependencies: material, application, functional stresses, use, and tolerances.

Continued research and development of spring materials have helped to improve spring technology. The spring materials most commonly used include high-carbon spring steels, alloy spring steels, stainless spring steels, music wire, oil-tempered steel, copper-based alloys, and nickel-based alloys. Spring materials, depending on use, may have to withstand high operating temperatures, and high stresses under repeated loading.

Spring design criteria are generally based on material gage, kind of material, spring index, direction of the helix, type of ends, and

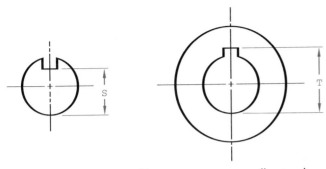

Figure 8-49 Keyseats and keyways are generally sized as related to shaft size.

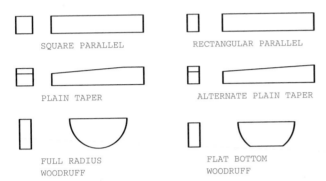

SQUARE PARALLEL  RECTANGULAR PARALLEL

PLAIN TAPER  ALTERNATE PLAIN TAPER

FULL RADIUS WOODRUFF  FLAT BOTTOM WOODRUFF

Figure 8-50 Types of keys.

function. Spring wire gages are available from several different sources ranging in diameter from number 7/0 (.490 inch) to number 80 (.013 inch). The most commonly used spring gages range from 4/0 to 40. There are a variety of spring materials available in round or square stock for use depending on spring function and design stresses. The spring index is a ratio of the average coil diam-

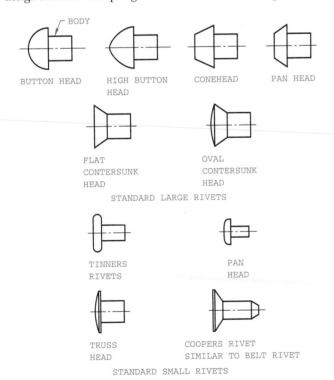

BUTTON HEAD  HIGH BUTTON HEAD  CONEHEAD  PAN HEAD

FLAT CONTERSUNK HEAD  OVAL CONTERSUNK HEAD

STANDARD LARGE RIVETS

TINNERS RIVETS  PAN HEAD

TRUSS HEAD  COOPERS RIVET SIMILAR TO BELT RIVET

STANDARD SMALL RIVETS

Figure 8-51 Common rivets.

eter to the wire diameter. The index is a factor in determining spring stress, deflection, and the evaluation of the number of coils needed and the spring diameter. Recommended index ratios range between 7 and 9, although other ratios commonly used range from 4 to 16. The direction of the helix is a design factor when springs must operate in conjunction with threads or with one spring inside of another. In such situations the helix of one feature should be in the opposite direction of the helix for the other feature. Compression springs are available with ground or unground ends. Unground, or rough, ends are less expensive than ground ends. If the spring is required to rest flat on its end, ground ends should be used. Spring function depends on two basic factors, compression or extension. Compression springs release their energy and return to the normal form when compressed. Extension springs release their energy and return to the normal form when extended (see Figure 8–52).

COMPRESSION SPRING  EXTENSION SPRING

Figure 8-52 Compression and extension springs.

## Spring Terminology

The springs shown in Figure 8–53 show some common characteristics.

*Ends.* Compression springs have four general types of ends: open or closed ground ends or open or closed unground ends, shown in Figure 8–54. Extension springs have a large variety of optional ends, a few of which are shown in Figure 8–55.

*Helix Direction.* The helix direction may be specified as right hand or left hand. (See Figure 8–54.)

*Free Length.* The length of the spring when there is no pressure or stress to affect compression or extension is known as free length.

*Compression Length.* The compression length is the maximum recommended design length for the spring when compressed.

*Solid Height.* The solid height is the maximum compression possible. The design function of the spring should not allow the spring to reach solid height when in operation unless this factor is a function of the machinery.

*Loading Extension.* The extended distance to which an extension spring is designed to operate is the loading extension length.

*Pitch.* The pitch is one complete helical revolution, or the distance from a point on one coil to the same corresponding point on the next coil.

## Torsion Springs

Torsion springs are designed to transmit energy by a turning or twisting action. Torsion is defined as a twisting action which tends to turn one part or end around a longitudinal axis, while the other

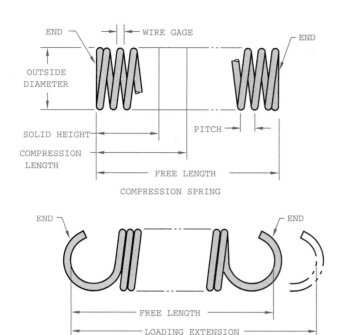

Figure 8–53 Spring characteristics.

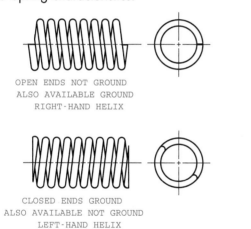

Figure 8–54 Helix direction and compression spring end types.

part or end remains fixed. Torsion springs are often designed as antibacklash devices or as self-closing or self-reversing units (see Figure 8–56).

## Flat Springs

Flat springs are arched or bent flat-metal shapes designed so that when placed in machinery they cause tension on adjacent parts. The tension may be used to level parts, provide a cushion, or position the relative movement of one part to another. One of the most common examples of flat springs are leaf springs on an automobile.

## Spring Representations

There are three types of spring representations: detailed, schematic, and simplified, as seen in Figure 8–57. Detailed spring drawings are used in situations that requir0e a realistic representation, such as vendors' catalogs, assembly instructions, or detailed assemblies.

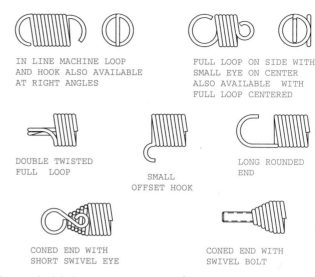

Figure 8–55 Extension spring end types.

**Material: 302 Stainless Steel (Spring Tamper) Passivated**

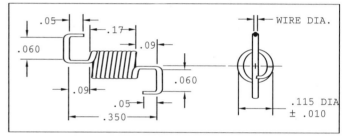

Figure 8–56 Torsion spring, also called antibacklash spring. *Courtesy NORDEX, Inc.*

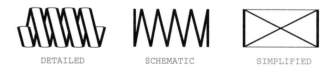

Figure 8–57 Spring representations.

Schematic spring representations are commonly used on prints. The single-line schematic symbols clearly represent springs without taking the additional time required to draw a detailed spring. The use of simplified spring drawings is limited to situations where the clear resemblance of a spring is not necessary. The simplified spring symbol is accompanied by clearly written spring specifications.

## Spring Specifications

No matter which representation is used, several important specifications accompany the spring symbol. Spring information is generally given in the form of a specific or general note.

Spring specifications include outside or inside diameter, wire gage, kind of material, type of ends, surface finish, free and compressed length, and number of coils. Other information, when required, may include spring design criteria and heat treatment specifications. The information is often provided on a drawing as shown in Figure 8–58. The material note is usually found in the title block.

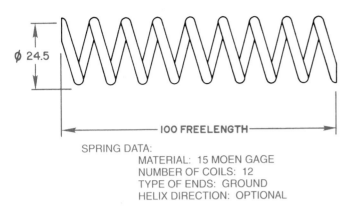

SPRING DATA:
          MATERIAL:  15 MOEN GAGE
          NUMBER OF COILS:  12
          TYPE OF ENDS:  GROUND
          HELIX DIRECTION:  OPTIONAL

Figure 8–58 Detailed spring drawing with spring data chart.

# CHAPTER 8 TEST

Multiple choice: Respond to the following by selecting a, b, c, or d to best answer the question or complete the statement.

1. These are a helix or conical spiral form on the external surface of a shaft or internal surface of a cylindrical hole:
   a. screw thread
   b. tap
   c. die
   d. bolt and nut

2. This is a machine tool used for cutting external threads:
   a. die
   b. tap
   c. tap drill
   d. thread tool

3. This is a machine tool used to form internal threads:
   a. die
   b. tap
   c. tap drill
   d. thread tool

4. This is the screw thread dimension from root to root through the axis:
   a. pitch
   b. thread form
   c. minor diameter
   d. major diameter

5. This is the screw thread dimension from crest to crest through the axis:
   a. pitch
   b. thread form
   c. minor diameter
   d. major diameter

6. This is used to make a hole in the material before tapping:
   a. die
   b. tap
   c. tap drill
   d. thread cutter

7. This is the distance measured parallel to the axis from a point on one thread to the same corresponding point on the next thread:
   a. pitch
   b. thread form
   c. minor diameter
   d. major diameter

8. This is a thread that engages with a mating thread by rotating clockwise:
   a. right-hand thread
   b. left-hand thread
   c. clockwise thread
   d. counterclockwise thread

9. This is a thread that engages with a mating thread by rotating counterclockwise:
   a. right-hand thread
   b. left-hand thread
   c. clockwise thread
   d. counter clockwise thread

10. This is an angular relief at the last thread to help allow the thread to more easily engage with a mating part:
   a. lead
   c. crest
   b. chamfer
   d. thread

11. This is the lateral distance a thread travels during one complete revolution:
   a. lead
   b. chamfer
   c. crest
   d. thread

12. The underlined part of this thread note, M10X1.5–6H, means:
   a. multithread
   b. metric thread
   c. major diameter
   d. minor diameter

13. The underlined part of this thread note, M10X1.5–6H, means:
   a. threads per inch
   b. threads per meter
   c. major diameter
   d. thread pitch

14. The underlined part of this thread note, M10X1.5–6H, means:
   a. threads per inch
   b. threads per meter
   c. tolerance grade
   d. thread pitch

15. The underlined part of this thread note, M10X1.5–6H, means:
   a. threads per inch
   b. threads per meter
   c. tolerance grade
   d. thread pitch

16. The underlined part of this thread note, <u>1/2</u>–13UNC–2A, means:
    a. threads per inch
    b. major diameter
    c. thread series
    d. class of fit

17. The underlined part of this thread note, 1/2–13<u>UNC</u>–2A, means:
    a. threads per inch
    b. major diameter
    c. thread series
    d. class of fit

18. The underlined part of this thread note, 1/2–<u>13</u>UNC–2A, means:
    a. threads per inch
    b. major diameter
    c. thread series
    d. class of fit

19. The underlined part of this thread note, 1/2–13UNC–<u>2</u>A, means:
    a. threads per inch
    b. major diameter
    c. thread series
    d. class of fit

20. The underlined part of this thread note, 1/2–13UNC–2<u>A</u>, means:
    a. internal thread
    b. external thread
    c. thread series
    d. class of fit

21. The underlined part of this thread note, 1/2–13UNC–2B <u>LH</u>, means:
    a. light helix
    b. limited hold
    c. long helix
    d. left hand

22. The underlined part of this thread note, 3/4–13 <u>NPT</u> means National Pipe Thread:
    a. true
    b. false

23. This is a screw-threaded fastener used for general assembly of machine parts:
    a. cap screws
    b. machine screws
    c. set screws
    d. self-tapping screws

24. This screw is designed for use in situations where the mating thread is created by the fastener:
    a. cap screws
    b. machine screws
    c. set screws
    d. self-tapping screws

25. These screws are used to help prevent rotary motion and to transmit power between two parts:
    a. cap screws
    b. machine screws
    c. set screws
    d. self-tapping screws

26. These are fine-finished machine screws which are generally used without a nut:
    a. cap screws
    b. machine screws
    c. set screws
    d. self-tapping screws

27. These are helically formed coils of diamond-shaped wire which are screwed into a threaded hole to form a mating internal thread for a threaded fastener:
    a. cap screws
    b. machine screws
    c. threaded insert
    d. self-tapping screws

28. This is used as a fastening device in combination with a bolt to hold two or more pieces of material together:
    a. threaded insert
    b. nut
    c. washer
    d. dowel pin

29. This is used to retain parts in a fixed position to keep parts aligned:
    a. threaded insert
    b. nut
    c. washer
    d. dowel pin

30. These are used under a nut or bolt head, or at other machinery wear points, to serve as a cushion or a bearing surface:
    a. threaded insert
    b. nut
    c. washer
    d. dowel pin

31. These are used for perfect alignment of accurately constructed parts, and for parts that have to be taken apart frequently:
    a. taper pin
    b. rivet
    c. key
    d. dowel pin

32. This is a machine element that is used to provide a positive connection for transmitting torque between a shaft and a hub:
    a. taper pin
    b. retaining ring
    c. key
    d. dowel pin

33. This is a metal pin with a head used to fasten two or more materials together:
    a. taper pin
    b. rivet
    c. key
    d. dowel pin

34. This is a mechanical device, often in the form of a helical coil, that yields by expansion or contraction due to pressure, force, or stress applied:
    a. screw thread
    b. retaining ring
    c. key
    d. spring

35. This is the condition of the spring when there is no pressure or stress to affect compression or extension:
    a. compression length
    b. free length
    c. solid height
    d. loading extension

36. The extended distance to which an extension spring is designed to operate is the:
    a. compression length
    b. free length
    c. solid height
    d. loading extension

37. This is the maximum recommended design lengths when a spring is compressed:
    a. compression length
    b. free length
    c. solid height
    d. loading extension

38. These springs are designed to transmit energy by a turning or twisting action:
    a. compression spring
    b. extension spring
    c. torsion spring
    d. flat spring

39. These springs are arched or bent shapes designed so that when placed in machinery they cause tension on adjacent parts:
    a. compression spring
    b. extension spring
    c. torsion spring
    d. flat spring

40. This is one complete helical revolution, or the distance from one point on one coil to the same corresponding point on the next coil:
    a. compression length
    b. free length
    c. solid height
    d. spring pitch

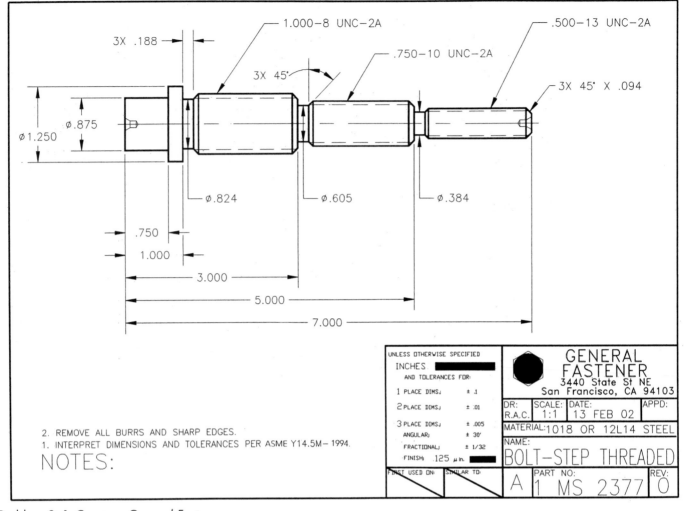

Problem 8–1 *Courtesy General Fastener.*

## CHAPTER 8 PROBLEMS

**PROBLEM 8—1** Answer the following questions as you refer to the print shown on this page.

1. How is it possible to represent this part with only one view?

2. What type of thread representation is used on this print?

3. List the thread specifications given on this part.

4. Give the length, including the chamfers, of the largest thread.

5. Give the length, including the chamfers, of the middle thread.

6. Give the length, including the chamfers, of the smallest thread.

7. Explain the meaning of the note: 3 **X** 45° **X** .094.

8. What determines the width of the chamfer labeled 3 **X** 45°?

9. What does the underlined part of this thread mean: 750–10UNC–2A?

10. What does the underlined part of this thread mean: 750–10UNC–2A?

11. What does the underlined part of this thread mean: 750–10UNC–2A?

12. What does the underlined part of this thread mean: 750–10UNC–2A?

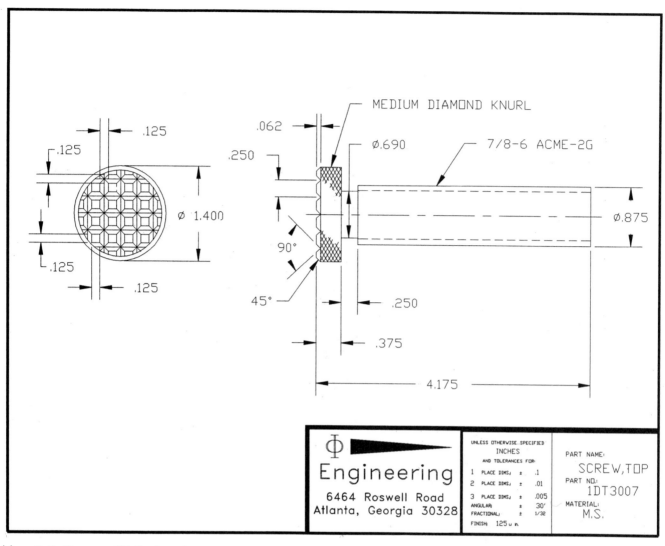

Problem 8–2 *Courtesy IO Engineering.*

**PROBLEM 8–2** Answer the following questions as you refer to the print shown on this page.

1. Describe the views shown on this print.
   _____
   _____

2. Give the thread specification.
   _____
   _____

3. What is the length of the thread?
   _____
   _____

4. Give the knurl specification.
   _____
   _____

5. What does the 7/8 on the thread specification mean?
   _____
   _____

6. What does the 6 in the thread specification mean?
   _____

7. Give the width and depth of the 90° angle feature.
   _____
   _____

8. What is the diameter and length of the neck?
   _____
   _____

9. Give the diameter of the head.
   _____
   _____

10. Give the full name of the material used to make this part.
    _____
    _____

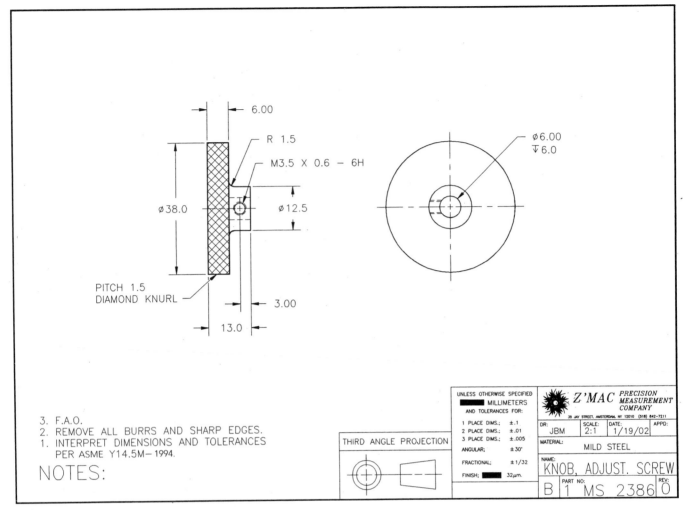

NOTES:
1. INTERPRET DIMENSIONS AND TOLERANCES
   PER ASME Y14.5M- 1994.
2. REMOVE ALL BURRS AND SHARP EDGES.
3. F.A.O.

| UNLESS OTHERWISE SPECIFIED ■ MILLIMETERS AND TOLERANCES FOR: | | | | |
|---|---|---|---|---|
| 1 PLACE DIMS.; ±.1 | | | | |
| 2 PLACE DIMS.; ±.01 | DR: JBM | SCALE: 2:1 | DATE: 1/19/02 | APPD: |
| 3 PLACE DIMS.; ±.005 | MATERIAL: MILD STEEL | | | |
| ANGULAR; ±30' | | | | |
| FRACTIONAL; ±1/32 | NAME: KNOB, ADJUST. SCREW | | | |
| FINISH; ■ 32μm. | B | 1 | PART NO: MS 2386 | REV: 0 |

Z'MAC PRECISION MEASUREMENT COMPANY
35 JAY STREET, AMSTERDAM, NY 12010 (518) 842-7211

THIRD ANGLE PROJECTION

Problem 8–3   *Courtesy Jim B. MacDonald.*

**PROBLEM 8–3** Answer the following questions as you refer to the print shown on this page.

1. Is the print in inches or millimeters?

2. Give the specifications for the end hole.

3. What is the diameter of the knurled feature?

4. Give the location dimension to the thread.

5. What does the underlined part of this thread mean: M3.5 X 0.6–6H?

6. What does the underlined part of this thread mean: M3.5 X 0.6–6H?

7. What does the underlined part of this thread mean: M3.5 X 0.6–6H?

8. What does the underlined part of this thread mean: M3.5 X 0.6–6H?

9. What does the underlined part of this thread mean: M3.5 X 0.6–6H?

10. Give the pitch of the thread shown on this print.

11. Define *thread pitch*.

12. Give the knurl specification.

13. What is the finish for the part shown on this print?

14. Give the tolerance of the Ø38.0 dimension.

15. Give the tolerance of the 6.00 dimension.

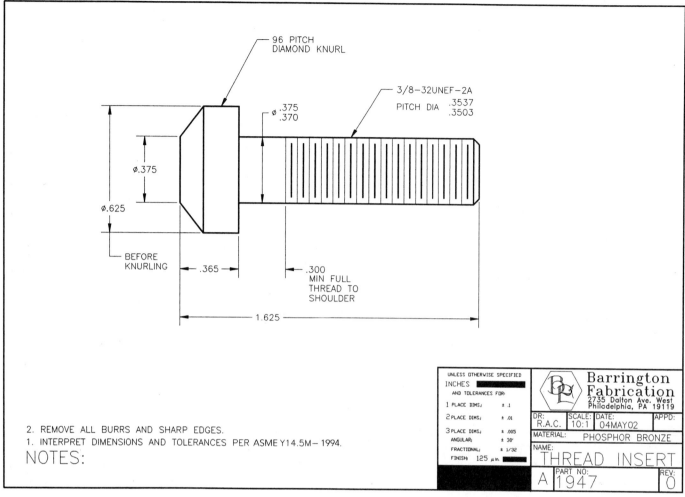

Problem 8–4 *Courtesy Richard Clouser.*

**PROBLEM 8–4** Answer the following questions as you refer to the print shown on this page.

1. What type of thread representation is used on this part?

2. Give the complete thread specification.

3. What is the diameter of the knurled features before knurling?

4. What information are you given about the length of thread?

5. Give the shoulder diameter?

6. What is the tolerance of the pitch diameter?

7. Give the tolerance of the shoulder.

8. What is the tolerance of the 1.625 dimension?

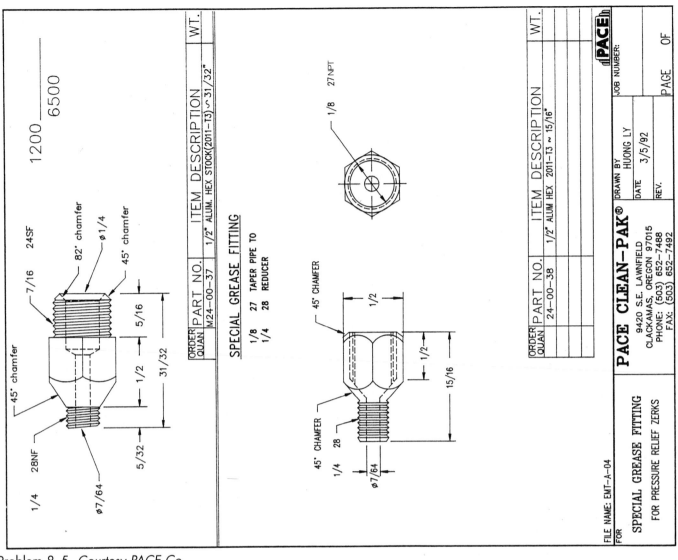

Problem 8–5  *Courtesy PACE Co.*

**PROBLEM 8–5** Answer the following questions as you refer to the print shown on this page.

1. What type of thread representations are used on this print?

   _____

   _____

2. Give the part numbers and item descriptions of the parts displayed on the print.

   _____

   _____

   _____

3. List the different thread specifications found on this print.

   _____

   _____

   _____

   _____

   _____

   _____

   _____

   _____

   _____

   _____

   _____

4. Give the name of the parts.

   _____

   _____

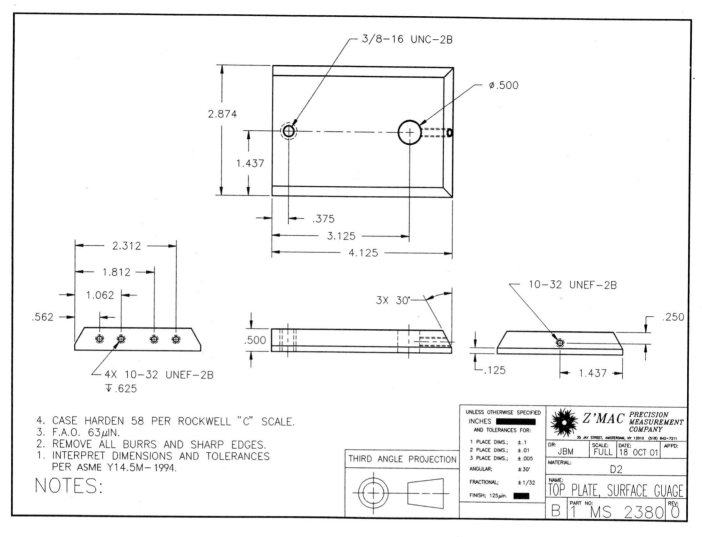

Problem 8–6 *Courtesy Jim B. MacDonald.*

**PROBLEM 8–6** Answer the following questions as you refer to the print shown on this page.

1. Is the print in inches or millimeters?

2. List each of the different thread specifications given on this print.

3. What type of dimensioning practice is used to dimension the location of the threads in the left-side view?

4. How many sides of the part are angled?

5. What is the angle, from vertical, on the sides of the part?

6. What does the underlined part of this thread mean: 4 **X** 10–32UNEF–2B?

7. What does the underlined part of this thread mean: 4 **X** <u>10</u>–32UNEF–2B?

8. What does the underlined part of this thread mean: 4 **X** 10–<u>32</u>UNEF–2B?

9. What does the underlined part of this thread mean: 4 **X** 10–32<u>UNEF</u>–2B?

10. What does the underlined part of this thread mean: 4 **X** 10–32UNEF–<u>2</u>B?

11. What does the underlined part of this thread mean: 4 **X** 10–32UNEF–2<u>B</u>?

12. Give the case hardening specification for this part.

# Welding Processes and Reading Welding Representations

## LEARNING OBJECTIVES

After completing this chapter you will be able to:

■ Answer questions about a variety of welding processes.

■ Identify the elements of welding symbols.

■ Identify types of welds on prints.

■ Read symbols and information related to destructive (DT) and nondestructive tests (NDT).

■ Read welding specifications.

Welding is a process of joining two or more pieces of like metals by heating the material to a temperature high enough to cause softening or melting. The location of the weld is where the materials actually combine the grain structure from one piece to the other. The parts that are welded become one and the properly welded joint is as strong or stronger than the original material. Welding can be performed with or without pressure applied to the materials. Some materials may actually be welded together by pressure alone. Most welding operations, however, are performed by filing a heated joint between pieces with molten metal.

Welding is actually a method of fastening parts. Welding was not discussed in the section about fasteners in Chapter 8 because the weld is a more permanent fastening application than screw threads or pins, for example. Welding is a common fastening method used in many manufacturing applications and industries from automobile to aircraft manufacturing and from computers to ship building. Some of the advantages of welding over other fastening methods include better strength, better weight distribution and reduction, a possible decrease in the size of castings or forgings needed in an assembly, and a potential savings of time and manufacturing costs.

## WELDING PROCESSES

A large number of welding processes are available for use in industry, as shown in Figure 9–1. The most common welding processes include oxygen gas welding, shielded metal arc welding, gas tungsten arc welding, and gas metal arc welding.

### Oxygen Gas Welding

*Oxygen gas welding*, commonly known as *oxyfuel* or *oxyacetylene welding*, can also be performed with such fuels as natural gas, propane, or propylene. Oxyfuel welding is most typically used to fabricate thin materials, such as sheet metal and thin-wall pipe or tubing. Oxyfuel processes are also used for repair work and metal cutting. One advantage of oxyfuel welding is that the equipment and operating costs are less than with other methods. But other welding methods have advanced over the oxyfuel process because they are faster, cleaner, and cause less material distortion. Common oxyfuel welding and cutting equipment is shown in Figure 9–2.

Also associated with oxyfuel applications are soldering, brazing, and braze welding. These methods are more of a bonding process than welding, as the base material remains solid, while a filler metal is melted into a joint. *Soldering* and brazing differ in application temperature. Soldering is done below 450°C and brazing above 450°C. Like alloys may be used depending upon their melting temperatures. The filler generally associated with soldering is solder. Solder is an alloy of tin and lead. The filler metal associated with brazing is an alloy of copper and zinc. *Brazing* is a process of joining two very closely fitting metals by heating the pieces, causing the filler metal to be drawn into the joint by capillary action. *Braze welding* is more of a joint filling process that does not rely on capillary action. Another process that uses an oxyfuel mixture is *flame cutting*. This process uses a high-temperature gas flame to preheat the metal to a kindling temperature, at which time a stream of pure oxygen is injected to cause the cutting action.

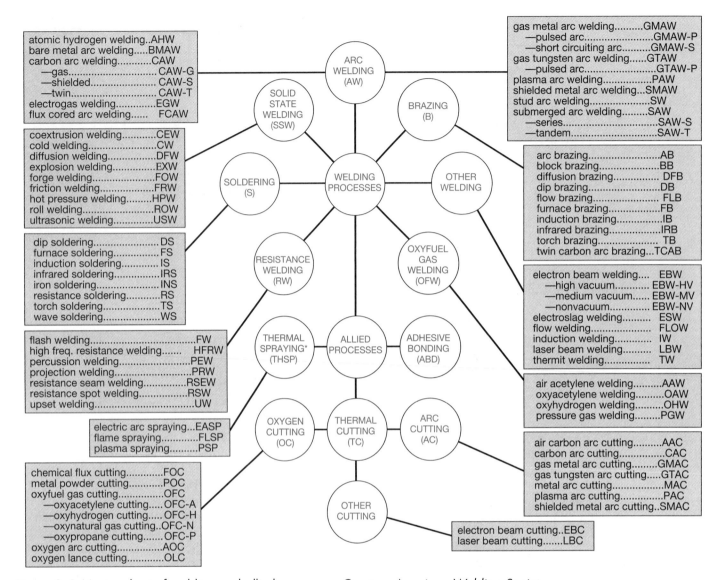

atomic hydrogen welding..AHW
bare metal arc welding.....BMAW
carbon arc welding............CAW
—gas.................CAW-G
—shielded..................CAW-S
—twin.....................CAW-T
electrogas welding.............EGW
flux cored arc welding......   FCAW

coextrusion welding............CEW
cold welding......................CW
diffusion welding.................DFW
explosion welding..............EXW
forge welding......................FOW
friction welding.................FRW
hot pressure welding........HPW
roll welding..........................ROW
ultrasonic welding...............USW

dip soldering.........................DS
furnace soldering...............FS
induction soldering............IS
infrared soldering...............IRS
iron soldering.....................INS
resistance soldering............RS
torch soldering....................TS
wave soldering...................WS

flash welding...........................FW
high freq. resistance welding.......   HFRW
percussion welding.........................PEW
projection welding..........................PRW
resistance seam welding...............RSEW
resistance spot welding..................RSW
upset welding.............................UW

electric arc spraying...EASP
flame spraying.............FLSP
plasma spraying..........PSP

chemical flux cutting.............FOC
metal powder cutting............POC
oxyfuel gas cutting...............OFC
—oxyacetylene cutting.....OFC-A
—oxyhydrogen cutting......OFC-H
—oxynatural gas cutting..OFC-N
—oxypropane cutting.......OFC-P
oxygen arc cutting................AOC
oxygen lance cutting.............OLC

gas metal arc welding..........GMAW
—pulsed arc....................GMAW-P
—short circuiting arc.........GMAW-S
gas tungsten arc welding....GTAW
—pulsed arc.....................GTAW-P
plasma arc welding..............PAW
shielded metal arc welding...SMAW
stud arc welding....................SW
submerged arc welding.........SAW
—series.........................SAW-S
—tandem........................SAW-T

arc brazing.......................AB
block brazing....................BB
diffusion brazing................DFB
dip brazing.......................DB
flow brazing...................... FLB
furnace brazing.................FB
induction brazing...............IB
infrared brazing................IRB
torch brazing.................... TB
twin carbon arc brazing...TCAB

electron beam welding.... EBW
—high vacuum........... EBW-HV
—medium vacuum...... EBW-MV
—nonvacuum............. EBW-NV
electroslag welding.......... ESW
flow welding.................... FLOW
induction welding............. IW
laser beam welding......... LBW
thermit welding................ TW

air acetylene welding..........AAW
oxyacetylene welding..........OAW
oxyhydrogen welding..........OHW
pressure gas welding.........PGW

air carbon arc cutting..........AAC
carbon arc cutting................CAC
gas metal arc cutting.........GMAC
gas tungsten arc cutting.....GTAC
metal arc cutting.................MAC
plasma arc cutting..............PAC
shielded metal arc cutting..SMAC

electron beam cutting..EBC
laser beam cutting.......LBC

Figure 9–1 Master chart of welding and allied processes. *Courtesy American Welding Society.*

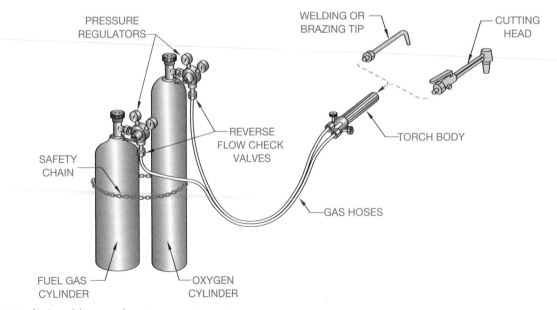

Figure 9–2 Oxyfuel welding and cutting equipment.

## Shielded Metal Arc Welding

*Shielded metal arc* or *stick electrode welding* is the most traditionally used welding methods. High-quality welds on a variety of metals and thicknesses can be made rapidly with excellent uniformity. This method uses a flux-covered, metal electrode to carry an electrical current forming an arc that melts the work and the electrode. The molten metal from the electrode mixes with the melting base material forming the weld. Shielded metal arc welding is popular because of low-cost equipment and supplies, flexibility, portability, and versatility. Figure 9–3 shows a shielded metal arc welding setup.

## Gas Tungsten Arc Welding

The *gas tungsten arc welding* process is sometimes referred to as TIG (*tungsten inert gas*) welding, or as *Heliarc®*, which is a trademark of the Union Carbide Corporation. Gas tungsten arc welding can be performed on a wider variety of materials than shield metal arc welding, and produces clean, high-quality welds. This welding process is useful for certain materials and applications. Gas tungsten arc welding is generally limited to thin materials, high-integrity joints, or small parts, because of its slow welding speed and high cost of equipment and materials (see Figure 9–4).

## Gas Metal Arc Welding

Another welding process that is extremely fast, is economical, and produces a very clean weld is *gas metal arc welding*. This process is used to weld thin material or heavy plate. It was used previously for welding aluminum using a metal inert gas shield, a process which was referred to as MIG. The present application employs a current-carrying wire that is fed into a joint between pieces to form the weld. This welding process is used in industry with automatic or robotic welding machines to produce rapidly made, high-quality welds in any welding position. While the expense of the equipment remains high, the cost is declining because of its popularity. Figure 9–5 shows a gas metal arc welding equipment.

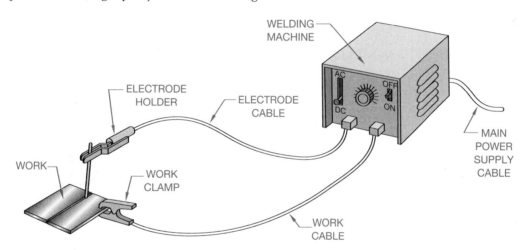

Figure 9–3 Shielded metal arc welding equipment.

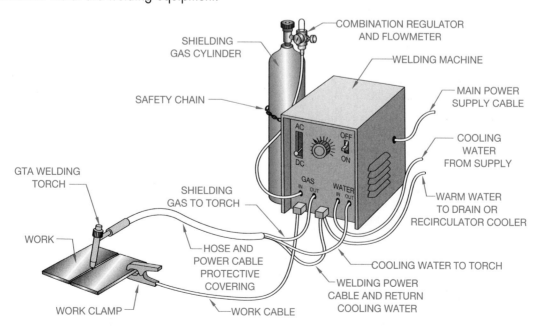

Figure 9–4 Gas tungsten arc welding equipment.

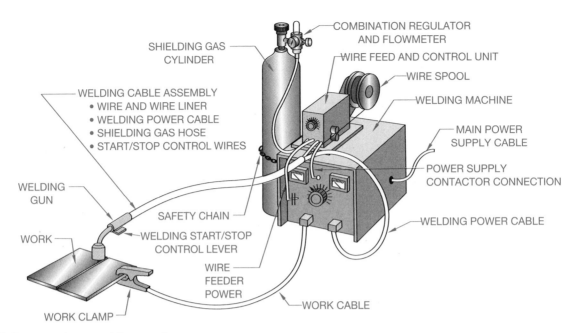

Figure 9–5 Gas metal arc welding equipment.

## ELEMENTS OF WELDING PRINTS

Welding prints are made up of several parts to be welded together. These prints are usually called *weldments*, or *welding assemblies* or *subassemblies*. The welding assembly typically shows the parts together in multiview with all the fabrication dimensions, types of joints, and weld symbols. Welding symbols identify the type of weld, the location of the weld, the welding process, the size and length of the weld, and other weld information. The welding assembly has a list of materials which generally provides a key to the assembly, the number of each part, part size, and material. Figure 9–6 shows a welding subassembly. When additional clarity of components parts must be given, then detailed drawings of each part are prepared, as shown in Figure 9–7.

### Welding Symbols

*AWS* The welding symbol represents complete information about the weld. The welding symbols discussed here are in accordance with the American Welding Society document AWS A2.4.

A welding symbol contains a few basic components, beginning with the reference line, tail, and leader, as shown in Figure 9–8. Figure 9–9 shows the use of multiple leaders and specific notes with the welding symbol.

After establishing the reference line, place additional information on the reference line to continue the weld specification. Figure 9–10 shows the standard location of welding symbol elements as related to the reference line, tail, and leader.

### Types of Welds

The next information that is applied to the reference line is the type of weld. The type of weld is associated with the weld shape and/or the type of groove to which the weld is applied. Figure 9–11 shows the information that is associated with the types of welds.

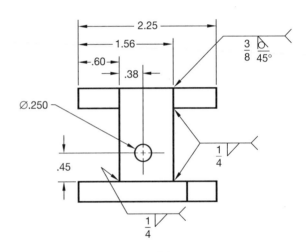

Figure 9–6 Welded subassembly.

*Fillet Weld.* A fillet weld is formed in the internal corner of the angle formed by two pieces of metal. The size of the fillet weld is shown on the same side of the reference line as the weld symbol and to the left of the symbol. When both legs of the fillet weld are the same, the size is given once, as in Figure 9–12. When the leg lengths are different in size, the vertical dimension is followed by the horizontal dimension (see Figure 9–13).

*Square Groove Weld.* A square groove weld is applied to a butt joint between two pieces of metal. The two pieces of metal are spaced apart a given distance, known as the root opening. If the root opening distance is a standard in the company, this dimension is assumed. If the root opening is not standard, the specified dimension is given to the left of the square groove symbol, as shown in Figure 9–14.

*V Groove Weld.* A V groove weld is formed between two adjacent parts when the side of each part is beveled to form a groove between the parts in the shape of a V. The included angle of the V may be given with or without a root opening, as shown in Figure 9–15.

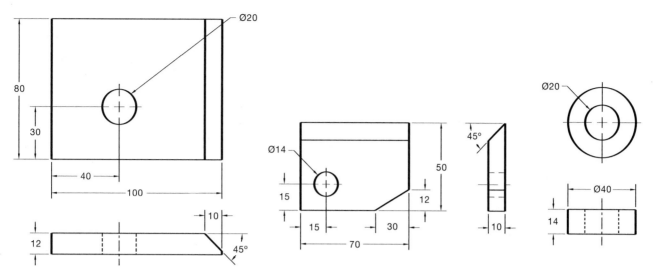

Figure 9–7 Drawings of each part of the welded subassembly.

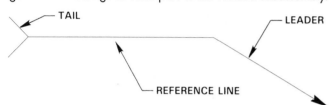

Figure 9–8 Welding symbols reference line, tail, and leader.

*Bevel Groove Weld.* The bevel groove weld is created when one piece is square and the other piece has a beveled surface. The bevel weld may be given with a bevel angle and a root opening, as shown in Figure 9–16.

*U Groove Weld.* A U groove weld is created when the groove between two parts is in the form of a U. The angle formed by the sides of the U shape, the root, and the weld size are generally given (see Figure 9–17).

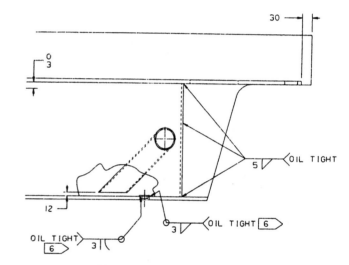

Figure 9–9 Welding symbol leader use. *Courtesy Hyster Company.*

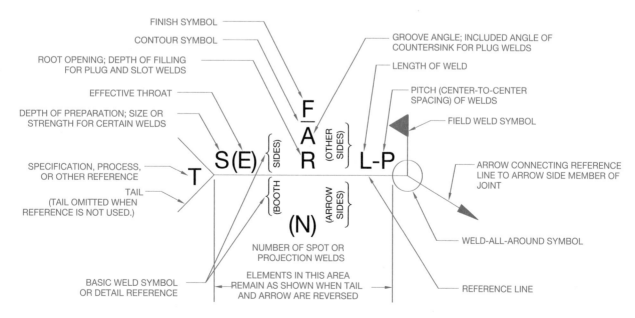

Figure 9–10 Standard location of elements of a welding symbol. *Courtesy American Welding Society.*

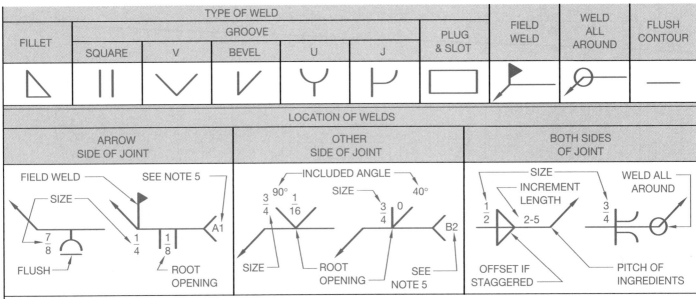

| | TYPE OF WELD | | | | | | | | |
|---|---|---|---|---|---|---|---|---|---|
| FILLET | GROOVE | | | | | PLUG & SLOT | FIELD WELD | WELD ALL AROUND | FLUSH CONTOUR |
| | SQUARE | V | BEVEL | U | J | | | | |

| LOCATION OF WELDS | | |
|---|---|---|
| ARROW SIDE OF JOINT | OTHER SIDE OF JOINT | BOTH SIDES OF JOINT |

1. THE SIDE OF THE JOINT TO WHICH THE ARROW POINTS IS THE ARROW SIDE.
2. BOTH-SIDES WELDS OF SAME TYPE ARE OF SAME SIZE UNLESS OTHERWISE SHOWN.
3. SYMBOLS APPLY BETWEEN ABRUPT CHANGES IN DIRECTION OF JOINT OR AS DIMENSIONED (EXCEPT WHERE ALL AROUND SYMBOL IS USED).
4. ALL WELDS ARE CONTINUOUS AND OF USER'S STANDARD PROPORTIONS, UNLESS OTHERWISE SHOWN.
5. TAIL OF ARROW USED FOR SPECIFICATION REFERENCE. (TAIL MAY BE OMITTED WHEN REFERENCE NOT USED.)
6. DIMENSIONS OF WELD SIZES, INCREMENT LENGTHS, AND SPACING IN INCHES.

Figure 9–11 Standard welding symbols. *Courtesy American Welding Society.*

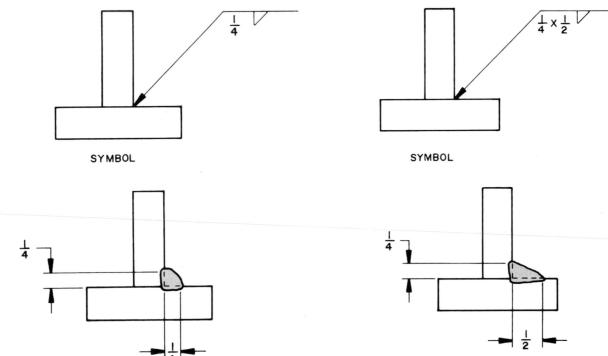

Figure 9–12 Symbol for fillet weld with equal legs.

Figure 9–13 Symbol for fillet weld with different leg lengths.

*J Groove Weld.* The J groove weld is necessary when one piece is a square cut and the other piece is in a J-shaped groove. The included angle, the root opening, and the weld size are given, as shown in Figure 9–18.

*Plug Welds.* A plug weld is made in a hole in one piece of metal that is lapped over another piece of metal. These welds are spec-

ified by giving the weld size, angle, depth, and pitch, as shown in Figure 9–19(a). The same type of weld may be applied to a slot. This is referred to as a slot weld, as shown in Figure 9–19(b).

*Field Weld.* A field weld is a weld that will be performed in the field rather than in a fabrication shop. The field weld symbol is used when two or more subassemblies must be welded together

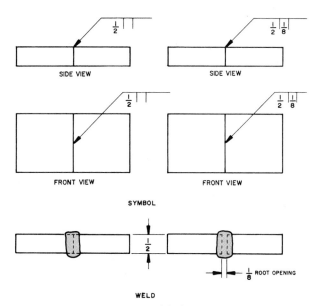

Figure 9–14 Square groove weld showing root opening.

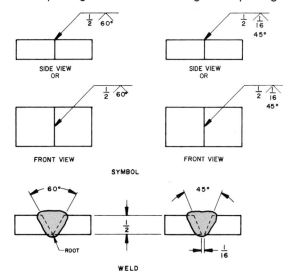

Figure 9–15 V groove weld.

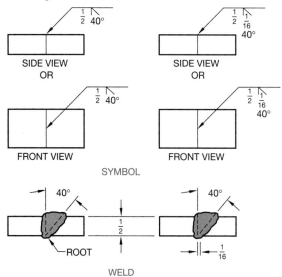

Figure 9–16 Bevel groove weld.

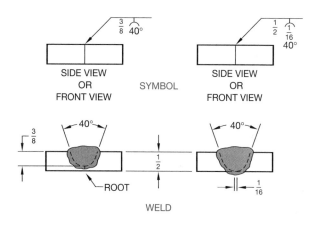

Figure 9–17 U groove weld.

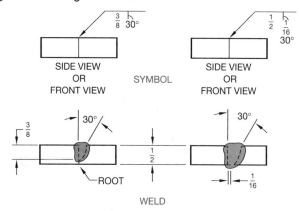

Figure 9–18 J groove weld.

on the job site rather than in a shop. The reason for this application is that the individual components may be easier to transport disassembled, or the mounting procedure may require job site installation. The field weld symbol is a flag attached to the reference line at the leader intersection, as shown in Figure 9–10 or Figure 9–11.

*Weld-all-around.* When a welded connection must be performed all around a feature, the weld-all-around symbol is attached to the reference line at the junction of the leader. This makes it clear that the weld surrounds the feature rather than attaches at a certain increment or length (see Figure 9–20).

*Flush Contour Weld.* Generally the surface contour of a weld is raised above the surface face. If this is undesirable, a flush surface symbol must be applied to the weld symbol. When the flush contour weld symbol is applied without any further consideration, the welder must perform this effect without any further finishing. The other option is to specify a flush finish using another process. The letter designating the other process is placed above the flush contour symbol for another side application, or below the flush contour symbol for an arrow-side application. The options include: C = chipping, G = grinding, M = machining, R = rolling, or H = hammering (see Figure 9–21).

*Weld Length and Increment.* When a weld is not continuous along the length of a part, the weld length should be given. In some situations, the weld along the length of a feature is given in lengths spaced a given distance apart. The distance from one

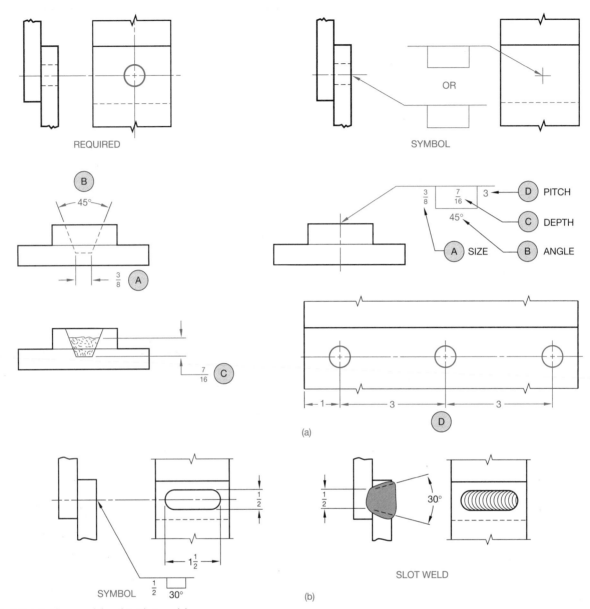

(a)

(b)

Figure 9–19 (a) Plug welds. (b) Slot welds.

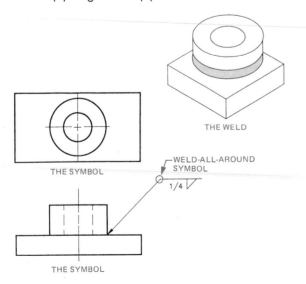

Figure 9–20 Weld-all-around.

point on a weld length to the same corresponding point on the next weld is called the *pitch*; generally, from center to center of the welds. The weld length and increment are shown to the right of the weld symbol, as shown in Figure 9–22.

*Spot Weld.* Spot welding is a process of resistance welding, where the base materials are clamped between two electrodes and a momentary electric current produces the heat for welding at the contact spot. Spot welding is generally associated with welding sheet metal lap seams. The size of the spot welding is given as a diameter to the left of the symbol. The center-to-center pitch is given to the right of the symbol (see Figure 9–23). The strength of the spot welds can be given as minimum shear strength in pounds per spot to the left of the symbol, as shown in Figure 9–24(a). When a specific number of spot welds is required in a seam or joint, the quantity is placed above or below the symbol in parentheses, as shown in Figure 9–24(b).

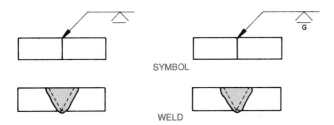

Figure 9–21 Flush contour weld.

*Seam Weld.* A seam weld is a continuous weld made between or upon overlapping members. The continuous weld may consist of a single weld bead or a series of overlapping spot welds. The dimensions of seam welds, shown on the same side of the reference line as the weld symbol, relate to either size or strength. The weld size width is shown to the left of the symbol in fractional or decimal inches, or millimeters, as shown in Figure 9–25(a). The weld length, when specified, is provided to the right of the symbol, as shown in Figure 9–25(b). The strength of a seam weld is expressed as minimum acceptable

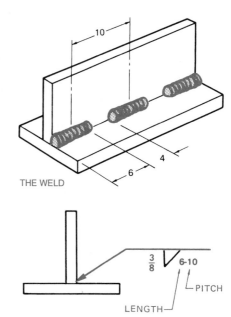

THE WELD

Figure 9–22 Intermittent fillet weld.

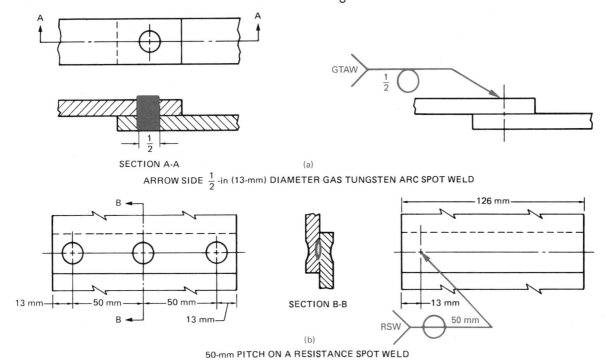

SECTION A-A

(a)

ARROW SIDE $\frac{1}{2}$-in (13-mm) DIAMETER GAS TUNGSTEN ARC SPOT WELD

SECTION B-B

(b)

50-mm PITCH ON A RESISTANCE SPOT WELD

Figure 9–23 Designating spot welds.

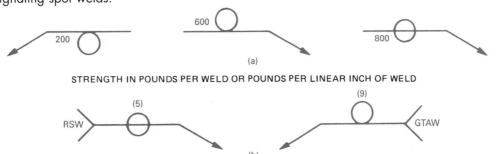

(a)

STRENGTH IN POUNDS PER WELD OR POUNDS PER LINEAR INCH OF WELD

(b)

QUANTITY OF SPOT WELDS

Figure 9–24 (a) Indicating strength in pounds per weld or pounds per linear inch of weld. (b) Designating strength and number of spot welds.

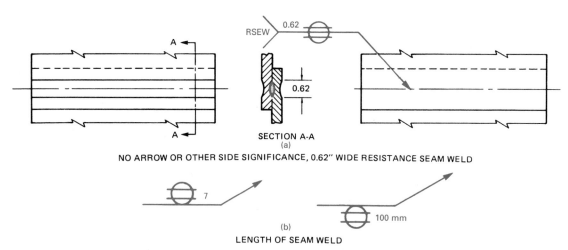

Figure 9–25 Indicating the length of a seam weld.

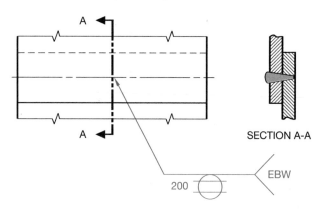

Figure 9–26 Indicating the strength of a seam weld.

shear strength in pounds per linear inch placed to the left of the weld symbol, as shown in Figure 9–26.

## Weld Symbol Leader Arrow Related to Weld Location

Welding symbols are applied to the joint as the basic reference. All joints have an *arrow side* and an *other side*. When fillet and groove welds are used, the welding symbol leader arrows connect the symbol reference line to one side of the joint known as the arrow side. The side opposite the location of the arrow is called the other side. If the weld is to be deposited on the arrow side of the joint, the proper weld symbol is placed below the reference line, as shown in Figure 9–27(a). If the weld is to be deposited on the side of the joint opposite the arrow, the weld symbol is placed *above* the reference line, as you can see in Figure 9–27(b). When welds are to be deposited on both sides of the joint, the same weld symbol is shown *above and below* the reference line, as shown in Figure 9–27(c) and Figure 9–27(d).

For plug, spot, seam, or resistance welding symbols, the leader arrow connects the welding symbol reference line to the outer surface of one of the members of the joint at the center line of the desired weld. The member that the arrow points to is considered the arrow-side member. The member opposite of the arrow is considered the other-side member.

## Additional Weld Characteristics

*Weld Penetration.* Unless otherwise specified, a weld penetrates through the thickness of the parts at the joint. The size of the groove weld remains to the left of the weld symbol. Figure 9–28(a) shows the size of grooved welds with partial penetration. Notice in Figure 9–28(b) that a weld with partial penetration may specify the depth of the groove followed by the depth of weld penetration in parentheses, with both items placed to the left of the weld symbol.

When single-groove and symmetrical double-groove welds penetrate completely through the parts being joined, the weld size may be omitted, as shown in Figure 9–29. The depth of penetration of flare-formed groove welds is assumed to extend to the tangent points of the members, as shown in Figure 9–30.

*Flange Welds.* Flange welds are used on light-gage metal joints where the edges to be joined are flanged or flared. Dimensions of flange welds are placed to the left of the weld symbol. Further, the radius and height of the weld above the point of tangency are indicated by showing both the radius and the height separated by a plus (+) symbol. The size of the flange weld is then placed outward of the flange dimensions (see Figure 9–31).

*Welding Process Designation.* The tail is added to the welding symbol when it is necessary to designate the welding specification, procedures, or other supplementary information needed to fabricate the weld (see Figure 9–32).

*Weld Joints.* The types of weld joints are often closely associated with the types of weld grooves already discussed. The weld grooves may be applied to any of the typical joint types. The weld joints used in most weldments are the butt, lap, tee, outside corner, and edge joints shown in Figure 9–33.

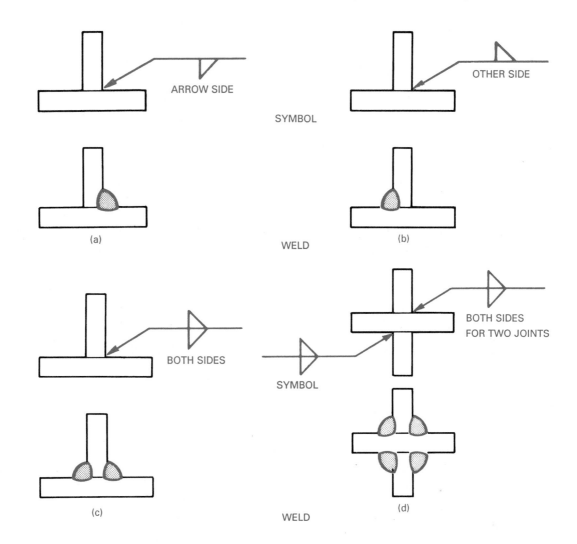

Figure 9–27 Designating weld locations. *Courtesy American Welding Society.*

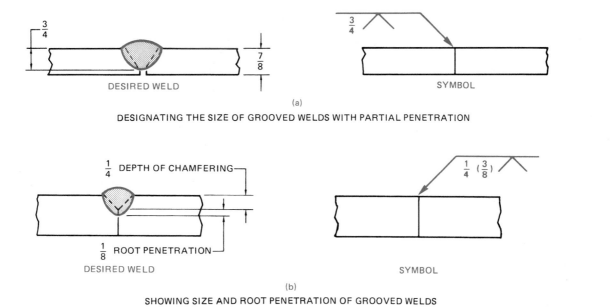

Figure 9–28 (a) Designating the size of grooved welds with partial penetration. (b) Showing size and penetration of grooved welds. *Courtesy American Welding Society.*

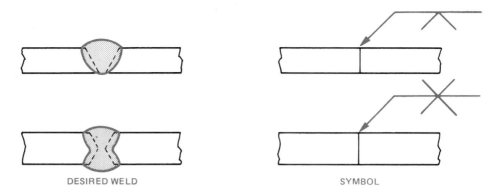

Figure 9–29 Designating single- and double-groove welds with complete penetration. *Courtesy American Welding Society.*

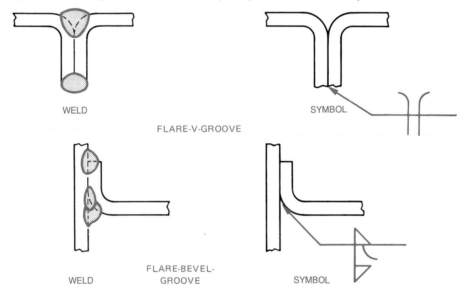

Figure 9–30 Designating flare V and flare bevel groove welds. *Courtesy American Welding Society.*

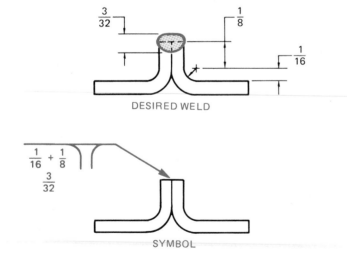

Figure 9–31 Applying dimensions to flange welds. *Courtesy American Welding Society.*

## WELDING TESTS

There are two types of welding tests—destructive and nondestructive tests.

### Destructive Tests (DT)

*Destructive tests (DT)* use the application of a specific force on the weld until the weld fails. These types of tests may include the analysis of tensile compression, bending, torsion, or shear strength. Figure 9–34 shows the relationship of forces that may be applied to a weld. The continuity of a weld is when the desired characteristics of the weld exist throughout the weld length. Discontinuity or lack of continuity exists when a change in the shape or structure exists. The types of problems that alter the desired weld characteristic can include cracks, bumps, seams or laps, or changes in density. The intent of destructive testing is to determine how much of a discontinuity can exist in a weld before the weld is considered to be flawed. Parts may be periodically selected for destructive testing. The weld that is tested is unfit for any further use.

### Nondestructive Tests (NDT)

*Nondestructive Tests (NDT)* are tests for potential defects in welds that are performed without destroying or damaging the weld or the part. The types of nondestructive tests and the corresponding symbol for each test are shown in Figure 9–35. The increased use of

A-2 REFERENCE

SAW PROCESS

BMAW-MA PROCESS AND METHOD

NO SPECIFICATIONS REQUIRED

(a)

|  | Welding Process | Letter Designation |
|---|---|---|
| Brazing | Torch brazing | TB |
|  | Induction brazing | IB |
|  | Resistance brazing | RB |
| Flow Welding | Flow welding | FLOW |
| Induction Welding | Induction welding | IW |
| Arc Welding | Bare metal arc welding | BMAW |
|  | Submerged arc welding | SAW |
|  | Shielded metal arc welding | SMAW |
|  | Carbon arc welding | CAW |
|  | Oxyhydrogen welding | OHW |
| Gas Welding | Oxyacetylene welding | OAW |

The following suffixes may be added if desired to indicate the method of applying the above processes:

| Automatic welding | -AU |
|---|---|
| Machine welding | -ME |
| Manual welding | -MA |
| Semiautomatic welding | -SA |

(b)

Figure 9–32 (a) Locations for weld specifications, processes. (b) Other references about weld symbols.

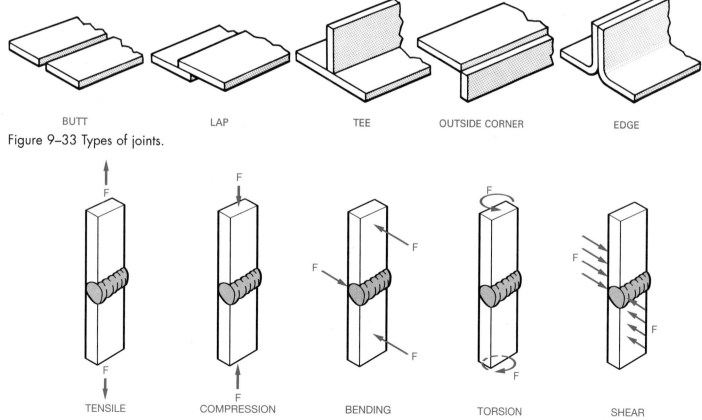

BUTT          LAP          TEE          OUTSIDE CORNER          EDGE

Figure 9–33 Types of joints.

TENSILE          COMPRESSION          BENDING          TORSION          SHEAR

Figure 9–34 Forces on a weld.

| Type of Nondestructive Test | Symbol |
| --- | --- |
| Visual | VT |
| Penetrant | PT |
| Dye penetrant | DPT |
| Fluorescent penetrant | FPT |
| Magnetic particle | MT |
| Eddy current | ET |
| Ultrasonic | UT |
| Acoustic emission | AET |
| Leak | LT |
| Proof | PRT |
| Radiographic | RT |
| Neutron radiographic | NRT |

Figure 9–35 Standard nondestructive testing symbols.

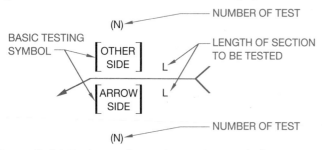

Figure 9–36 Basic nondestructive testing symbol.

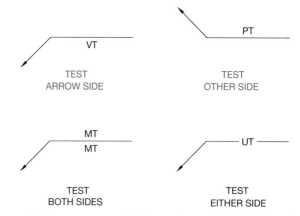

Figure 9–37 Testing symbols used to indicate what side is to be tested.

NDT as a means of fabrication quality control has resulted in the development of standard NDT symbols. The testing symbols are used in conjunction with the weld symbol to denote the area to be tested and the type of test to be used (see Figure 9–36).

The location of the testing symbol above, below, or placed in a break on the reference line has the same reference to the weld joint as the weld symbol application. Test symbols below the reference line mean arrow-side tests. Symbols above the reference line are for other side tests, and a test symbol placed in a break on the line indicates no preference of side to be tested. Test symbols placed on both sides of the reference line require the weld to be tested on both sides of the joint (see Figure 9–37).

Two or more different tests may be required on the same section or length of weld. Methods of combining welding test symbols to indicate more than one test procedure are shown in Figure 9–38. The length of the weld to be tested can be shown to the right of the test symbol, or may be provided as a dimension line giving the extent of the test length, as shown in Figure 9–39. The number of tests to be made may be identified in parentheses below the test symbol for arrow-side tests, or

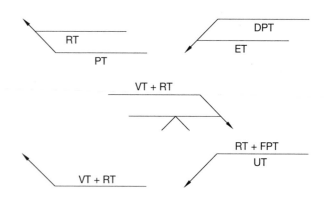

Figure 9–38 Methods of combining testing symbols.

above the symbol for other-side tests, as shown in Figure 9–40. The welding symbols and nondestructive testing symbols can be combined, as shown in Figure 9–41. The combination symbol is appropriate to help the welder and inspector identify welds that require special attention. When a radiograph test is needed, a special symbol and the angle of radiation may be specified, as shown in Figure 9–42.

## WELDING SPECIFICATIONS

A welding specification is a detailed statement of the legal requirements for a specific classification or type of product. Products manufactured to code or specification requirements commonly must be inspected and tested to assure compliance.

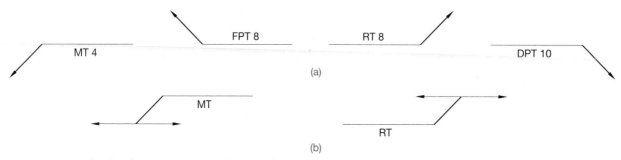

Figure 9–39 Two methods of designating the length of weld to be tested.

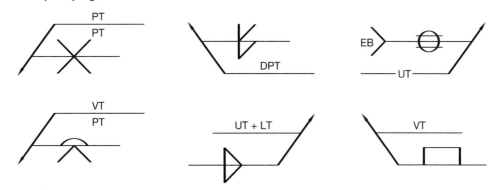

Figure 9–40 Method of specifying number of tests to be made.

Figure 9–41 Combination welding and nondestructive testing symbols.

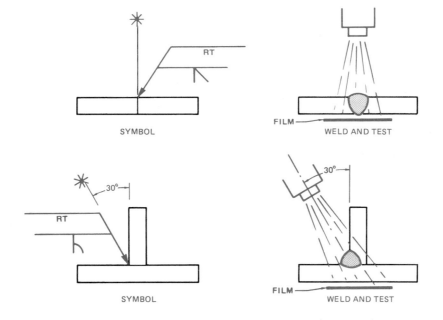

Figure 9–42 Combination symbol for welding and radiation location for testing.

There are a number of agencies and organizations that publish welding codes and specifications. The application of a particular code or specification to a weldment can be the result of one or more of the following requirements:

■ Local, state, or federal government regulations

■ Bonding or insurance company requirements

■ Customer requirements

■ Standard industrial practice

Commonly used codes include:

■ No. 1104, American Petroleum Institute (API). Used for pipeline specifications.

■ Section IX, American Society of Mechanical Engineers (ASME). Used to specify welds for pressure vessels.

■ D1.1, American Welding Society (AWS). Welding specifications for bridges and buildings.

■ AASHT, American Association of State Highways and Transportation Officials.

■ AIAA, Aerospace Industries Association of America.

■ AISC, American Institute of Steel Construction.

■ ANSI, American National Standard Institute.

■ AREA, American Railway Engineering Association.

■ AWWA, American Water Works Association.

■ AAR, Association of American Railroads.

■ MILSTD, Military Standards, Department of Defense.

■ SAE, Society of Automotive Engineers.

Note in the problem assignments that the welding specifications and notes are provided under SPECIFICATIONS and in the general notes.

# CHAPTER 9 TEST

Multiple choice: Respond to the following by selecting a, b, c, or d to best answer the question or complete the statement.

1. This is the most traditionally used welding method, which uses a flux-covered metal electrode to carry an electrical current forming an arc that melts the work and the electrode.
   a. oxyacetylene welding
   b. shielded metal arc welding
   c. gas tungsten arc welding
   d. gas metal arc welding

2. This low-equipment and operating-cost welding process is most typically used to fabricate thin materials or to do repair work and metal cutting.
   a. oxyacetylene welding
   b. shielded metal arc welding
   c. gas tungsten arc welding
   d. gas metal arc welding

3. This welding process is extremely fast, economical, and produces a very clean weld, where a current carrying wire is fed into a joint between pieces to form the weld. It was previously used for welding aluminum using a metal inert gas known as MIG.
   a. oxyacetylene welding
   b. shielded metal arc welding
   c. gas tungsten arc welding
   d. gas metal arc welding

4. This welding process is useful on certain materials and applications, but is generally limited to thin materials, high-integrity joints, or small parts, because of its slow welding speed and high cost of equipment and materials.
   a. oxyacetylene welding
   b. shielded metal arc welding
   c. gas tungsten arc welding
   d. gas metal arc welding

5. Welding drawings are usually called:
   a. welders
   b. weld drawings
   c. welding drawings
   d. weldments

6. This type of weld is formed between two adjacent parts when the side of each part is beveled to form a groove between the parts.
   a. fillet weld
   b. square groove weld
   c. V groove weld
   d. bevel groove word

7. This type of weld is formed in the internal corner of the angle formed by two pieces of metal.
   a. fillet weld
   b. square groove weld
   c. V groove weld
   d. bevel groove weld

8. This type of weld is applied to a butt joint between two pieces of metal.
   a. fillet weld
   b. square groove weld
   c. V groove weld
   d. bevel groove weld

9. This type of weld is created when one piece is square and the other piece has a beveled surface.
   a. fillet weld
   b. square groove weld
   c. V groove weld
   d. bevel groove weld

10. This type of weld is made in a hole in one piece of metal that is lapped over another piece of metal.
    a. lap weld
    b. spot weld
    c. plug weld
    d. seam weld

11. This is a process in resistance welding where the base materials are clamped between two electrodes and a moment electric current produces the heat for welding at the contact location.
    a. lap weld
    b. spot weld
    c. plug weld
    d. seam weld

12. This is a continuous weld made between or on overlapping members.
    a. lap weld
    b. spot weld
    c. plug weld
    d. seam weld

13. If the welding symbol is above the reference line, this is known as:
    a. arrow side
    b. other side

14. If the welding symbol is below the reference line, this is known as:
    a. arrow side
    b. other side

15. When welds are given in lengths spaced a given distance apart, the length of each weld increment is known as:

    a. length

    b. increment

    c. pitch

    d. distance

16. When welds are given in lengths spaced a given distance apart, the distance from a point on one weld to the same corresponding point on the next weld is called:

    a. length

    b. increment

    c. pitch

    d. distance

17. This type of weld is used on light-gage metal joints where the edges to be joined are flared.

    a. lap weld

    b. flange weld

    c. flared weld

    d. seam weld

18. This type of weld test uses the application of specific force on the weld until the weld fails.

    a. destructive test

    b. nondestructive test

19. This type of weld test is performed by visual inspection, the use of penetrant, magnetic particles, sound, liquid, radiographic, or neutron procedures to ensure weld quality.

    a. destructive test

    b. nondestructive test

20. These are detailed statements of the legal requirements for a specific classification or type of product manufactured and commonly indicate inspection and testing requirements to assure compliance with codes.

    a. welding symbols

    b. welding codes

    c. welding specifications

    d. welding laws

## CHAPTER 9 PROBLEMS

Given the following welding symbols with letters (a, b, c, etc.) pointing to various components, identify the component related to each letter in the blanks provided.

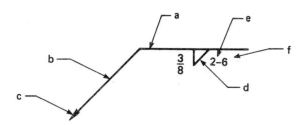

**PROBLEM 9–1**

a. _____
b. _____
c. _____
d. _____
e. _____
f. _____

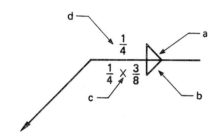

**PROBLEM 9–2**

a. _____
b. _____
c. _____
d. _____

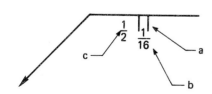

**PROBLEM 9–3**

a. _____
b. _____
c. _____

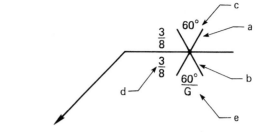

**PROBLEM 9–4**

a. _____
b. _____
c. _____
d. _____
e. _____

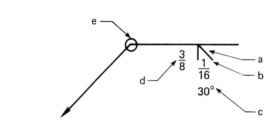

**PROBLEM 9–5**

a. _____
b. _____
c. _____
d. _____
e. _____

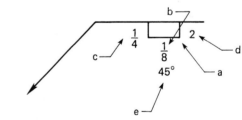

**PROBLEM 9–6**

a. _____
b. _____
c. _____
d. _____
e. _____

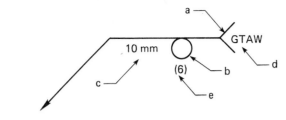

**PROBLEM 9–7**

a. _____
b. _____
c. _____
d. _____
e. _____

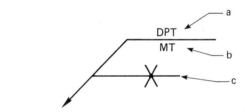

**PROBLEM 9–8**

a. _____
b. _____
c. _____

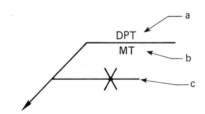

**PROBLEM 9–9**

a. _____
b. _____
c. _____

**PROBLEM 9–10**

a. _____
b. _____
c. _____

**PROBLEM 9–11** Answer the following questions based on the following print.

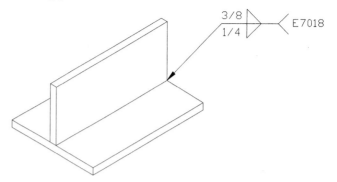

1. What is the type of weld?

   _____

2. Give the arrow-side weld size.

   _____

3. What is the other-side weld size?

   _____

4. What are the specifications given with this welding symbol?

   _____

**PROBLEM 9–12** Answer the following questions based on the print shown on the next page.

1. What is the type of weld at A?

   _____

2. Is the weld at A arrow side or other side?

   _____

3. What is the weld size at A?

   _____

4. Identify what the O.A.W. specification means.

   _____

5. Is the weld at B arrow side or other side?

   _____

6. What does it mean when there is a symbol on both sides of the reference line as in example C?

   _____

7. What is the weld groove type at D?

   _____

8. What does the 1/8 mean when placed inside the weld groove symbol, as in example D?

   _____

9. Identify the meaning of the specification given at welding symbol E.

   _____

10. Is the weld at E arrow side or other side?

    _____

Problem 9–12

**PROBLEM 9–13** Answer the following questions based on the print shown on the next page.

1. What type of weld symbol is shown on this print?

   _____

2. Is the weld an arrow-side or other-side weld?

   _____

3. What does the line under the weld symbol mean?

   _____

4. Why is there a note 2 PLACES below the weld symbol?

   _____

5. Give the dimension from the center of the Ø.781 hole to the bottom of the welded pieces.

   _____

6. What is the distance between the welded pieces?

   _____

**PROBLEM 9–14** Answer the following questions based on the print shown on the next page.

1. Describe the weld symbol at A.

   _____

2. What is the type of weld on the arrow side at B?

   _____

3. What is the type of weld on the other side at B?

   _____

4. Explain the meaning of the circle located where the reference line and leader meet at C.

   _____

5. What is the type of weld on the arrow side at D?

   _____

6. What is the type of weld on the other side at B?

   _____

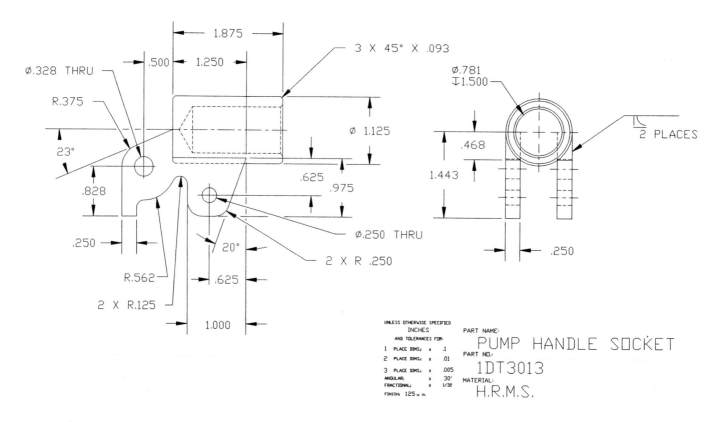

UNLESS OTHERWISE SPECIFIED
INCHES
AND TOLERANCES FOR:

| | | |
|---|---|---|
| 1 PLACE DIMS; | ± | .1 |
| 2 PLACE DIMS; | ± | .01 |
| 3 PLACE DIMS; | ± | .005 |
| ANGULAR; | ± | 30' |
| FRACTIONAL; | ± | 1/32 |
| FINISH; 125 μ in. | | |

PART NAME:
PUMP HANDLE SOCKET
PART NO.:
1DT3013
MATERIAL:
H.R.M.S.

Problem 9–13

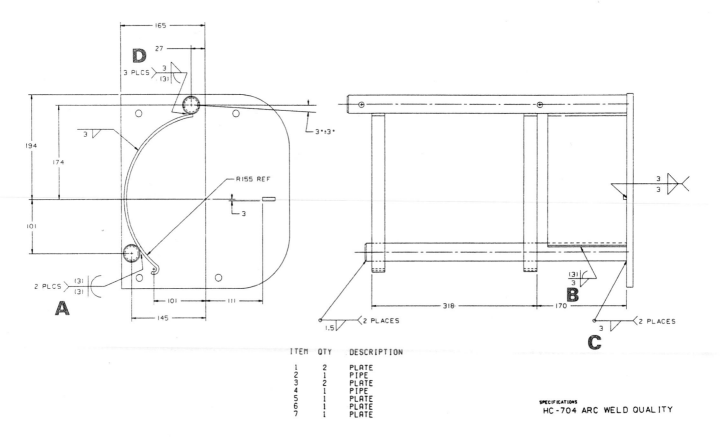

| ITEM | QTY | DESCRIPTION |
|---|---|---|
| 1 | 2 | PLATE |
| 2 | 1 | PIPE |
| 3 | 2 | PLATE |
| 4 | 1 | PIPE |
| 5 | 1 | PLATE |
| 6 | 1 | PLATE |
| 7 | 1 | PLATE |

SPECIFICATIONS
HC-704 ARC WELD QUALITY

Problem 9–14 *Courtesy Hyster Company.*

# Reading Sections, Revolutions, and Conventional Breaks

## LEARNING OBJECTIVES

After completing this chapter you will be able to:

- Identify the ANSI standard for multiview and sectional view drawings.
- Identify types and read prints with sections, revolutions, and conventional breaks.
- Read and identify section line symbols.

*ANSI/ASME* The American National Standards Institute / American Society of Mechanical Engineers document that governs sectioning practices is titled *Multiview and Sectional View Drawings* ASME Y14.3M. Lines associated with sectional view drawings include cutting-plane lines, section lines, and break lines. The standard for these and other lines are found in Line Conventions and Lettering ASME Y14.2M. The content of this chapter is based on the ANSI/ASME standard and provides in-depth analysis of the techniques and methods of sectional view presentation.

## SECTIONING

*Sections*, or *sectional* views, are used to describe the interior portions of an object that would otherwise be difficult to visualize. Interior features that are described using hidden lines may not appear as clear as if they were exposed for viewing as visible features. Figure 10–1 shows an object in conventional multiview representation and using a sectional view. Notice how the hidden features are clarified in the sectional view.

## CUTTING-PLANE LINES

The sectional view is found by placing an imaginary cutting plane through the object as if the area to be exposed were cut away. The next view then becomes the sectional view by removing the portion of the object between the viewer and the cutting plane (see Figure 10–2).

The sectional view is normally projected from the view that has the cutting plane the same as a view in multiview. The cutting-plane line is a thick line that represents the cutting plane, as shown

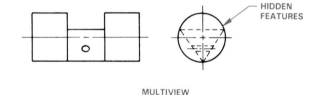

MULTIVIEW

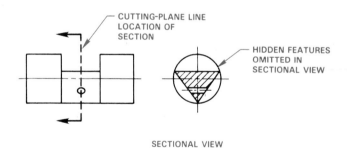

SECTIONAL VIEW

Figure 10–1 Conventional multiview compared to a sectional view.

in Figure 10–2. The *cutting-plane line* is capped on the ends with arrowheads that show the direction of sight of the sectional view. When the extent of the cutting plane is obvious, only the ends of the cutting-plane line may be used, as shown in Figure 10–3.

If lack of space restricts the normal placement of a sectional view, the view may be placed in an alternate location. When this is done, the sectional view is not rotated, but remains in the same orientation as if it were a direct projection from the cutting plane. The cutting planes and related sectional views are labeled with letters beginning with A, as shown in Figure 10–4.

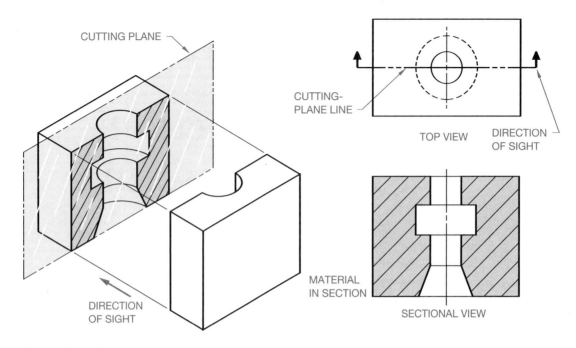

Figure 10–2 Cutting-plane line.

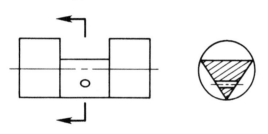

Figure 10–3 Simplified cutting-plane line.

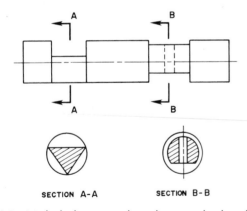

SECTION A-A          SECTION B-B

Figure 10–4 Labeled cutting-plane lines and related sectional views.

The cutting-plane line may be completely omitted when the location of the cutting plane is clearly obvious (see Figure 10–5).

## SECTION LINES

*Section lines* are thin lines used in the view of the section to show where the cutting-plane line has cut through material (see Figure 10–6). Section lines are usually drawn equally spaced at 45°, but any convenient angle may be used. Section lines are drawn in

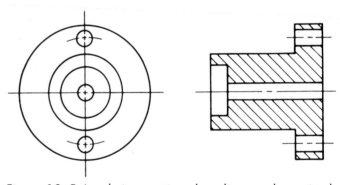

Figure 10–5 An obvious cutting-plane line can be omitted.

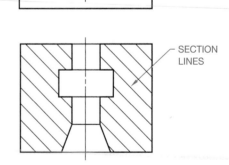

Figure 10–6 Sections lines.

opposite directions on adjacent parts, and when several parts are next to each other, any suitable angle is used to make the parts appear clearly separate. When a very large area requires section lining, you can use outline section lining.

Equally spaced section lines mean either a general material designation or cast iron, with the actual material identification located in the drawing title block. The other option is found on a print's coded section lining. Coded section lining is often used when a section is taken through an assembly of different materials. Look at Figure 4–25 to see the variety of section lines that may be found on a print. Very thin features in section are represented as an outline (see Figure 10–7). General section lines are evenly spaced. The amount of space between lines is dependent on the size of the part. Very large parts have larger spacing than very small parts.

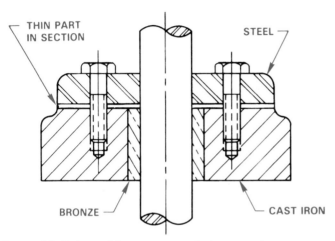

Figure 10–7 Assembly section, coded section lines.

## FULL SECTIONS

A *full section* is drawn when the cutting plane extends completely through the object, usually along a center plane, as shown in Figure 10–8. The object shown in Figure 10–8 could have used two full sections to further clarify hidden features. In that case, the cutting planes and related views are labeled. The cutting planes are labeled near the arrowhead with letters, such as AA and BB, as in Figure 10–9. Notice how the related sections are labeled to match the cutting-plane lines. SECTION A–A correlates with cutting-plane line AA, and SECTION B–B correlates with cutting-plane line BB (see Figure 10–9).

## HALF SECTIONS

A *half section* can be used when a symmetrical object requires sectioning. The cutting-plane line of a half section actually removes one quarter of the object. The advantage of a half section is that the sectional view shows half of the object in section and the other half of the object as it would normally appear. Thus the name half section (see Figure 10–10). Notice that a centerline is used in the sectional view to separate the sectioned portion from the unsectioned portion. Hidden lines are generally omitted from sectional views unless their use improves clarity.

## OFFSET SECTIONS

You can section staggered interior features of an object by allowing the cutting-plane line to *offset* through the features, as shown in

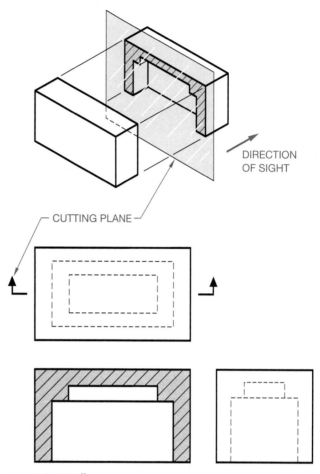

Figure 10–8 Full section.

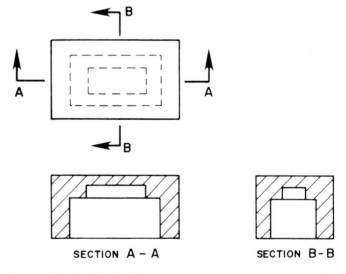

Figure 10–9 Multiple full sections. The cutting planes and related sections are labeled to correlate.

Figure 10–11. Notice in Figure 10–11, there is no line in the sectional view, indicating a change in direction of the cutting-plane line. Normally the cutting-plane line in an offset section extends completely through the object to clearly show the location of the section.

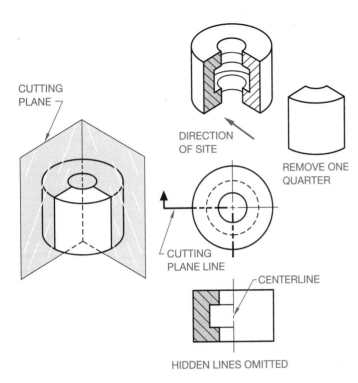

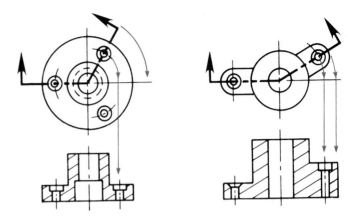

Figure 10–12 Aligned section.

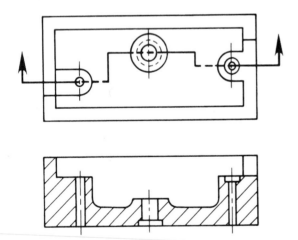

Figure 10–10 Half section.

Figure 10–11 Offset section.

## ALIGNED SECTIONS

Similar to the offset section, the *aligned section* cutting-plane line also staggers to pass through offset features of an object. Normally the change in direction of the cutting-plane line is less than 90° in an aligned section. When this section is taken, the sectional view is drawn as if the cutting plane is rotated to a plane perpendicular to the line of sight, as shown in Figure 10–12.

## UNSECTIONED FEATURES

Specific features of an object are commonly left unsectioned in a sectional view if the cutting-plane line passes through the feature and parallel to it. The types of features that are left unsectioned for clarity are bolts, nuts, rivets, screws, shafts, ribs, webs, spokes,

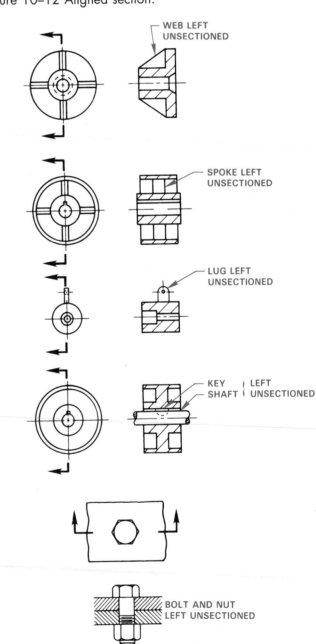

Figure 10–13 Certain features are not sectioned when a cutting-plane line passes parallel to their axis.

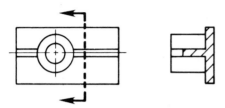

Figure 10–14 Cutting plane perpendicular to normally unsectioned features.

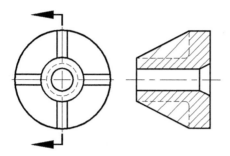

Figure 10–15 Alternate section-line method.

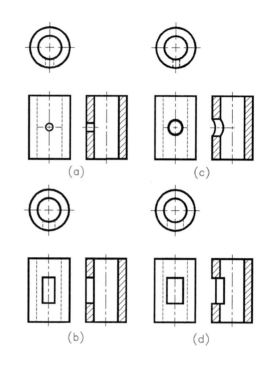

(a)  (c)

(b)  (d)

Figure 10–16 Intersections in section.

bearings, gear teeth, pins, and keys (see Figure 10–13). When the cutting-plane line passes through the previously described features perpendicular to their axes, section lines are shown as seen in Figure 10–14.

It is most common to draw the outline of the previously discussed features, such as webs, lugs, spokes, keys, and shafts, without section lines, but a less common method called *alternate section lines* can be used. With this method, you draw the normally unsectioned feature using hidden lines, and you draw every other section line through the feature, as shown in Figure 10–15.

## INTERSECTIONS IN SECTION

When a section is drawn through a small intersecting shape, the true projection is ignored. The detail is too complex to represent. See Figure 10–16(a) and (b). However, larger intersecting features are drawn as their true representation. See Figure 10–16(c) and (d). The professional decision is up to you.

## CONVENTIONAL REVOLUTIONS

When the true projection of a feature results in foreshortening, the feature is revolved onto a plane perpendicular to the line of sight, as in Figure 10–17. The revolved spoke shown in Figure 10–17 gives a clear representation. Figure 10–18 shows another illustration of conventional revolution compared to true projection. Notice how the true projection results in a distorted and foreshortened representation of the spoke. The revolved spoke in the preferred view is clear and easy to read. The practice illustrated in Figure 10–17 and Figure 10–18 also applies to features of unsectioned objects in multiview, as shown in Figure 10–19.

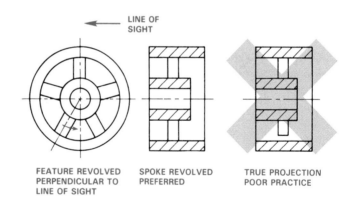

FEATURE REVOLVED PERPENDICULAR TO LINE OF SIGHT

SPOKE REVOLVED PREFERRED

TRUE PROJECTION POOR PRACTICE

Figure 10–17 Conventional revolution

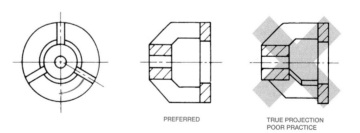

PREFERRED

TRUE PROJECTION POOR PRACTICE

Figure 10–18 Conventional revolution in section.

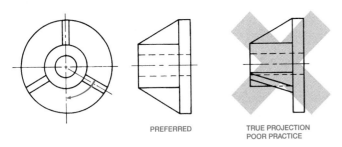

PREFERRED

TRUE PROJECTION
POOR PRACTICE

Figure 10–19 Conventional revolution in multiview.

## BROKEN-OUT SECTIONS

Often a small portion of a part may be broken away to expose and clarify an interior feature. This technique is called a *broken-out* section. There is no cutting-plane line used, as you can see in Figure 10–20. A short break line is generally used with a broken-out section.

## AUXILIARY SECTIONS

A section that appears in an auxiliary view is known as an *auxiliary section*. Auxiliary sections are generally projected directly from the view of the cutting plane. If these sections must be moved to other locations on the drawing sheet, they remain in the same relationship (not rotated) as if taken directly from the view of the cutting plane (see Figure 10–21).

## CONVENTIONAL BREAKS

When a long object of constant shape throughout its length requires shortening, *conventional breaks* may be used. You can use these breaks effectively to save paper or space, or to increase the scale of an otherwise very long part. Figure 10–22 shows typical conventional breaks.

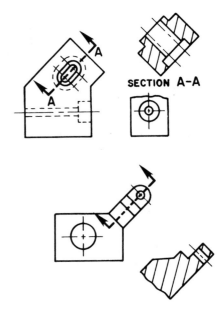

SECTION A-A

Figure 10–21 Auxiliary sections.

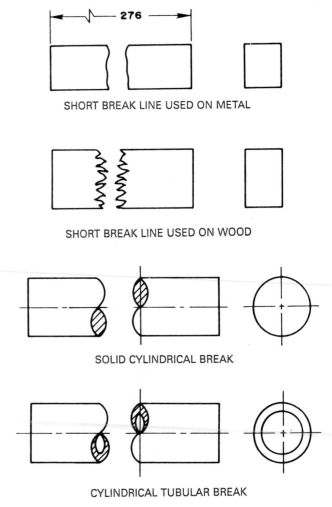

276

SHORT BREAK LINE USED ON METAL

SHORT BREAK LINE USED ON WOOD

SOLID CYLINDRICAL BREAK

CYLINDRICAL TUBULAR BREAK

Figure 10–22 Conventional breaks for various materials and shapes.

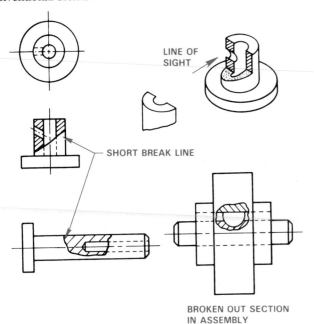

LINE OF SIGHT

SHORT BREAK LINE

BROKEN OUT SECTION IN ASSEMBLY

Figure 10–20 Broken-out section.

## REVOLVED SECTIONS

When a feature has a constant shape throughout the length that cannot be shown in an external view, a *revolved shape* may be used. The desired section is revolved 90° onto a plane perpendicular to the line of sight, as shown in Figure 10–23. Revolved sections can be represented on a drawing in one of two ways, as shown in Figure 10–24. In Figure 10–24(a) the revolved section is drawn on the part. The revolved section can also be broken away, as seen in Figure 10–24(b). The surrounding space can be used for dimensions, as shown in Figure 10–25(a). Notice, in Figure 10–25(b), that very thin features less than 4 mm thick in section can be left without section lines. This practice is also common when sectioning a gasket or other thin material.

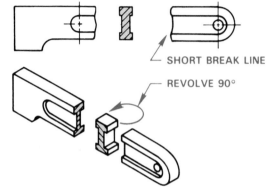

Figure 10–23 Revolved section.

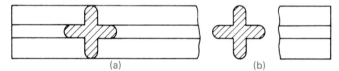

Figure 10–24 (a) Revolved section not broken-away. (b) Revolved section broken away.

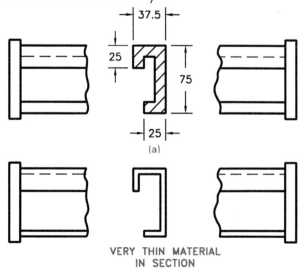

Figure 10–25 (a) Dimensioning a broken-away revolved section. (b) A revolved section through thin material. Section lines can be omitted when material less than 4 mm thick is sectioned.

## REMOVED SECTIONS

*Removed sections* are similar to revolved sections except that they are removed from the view. A cutting-plane line is placed through the object where the section is taken. Removed sections are not generally placed in direct alignment with the cutting-plane line, but are placed in a surrounding area, as shown in Figure 10–26.

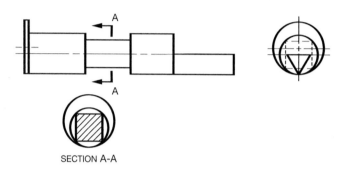

Figure 10–26 Removed section.

The removed section is drawn at a larger scale so that close detail can be more clearly identified, as shown in Figure 10–27. The sectional view is labeled, as shown, with Section A-A and the revised scale, which in this example is 2:1. The predominant scale of the principal views is shown in the title block.

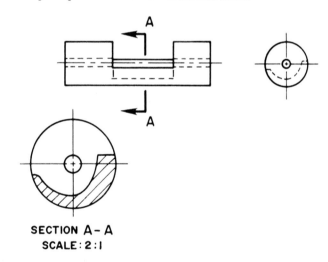

Figure 10–27 Enlarged removed section.

Multiple removed sections are generally arranged on the sheet in alphabetical order from left to right and top to bottom (see Figure 10–28). Notice that hidden lines have been omitted for clarity. The cutting planes and related sections are labeled alphabetically, excluding the letters I, O, and Q as they may be mistaken for numbers. When the entire alphabet has been used, sections are labeled with double letters beginning with AA, BB, and so on.

Another method of drawing a removed section is to extend a centerline next to a symmetrical feature and revolve the section on the centerline, as shown in Figure 10–29. The removed section can be drawn to the same scale or enlarged as necessary to clarify detail.

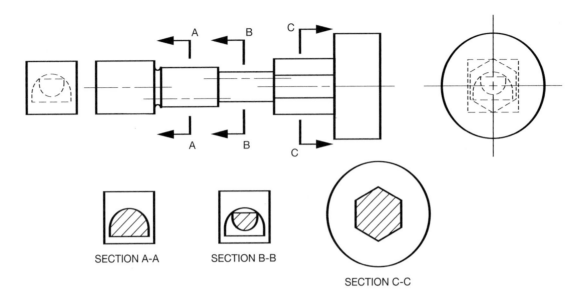

SECTION A-A    SECTION B-B

SECTION C-C

Figure 10–28 Multiple removed sections. Hidden lines in the profile views help illustrate importance of sectional views.

## PULL IT ALL TOGETHER

Now that you have learned about all the different sectioning techniques and conventional revolution methods, take some time to look at as many prints as you can in an effort to identify cutting-plane lines and related sectional views. Start with the complex print shown in Figure 10-30. This print shows a front view, left-side view, rear view, a full section, and an auxiliary section.

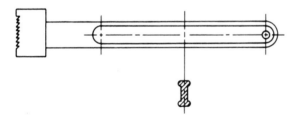

Figure 10–29 Alternate removed section method.

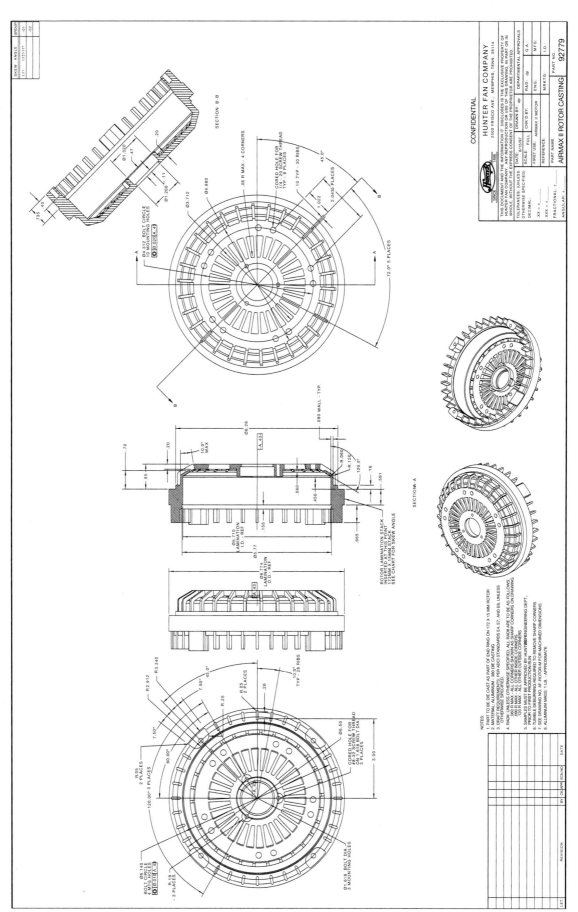

Figure 10—30 An actual industry drawing of a complex part, using front, left-side, and rear views, along with a full and auxiliary section. Compliments of Hunter Fan Company, Memphis, TN.

# CHAPTER 10 TEST

Multiple choice: Respond to the following by selecting a, b, c, or d to best answer the question or complete the statement.

1. This type of view is used to describe the interior portions of an object that would otherwise be difficult to visualize.
   a. multiview
   b. auxiliary view
   c. sectional view
   d. view enlargement

2. This type of line is capped on the ends with arrows that show the direction of sight of the sectional view.
   a. viewing-plane line
   b. cutting-plane line
   c. section line
   d. phantom line

3. These are thin lines used in the view of the section to show where the material has been cut.
   a. viewing-plane line
   b. cutting-plane line
   c. section line
   d. phantom line

4. This type of section may be used when a symmetrical object requires sectioning and the cutting-plane line removes one quarter of the object.
   a. full section
   b. half section
   c. offset section
   d. aligned section

5. This type of section is drawn when the cutting-plane line extends completely through the object, usually along a center plane.
   a. full section
   b. half section
   c. offset section
   d. aligned section

6. This type of section has a cutting-plane line that staggers to pass through offset features of an object where the change in cutting-plane direction is usually less than 90°.
   a. full section
   b. half section
   c. offset section
   d. aligned section

7. This type of section has a cutting-plane line that turns at 90° angles to cut through staggered features of an object.
   a. full section
   b. half section
   c. offset section
   d. aligned section

8. This type of section is used to remove a small portion of an object to expose and clarify an internal feature.
   a. broken-out section
   b. auxiliary section
   c. revolved section
   d. removed section

9. This type of section is placed in a convenient location on the drawing and is correlated with the cutting-plane line.
   a. broken-out section
   b. auxiliary section
   c. revolved section
   d. removed section

10. This type of section is not aligned with the regular multi-views and is used to show the true size and shape of interior features that would otherwise be foreshortened in another type of section.
    a. broken-out section
    b. auxiliary section
    c. revolved section
    d. removed section

11. This type of section is commonly used on an object that is consistent in shape throughout its length and where the shapes cannot be clearly shown in a regular multi-view, and no cutting-plane line is typically used.
    a. broken-out section
    b. auxiliary section
    c. revolved section
    d. removed section

12. This type of symbol is used when a long shape that is constant in shape throughout requires shortening.
    a. broken-out section
    b. conventional break
    c. revolved section
    d. conventional section

13. This practice occurs when a foreshortened feature is revolved onto a plane perpendicular to the line of sight.
    a. conventional revolution
    b. conventional break
    c. drafting convention
    d. conventional section

14. Pick the following feature that remains unsectioned even if the cutting-plane line passes through parallel to its axis.
    a. shaft
    b. bolt
    c. spoke
    d. all of the above

15. This is the ANSI standard that governs sectioning techniques.

    a. Sectional View Drawings

    b. Sectional View Conventions

    c. Multiview and Sectional View Drawings

    d. Multiview and Sectional View Standards

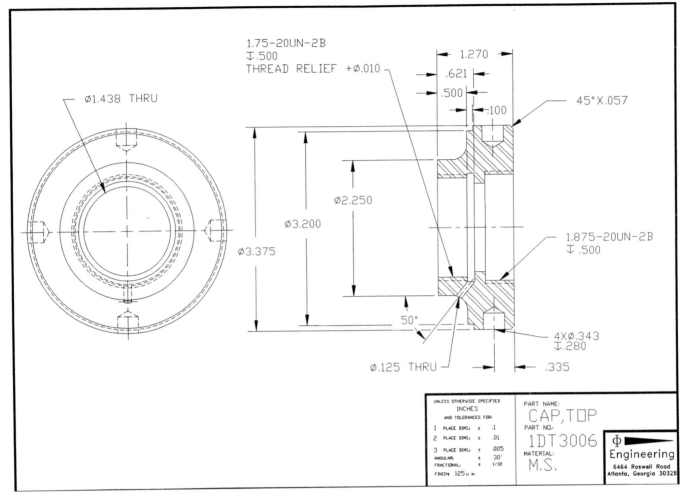

Problem 10–1 *Courtesy IO Engineering.*

## CHAPTER 10 PROBLEMS

**PROBLEM 10–1** Answer the following questions as you refer to the print shown on this page.

1. Describe the views shown on this print.

2. What type of section is represented on this print?

3. Why is the cutting-plane line omitted from this print?

4. Give the largest thread specification shown on this part.

5. What is the angle from horizontal of the Ø.125 hole?

6. What is the angular relationship between the 4 x Ø.343 holes?

7. Give the full name of the material used in this part.

8. What are the overall dimensions of the part?

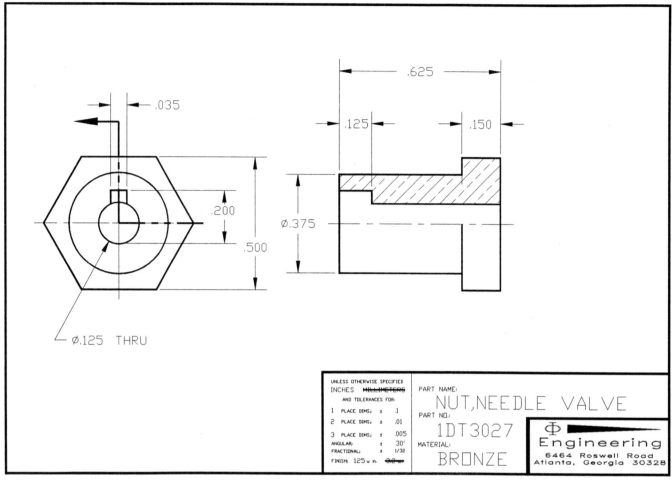

Problem 10–2 *Courtesy IO Engineering.*

**PROBLEM 10–2** Answer the following questions as you refer to the print shown on this page.

1. Describe the views shown on this print.
   _____

2. What type of section is represented on this print?
   _____

3. What is the dimension across the flats of the hexagon?
   _____

4. What is the specification of the center hole?
   _____

5. Name the material used in this part.
   _____

6. What are the overall dimensions of the part?
   _____

7. Give all dimensions of the keyway.
   _____

8. What is the diameter of the main cylinder?
   _____

9. What is the width (thickness) of the hexagonal feature?
   _____

10. What is the tolerance of the .500 dimension?
    _____

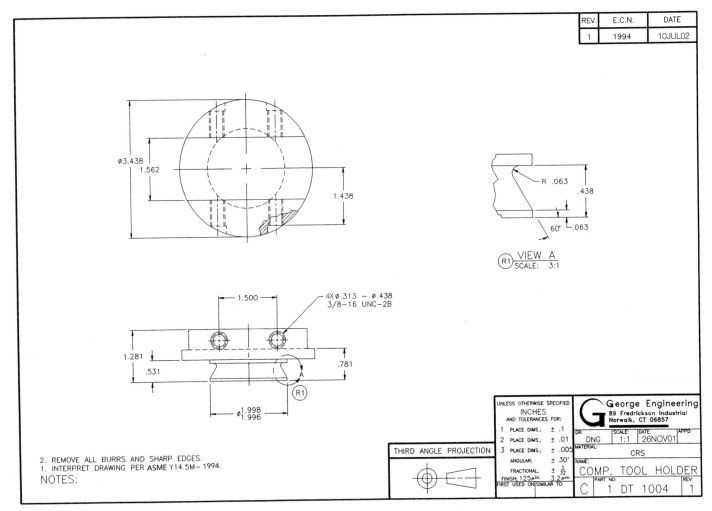

Problem 10–3 *Courtesy David George.*

**PROBLEM 10–3** Answer the following questions as you refer to the print shown on this page.

1. Describe the views shown on this print.

2. What type of section is used on this print?

3. Explain the purpose of the section used on this print.

4. What is the purpose of VIEW A?

5. How many times has this print been revised?

6. What is the tolerance of the Ø1.998/1.996 dimension?

7. What is the diameter of the four holes prior to taping?

8. Give the dimension from the center of the part to the bottom of the counterbore.

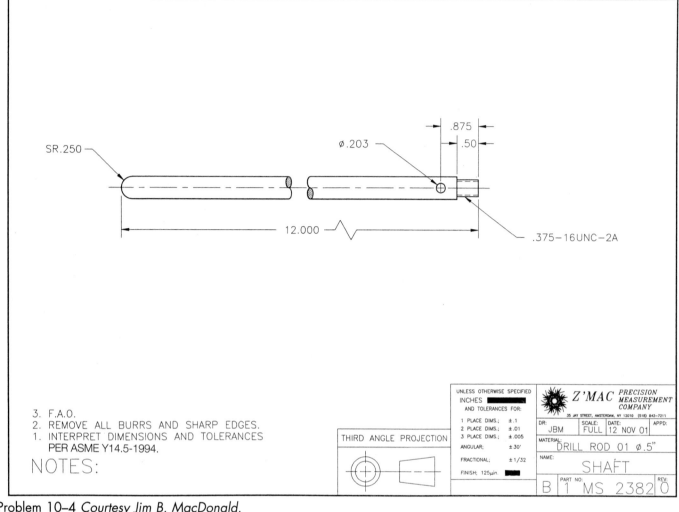

3. F.A.O.
2. REMOVE ALL BURRS AND SHARP EDGES.
1. INTERPRET DIMENSIONS AND TOLERANCES
   PER ASME Y14.5-1994.
NOTES:

Problem 10–4 *Courtesy Jim B. MacDonald.*

**PROBLEM 10–4** Answer the following questions as you refer to the print shown on this page.

1. Describe the conventional break used on this print.

   _____

2. What is the purpose of the conventional break used on this print?

   _____

3. Give the total length of the part.

   _____

4. What is the principal diameter of the part?

   _____

5. Explain the meaning of SR .250.

   _____

6. Give the thread specification.

   _____

7. What is the thread length?

   _____

8. Give the location dimension to the Ø.203 hole.

   _____

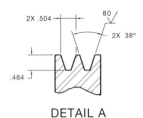

DETAIL A

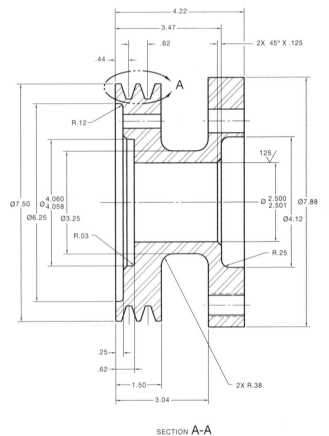

SECTION A-A

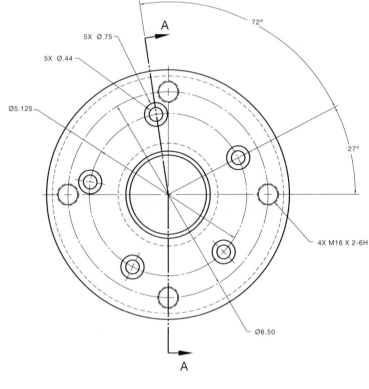

Problem 10–5 Part Name: Crankshaft Adapter.    Material: CI.    *Courtesy American Hoist and Derrick Company.*

**PROBLEM 10–5** Answer the following questions as you refer to the print shown on this page.

1. Name the type of section found at AA.

2. Give the depth and included angle of the v grooves.

3. Identify the depth of the Ø.

4. Give the depth of the Ø4.060/4.058 from the inside face of the Ø.

5. Calculate the thickness of material where the Ø.44 holes are located.

6. Calculate the thickness of material where the Ø.75 holes are located.

7. Calculate the length of the 2.500/2.501 hole, excluding the chamfer.

8. Identify the thread specification found on this print.

9. Give the overall diameter and depth of the part.

10. Identify the surface finish on the V grooves.

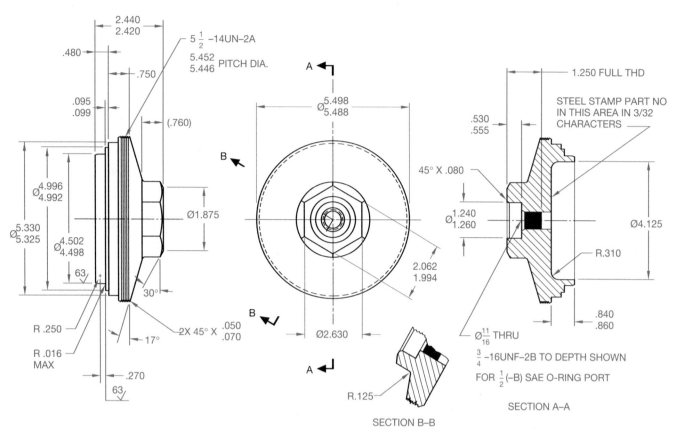

Problem 10–6 Part Name: Accumulator Plug,     Material: Phosphor Bronze.     *Courtesy Stanley Hydraulic Tools, a Division of the Stanley Works.*

**PROBLEM 10–6** Answer the following questions as you refer to the print shown on this page.

1. Identify the thickness of the part.

   _____

2. Give the specification found on the overall diameter.

   _____

3. Identify the dimension found on the hexagon.

   _____

4. Name the type of section found at AA.

   _____

5. Identify the type of section found at BB.

   _____

6. Give the specification for the o-ring port.

   _____

7. Identify the specifications for the part number.

   _____

8. Give the surface finish found on the Ø4.502/4.498.

   _____

9. Give the angle and any related specification of the three angled surfaces on SECTION B-B, starting with the smallest dimension.

   _____

10. Give the depth of the Ø1.240/1.260 from the face of the hexagon.

   _____

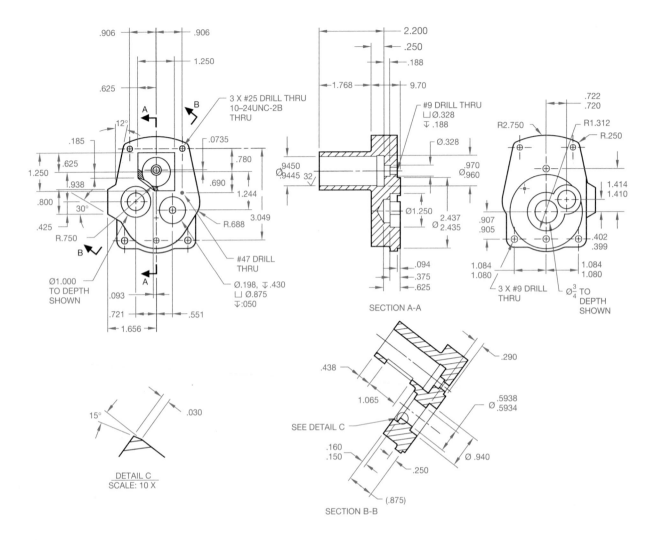

SECTION A-A

SECTION B-B

DETAIL C
SCALE: 10 X

Problem 10–7 Part Name: Bulk Head.    Material: Cast Iron.    *Courtesy Stanley Hydraulic Tools, a Division of the Stanley Works.*

**PROBLEM 10–7** Answer the following questions as you refer to the print shown on this page.

1.  Name the type of section found at AA.

2.  Name the type of section found at BB.

3.  Give the depth of the Ø1.000.

4.  Identify the length of the R.750.

5.  Give the depth of the 3/4 DRILL.

6.  Give the chamfer specifications found on the Ø.5938/.5934.

7.  Give the depth of the Ø.940.

8.  Give the diameter of the .094 boss.

9.  Identify the surface finish on the Ø.9450/.9445.

10. Give the specification of the threads.

# *Reading Geometric Tolerancing*

## LEARNING OBJECTIVES

After completing this chapter you will be able to:

- Identify the ASME standard for dimensioning and tolerancing.
- Read prints containing geometric dimensioning applications.
- Provide datum identification as given on actual prints.
- Read datum target points, lines, area, and related datum target symbols.

- Calculate the geometric tolerance at given produced sizes based on the material condition symbol.
- Read and explain the information given in feature control frames presented on prints.
- Calculate the virtual condition for given applications.

*Geometric tolerancing (GT)* is the dimensioning and tolerancing of individual features of a part where the permissible variations relate to characteristics of form, profile, orientation, runout, or the relationship between features. This subject is commonly referred to as geometric dimensioning and tolerancing (GD & T).

*ASME* The standards for geometric dimensioning and tolerancing (GD & T) are governed by the American National Standards document ASME Y14.5M–1994 titled *Dimensioning and Tolerancing*, published by the American Society of Mechanical Engineers.

## GENERAL TOLERANCING

Tolerancing was introduced in Chapter 7 with definitions and examples of how dimensions are presented on a print. Review Chapter 7 to establish a solid understanding of dimensioning and tolerancing terminology.

The term *general tolerancing* as used here implies dimensioning without the use of geometric tolerancing. General tolerancing applies a degree of form and location control by increasing or decreasing the tolerance. The *limits of size* of a feature control the amount of variation in size and geometric form. This is the boundary between maximum material condition (MMC) and least material condition (LMC). MMC and LMC are defined later in this chapter. The form of the feature may vary between the upper limit and lower limit of a size dimension. This is known as *extreme form variation* (see Figure 11–1). The addition of GT is necessary to control form, but still permits a relatively large-size tolerance.

## SYMBOLOGY

Symbols on prints represent specific information that would otherwise be difficult and time-consuming to duplicate in note form. Geometric tolerancing symbols are divided into the following types:

1. Datum symbols

2. Geometric characteristic symbols

3. Material condition (modifying) symbols

4. Feature control frame

5. Supplementary symbols

## DATUM FEATURE SYMBOLS

*Datums* are considered to be theoretically perfect surfaces, planes, points, or axes that are established from the true geometric counterpart of the datum feature. The datum feature, as shown in Figure 11–2, is the actual feature of the part that is used to establish the datum. Datums are used to originate size and location dimensions. Without the use of GT, datums are often implied. When dimensions originate at a common surface, for example, that surface is assumed to be the datum. The only problem with this method is that the manufacturing operation may not read the datum in the same way as the engineering department, or the implied datum may be ignored all together. With the use of defined datums, each part is made with dimensions that begin from the same original; there is no variation.

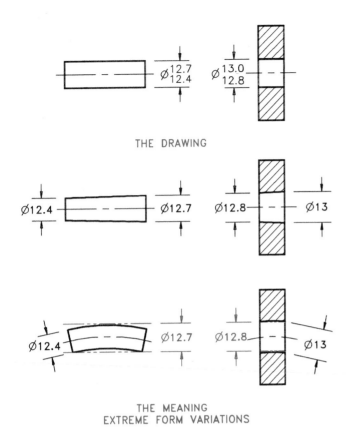

THE DRAWING

THE MEANING
EXTREME FORM VARIATIONS

Figure 11–1 Extreme form variations of given tolerances.

Datum feature symbols are shown in Figure 11–3. The triangular base can be filled or unfilled. The filled triangle helps the symbol show up better on a drawing.

Datum feature symbols are placed in the view that shows the edge of a surface, or they are attached to a diameter or symmetrical dimension when associated with a centerline or center plane (see Figure 11–4).

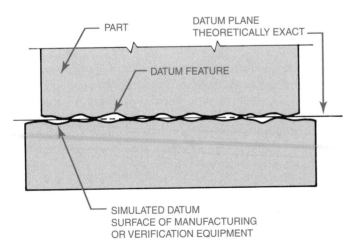

Figure 11–2 Magnified representation of a datum feature.

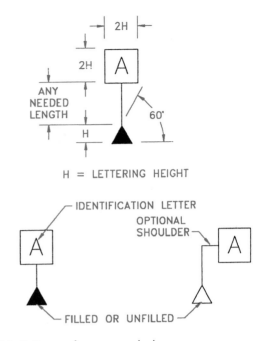

H = LETTERING HEIGHT

Figure 11–3 Datum feature symbol.

## DATUM REFERENCE FRAME

Datum referencing is used to relate features of a part to an appropriate datum or datum reference frame. A datum indicates the origin of a dimensional relationship between a toleranced feature and a designated feature or features on a part. The datum identification symbol on a drawing relates to an actual feature on the part known as the *datum feature*. The part is then placed on the surface of an inspection table or manufacturing verification equipment. This inspection table or equipment is referred to as the *simulated datum*. The *datum* is the theoretically exact plane, axis, or point that is established by the true geometric counterpart of the specified datum feature. Measurements cannot be made from the true geometric counterpart, because it is theoretical or assumed to exist. The machine tables, surface plates, or inspection tables are of such high quality that they are used to simulate the datums from which measurements are taken and dimensions are verified. In this way, each dimension always originates from the same reliable location. Dimensions are never taken or verified from one surface of the part to another.

Sufficient datum features, those most important to the design of a part, are chosen to position the part in relation to a set of three mutually perpendicular planes, jointly called a *datum reference frame*. This reference frame exists in theory only and not on the part. Therefore, it is necessary to establish a method for simulating the theoretical reference frame from the actual features of the part. This simulation is accomplished by positioning the part on appropriate datum features to adequately relate the part to the reference frame and to restrict the motion of the part in relation to it (see Figure 11–5). These planes are simulated in a mutually perpendicular relationship to provide direction, as well as the origin for related dimensions and measurements. Thus, when a part is positioned on the datum reference frame, dimensions related to the

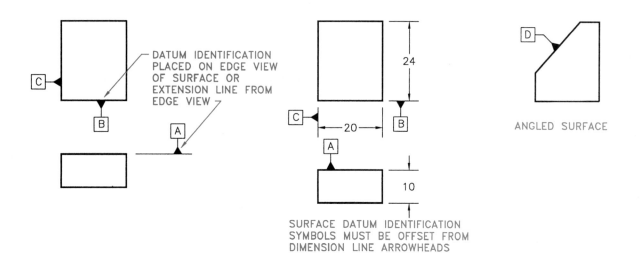

SURFACE DATUMS

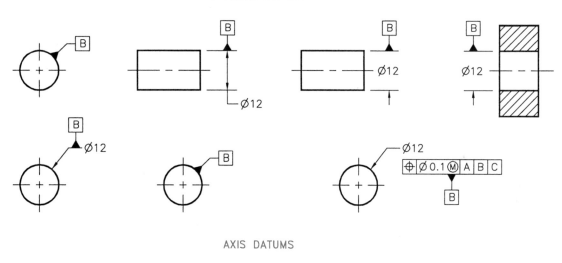

AXIS DATUMS

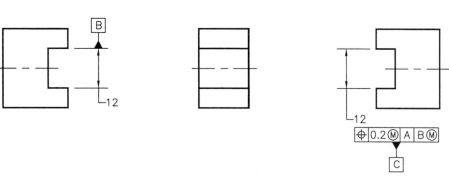

AXIS AND CENTER PLANE DATUM IDENTIFICATION
SYMBOLS MUST ALIGN WITH OR REPLACE THE
DIMENSION LINE ARROWHEAD OR BE PLACED ON
THE FEATURE, LEADER SHOULDER, OR FEATURE
CONTROL FRAME.

CENTER PLANE DATUMS

Figure 11–4 Datum identification.

datum reference frame by a feature control frame or note are thereby mutually perpendicular. This theoretical reference frame constitutes the three-plane dimensioning system used for datum referencing.

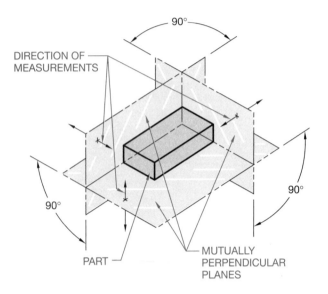

FIGURE 11–5 Datum reference frame.

## DATUM FEATURES

A datum feature is selected on the basis of its geometric relationship to the toleranced feature and the requirements of the design. To ensure proper part interface and assembly, corresponding features of mating parts are also selected as a datum feature where practical. Datum features must be readily recognizable on the part. Therefore, in the case of symmetrical parts or parts with identical features, physical identification of the datum feature on the part may be necessary. A datum feature should be accessible on the part and be of sufficient size to permit processing operations.

The three reference planes are mutually perpendicular unless otherwise specified. Planes in the datum reference frame are placed in order of importance. For example, the plane that originates most location dimensions, or has the most functional importance, is the *primary*, or first, datum plane. The next datum is the *secondary*, or second, datum plane, and the third element of the frame is called the *tertiary*, or third, datum plane. The desired order of precedence is indicated by entering the appropriate datum reference letters from left to right in the feature control frame. For instructional purposes it may be convenient to label datums as A, B, and C; although, in industry, other letters are also used to identify datums, such as D, E, and F, or X, Y, and Z. The letters that should be avoided are O, Q, and I. Figure 11–6(a) shows a part and

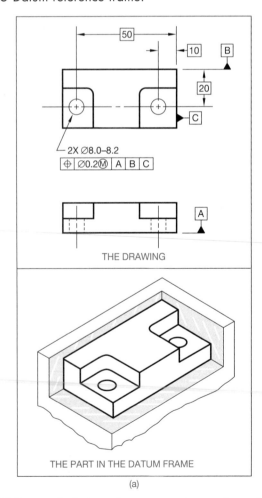

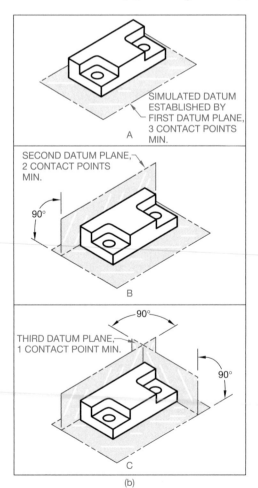

(a)            (b)

Figure 11–6 (a) Part in which datum features are plane surfaces; (b) sequence of datum features relates part to datum reference frame.

the plane that are chosen as datum features. Notice, in Figure 11–6(b), how the order of precedence of datum features relates the part to the datum reference frame. The datum features are identified as surfaces A, B, and C. These surfaces are most important to the design and function of the part. Surfaces A, B, and C are the primary, secondary, and tertiary datum features, respectively, because they appear in that order in the feature control frame.

Multiple datum frames can be established for some parts, depending on the complexity and function of the part. The relationship between datum frames is often controlled by a representative angle (see Figure 11–7).

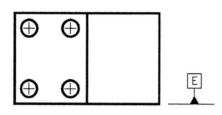

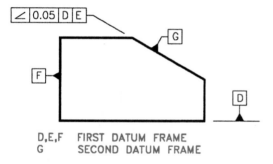

D,E,F    FIRST DATUM FRAME
G        SECOND DATUM FRAME

Figure 11–7 Multiple datum frames.

## Partial Surface Datums

In some situations it may be more realistic to apply a datum feature symbol to a portion of a surface rather than the entire surface. For example, when a long part has related features located in one or more concentrated places, the datum may be located next to the important features. When you do this, a chain line is used to identify the extent of the datum feature. The location and length of the chain line must be dimensioned (see Figure 11–8).

## Coplanar Surface Datums

Coplanar surfaces are two or more surfaces that are in the same plane. The relationship of coplanar surface datums may characterize the surfaces as one plane or datum in correlated geometric tolerance callouts, as shown in Figure 11–9.

# DATUM TARGET SYMBOLS

In many situations it is not possible to establish an entire surface or surfaces as datums, or it may not be practical to coordinate a partial datum surface with the part features. When this happens, because of the size or shape of the part, datum targets can be used

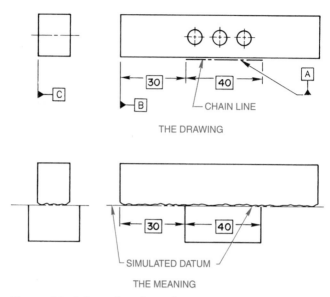

Figure 11–8 Partial surface datums.

to establish datum planes. Datum targets are especially useful on parts with surface or contour irregularities, such as sand castings or forgings, on some sheet metal parts subject to bowing or warpage, or on weldments where heat can induce warpage. Datum targets are designated points, lines, or surface areas you can use to establish the datum reference frame. The primary datum is established by three points or contact locations. The three locations are placed on the part so they form a stable, well-spaced pattern and are not in a line. The secondary datum is created by two points or locations. The tertiary datum is established by one point or location. Remember, the purpose of the datum frame is to prepare a stable position for the part dimensions to be established in relationship to corresponding datums. For this reason, the three points on the primary datum are used to provide stability similar to a three-legged stool. The two points on the secondary datum

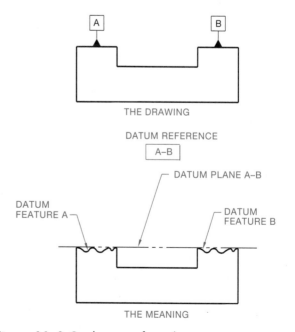

Figure 11–9 Coplanar surface datums.

provide the needed amount of stability when the object is placed against the secondary datum. Finally, the tertiary datum requires only one point of contact to complete the stability between the three datum frame elements (see Figure 11–10).

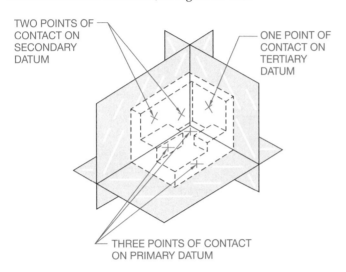

TWO POINTS OF CONTACT ON SECONDARY DATUM

ONE POINT OF CONTACT ON TERTIARY DATUM

THREE POINTS OF CONTACT ON PRIMARY DATUM

Figure 11–10 Datum frame established by datum target symbols.

The datum target symbol is connected with a leader that points to a target point, line, or surface area. The datum target symbol is divided into two halves by a horizontal line. The top half of the symbol is reserved for identification of the datum target area size when used. The bottom half of the symbol is used for datum target identification. For example, if there are three datum target points on a datum, the first point may be labeled A1, the second A2, and the third A3. The datum target point is located on the surface or edge view from adjacent datums with basic dimensions (see Figure 11–11). Chain dimensioning is commonly used to locate datum target points and target areas. The location dimensions originate from datums. Datum target areas are located to their centers (see Figure 11–12). When the surface of a part has an irregular contour or different levels, the contact points, lines, or areas may lie on the same plane and different lengths of locating pins can be used. The pins used to make point contact are usually rounded. The pins for target area contact have a flat surface equal in diameter to the specified target area.

## DATUM AXIS

The datum frame established by a cylindrical object is the base of the part. The two theoretical planes, represented in Figure 11–13 by the X and Y center planes, cross to establish the datum axis. The actual secondary datum is the cylindrical surface. The center planes located at X and Y are used to indicate the direction of dimensions that originate from the datum axis (see Figure 11–14).

Datum target points, lines, or surface areas can also be used to establish a datum axis. A primary datum axis can be established by two sets of three equally spaced targets: a set at one end of the cylinder and the other set near the other end, as shown in Figure 11–15. When two cylindrical features of different diameters are used to establish a datum axis, the datum target points are iden-

tified in correspondence to the adjacent cylindrical datum feature, as shown in Figure 11–16. Cylindrical datum target areas and circular datum target lines may also be used to establish the datum axis of cylindrically shaped parts, as shown in Figure 11–17. A secondary datum axis is established by placing three equally spaced targets on the cylindrical surface (see Figure 11–18).

## FEATURE CONTROL FRAME

The feature control frame is used to relate a geometric tolerance to a part feature. The elements in a feature control frame must always be in the same order. The most basic format is when a geometric characteristic and related tolerance are applied to an individual feature, as shown in Figure 11–19. The next expanded format is when a geometric characteristic, tolerance zone descriptor, and material condition symbol are used in the feature control frame (see Figure 11–20). One, two, or three datum references can be included in a feature control frame, as shown in Figure 11–21. The material condition symbol that applies to the feature tolerance is always placed after the geometric tolerance. When a material condition symbol is applied to a datum reference, it is placed after the datum reference and in the same compartment, as shown in Figure 11–22.

The feature control frame is connected to the feature with a leader or extension line (see Figure 11–23). The feature control frame can be combined with a datum feature symbol when the feature controlled by the geometric tolerance also serves as a datum. The datum feature symbol and the datum reference in the feature control frame are considered separately. When a datum feature symbol and a feature control frame are combined, the feature control frame is normally shown first. The datum feature symbol normally is centered on the base of the feature control frame (see Figure 11–24).

## BASIC DIMENSIONS

A *basic dimension* is defined as any size or location dimension that is used to identify the theoretically exact size, profile, orientation, or location of a feature or datum target. Basic dimensions are the basis from which permissible variations are established by tolerances on other dimensions in notes or in feature control frames. A basic dimension is described on a print by a box around the dimension. Metric and inch basic dimension numerals follow the same rules applied to other dimension numerals. Refer to Chapter 7. Metric basic dimensions contain only the number of decimal places needed to display the intended control. For inch basic dimensions, the basic dimension value is expressed in the same number of decimal places as the related tolerance (see Figure 11–25).

## GEOMETRIC TOLERANCES

Geometric tolerances are divided into five types: form, profile, orientation, runout, and location. These tolerances are subdivided into 13 characteristics and modifying terms, all of which are discussed in this section.

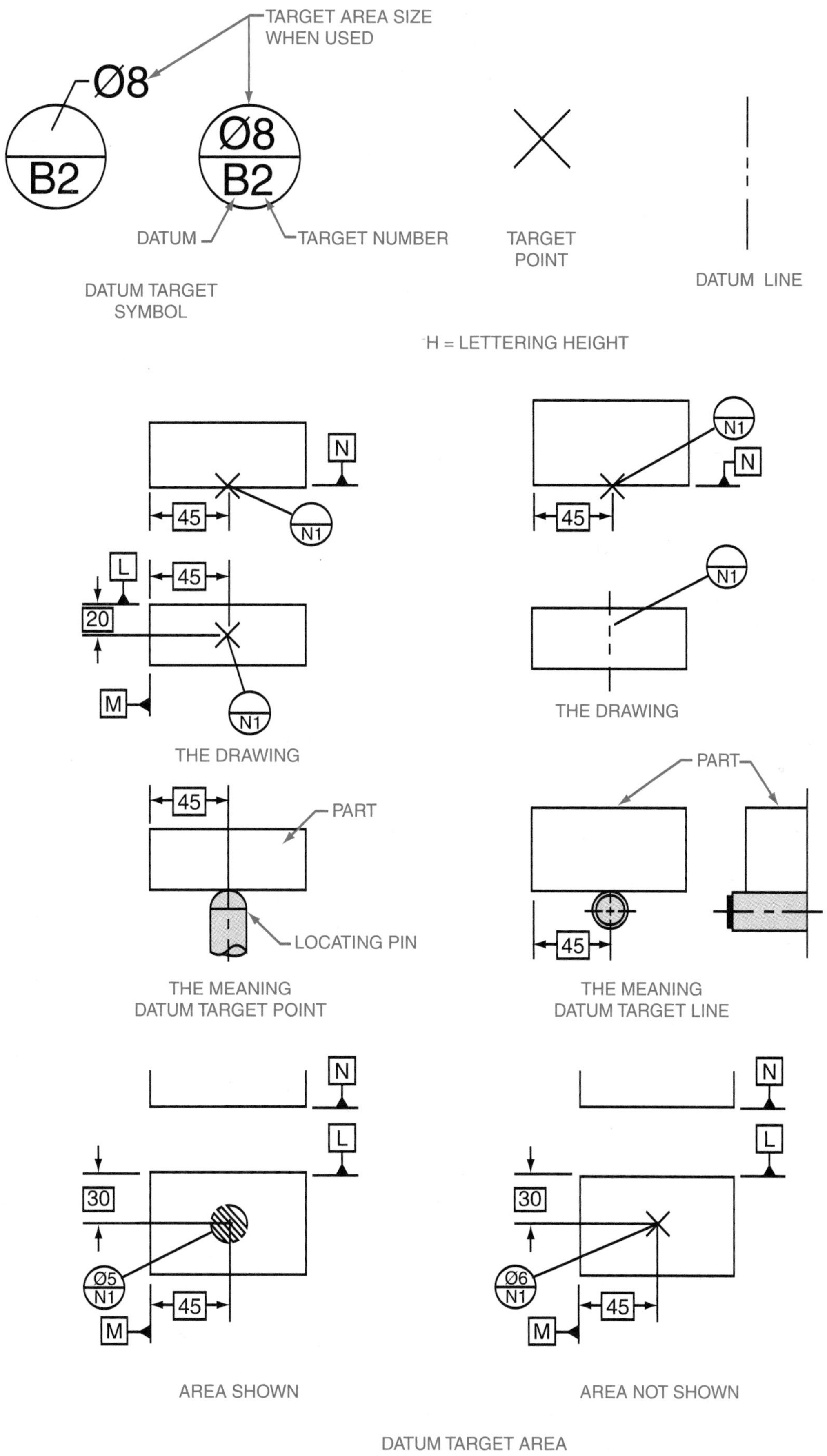

Figure 11–11 Datum target symbol, target point, datum line, and target area.

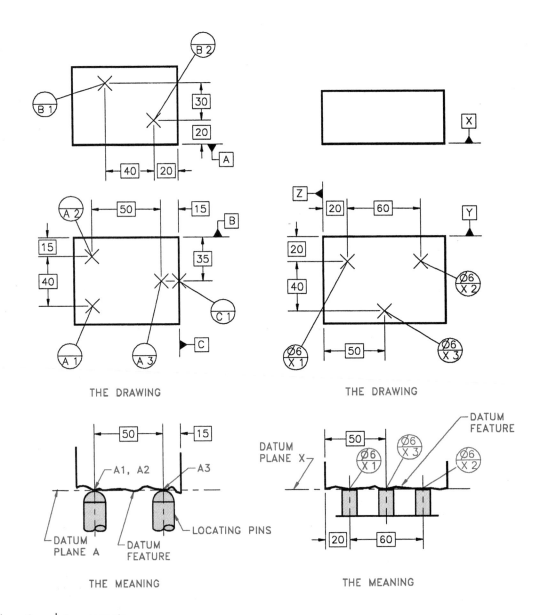

Figure 11-12 Locating datum targets.

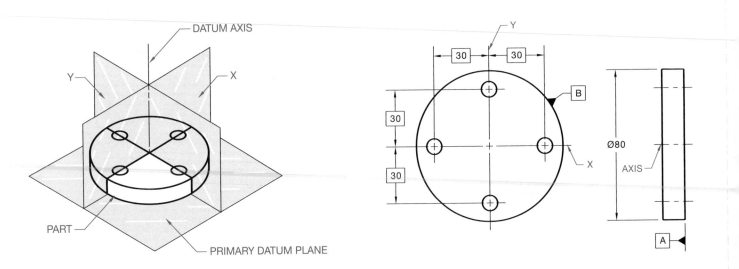

Figure 11-13 Datum frame for cylindrical object.

Figure 11-14 Datum axis.

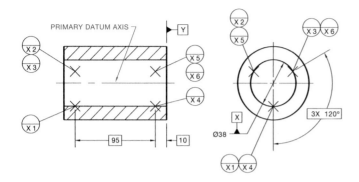

Figure 11–15 Establishing a primary datum axis with two sets of three equally spaced targets.

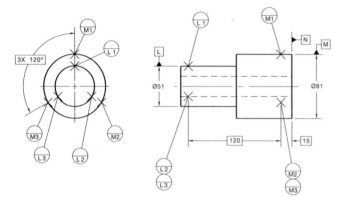

Figure 11–16 Datum targets identified on adjacent cylindrical features.

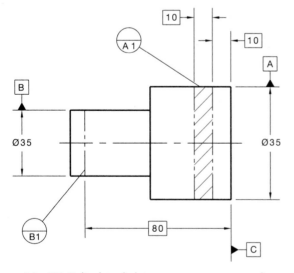

Figure 11–17 Cylindrical datum target areas and circular datum target lines.

## Form Tolerance

A tolerance of form is commonly applied to individual features or elements of single features and is not related to datums. The amount of given form variation must fall within the specified size tolerance zone.

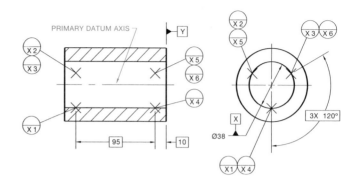

Figure 11–18 Establishing a secondary datum on a cylindrical object with three equally spaced targets.

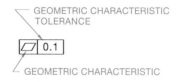

Figure 11–19 Feature control frame with geometric characteristics and related tolerance.

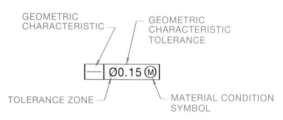

Figure 11–20 Feature control frame with geometric characteristic, tolerance zone descriptor, and material condition symbol.

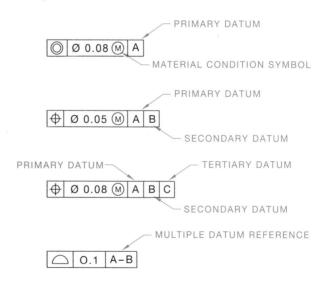

Figure 11–21 Applying one, two, or three datum references to the feature control frame.

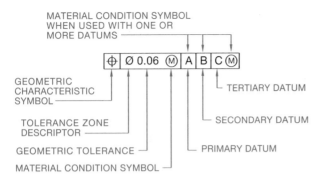

MATERIAL CONDITION SYMBOL
WHEN USED WITH ONE OR
MORE DATUMS

GEOMETRIC
CHARACTERISTIC
SYMBOL

TOLERANCE ZONE
DESCRIPTOR

GEOMETRIC TOLERANCE

MATERIAL CONDITION SYMBOL

TERTIARY DATUM

SECONDARY DATUM

PRIMARY DATUM

⊕ Ø 0.06 Ⓜ A B C Ⓜ

Figure 11–22 Elements of a feature control frame.

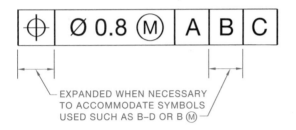

⊕ Ø 0.8 Ⓜ A B C

EXPANDED WHEN NECESSARY
TO ACCOMMODATE SYMBOLS
USED SUCH AS B–D OR B Ⓜ

Figure 11–23 Detailed feature control frame and application.

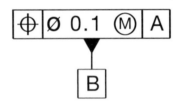

⊕ Ø 0.1 Ⓜ A

B

Figure 11–24 Combination datum feature symbol and feature control frame.

45.5

30°

BASIC DIMENSION SYMBOLS

## Straightness

The straightness symbol is detailed in Figure 11–26. Perfect straightness exists when a surface element or the axis of a part is a straight line. A straightness tolerance allows for a specified amount of variation from a straight line. A straightness tolerance allows for a specified amount of variation from a straight line. The straightness feature control frame can be attached to the surface of the object with a leader or combined with the diameter dimension of the part. When straightness is connected to the surface of the feature with a leader, *surface straightness* is implied, as shown in Figure 11–27. When the straightness feature control frame is combined with the diameter dimension of the part, *axis straightness* is specified (see Figure 11–28). The tolerance zone shape is cylindrical for this application. While a straightness tolerance is common on cylindrical parts, straightness may also be applied to the surface of noncylindrical parts. When specifying center-plane feature straightness, a center plane is implied and the diameter tolerance zone descriptor is omitted. The straightness geometric tolerance must always be less than the size tolerance, thereby refining the size tolerance.

Unit straightness is a situation where a specified tolerance is given per unit of length, for example 25 mm units, and a greater amount of tolerance is provided over the total length of the part (see Figure 11–29). This is intended to control excessive waviness in a long part.

## Flatness

The flatness tolerance symbol is detailed in relationship to the feature control frame in Figure 11–30. A surface is considered *flat* when all of the surface elements lie in one plane. A flatness geometric tolerance callout allows a specified amount of surface variation from a flat plane. The flatness tolerance zone establishes two

| 14 | 2.000 |
|---|---|
| ASSOCIATED WITH | ASSOCIATED WITH |

⊕ Ø 0.15 Ⓜ A B C          ⊕ Ø .005 Ⓜ A B C

METRIC APPLICATIONS          INCH APPLICATIONS

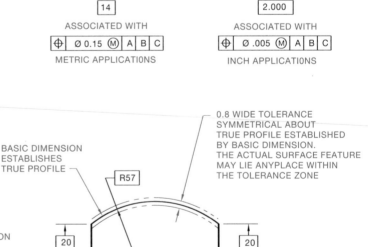

Figure 11–25 Basic dimensions.

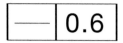

Figure 11–26 Straightness feature control frame.

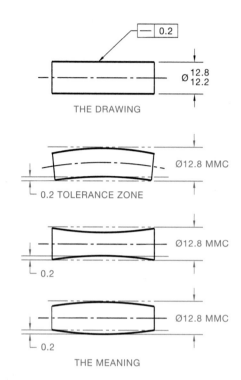

Figure 11–27 Surface straightness.

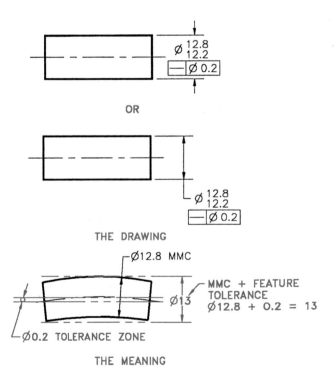

Figure 11–28 Axis straightness.

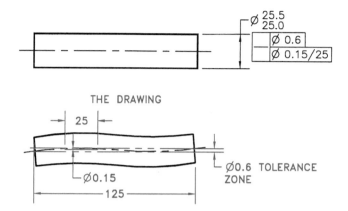

Figure 11–29 Unit straightness.

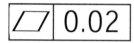

Figure 11–30 Flatness feature control frame.

parallel planes. The actual surface of the object may not extend beyond the boundary of the size tolerance zone and, when associated with the size dimension, the flatness tolerance must be smaller than the size tolerance. The flatness feature control flame can be connected to the edge view of the surface with a leader or with an extension line (see Figure 11–31).

*Unit flatness* is specified when it is desirable to control the flatness of surface units. The unit size can be 25 x 25 mm or 1.00 x 1.00 in. This is used to prevent abrupt surface variation. You can present a unit flatness callout with or without a separate tolerance for the total area. Unit flatness, just as unit straightness, often has a total tolerance to avoid a situation where the unit tolerance gets out of control. The size of the unit area is given after the unit tolerance specification. The feature control frame is doubled in height and the total area flatness is given first, as shown in Figure 11–32.

## Circularity

The circularity geometric characteristic symbol is shown in a feature control frame in Figure 11–33. Circularity can be applied to cylindrical, conical, or spherical shapes. Circularity exists when all of the elements of a circle are the same distance from the center. *Circularity* is a cross-sectional evaluation of the feature to determine if the circular surface lies between a tolerance zone that is made up of two concentric circles. The cross-sectional tolerance zone is established perpendicular to the axis of the feature. The term *circular* or *line element* is used to refer to a cross-sectional or single-line tolerance zone, rather than a blanket or entire-surface tolerance zone. The circularity geometric tolerance zone is a radius dimension. The circularity feature control frame may be connected to the part with a leader in the circular or rectangular view, as shown in Figure 11–34. The circularity tolerance must always be less than the size tolerance, except for parts subject to free state variations. *Free state variation* is a term used to describe distortion of a part after removal of forces applied during

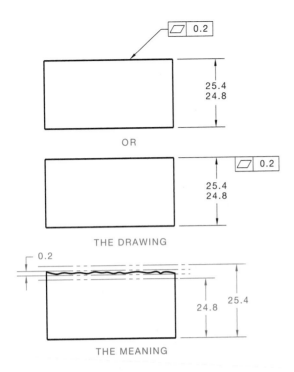

Figure 11–31 Flatness representation.

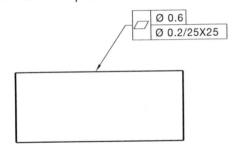

Figure 11–32 Unit flatness.

manufacturing, such as with a rubber gasket or sheet metal. The part may have to meet the tolerance specifications while in free state, or it may be necessary to hold features in a simulated mating part to verify dimensions. The free state symbol, shown in Figure 11–35, is placed in the feature control frame after the geometric tolerance and material condition symbol, if any.

### Cylindricity

The cylindricity geometric characteristic symbol is shown in Figure 11–36. Cylindricity is similar to circularity in that both have a radius tolerance zone. The difference is that circularity is a cross-sectional tolerance that results in a feature that must lie between two concentric circles, while cylindricity is a blanket tolerance that results in a feature lying between two concentric cylinders. The cylindricity feature control frame may be connected with a leader to either the circular or rectangular view (see Figure 11–37). The cylindricity tolerance must be less than the size tolerance.

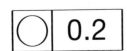

Figure 11–33 Circularity feature control frame.

### Profile Tolerance

*Profile* is used to control form or a combination of size, form, and orientation. Profile callouts are commonly used on arcs, curves, or irregular-shaped features, but can also be applied to plane surfaces. You define the shape and size of the profile with basic dimensions or tolerance dimensions. When tolerance dimensions are used to establish profile, the profile tolerance zone must be within the size tolerance. The profile feature control frame is connected to the longitudinal view with a leader line. The two types of profile are profile of a line and profile of a surface (see Figure 11–38).

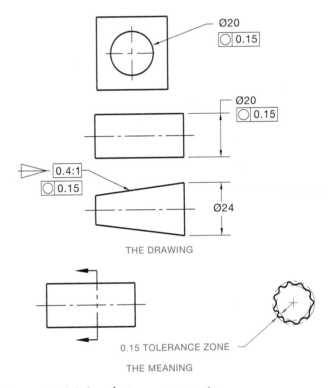

Figure 11–34 Circularity representation.

Figure 11–35 Using the free state symbol.

The *profile of a line* is cross-sectional or single-line tolerance that extends through the specified feature. The profile of a line may be controlled in relation to datums or without datums. The profile of a line may be all around the part by using the all around symbol on the leader, as seen in Figure 11–39, or between two specified points on the part by using the between symbol and identifying the points, as shown in Figure 11–40.

The *profile of a surface* is a blanket tolerance zone that affects all the surface elements of a feature equally. The profile of a surface is generally referenced to one or more datums so proper orientation of the profile boundary can be maintained. The profile of a surface can be applied all around a part, as in Figure 11–41, or between two specified points, as shown in Figure 11–42.

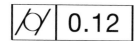

Figure 11–36 Cylindricity feature control frame.

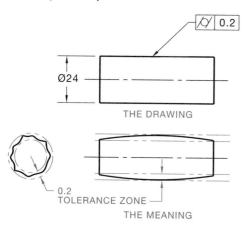

Figure 11–37 Cylindricity representation.

The profile of a line or the profile of a surface is a bilateral tolerance zone when the leader line points to the surface of the part without any additional symbology, as seen in Figures 11–38 and 11–39. A *bilateral profile tolerance* means that the tolerance zone is split equally on each side of the specified perfect form. Either type of profile callout may also have a unilateral tolerance zone specified. A unilateral tolerance zone places the entire zone on only one side of the true profile or perfect form. This is accomplished on a print with a phantom line parallel to the surface where the leader arrowhead touches the part. The phantom line is placed any clear distance from the part surface and is either inside or outside depending on the specified direction of the unilateral tolerance from the true profile. The actual feature is confined between the true profile and the given tolerance reference (see Figure 11–43).

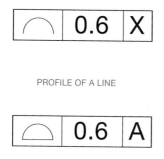

PROFILE OF A LINE

PROFILE OF A SURFACE

Figure 11–38 Profile feature control frames.

The profile of coplanar surfaces can also be specified by placing a phantom line between the surfaces in the view where they appear as edges. You identify the number of coplanar surfaces below the feature control frame with a note, such as 2 SURFACES. The profile feature control frame is then connected to the phantom line with a leader (see Figure 11–44). The profile is also used to control the angle of an inclined surface in relationship to a datum (see Figure 11–45).

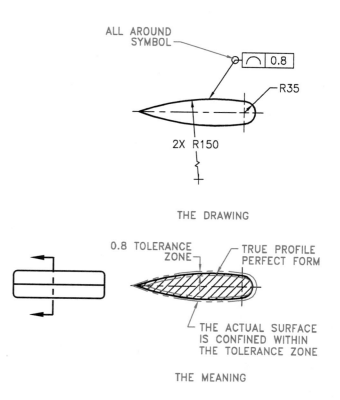

Figure 11–39 Profile of a line all around, and the all around symbol.

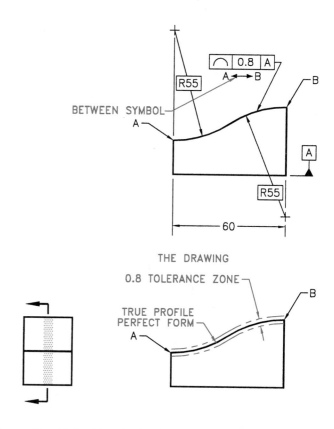

Figure 11–40 Profile of a line between two given points, and the between symbol.

## Orientation Tolerance

*Orientation* tolerances refer to a specific group of geometric characteristics that establish a relationship between the features of an object. The tolerances that orient one feature to another are parallelism, perpendicularity, angularity, and, in some applications, profile. Orientation tolerances require that you use one or more datums to establish the relationship between features. Parallelism, perpendicularity, and angularity also control flatness or cylindricity, depending on the shape of the feature being controlled. The geometric tolerance must fall within the size tolerance of the part.

### Parallelism

The parallelism geometric characteristic symbol is shown in Figure 11–46. A *parallelism* tolerance zone requires that the actual feature is between two parallel planes or lines that are parallel to a datum. The parallelism feature control frame can be attached to the feature with a leader or on an extension line of the surface (see Figure 11–47). Unless otherwise specified, a parallelism tolerance zone is a total tolerance that covers the entire surface. If a single-line element is to be specified instead of the surface, the note EACH ELEMENT, or EACH RADIAL ELEMENT for an arc-shaped surface, must be added below the feature control frame (see Figure 11–48). This technique also applies to perpendicularity and angularity.

In certain instances parallelism can also be applied to a cylindrical feature. The tolerance zone is a cylindrical shape that is parallel to a datum axis reference. The feature control frame that relates to an axis specification is attached to the diameter dimension, and a diameter zone descriptor should precede the geometric tolerance (see Figure 11–49).

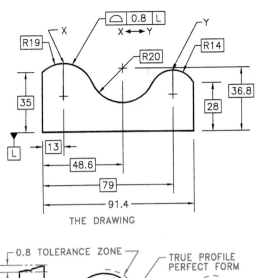

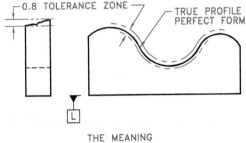

Figure 11–42 Profile of a surface between two given points.

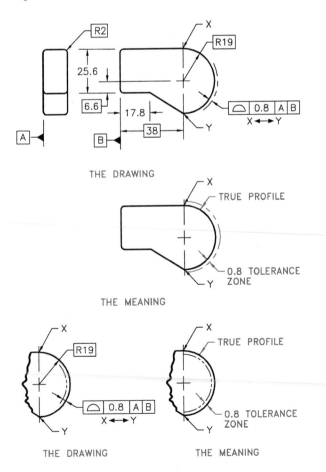

Figure 11–43 A unilateral profile representation.

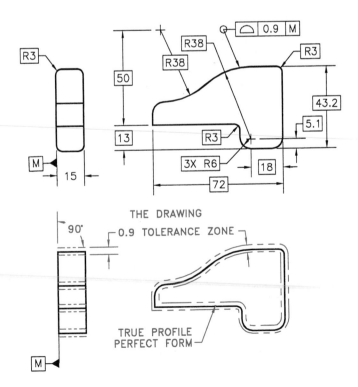

Figure 11–41 Profile of a surface all around.

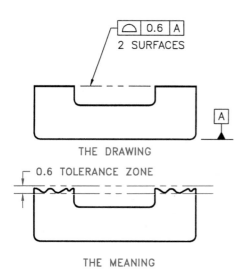

Figure 11–44 Profile of coplanar surfaces.

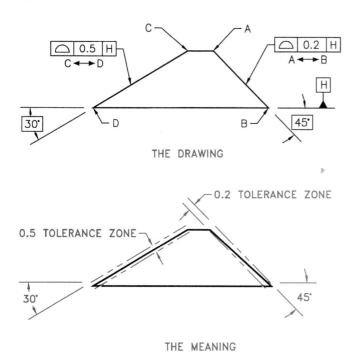

Figure 11–45 Profile of an inclined surface.

Figure 11–46 Parallelism feature control frame.

## Perpendicularity

The perpendicularity geometric characteristic symbol is shown in Figure 11–50. A *perpendicularity* geometric tolerance requires that a given feature be located between two parallel planes or lines, or within a cylindrical tolerance zone that is a basic 90° to a datum. The perpendicularity feature control frame can be connected to the feature surface with a leader or an extension line, or attached to the diameter dimension for axis perpendicularity (see Figure 11–51).

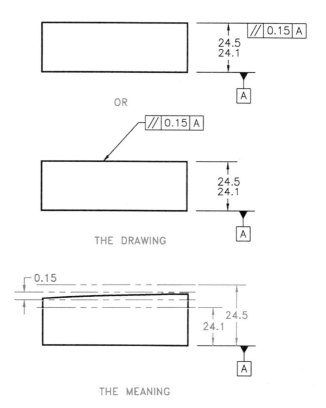

Figure 11–47 Parallelism representation.

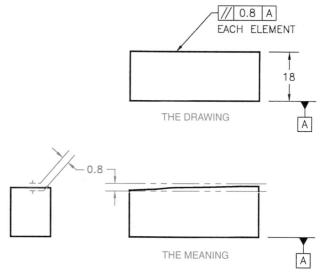

Figure 11–48 Single-line element parallelism.

## Angularity

The angularity geometric characteristic symbol is represented in Figure 11–52. An *angularity* tolerance zone places a given feature between two parallel planes that are at a specified basic angular dimension from a datum. The basic angle from the datum can be any amount except 90°. The angularity feature control frame can be connected to the surface with a leader or from an extension line, or attached to the diameter dimension for axis angularity (see Figure 11–53).

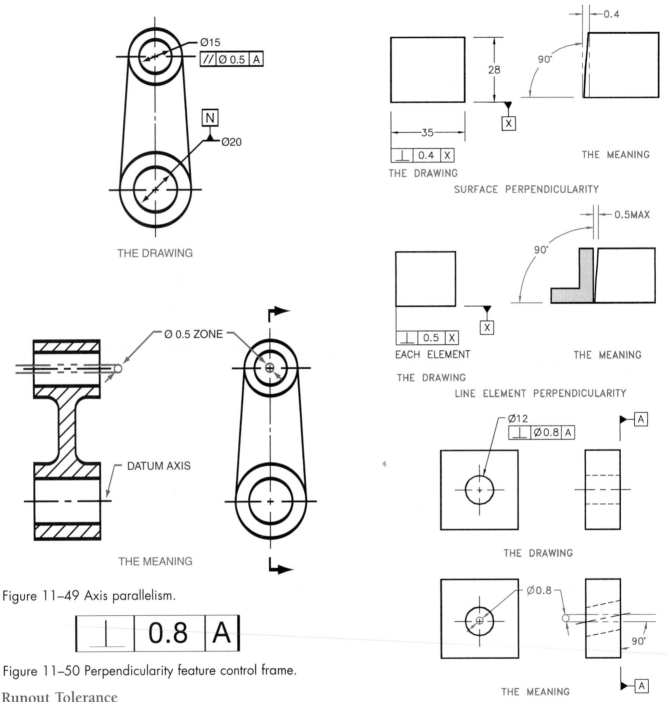

Figure 11–49 Axis parallelism.

Figure 11–50 Perpendicularity feature control frame.

## Runout Tolerance

*Runout* is used to control the relationship of radial features to a datum axis and features that are 90° to a datum axis. These types of features include cylindrical, tapered, and curved shapes, as well as plane surfaces, that are at right angles to a datum axis. A runout tolerance is determined when a dial indicator is placed on the surface to be inspected and the part is rotated 360°. The *full indicator movement (FIM)* on the dial indicator must not exceed the amount of specified tolerance. The runout feature control frame is attached to the required surface with a leader. The runout tolerance applies to the length of the intended surface or until there is a break, or a change in shape or diameter of the surface. There are two types of runout: circular and total (see Figure 11–54).

Figure 11–51 Perpendicularity representations.

## Circular Runout

*Circular runout* provides control of single, circular elements of a surface. Circular runout controls circularity and the relationship of the common axes (coaxial) of parts. This tolerance, established by the FIM of a dial indicator, is placed at one location on the feature as the part is rotated 360° (see Figure 11–55).

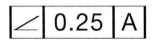

Figure 11–52 Angularity feature control frame.

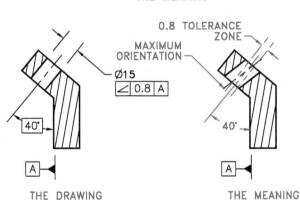

THE DRAWING

THE MEANING

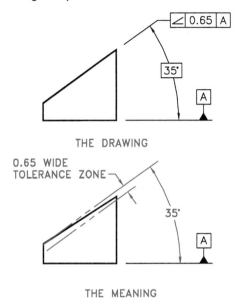

Figure 11–53 Angularity representations.

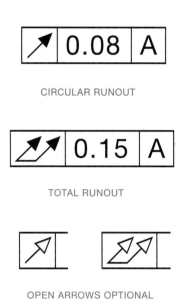

CIRCULAR RUNOUT

TOTAL RUNOUT

OPEN ARROWS OPTIONAL

Figure 11–54 Runout feature control frames.

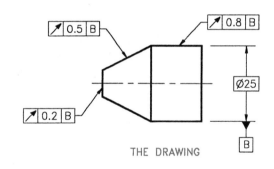

THE DRAWING

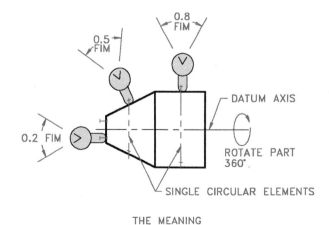

THE MEANING

Figure 11–55 Circular runout.

**Total Runout**

*Total runout* provides composite control of all surface elements. The tolerance is applied simultaneously to all circular and profile measuring positions as the part is rotated 360° (see Figure 11–56). Where applied to surfaces constructed around a datum axis, total runout is used to control cumulative variations of circularity, straightness, coaxiality, angularity, taper, and profile of a surface. Where applied to surfaces constructed at right angles to a datum axis, total runout controls cumulative variations of perpendicularity to detect wobble and flatness, and to detect concavity or convexity.

A portion of a surface can have a specified runout tolerance if you don't want to control the entire surface. You do so with a chain line placed in the linear view adjacent to the desired location. The chain line is located with basic dimension, as shown in Figure 11–57. When a part has compound datum features—that is, where more than one datum feature is used to establish a common datum—the combined datum features are shown separated by a

dash in the feature control frame. The datums are of equal importance (see Figure 11–57).

## MATERIAL CONDITION SYMBOLS

Material condition symbols are used in conjunction with the feature tolerance or datum reference in the feature control frame. The *material condition symbols* are required to establish the relationship between the size or location of the feature and the geometric tolerance. The use of different material condition symbols alters the effect of this relationship. The material condition modifying elements are maximum material condition, MMC; regardless of feature size, RFS; and least material condition, LMC. There is no symbol for RFS, because it is assumed for all geometric tolerances and datum references, unless another material condition symbol is specified. The standard material condition symbols are shown in Figure 11–58.

### Perfect Form Envelope

The form of a feature is controlled by the size tolerance limits. The envelope, or boundary, of these size limits is established at MMC. Remember from the discussion in Chapter 7 that MMC is the largest limit for an external feature and the smallest limit for an internal feature. The key is *most material*. The true geometric form of the feature is at MMC. This is known as the *perfect form boundary*. This is also referred to as the perfect form envelope. If the part feature is produced at MMC, it is considered to be at perfect form. When you want to permit a surface or surfaces of a feature to exceed the boundary of perfect form at MMC, a note, such as PERFECT FORM AT MMC NOT REQUIRED, is specified, exempting the pertinent size dimension. If a feature is produced at LMC, the opposite of MMC, the form is allowed to vary within the geometric tolerance zone or to the extent of the MMC envelope.

### Regardless of Feature Size

*Regardless of feature size (RFS)* means that the geometric tolerance applies at any produced size. The tolerance remains as the specified value regardless of the actual size of the feature. Regardless of feature size is implied (assumed) for all geometric characteristics and related datums, unless otherwise specified.

### Effect of RFS on Surface Straightness

RFS is implied for the straightness geometric characteristic. When a surface straightness tolerance is used by connecting a leader from the feature control frame to the surface of the part, the geometric tolerance remains the same regardless of the feature size, and the actual size may not exceed the perfect form envelope at MMC. An acceptable part can be produced between the given size tolerance, and any straightness irregularity may not be greater than the specified geometric tolerance. Perfect form is required at MMC (see Figure 11–59).

### Effect of RFS on Axis Straightness

When the straightness tolerance is applied to the axis of the feature by a relationship with the diameter dimension, RFS is implied unless otherwise specified, and the actual feature size plus the geometric tolerance may exceed the MMC perfect form envelope (see Figure 11–60).

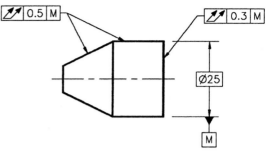

THE DRAWING

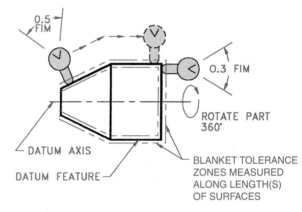

THE MEANING

Figure 11–56 Total runout.

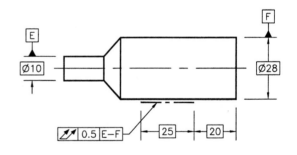

Figure 11–57 Partial surface runout.

MAXIMUM MATERIAL CONDITION
MMC

**NO SYMBOL.**
RFS IS IMPLIED UNLESS
OTHERWISE SPECIFIED

REGARDLESS OF FEATURE SIZE
RFS

LEAST MATERIAL CONDITION
LMC

Figure 11–58 Material condition symbols.

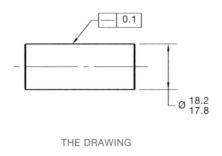

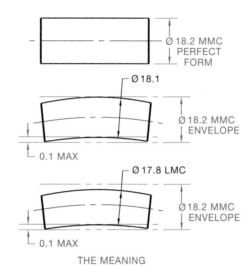

THE DRAWING

| ACTUAL SIZE | GEOMETRIC TOLERANCE |
|-------------|---------------------|
| 18.2 MMC | 0–PERFECT FORM |
| 18.1 | 0.1 |
| 18.0 | 0.1 |
| 17.9 | 0.1 |
| 17.8 LMC | 0.1 |

Figure 11–59 Effect of regardless of feature size, RFS, on surface straightness. Perfect form is required at MMC.

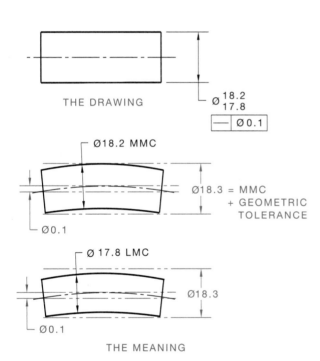

THE DRAWING

THE MEANING

| ACTUAL SIZE | GEOMETRIC TOLERANCE |
|-------------|---------------------|
| 18.2 MMC | 0.1 |
| 18.1 | 0.1 |
| 18.0 | 0.1 |
| 17.9 | 0.1 |
| 17.8 LMC | 0.1 |

Figure 11–60 Effect of RFS on axis straightness.

## Datum Feature RFS

Datum features that are influenced by size variations, such as diameters and widths, are also subject to variations in form. RFS is implied, unless otherwise specified. When a datum feature has a size dimension and a geometric form tolerance, the size of the simulated datum is the MMC size limit. This rule applies except for axis straightness where the envelope is allowed to exceed MMC. Figure 11–61 shows the effect of RFS on the primary datum feature with axis and center plane datums. When the datum features are secondary or tertiary, the axis or center plane also has an angular relationship to the primary datum (see Figure 11–62).

## Maximum Material Condition

The use of MMC in conjunction with the geometric tolerance in a feature control frame means that the given tolerance applies at the MMC-produced size. As the feature dimension departs from MMC, the geometric tolerance is allowed to increase equal to the change. The maximum amount of change is at the LMC-produced size. MMC must be specified for any geometric characteristic where its application is desired.

## Effect of MMC on Form Tolerance

When MMC is specified in conjunction with an axis straightness callout, the MMC feature size envelope is exceeded by the given geometric tolerance. The given geometric tolerance is held at the MMC-produced size; and as the actual produced size departs from MMC, the geometric tolerance is allowed to increase equal to the change from MMC to a maximum amount of departure at LMC (see Figure 11–63). The same situation can also be demonstrated with a perpendicularity tolerance where a datum axis is perpendicular to a given datum, as in Figure 11–64.

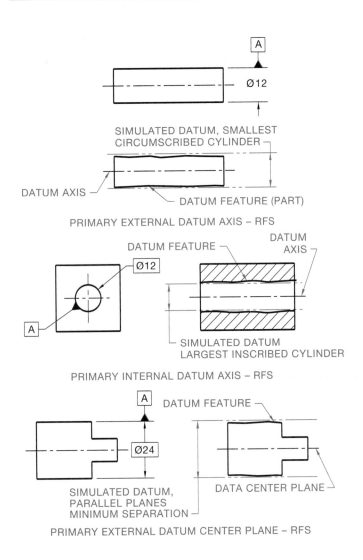

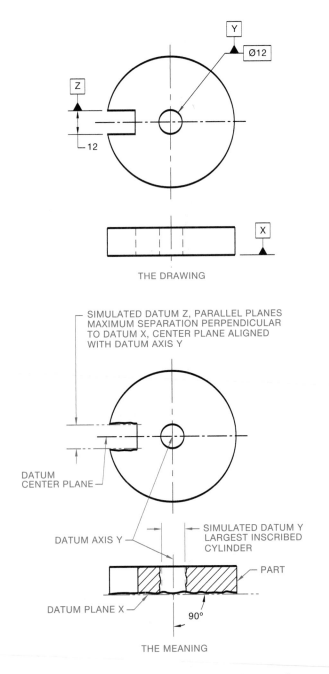

Figure 11–61 Effect of RFS on the primary datum feature with axis and center-plane datums.

### Least Material Condition

The application of LMC is not as prevalent as MMC. LMC can be used when it is desirable to control minimum wall thickness. This is discussed with positional geometric tolerance at LMC later. When an LMC material condition symbol is used in conjunction with a geometric tolerance, the specified tolerance applies at the LMC-produced size. GDT controls are refinement of size controls. As the actual produced size deviates from LMC toward MMC, the tolerance is allowed an increase equal to the amount of change from LMC (see Figure 11–65).

## LOCATION TOLERANCE

Location tolerances include concentricity, symmetry, and position. *True position* is the theoretically exact location of the axis or center plane of a feature. The true position is located using basic dimensions. The locational tolerance specifies the amount that the axis of the feature is allowed to deviate from true position. All geometric tolerances imply RFS. MMC or LMC must be specified if

Figure 11–62 The secondary or tertiary datum relationship to the primary datum.

desired. Reference to true position dimensions must be provided as basic dimensions at each required location on the drawing, or a general note can be used to specify that UNTOLERANCED DIMENSIONS LOCATING TRUE POSITION ARE BASIC. Basic dimensions are theoretically perfect, so that datum or chain dimensioning may be used equally to locate true position because there is no tolerance buildup. The location feature control frame is generally added to the note or dimension of the related feature.

### Concentricity

The concentricity geometric characteristic symbol is shown in Figure 11–66. *Concentricity* is the relationship of the axes of

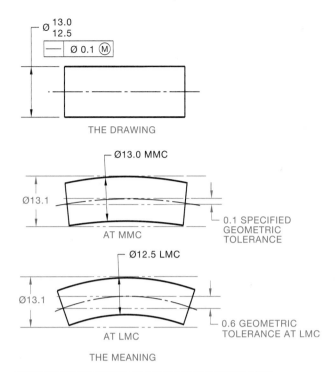

THE DRAWING

Ø13.0 MMC

Ø13.1

AT MMC

0.1 SPECIFIED
GEOMETRIC
TOLERANCE

Ø12.5 LMC

Ø13.1

AT LMC

0.6 GEOMETRIC
TOLERANCE AT LMC

THE MEANING

| ACTUAL SIZE | GEOMETRIC TOLERANCE |
|---|---|
| 13.0 MMC | 0.1 |
| 12.9 | 0.2 |
| 12.8 | 0.3 |
| 12.7 | 0.4 |
| 12.6 | 0.5 |
| 12.5 LMC | 0.6 |

Figure 11–63 Effect of maximum material condition, MMC, on form tolerance and axis straightness.

cylindrical shapes. Perfect concentricity exists when the axes of two or more cylindrical features are in perfect alignment (see Figure 11–67). A concentricity geometric tolerance allows a specified amount of deviation of the axes of concentric cylinders, as shown in Figure 11–68. The geometric tolerance and related datums imply RFS. It is difficult to control the axis relationship specified by concentricity, so runout is often used. Where the balance of a shaft is critical, concentricity can be used.

## Position Tolerance of Symmetrical Features

You can use position tolerancing to locate the center plane of one or more features relative to a datum center plane. Figure 11–69 shows optional feature control frames, where the position geometric tolerance and related datum reference are controlled on an RFS or MMC basis. When the position tolerance is at MMC, the related datum feature may be RFS, MMC, or LMC, depending on design requirements. If the position tolerance is held RFS, the related datum reference is RFS. RFS is implied unless otherwise specified.

## Symmetry

Symmetry is the center-plane relationship between two or more features. Perfect symmetry exists when the center plane of two or

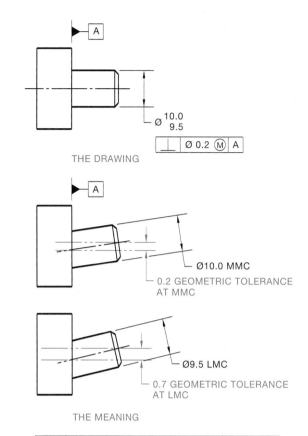

THE DRAWING

A

Ø10.0 MMC

0.2 GEOMETRIC TOLERANCE
AT MMC

Ø9.5 LMC

0.7 GEOMETRIC TOLERANCE
AT LMC

THE MEANING

| ACTUAL SIZE | GEOMETRIC TOLERANCE |
|---|---|
| 10.0 MMC | 0.2 |
| 9.9 | 0.3 |
| 9.8 | 0.4 |
| 9.7 | 0.5 |
| 9.6 | 0.6 |
| 9.5 LMC | 0.7 |

Figure 11–64 Effect of MMC on form tolerance and axis perpendicularity.

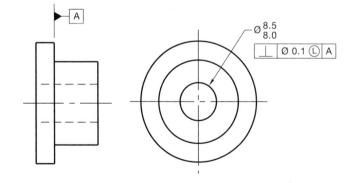

| ACTUAL SIZE | GEOMETRIC TOLERANCE |
|---|---|
| 8.0 MMC | 0.6 |
| 8.1 | 0.5 |
| 8.2 | 0.4 |
| 8.3 | 0.3 |
| 8.4 | 0.2 |
| 8.5 LMC | 0.1 |

Figure 11–65 Application of least material condition, LMC.

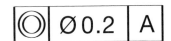

Figure 11–66 Concentricity feature control frame.

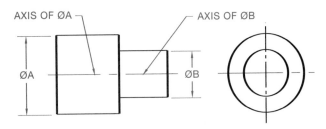

Figure 11–67 Perfect concentricity when both axes coincide.

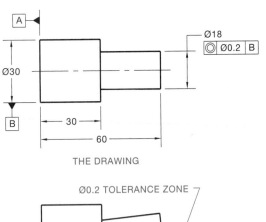

THE DRAWING

Ø0.2 TOLERANCE ZONE

EXTREME POSITION OF FEATURE AXIS

DATUM AXIS

THE MEANING

Figure 11–68 Concentricity represented.

more features is in alignment. The geometric characteristic symbol used to specify symmetry is shown in the feature control frame in Figure 11–70. The *symmetry* geometric tolerance is a zone in which the median points of opposite symmetrical surfaces align with the datum center plane. The symmetry geometric tolerance and related datum reference are applied only on an RFS basis (see Figure 11–71). A zero position tolerance at MMC is used when you need to control the symmetrical relationship of features within their size limits. An explanation of zero position tolerance at MMC follows.

## Position

The position geometric characteristic symbol, also used for symmetry, is shown in Figure 11–72. RFS is implied for the position geometric tolerance and related datum reference, unless otherwise specified. MMC or LMC is applied to the position tolerance and datum reference as needed for design. The maximum material condition application is common, although some cases require the use of it regardless of feature size or least material condition. The feature control frame is applied to the note or size dimension of the

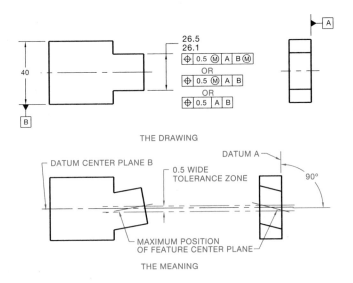

THE DRAWING

DATUM CENTER PLANE B

0.5 WIDE TOLERANCE ZONE

DATUM A

90°

MAXIMUM POSITION OF FEATURE CENTER PLANE

THE MEANING

Figure 11–69 Position geometric characteristic specifying symmetry.

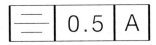

Figure 11–70 The symmetry symbol detailed in a feature control frame.

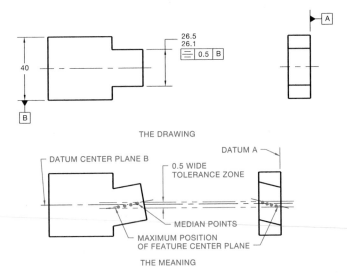

THE DRAWING

DATUM CENTER PLANE B

0.5 WIDE TOLERANCE ZONE

DATUM A

MEDIAN POINTS

MAXIMUM POSITION OF FEATURE CENTER PLANE

THE MEANING

Figure 11–71 Symmetry application.

Figure 11–72 Position feature control frame.

feature. A *positional* tolerance defines a zone within which the center, axis, or center plane of a feature or size is permitted to vary from the true, theoretically exact, position. Basic dimensions establish the true position from specified datum features and between interrelated features. A positional tolerance is indicated by the position symbol, a tolerance, and appropriate datum references placed in a feature control frame. The location of each feature is given by

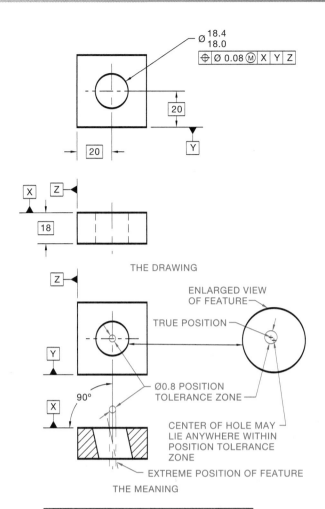

| ACTUAL SIZE | POSITIONAL TOLERANCE |
|---|---|
| 18.0 MMC | 0.08 |
| 18.1 | 0.18 |
| 18.2 | 0.28 |
| 18.3 | 0.38 |
| 18.4 LMC | 0.48 |

Figure 11–73 Positional tolerance at maximum material condition, MMC.

basic dimensions. Dimensions locating true position must be excluded from the drawing's general tolerances by applying the basic dimension symbol to each basic dimension or by specifying on the drawing or drawing reference the general note UNTOLERANCED DIMENSIONS LOCATING TRUE POSITION ARE BASIC.

### Positional Tolerance at Maximum Material Condition (MMC)

Positional tolerance at MMC means that the given tolerance is held at the MMC-produced size. Then, as the feature size departs from MMC, the position tolerance increases equal to the amount of change from MMC to the maximum change at LMC. Positional tolerance at MMC may be defined by the feature axis or surface. The datum references commonly establish true position perpendicular to the primary datum and the coordinate location dimensions to the secondary and tertiary datums. The position tolerance zone is a cylinder equal in diameter to the given position tolerance. The cylindrical tolerance zone extends through the thickness of the part, unless otherwise specified. The actual centerline of the

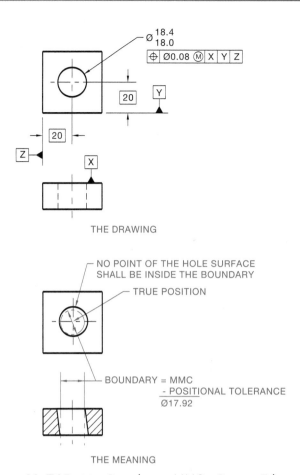

Figure 11–74 Position Boundary = MMC – Position Tolerance

feature may be anywhere within the cylindrical tolerance zone (see Figure 11–73). Another explanation of position may be related to the surface of a hole. The hole surface cannot be inside a cylindrical tolerance zone established by the MMC diameter of the hole less the position tolerance (see Figure 11–74).

### Zero Positional Tolerance at MMC

When the positional tolerance is associated with the MMC symbol, the tolerance is allowed to exceed the specified amount when the actual feature size departs from MMC. When the actual sizes of features are manufactured very close to MMC, it is critical that the feature axis or surface not exceed the boundaries discussed previously. When parts are rejected because of this situation, it is possible to increase the acceptability of mating parts by reducing the MMC size of the feature to minimum allowance with the mating part (virtual condition) and providing a zero positional tolerance at MMC. The positional tolerance is dependent on the feature size. When zero positional tolerance is used, no positional tolerance is allowed when the part is produced at MMC. True position is required at MMC. As the actual size departs from MMC, the positional tolerance increases equal to the amount of change to the maximum tolerance at LMC (see Figure 11–75).

### Positional Tolerance at RFS

You can apply RFS to the positional tolerance when it is desirable to maintain the given tolerance at any produced size. This application results in close positional control (see Figure 11–76). No additional tolerance is permitted based on size of feature.

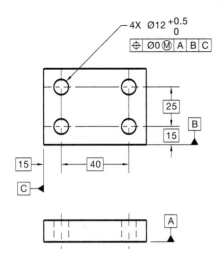

Figure 11–75 Zero positional tolerances at MMC.

| ACTUAL SIZE | POSITIONAL TOLERANCE |
|---|---|
| 12.0 MMC ZERO ALLOWANCE 12mm FASTENER AT MMC | 0 |
| 12.1 | 0.1 |
| 12.2 | 0.2 |
| 12.3 | 0.3 |
| 12.4 | 0.4 |
| 12.5 | 0.5 |

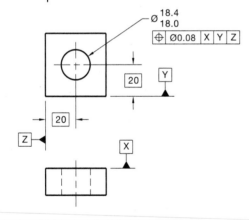

Figure 11–76 Positional tolerance at RFS.

| ACTUAL SIZE | POSITIONAL TOLERANCE |
|---|---|
| 18.0 MMC | 0.08 |
| 18.1 | 0.08 |
| 18.2 | 0.08 |
| 18.3 | 0.08 |
| 18.4 LMC | 0.08 |

### Positional Tolerance at LMC

Positional tolerance at LMC is used to control the relationship of the feature surface and the true position at largest hole size. The function of the LMC specification is generally for the control of minimum-edge distances. When LMC is used, the given positional tolerance is held at the LMC-produced size where perfect form is required. As the actual size departs from LMC toward MMC, the positional tolerance zone is allowed to increase equal to the amount of change. The maximum positional tolerance is at the MMC-produced size (see Figure 11–77).

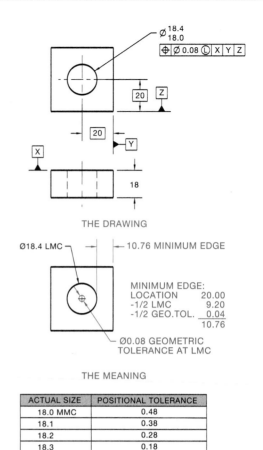

Figure 11–77 Positional tolerance at LMC.

| ACTUAL SIZE | POSITIONAL TOLERANCE |
|---|---|
| 18.0 MMC | 0.48 |
| 18.1 | 0.38 |
| 18.2 | 0.28 |
| 18.3 | 0.18 |
| 18.4 LMC | 0.08 |

## DATUM PRECEDENCE AND MATERIAL CONDITION

The effect of material condition on the datum and related feature may be altered by changing the datum precedence and the applied material condition symbol. The datum precedence is established by the order of placement in the feature control frame. The first datum listed is the primary datum; subsequent datums are secondary and tertiary. Figure 11–78 shows the effect of altering datum precedence and material condition.

## POSITION OF MULTIPLE FEATURES

The location of multiple features is handled in a manner similar to the location of a single feature. The true positions of the features are located with basic dimensions using rectangular or polar coordinates. The features are then identified by quantity, size, and position (see Figure 11–79). When two or more separate patterns are referenced to the same datums and with the same datum precedence, the patterns are functionally the same. Verification of location and size is performed together. If this situation occurs and the interrelationship between patterns is not desired, the specific note SEP REQT, meaning separate requirement, shall be placed beneath each affected feature control frame.

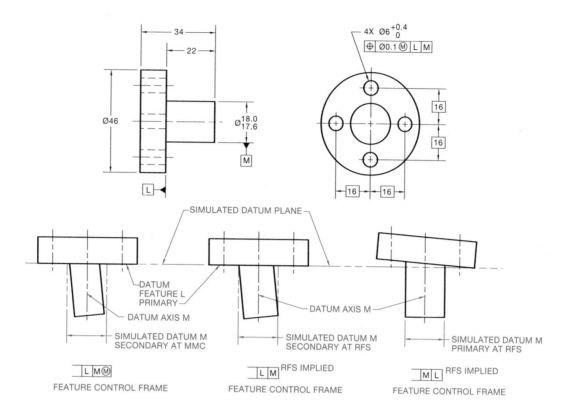

Figure 11–78 The effect of altering datum precedence and material condition.

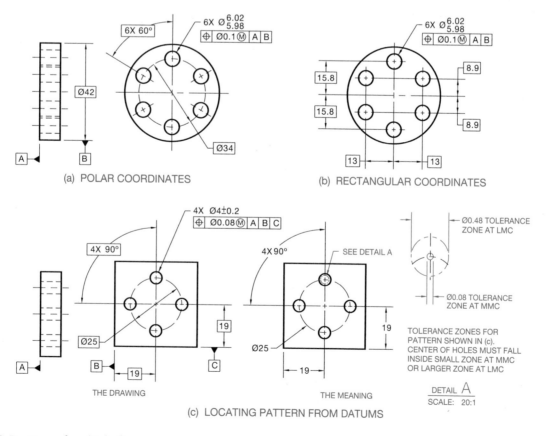

Figure 11–79 Position of multiple features.

## COMPOSITE POSITIONAL TOLERANCING

In some situations it is permissible to have the location of individual features in a pattern differ from the tolerance related to the items as a group. When you do this, the group of features is given a positional tolerance that is greater in diameter than the zone specified for the individual features. The tolerance zone of the individual features must fall within the group zone. The positional tolerance of the individual elements controls the perpendicularity of the features. An individual tolerance zone may extend partly beyond the group zone only if the feature axis does not fall outside the confines of both zones. When that is the case, the feature control frame is expanded in height and divided into two parts. The upper part, known as the *pattern locating control*, specifies the larger positional tolerance for the pattern of features as a group. The lower entry, commonly called the *feature relating control*, specifies the smaller positional tolerance for the individual features within the pattern. The pattern locating control is located first with basic dimensions. The feature relating control is established at the actual position of the feature center. Only the primary datum is represented in the feature relating control (see Figure 11–80).

## TWO SINGLE-SEGMENT FEATURE CONTROL FRAMES

The composite position tolerance is specified by a feature control frame doubled in height, with one position symbol shown in the first compartment and only a single datum reference for orientation given for the feature relating control. The *two single-segment feature control frame* is similar, except there are two position symbols, each displayed in a separate compartment, and a two-datum reference in the lower feature control frame. Look at Figure 11–81. The top feature control frame is the pattern locating control that works as previously discussed. The lower feature control frame is the feature relating control in which two datums control the orientation and alignment with the pattern locating control. This type of position tolerance provides a tighter relationship of the holes within the pattern than the composite position tolerance.

## POSITIONAL TOLERANCE OF TABS

Positional tolerancing of tabs may be accomplished by identifying related datums, dimensioning the relationship between the tabs, and providing the number of units followed by the size and feature control frame (see Figure 11–82).

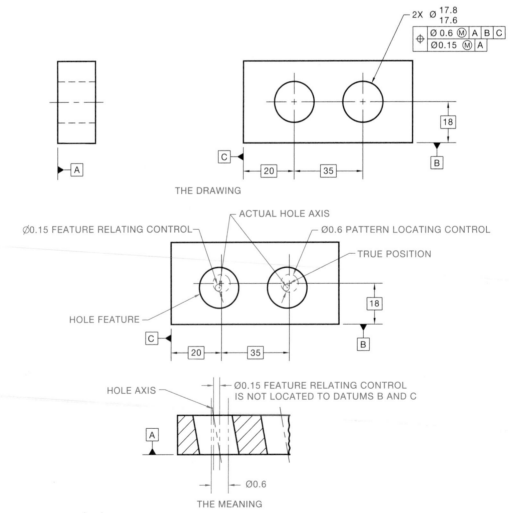

Figure 11–80 Composite positional tolerance.

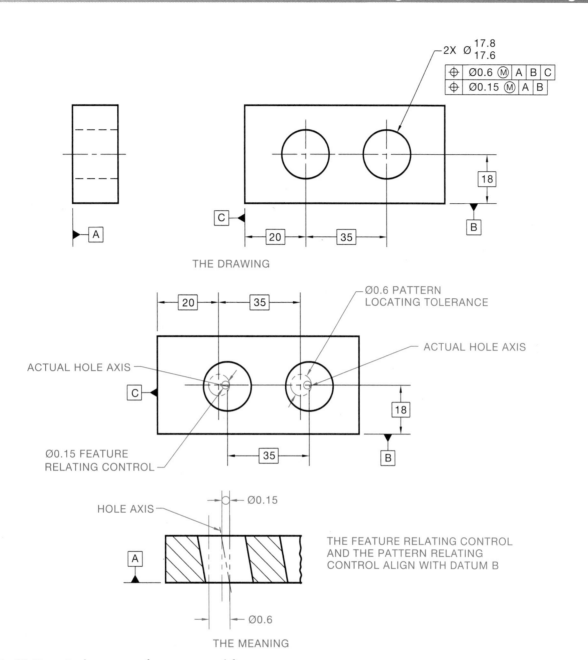

THE DRAWING

THE MEANING

Figure 11–81 Two single-segment feature control frame.

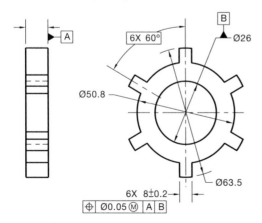

Figure 11–82 Positional tolerance of tabs; a similar practice for slots.

## POSITIONAL TOLERANCE OF SLOTTED HOLES

The positional tolerance of slotted holes may be accomplished by locating the slot centers with basic dimensions and providing a feature control frame to both the length and width of the slotted holes. (see Figure 11–83). The word BOUNDARY is placed below each feature control frame. This means that each slotted feature is controlled by a theoretical boundary of identical shape that is located at true position. The size of each slot must remain within the size limits, and no portion of the surface may enter the theoretical boundary, which is the size of the boundary calculated by MMC (each dimension, length, and width) – Position Tolerance. Normally there is a greater position tolerance for the length than the width. When the position tolerance is the same for the length

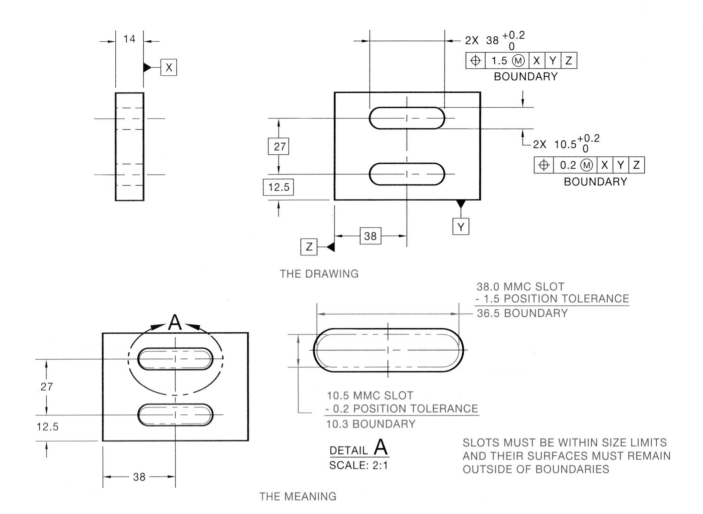

Figure 11–83 Positional tolerance for slotted holes.

and width, the feature control frame is separated from the size dimensions and connected to one of the slots with a leader.

## POSITIONAL TOLERANCE OF THREADED FEATURES AND GEARS

The orientation or positional tolerance of screw threads applies to the datum axis. The datum feature is established by a pitch diameter cylinder. If a different datum feature is required, the note should be lettered below the feature control frame, for example, MINOR DIA or MAJOR DIA. Similarly, the intended datum feature for gears or splines should be identified below the feature control frame. The options include MAJOR DIA, PITCH DIA, or MINOR DIA.

## POSITIONAL TOLERANCE OF COUNTER-BORED HOLES

Counterbored holes are a type of coaxial feature. *Coaxial* means that one or more features have the same axis. There are three different ways to provide positional tolerancing to counterbored holes, as shown in Figure 11–84.

## POSITIONAL TOLERANCING OF COAXIAL HOLES

Coaxial holes that are aligned through different features may be controlled with positional tolerancing using a feature-relating zone for each hole and a pattern-locating zone for the holes as a group. Some possible options are shown in Figure 11–85.

## POSITIONAL TOLERANCING OF NON-PARALLEL HOLES

Positional tolerancing can be applied to holes that are not parallel to each other. The axis may not be perpendicular to a surface, as shown in Figure 11–86. The feature is located with a base longitudinal and angular dimension.

## PROJECTED TOLERANCE ZONE

The standard application of a positional tolerance implies that the cylindrical zone extends through the thickness of the part or feature. In situations where there is the possibility of interference with

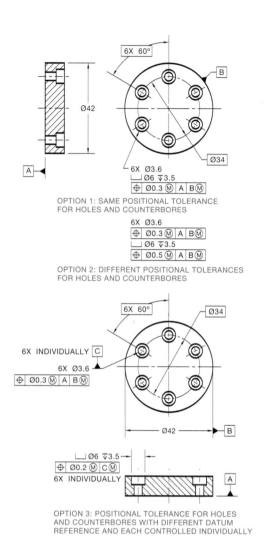

OPTION 1: SAME POSITIONAL TOLERANCE
FOR HOLES AND COUNTERBORES

OPTION 2: DIFFERENT POSITIONAL TOLERANCES
FOR HOLES AND COUNTERBORES

OPTION 3: POSITIONAL TOLERANCE FOR HOLES
AND COUNTERBORES WITH DIFFERENT DATUM
REFERENCE AND EACH CONTROLLED INDIVIDUALLY

Figure 11–84 Positional tolerance of counterbored holes.

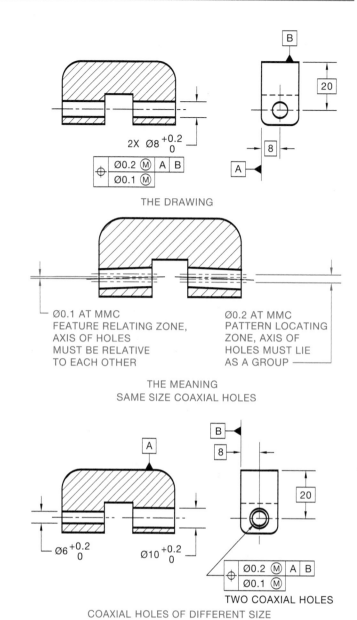

THE DRAWING

Ø0.1 AT MMC
FEATURE RELATING ZONE,
AXIS OF HOLES
MUST BE RELATIVE
TO EACH OTHER

Ø0.2 AT MMC
PATTERN LOCATING
ZONE, AXIS OF
HOLES MUST LIE
AS A GROUP

THE MEANING
SAME SIZE COAXIAL HOLES

TWO COAXIAL HOLES

COAXIAL HOLES OF DIFFERENT SIZE

Figure 11–85 Positional tolerance for coaxial holes.

mating parts, the tolerance zone could be extended or projected away from the primary datum controlling the axis of the related feature. The amount of projection is equal to the thickness of the mating part. This works because the axis of the threaded or press-fit feature is limited by the exact location and angle of the thread or hole within which it is assembled. This type of application is especially useful when the mating features are screws, pins, or studs. These types of conditions are referred to as *fixed fasteners*, because the fastener is fixed in the mating part and there is no clearance allowance. The projected tolerance zone can be handled in one of two ways, using the projected tolerance zone symbol (see Figure 11–87). When calculating the positional tolerance zones that are applied to the parts of a fixed fastener, you can use the following formula: MMC Hole – MMC Fastener (nominal thread size) ÷ 2 = Positional Tolerance Zone of Each Part. In some situations it may be desirable to provide more tolerance to one part than another. For example, 60% to the threaded part and 40% to the unthreaded part.

When parts are assembled with fasteners, such as bolts and nuts or rivets, and where all the parts have clearance holes to accommodate the fasteners, the application is referred to as a *floating fas-*

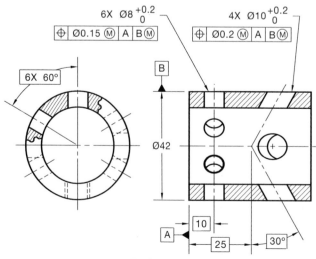

Figure 11–86 Positional tolerancing of nonparallel holes.

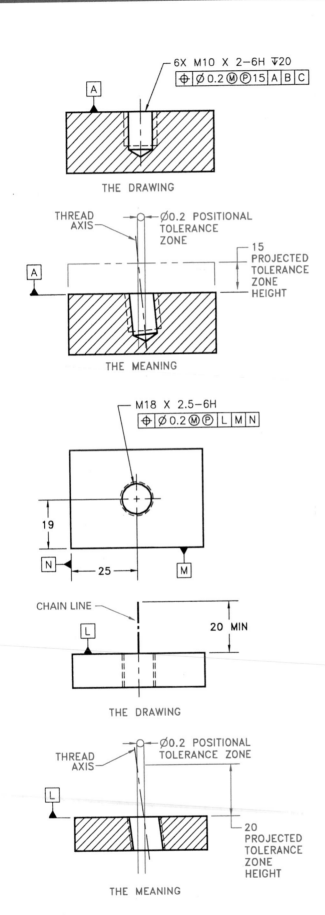

Figure 11–87 Projected tolerance zone.

*tener.* Floating fasteners require that the fastening device be secured on each side of the part, such as with a bolt and nut. With a fixed fastener, one of the parts is a fastening device. Greater tolerance flexibility with floating fasteners results from the fastener clearance at each part. When calculating the positional tolerance zone for floating fastener parts, the following formula is used: MMC Hole – MMC Fastener (nominal thread size) = Positional Tolerance Zone of Each Part.

## VIRTUAL CONDITION

The tolerances of a feature that relate to size, form, orientation, and location, including the possible application of MMC or RFS, are determined by the function of the part. Consideration must be given to the collective effect of these factors in determining the clearance between mating parts and in establishing gage feature sizes. The boundary created by the combined effects of size, MMC, and the geometric tolerance is known as the *virtual condition*. Virtual condition is the only condition where a feature size may be outside MMC. Controlling the clearance between mating parts is critical to the design process. When features are dimensioned using a combination of size and geometric tolerances, you need to consider the resulting effects of the specifications to ensure that parts will always fit together.

When a positional tolerance is applied to an internal feature, the Virtual Condition = MMC Hole – Positional Tolerance. This calculation determines the maximum feature size that should be allowed to fit within the hole (see Figure 11–88).

When perpendicularity is applied to an external diameter, such as a pin, the Virtual Condition = MMC Feature + Perpendicularity Geometric Tolerance. The virtual condition determines the smallest acceptable mating feature that fits over the given part while maintaining a positive connection between the surfaces at datum A, as shown in Figure 11–89.

## STATISTICAL TOLERANCING WITH GEOMETRIC CONTROLS

Methods of tolerancing for statistical process control (SPC) were introduced in Chapter 7. Statistical tolerances can also be used with geometric controls by placing the statistical tolerancing symbol in the feature control frame, as shown in Figure 11–90.

## COMBINATION CONTROLS

In some situations, compatible geometric characteristics may be combined in one feature control frame or separate frames associated with the same surface. This is normally done when the combined effect of two different geometric characteristics and tolerance zones is desired. The profile tolerance can be used to illustrate the combination of geometric characteristics. You can combine profile and parallelism to control the profile of a surface and the parallelism of each element to a datum. Combine profile and runout to control the line elements within the profile specification and circular elements within the runout tolerance, as shown in Figure 11–91.

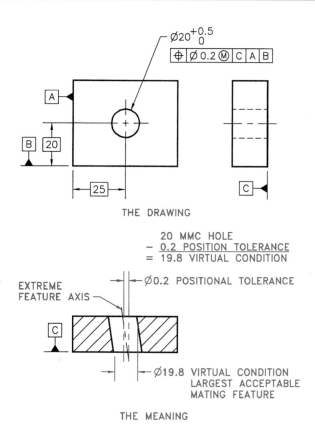

THE DRAWING

```
    20 MMC HOLE
 −  0.2 POSITION TOLERANCE
 =  19.8 VIRTUAL CONDITION
```

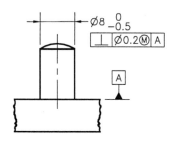

THE MEANING

Figure 11–88 Virtual Condition of Hole = MMC – Positional Tolerance.

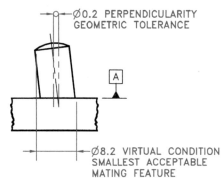

THE DRAWING

```
8 MMC FEATURE + 0.2 GEOMETRIC TOLERANCE
= 8.2 VIRTUAL CONDITION
```

THE MEANING

Figure 11–89 Virtual Condition Perpendicular Pin = MMC Feature + Geometric Tolerance.

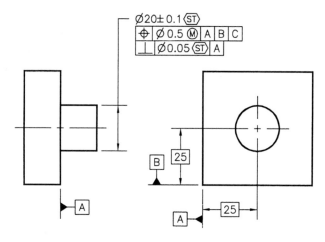

Figure 11–90 Statistical tolerancing with geometric controls.

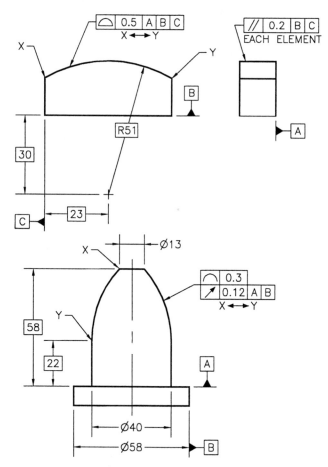

Figure 11–91 Combination controls (profile and runout).

The combined control of parallelism and perpendicularity is represented on the same feature, as shown in Figure 11–92. Other combined geometric characteristics are used when the callouts are compatible and the design function of the part requires such specific controls.

## PULL IT ALL TOGETHER

Now that you have learned the basics of GD & T, take some time to look at prints containing this information as you review the chapter. One of these prints is shown in Figure 11–93.

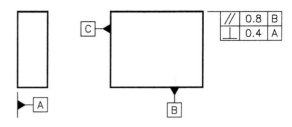

Figure 11–92 Combination control (parallelism and perpendicularity).

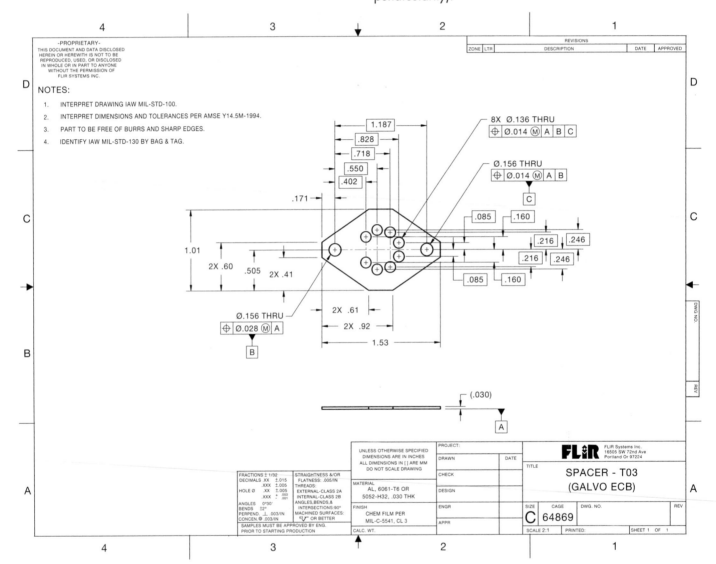

Figure 11–93 An actual industry drawing of a part with geometric tolerancing applied. *Courtesy Flir Systems, Inc.*

# CHAPTER 11 TEST

Multiple choice: Respond to the following by selecting a, b, c, or d to best answer the question or complete the statement.

1. This is the tolerancing of individual features of a part where the permissible variations relate to characteristics of form, profile, or the relationship between features.
   a. geometric tolerancing
   b. conventional tolerancing
   c. dimensioning and tolerancing
   d. form and location tolerancing

2. The ASME standard that governs dimensioning and tolerancing is:
   a. ASME Y14.2M
   b. ASME Y14.3
   c. ASME Y14.5M
   d. ASME Y14.6

3. The symbol used to identify datums on a drawing is called:
   a. datum identification symbol
   b. datum feature symbol
   c. datum symbol
   d. datum characteristic symbol

4. Sufficient datum features are chosen to position the part in relation to a set of three mutually perpendicular planes, which are jointly called a:
   a. coordinate dimension planes
   b. three-plane concept
   c. datum reference planes
   d. datum reference frame

5. Two or more surfaces that are on the same plane are called:
   a. common surfaces
   b. coordinate surfaces
   c. aligned surfaces
   d. coplanar surfaces

6. This symbol is used to relate a geometric tolerance to a part feature.
   a. feature control frame
   b. geometric characteristic symbol
   c. datum feature symbol
   d. geometric control frame

7. This is any size or location dimension that is used to identify the theoretically exact size, profile, orientation, or location of a feature or datum target.
   a. true position
   b. true dimension
   c. basic dimension
   d. specified dimension

8. This geometric tolerance allows a specified amount of variation from a straight line.
   a. straightness
   b. flatness
   c. profile
   d. parallelism

9. This geometric tolerance is a cross-sectional or single-line tolerance zone in the shape of two concentric circle.
   a. concentricity
   b. circularity
   c. cylindricity
   d. runout

10. This geometric tolerance allows a specified amount of surface variation from a flat plane.
    a. straightness
    b. flatness
    c. profile
    d. parallelism

11. This geometric tolerance is a blanket tolerance that results in a feature lying between two concentric cylinders.
    a. concentricity
    b. circularity
    c. cylindricity
    d. runout

12. This geometric tolerance is used to control form or a combination of size, form, and orientation.
    a. runout
    b. cylindricity
    c. concentricity
    d. profile

13. This geometric tolerance is used to control the relationship of radial features to a datum axis and features that are 90° to a datum axis.
    a. runout
    b. perpendicularity
    c. concentricity
    d. profile

14. This geometric tolerance requires that the actual feature be between two parallel planes or lines that are parallel to a datum.
    a. profile
    b. parallelism
    c. perpendicularity
    d. angularity

15. This geometric tolerance requires that the actual feature be between two parallel planes or lines that are at a basic angle to a datum.

    a. profile
    b. parallelism
    c. perpendicularity
    d. angularity

16. This geometric tolerance requires that the actual feature be between two parallel planes or lines that are perpendicular to a datum.

    a. profile
    c. perpendicularity
    b. parallelism
    d. angularity

17. The geometric tolerancing symbol that is placed in the first compartment of the feature control frame is called:

    a. geometric tolerancing symbol
    b. geometric characteristic symbol
    c. feature control symbol
    d. geometric dimensioning symbol

18. What is the MMC of a hole with the dimension ∅.745/.755?

    a. .745
    b. .755
    c. .750
    d. none of the above

19. What is the MMC of a shaft with the dimension ∅.745/.755?

    a. .745
    c. .750
    b. .755
    d. none of the above

20. The true geometric form of a feature at MMC is called:

    a. perfect form envelope
    c. boundary of perfect form
    b. perfect form boundary
    d. all of the above

21. Location tolerances include:

    a. circularity, concentricity
    c. position, symmetry, concentricity
    b. concentricity, profile, position
    d. true position, symmetry, runout

22. This geometric tolerance is the relationship of the axes of cylindrical shapes.

    a. cylindricity
    c. symmetry
    b. concentricity
    d. position

23. This geometric characteristic defines a zone within which the center, axis, or center plane of a feature is permitted to vary from a true, theoretically exact location.

    a. cylindricity
    b. concentricity
    c. symmetry
    d. position

24. This is the center plane relationship between two or more features.

    a. cylindricity
    b. concentricity
    c. symmetry
    d. position

25. What is it called when you have a situation where it may be permissible to allow the location of individual features in a pattern to differ from the tolerance related to the items as a group?

    a. composite positional tolerance
    c. virtual condition
    b. projected tolerance zone
    d. coaxial position tolerance

26. This takes into consideration the combined effect of the feature tolerance and the geometric tolerance at MMC.

    a. composite positional tolerance
    b. projected tolerance zone
    c. virtual condition
    d. combined geometric tolerance

27. This is where a positional tolerance zone is extended a given amount to accommodate the axis of a mating feature or part.

    a. composite positional tolerance
    b. projected tolerance zone
    c. virtual condition
    d. coaxial position tolerance

28. What is the virtual condition of a hole with a diameter of .870/.880 and a positional tolerance of .005?

    a. .865
    b. .870
    c. .875
    d. .885

29. What is the virtual condition of a pin with a diameter of 12.00 ± 0.25 mm and a perpendicularity geometric tolerance of .05?

    a. 11.70
    b. 12.25
    c. 12.30
    d. 12.75

30. The theoretically exact location of the center of a hole is known as:

    a. position
    b. exact position
    c. true position
    d. theoretical position

## Calculations

1. Given:

    a. Shaft Ø24.00/23.92.
    b. Straightness geometric tolerance 0.02.

    What is the geometric tolerance at the actual sizes specified below for the type of straightness and material condition shown?

    | Actual Size | Surface Straightness | | Axis Straightness | |
    |---|---|---|---|---|
    | | RFS | | RFS | MMC |
    | 24.00 | | | | |
    | 23.99 | | | | |
    | 23.98 | | | | |
    | 23.96 | | | | |
    | 23.94 | | | | |
    | 23.92 | | | | |

2. Given:

    a. Positional tolerance Ø0.02 at true position in reference to datums L, M, N.
    b. Hole size Ø8.50/8.40.

    What is the positional tolerance using different material condition symbols at the actual sizes shown in the table?

    | Actual Sizes | Material Condition Applied to Tolerance | | |
    |---|---|---|---|
    | | MMC | RFS | LMC |
    | 8.50 | | | |
    | 8.49 | | | |
    | 8.48 | | | |
    | 8.46 | | | |
    | 8.44 | | | |
    | 8.42 | | | |
    | 8.40 | | | |

3. If the positional tolerance of the hole in problem 2 above is zero at MMC, then what would the positional tolerance be at the actual produced sizes given below?

    | Actual Sizes | MMC |
    |---|---|
    | 8.50 | |
    | 8.48 | |
    | 8.46 | |
    | 8.44 | |
    | 8.42 | |
    | 8.40 | |

4. What is the virtual condition of a Ø12.2/12.0 hole that is located with a positional tolerance of Ø0.05 at MMC?

5. What is the virtual condition of a Ø6.0/5.9 pin established with perpendicularity to a datum A by Ø0.02 at MMC?

6. Calculate the positional tolerance for the location of holes with the following specifications.

    a. Floating fastener.

    b. Fastener: M20 x 2.5.

    c. Hole through two parts: Ø21.2/20.8.

    Positional tolerance for holes in part 1 equals _____, part 2 equals _____.

7. Calculate the positional tolerance for the location of holes with the following specifications.

    a. Fixed fastener.

    b. Part 1 hole: Ø9.0/8.6.

    c. Part 2 hole: M8 x 1.25.

    d. Equal positional tolerance for each part.

    Positional tolerance for holes in part 1 equals _____, part 2 equals _____.

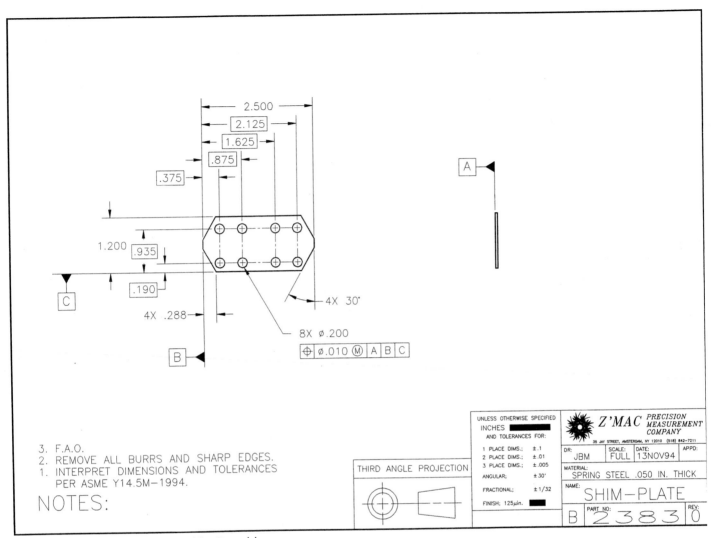

Problem 11–1 *Courtesy Jim B. MacDonald.*

## CHAPTER 11 PROBLEMS

**PROBLEM 11–1** Answer the following following questions as you refer to the print shown on this page.

1. How many datums are identified on this print?

2. What is the term associated with the rectangles around some of the dimensions?

3. Give the meaning of the term identified in question 2.

4. Name the geometric tolerancing symbol next to the 8 x Ø.200 dimension

5. What is the type of dimensioning on this print where dimensions to features originate from a common surface?

6. Interpret (explain the meaning of) each item in the symbol next to the 8 x Ø.200 dimension. Use the proper name when identifying symbols.

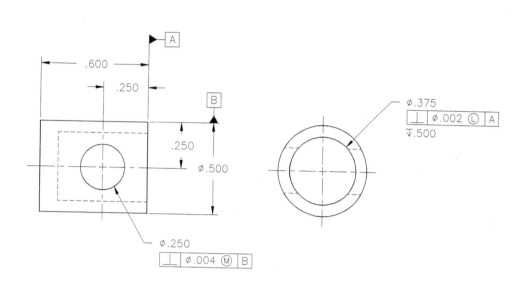

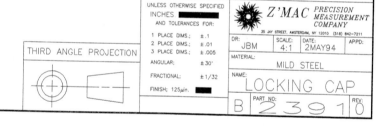

NOTES:
1. INTERPRET DIMENSIONS AND TOLERANCES
   PER ASME Y14.5M-1994.
2. REMOVE ALL BURRS AND SHARP EDGES.
3. F.A.O.

THIRD ANGLE PROJECTION

| UNLESS OTHERWISE SPECIFIED | | Z'MAC PRECISION MEASUREMENT COMPANY |
|---|---|---|
| INCHES | | 35 JAY STREET, AMSTERDAM, NY 12010 (518) 842-7211 |
| AND TOLERANCES FOR: | | |
| 1 PLACE DIMS.; ±.1 | DR: JBM | SCALE: 4:1 | DATE: 2MAY94 | APPD: |
| 2 PLACE DIMS.; ±.01 | MATERIAL: | |
| 3 PLACE DIMS.; ±.005 | MILD STEEL | |
| ANGULAR; ±30' | NAME: | |
| FRACTIONAL; ±1/32 | LOCKING CAP | |
| FINISH; 125μin. | B 23910 PART_NO: REV: 0 |

Problem 11–2 *Courtesy Jim B. MacDonald.*

**PROBLEM 11–2** Answer the following questions as you refer to the print shown on this page.

1. Give the overall dimensions of the part.

   _____

2. Give the limits of the Ø.375 dimension.

   _____

3. What is the maximum material condition of the Ø.375 dimension?

   _____

4. Given the following list of actual, produced sizes for the Ø.375 dimension, determine the geometric tolerance at each produced size.

| ACTUAL SIZE | GEOMETRIC TOLERANCE |
|---|---|
| .370 | _____ |
| .372 | _____ |
| .375 | _____ |
| .378 | _____ |
| .380 | _____ |

5. Give the limits of the Ø.250 dimension.

   _____

6. What is the maximum material condition of the Ø.250 dimension?

   _____

7. Given the following list of actual, produced sizes for the Ø.250 dimension, determine the geometric tolerance at each produced size.

| ACTUAL SIZE | GEOMETRIC TOLERANCE |
|---|---|
| .245 | _____ |
| .248 | _____ |
| .250 | _____ |
| .252 | _____ |
| .255 | _____ |

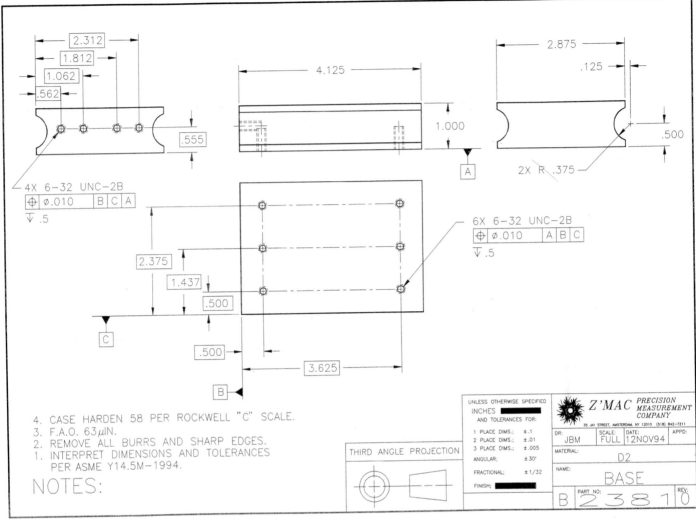

4. CASE HARDEN 58 PER ROCKWELL "C" SCALE.
3. F.A.O. 63 µIN.
2. REMOVE ALL BURRS AND SHARP EDGES.
1. INTERPRET DIMENSIONS AND TOLERANCES
   PER ASME Y14.5M–1994.
NOTES:

THIRD ANGLE PROJECTION

| UNLESS OTHERWISE SPECIFIED | | | |
|---|---|---|---|
| INCHES ▆▆▆▆ | | | |
| AND TOLERANCES FOR: | | | |
| 1 PLACE DIMS.; | ±.1 | | |
| 2 PLACE DIMS.; | ±.01 | | |
| 3 PLACE DIMS.; | ±.005 | | |
| ANGULAR; | ±30' | | |
| FRACTIONAL; | ±1/32 | | |
| FINISH; ▆▆▆ | | | |

*Z'MAC* PRECISION MEASUREMENT COMPANY
35 JAY STREET, AMSTERDAM, NY 12010 (518) 842-7211

| DR: JBM | SCALE: FULL | DATE: 12NOV94 | APPD: |
|---|---|---|---|
| MATERIAL: | | D2 | |
| NAME: | | BASE | |
| B | PART NO: 23810 | | REV: 0 |

Problem 11–3 *Courtesy Jim B. MacDonald.*

**PROBLEM 11–3** Answer the following questions as you refer to the print shown on this page.

1. Give the overall dimensions of the part.

2. How many datum feature symbols are found on this print?

3. What is the difference between the two feature control frames shown on this print?

4. What is the material condition application when there is no symbol found in the feature control frame after the geometric tolerance?

5. Describe the meaning of the material condition application identified in Question 4.

6. How many threaded holes are there in this part?

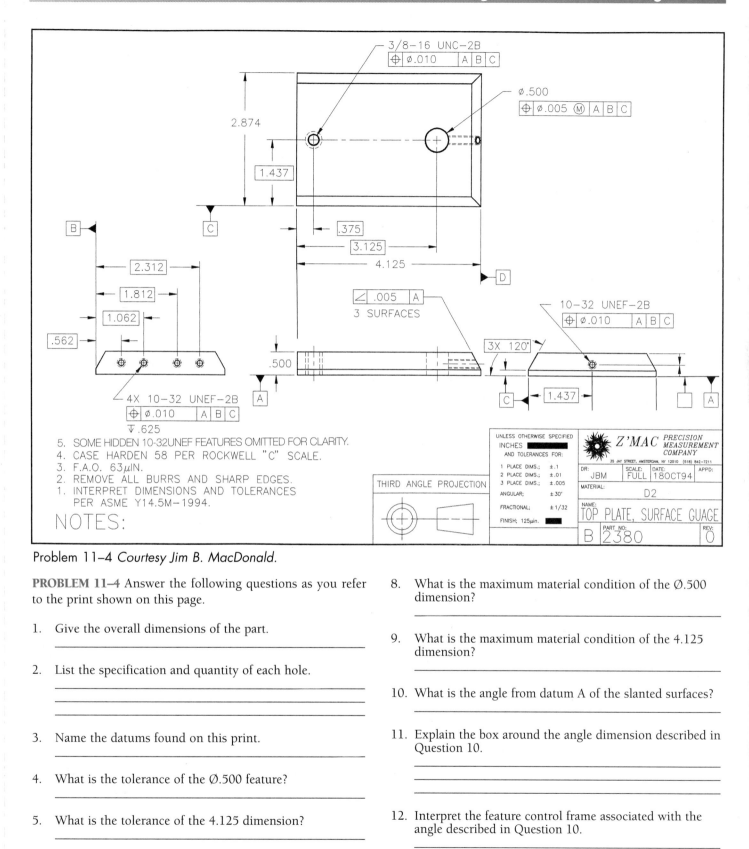

Problem 11–4 *Courtesy Jim B. MacDonald.*

**PROBLEM 11–4** Answer the following questions as you refer to the print shown on this page.

1. Give the overall dimensions of the part.

2. List the specification and quantity of each hole.

3. Name the datums found on this print.

4. What is the tolerance of the Ø.500 feature?

5. What is the tolerance of the 4.125 dimension?

6. Give the limits of the Ø.500 dimension.

7. Give the limits of the 4.125 dimension.

8. What is the maximum material condition of the Ø.500 dimension?

9. What is the maximum material condition of the 4.125 dimension?

10. What is the angle from datum A of the slanted surfaces?

11. Explain the box around the angle dimension described in Question 10.

12. Interpret the feature control frame associated with the angle described in Question 10.

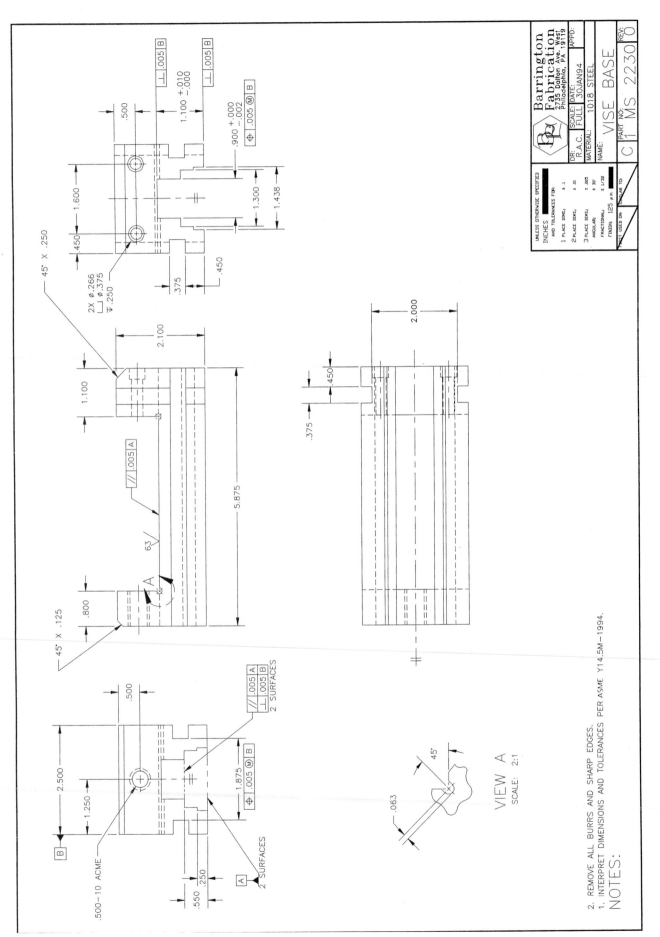

Problem 11–5 *Courtesy Richard Clouser.*

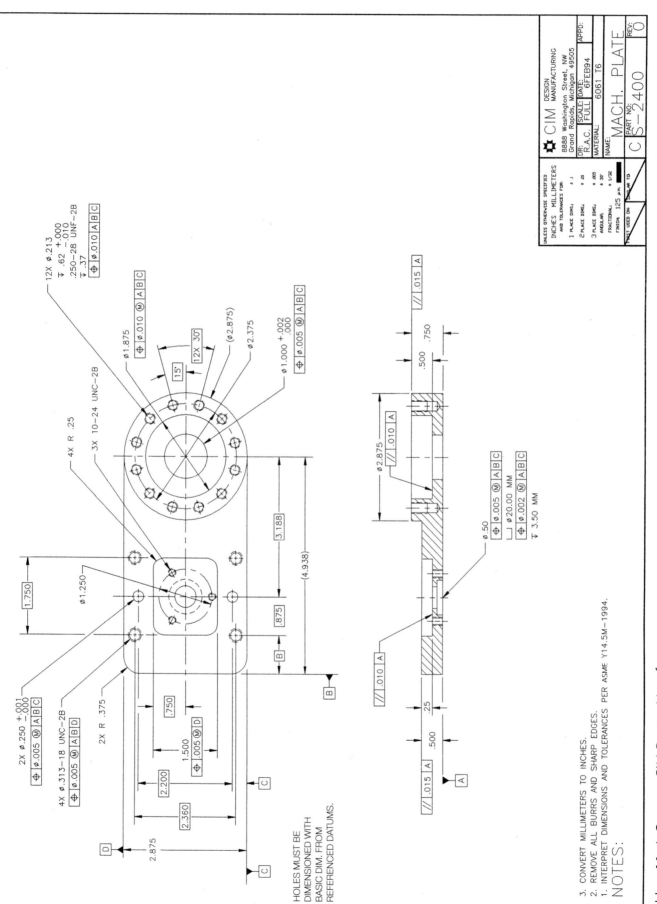

Problem 11–6 Courtesy CIM Design Manuf.

13. Why is there no material condition symbol in the feature control frame described in Question 12?

_____

_____

_____

_____

14. Interpret the feature control frame associated with the Ø.500 dimension.

_____

_____

_____

_____

15. Given the following list of actual, produced sizes, provide the geometric tolerance for the Ø.500 feature at each produced size.

| ACTUAL SIZE | GEOMETRIC TOLERANCE |
|---|---|
| .495 | _____ |
| .498 | _____ |
| .500 | _____ |
| .502 | _____ |
| .504 | _____ |
| .505 | _____ |

**PROBLEM 11–5** Answer the following questions as you refer to the print shown on page 318.

1. Give the overall dimensions of the part.

_____

2. Explain the meaning of datum feature symbol B.

_____

_____

3. Interpret the combination feature control frame.

_____

_____

_____

4. What does the note 2 SURFACES mean when placed under the feature control frames described in question 3?

_____

_____

5. Give the limits of the $.900 {}^{+.000}_{-.002}$ dimension.

_____

_____

6. What is the tolerance of the $.900 {}^{+.000}_{-.002}$ dimension?

_____

_____

7. What is the maximum material condition of the $.900 {}^{+.000}_{-.002}$ dimension?

_____

_____

8. Given the following list of actual produced sizes for the $.900 + .000/- .002$ dimension, determine the geometric tolerance at each produced size.

| ACTUAL SIZE | GEOMETRIC TOLERANCE |
|---|---|
| .898 | _____ |
| .899 | _____ |
| .900 | _____ |

**PROBLEM 11–6** Answer the following questions as you refer to the print shown on page 319.

1.  Give the overall dimensions of the part.

    _____

2.  List the datums found on this print.

    _____

3.  Give the limits of the .750 thickness dimension.

    _____

4.  Interpret the feature control frame associated with the top surface of the Ø2.875 feature and at the .750 thickness dimension.

    _____

    _____

5.  What is the material condition symbol associated with the feature control frame described in Question 4?

    _____

    _____

    _____

    _____

6.  Why is there no material condition symbol shown in the feature control symbol described in Question 4?

    _____

    _____

    _____

7.  Given the following list of actual, produced sizes for the .750 thickness dimension, give the parallelism geometric tolerance at each produced size.

    | ACTUAL SIZE | GEOMETRIC TOLERANCE |
    |---|---|
    | .755 | _____ |
    | .752 | _____ |
    | .750 | _____ |
    | .748 | _____ |
    | .745 | _____ |

8.  Describe datum feature D.

    _____

    _____

    _____

9.  Interpret the feature control frame associated with the 1.500 dimension in the top view.

    _____

    _____

    _____

10. What does the position of the feature control frame described in question 9 control?

    _____

    _____

    _____

11. What is the maximum material condition of the 1.500 feature described in Question 9?

    _____

12. Given the following list of actual produced sizes for the 1.500 feature described in Question 9 provide the geometric tolerance at each produced size.

    | ACTUAL SIZE | GEOMETRIC TOLERANCE |
    |---|---|
    | 1.495 | _____ |
    | 1.498 | _____ |
    | 1.500 | _____ |
    | 1.502 | _____ |
    | 1.505 | _____ |

13. How many .250-28UNF-2B threads are there, and what is the angle between each?

    _____

    _____

14. What is the diameter between the centers of the features described in Question 13?

    _____

    _____

15. What does the MM on the Ø20.00 and 3.50 DEEP dimension on the counterbore in the front view mean?

    _____

    _____

    _____

16. Why are there two position tolerance feature control frames associated with the Ø.50 hole and counterbore described in Question 15?

    _____

    _____

    _____

# Reading Cam, Gear, and Bearing Drawings

## LEARNING OBJECTIVES

After completing this chapter you will be able to:

- Define cam terminology and identify cam types and cam followers.
- Read information on plate cam prints.
- Read a drum cam print.
- Define gear terminology and identify gear types.
- Calculate gear train data.
- Read rack and pinion prints.

- Read worm gear prints.
- Identify bearing elements and types.
- Determine bearing specifications from bearing selection charts based on given information.
- Answer questions about bearing finish, lubrication, mountings, and gear and bearing assemblies.

## CAMS

A cam is a rotating mechanism that is used to convert rotary motion into a corresponding straight motion. The timing involved in the rotary motion is often the main design element of the cam. For example, a cam may be designed to make a follower rise a given amount in a given degree of rotation, dwell—which is to remain constant for an additional period of rotation—and, finally, fall back to the beginning in the last degree of rotation. The total movement of the cam follower happens in one 360° rotation of the cam. This movement is referred to as the displacement. Cams are generally in the shape of irregular plates, grooved plates, or grooved cylinders. The basic components of the cam mechanism are shown in Figure 12–1.

### Cam Types

There are basically three different types of cams: the plate cam, face cam, and drum cam (see Figure 12–2). The plate cam is the most commonly used type of cam.

### Cam Followers

There are several types of cam followers. The type used depends on the application. The most common type of follower is the roller follower. The roller follower works well at high speeds, reduces friction and heat, and keeps wear to a minimum. The arrangement of

the follower in relation to the cam shaft differs depending on the application. The roller followers, shown in Figure 12–3, include: the in-line follower where the axis of the follower is in line with the cam shaft; the offset roller follower; and the pivoted follower. The pivoted follower requires spring tension to keep the follower in contact with the cam profile.

Another type of cam follower is the knife-edged follower, shown in Figure 12–4(a). This follower is used for only low-speed and low-force applications. The knife-edged follower has a low resistance to wear, but is very responsive and can be used effectively in situations that require abrupt changes in the cam profile.

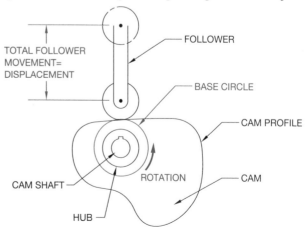

Figure 12–1 Elements of a cam mechanism.

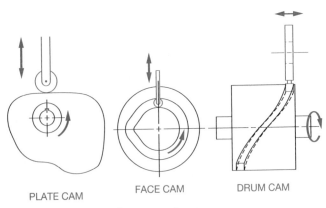

Figure 12–2 Types of cam mechanisms.

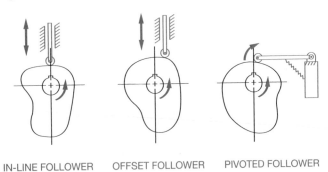

Figure 12–3 Types of cam roller followers.

The flat-faced follower, shown in Figure 12–4(b), is used in situations where the cam profile has a steep rise or fall. Designers often offset the axis of the follower. This practice causes the follower to rotate while in operation. This rotating action allows the follower surface to wear evenly and last longer.

## READING THE PLATE CAM DRAWING

The information found on the plate cam drawing (see Figure 12–5) includes:

- ■ Cam profile.
- ■ Hub dimensions, including cam shaft, outside diameter, width, keyway dimensions.
- ■ Roller follower placed in one convenient location, such as 60°, using phantom lines.
- ■ The drawing is set up as a chart drawing where A° equals the angle of the follower at each position, and R equals the radius from the center of the cam shaft to the center of the follower at each position.
- ■ A chart giving the values of the angles A and the radii R at each follower position.
- ■ Side view showing the cam plate thickness, and set screw location with thread specifications used.
- ■ Tolerances, unless otherwise specified.

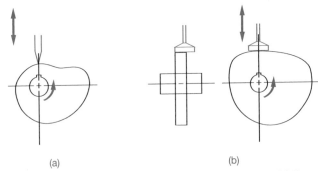

Figure 12–4 (a) Knife-edged cam follower. (b) Flat-faced follower.

## READING A DRUM CAM DRAWING

Drum cams are used when it is necessary for the follower to move in a path parallel to the axis of the cam. The drum cam is a cylinder with a groove machined in the surface in the shape of the cam profile. The cam follower moves along the path of the groove as the drum is rotated. The displacement diagram for a drum cam is actually the pattern of the drum surface as if it were rolled out flat. The height of the displacement diagram equals the height of the drum. The length of the displacement diagram is equal to the circumference of the drum. Refer to the drum cam drawing in Figure 12–6.

## GEARS

Gears are toothed wheels used to transmit motion and power from one shaft to another. Gears are rugged and durable and can transmit power with up to 98% efficiency with a long service life. Most gears are made of cast iron or steel, but brass and bronze alloys and plastic are also used for some applications. Gear selection is often done through vendors' catalogs or the use of standard formulas. A gear train exists when two or more gears are in combination for the purpose of transmitting power. Generally, two gears in mesh are used to increase or reduce speed or change the direction of motion from one shaft to another. When two gears are in mesh, the larger is called the gear and the smaller is called the pinion (see Figure 12–7).

*AGMA/ANSI* Gear selection generally follows the guidelines of the American Gear Manufacturers Association (AGMA) or the American National Standards Institute (ANSI).

## GEAR STRUCTURE

Gears are made in a variety of structures depending on the design requirements, but there is some basic terminology that can typically be associated with gear structure. The elements of the gear structure, shown in Figure 12–8, include the outside diameter, face, hub, bore, and keyway.

### Types of Hubs

The hubs are the lugs or shoulders projecting from one or both faces of some gears. These can be referred to as A, B, or C hubs. An A hub is also called a flush hub because there is no projection from

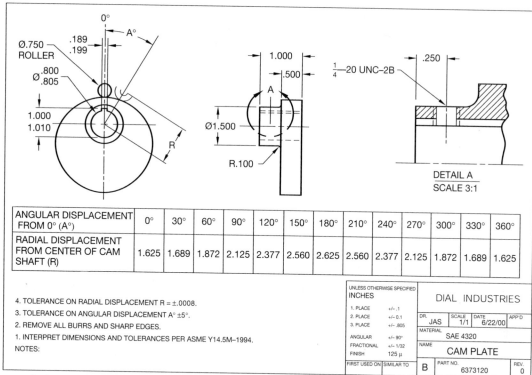

| ANGULAR DISPLACEMENT FROM 0° (A°) | 0° | 30° | 60° | 90° | 120° | 150° | 180° | 210° | 240° | 270° | 300° | 330° | 360° |
|---|---|---|---|---|---|---|---|---|---|---|---|---|---|
| RADIAL DISPLACEMENT FROM CENTER OF CAM SHAFT (R) | 1.625 | 1.689 | 1.872 | 2.125 | 2.377 | 2.560 | 2.625 | 2.560 | 2.377 | 2.125 | 1.872 | 1.689 | 1.625 |

NOTES:
1. INTERPRET DIMENSIONS AND TOLERANCES PER ASME Y14.5M–1994.
2. REMOVE ALL BURRS AND SHARP EDGES.
3. TOLERANCE ON ANGULAR DISPLACEMENT A° ±5°.
4. TOLERANCE ON RADIAL DISPLACEMENT R = ±.0008.

| UNLESS OTHERWISE SPECIFIED INCHES | | DIAL INDUSTRIES | | | |
|---|---|---|---|---|---|
| 1. PLACE | +/– .1 | | | | |
| 2. PLACE | +/– 0.1 | DR. JAS | SCALE 1/1 | DATE 6/22/00 | APP'D |
| 3. PLACE | +/– .805 | MATERIAL SAE 4320 | | | |
| ANGULAR | +/– 90° | NAME | | | |
| FRACTIONAL | +/– 1/32 | CAM PLATE | | | |
| FINISH | 125 μ | | | | |
| FIRST USED ON | SIMILAR TO | B | PART NO. 6373120 | | REV. 0 |

Figure 12–5 Formal plate cam CADD drawing. *Courtesy Dial Industries.*

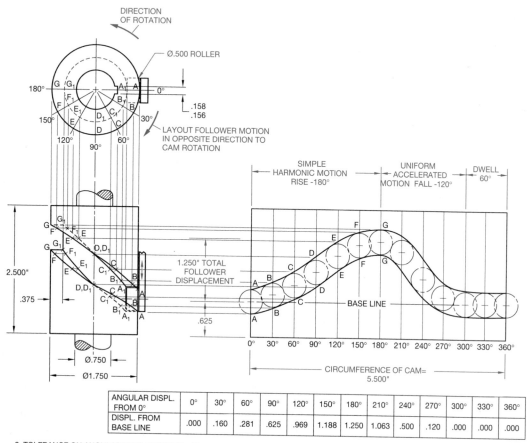

| ANGULAR DISPL. FROM 0° | 0° | 30° | 60° | 90° | 120° | 150° | 180° | 210° | 240° | 270° | 300° | 330° | 360° |
|---|---|---|---|---|---|---|---|---|---|---|---|---|---|
| DISPL. FROM BASE LINE | .000 | .160 | .281 | .625 | .969 | 1.188 | 1.250 | 1.063 | .500 | .120 | .000 | .000 | .000 |

NOTES:
1. TOLERANCE ON DISPLACEMENT FROM BASELINE ± .0008.
2. TOLERANCE ON ANGULAR DISPLACEMENT ± .5°.

Figure 12–6 Construction of a drum cam drawing.

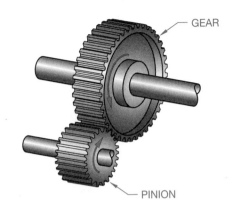

Figure 12–7 The gear and pinion.

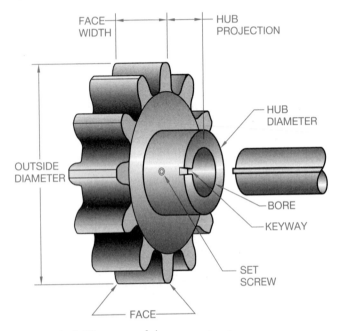

Figure 12–8 Elements of the gear structure.

the gear face. B hubs have a projection on one side of the gear, while C hubs have projections on both sides of the gear face (see Figure 12–9).

### Keyways, Keys, and Set Screws

Gears are usually held on the shaft with a key, keyway, and set screw. Refer to Chapter 6 for information about the manufacture of keyways, Chapter 7 for proper dimensioning, and Chapter 8 for screw thread specifications. One or more set screws are usually used to keep the key secure in the keyway (see Figure 12–10).

## SPLINES

Splines are teeth cut in a shaft and a gear or pulley bore and are used to prevent the gear or pulley from spinning on the shaft. Splines are often used when it is necessary for the gear or pulley to easily slide on the shaft. Splines may also be nonsliding and, in all cases, are stronger than keyways and keys. The standardization of splines is established by the Society of Automotive Engineers

(SAE), so that any two parts with the same spline specifications should fit together. The following is an example of an SAE spline specification:

$$\text{SAE } \underset{\text{(A)}}{2}\text{-}\underset{\text{(B)}}{1/2}\text{-}\underset{\text{(C)}}{10}\quad \underset{\text{(D)}}{\text{B SPLINE}}$$

The note components are described below:

(A)     Society of Automotive Engineers.

(B)     The outside diameter of the spline.

(C)     The number of teeth.

(D)     A = This is a fixed nonsliding spline.

B = This spline slides under no-load conditions.

C = This spline slides under load conditions.

### Involute Spline

Another spline standard is the involute spline. The teeth on the involute spline are similar to the curved teeth found on spur gears. The spline teeth generally have a shorter whole depth than standard spur gears, and the pressure angle is normally 30°.

## GEAR TYPES

The most common and simplest form of gear is the spur gear. This chapter discusses spur gears, bevel gears, and worm gears. Gear types are based on one or more of the following elements:

The relationship of the shafts; either parallel, intersecting, nonintersecting shafts, or rack and pinion.

- ■ Manufacturing cost.
- ■ Ease of maintenance in service.
- ■ Smooth and quiet operation.
- ■ Load-carrying ability.
- ■ Speed reduction capabilities.
- ■ Space requirements.

### Parallel Shafting Gears

Many different types of mating gears are designed with parallel shafts. These include spur and helical gears.

#### Spur Gears

There are two basic types of spur gears: external and internal spur gears. When two or more spur gears are cut on a single shaft, they are referred to as cluster gears. External spur gears are designed with the teeth of the gear on the outside of a cylinder (see Figure 12–11). External spur gears are the most common type of gear used in manufacturing. Internal spur gears have the teeth on the inside of the cylindrical gear (see Figure 12–12). The advantages of spur gears over other types is their low manufacturing cost, simple design, and ease of maintenance. The disadvantages include less load capacity and noise, which is greater than with other types of gears.

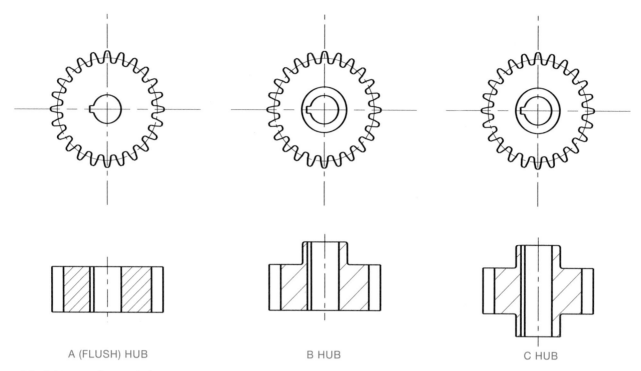

Figure 12–9 Types of gear hubs.

A (FLUSH) HUB      B HUB      C HUB

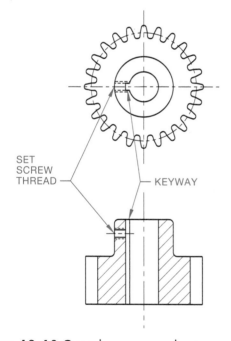

SET
SCREW
THREAD     KEYWAY

Figure 12–10 Gear, keyways, and set screw.

Figure 12–11 External spur gear.

### Helical Gears

Helical gears have their teeth cut at an angle that allows more than one tooth to be in contact (see Figure 12–13). Helical gears carry more load than equivalent-sized spur gears and operate more quietly and smoothly. The disadvantage of helical gears is that they develop end thrust. End thrust is a lateral force exerted on the end of the gear shaft. Thrust bearings are required to reduce the effect of this end thrust. Double helical gears are designed to eliminate the end thrust and provide long life under heavy loads. However,

they are more difficult and costly to manufacture. The herringbone gear shown in Figure 12–14 is a double helical gear without space between the two opposing sets of teeth.

## Intersecting Shafting Gears

Intersecting shafting gears allow for the change in direction of motion from the gear to the pinion. Different types of intersecting shafting gears include bevel and face gears.

### Bevel Gears

Bevel gears are conical in shape, allowing the shafts of the gear and pinion to intersect at 90° or any desired angle. The teeth on the bevel gear have the same shape as the teeth on spur gears except they taper toward the apex of the cone. Bevel gears provide for a speed change between the gear and pinion (see Figure 12–15). Miter gears are the same as bevel gears, except both the gear and pinion are the same size and are used when shafts must intersect at

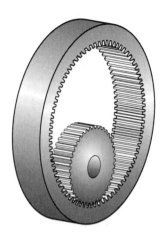

Figure 12–12 Internal spur gear.

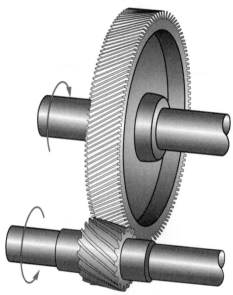

Figure 12–13 Helical gear.

Figure 12–14 Herringbone gear.

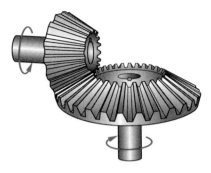

Figure 12–15 Bevel gears.

90° without speed reduction. Spiral bevel gears have the teeth cut at an angle, which provides the same advantages as helical gears over spur gears.

### Face Gears

The face gear is a combination of bevel gear and spur pinion, or bevel gear and helical pinion. This combination is used when the mounting accuracy is not as critical as with bevel gears. The load-carrying capabilities of face gears is not as good as that of bevel gears.

## Nonintersecting Shafting Gears

Gears with shafts that are at right angles but not intersecting are referred to as nonintersecting shafts. Gears that fall into this category are crossed helical, hypoid, and worm gears.

### Crossed Helical Gears

Also known as right angle helical or spiral gears, crossed helical gears provide for nonintersecting right angle shafts with low load-carrying capabilities (see Figure 12–16).

### Hypoid Gears

Hypoid gears have the same design as bevel gears except the gear shaft axes are offset and do not intersect (see Figure 12–17). The gear and pinion are often designed with bearings mounted on both

Figure 12–16 Crossed helical gears. Courtesy Browning Mfg. Division of Emerson Electric Co.

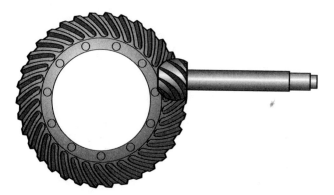

Figure 12–17 Hypoid gears.

sides for improved rigidity over standard bevel gears. Hypoid gears are very smooth, strong, and quiet in operation.

### Worm Gears

A worm and worm gear are shown in Figure 12–18. This type of gear is commonly used when a large speed reduction is required in a small space. The worm can be driven in either direction. When the gear is not in operation, the worm automatically locks in place. This is a particular advantage when it is important for the gears to have no movement of free travel when the equipment is shut off.

Figure 12–18 Worm and worm gear.

### Rack and Pinion

A rack and pinion is a spur pinion operating on a flat straight bar rack (see Figure 12–19). The rack and pinion is used to convert rotary motion into straight line motion.

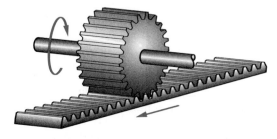

Figure 12–19 Rack and pinion.

## SPUR GEAR TERMINOLOGY

Spur gear teeth are straight and parallel to the gear shaft axis. The tooth profile is designed to transmit power at a constant rate, and with a minimum of vibration and noise. To achieve these requirements, an involute curve is used to establish the gear tooth profile. An involute curve is a spiral curve generated by a point on a chord as it unwinds from the circle. The contour of a gear tooth, based on the involute curve, is determined by a base circle, the diameter of which is controlled by a pressure angle. The pressure angle is the direction of push transmitted from a tooth on one gear to a tooth on the mating gear or pinion (see Figure 12–20). Two standard pressure angles, 14.5° and 20°, are used in spur gear design. The most commonly used pressure angle is 20° because it provides a stronger tooth for quieter running and heavier load-carrying characteristics. One of the basic rules of spur gear design is to have no fewer than 13 teeth on the running gear and 26 teeth on the mating gear.

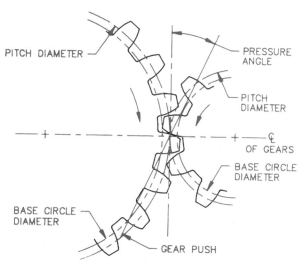

Figure 12–20 Spur gear pressure angle and related terminology.

Standard terminology and formulas control the requirements for spur gear design and specifications. Figure 12–21 shows a pictorial representation of the spur gear teeth with the components labeled.

### Diametral Pitch

The diametral pitch actually refers to the tooth size and has become the standard for tooth-size specifications. As you look at Figure 12–22, notice how the tooth size increases as the diametral pitch decreases. One of the most elementary rules of gear tooth design is that mating teeth must have the same diametral pitch.

## READING SPUR GEAR PRINTS

*ANSI* The standard that governs gear drawings is the American National Standards document ANSI Y14.7.1 Gear Drawing Standards — Part I. Because gear teeth are complex and time consuming to draw, simplified representations are used to make the

| TERM | DESCRIPTION | FORMULA |
|---|---|---|
| Pitch Diameter (D) | The diameter of an imaginary pitch circle on which a gear tooth is designed. Pitch circles of two spur gears are tangent. | $D = N/P$ |
| Diametral Pitch (P) | A ratio equal to the number of teeth on a gear per inch of pitch diameter. | $P = N/D$ |
| Number of Teeth (N) | The number of teeth on a gear. | $N = D \times P$ |
| Circular Pitch (p) | The distance from a point o one tooth to the corresponding point on the adjacent tooth, measured on the pitch circle. | $p = 3.1416 \times D/N$ $p = 3.1416/P$ |
| Center Distance (C) | The distance between the axis of two mating gears. | $C = $ sum of pitch DIA/2 |
| Addendum (a) | The radial distance from the pitch circle to the top of the tooth. | $a = 1/P$ |
| Dedendum (b) | The radial distance from the pitch to the bottom of the tooth. (This formula is for 20° teeth only.) | $b = 1.250/P$ |
| Whole Depth ($h_t$) | The full height of the tooth. It is equal to the sum of the addendum and the dedendum. | $h_t = a + b$ $h_t = 2.250/P$ |
| Working Depth ($h_k$) | The full height of the tooth. It is equal to the sum of the addendum and the dedendum. | $h_k = 2a$ $h_k = 2.000/P$ |
| Clearance (c) | The radial distance between the top of a tooth and the bottom of the mating tooth space. It is also the difference between the addendum and dedendum. | $c = b - a$ $c = .250/P$ |
| Outside Diameter ($D_o$) | The overall diameter of the gear. It is equal to the pitch diameter plus two dedendums. | $D_o = D + 2b$ |
| Root Diameter ($D_r$) | The diameter of a circle coinciding with the bottom of the tooth spaces. | $D_r = D - 2b$ |
| Circular Thickness (t) | The length of an arc between the two sides of a gear tooth measured on the pitch circle. | $t = 1.5708/P$ |
| Chordal Thickness ($t_c$) | The straight line thickness of a gear tooth measured on the pitch circle. | $t_c = D \sin (90°/N)$ |
| Chordal Addendum ($a_c$) | The height from the top of the tooth to the line of the chordal thickness. | $a_c = a + t^2/4D$ |
| Pressure Angle (Ø) | The angle of direction of pressure between contacting teeth. It determines the size of the base circle and the shape of the involute teeth. | |
| Base Circle Dia. ($D_a$) | The diameter of a circle from which the involute tooth form is generated. | $D_a = D \cos Ø$ |

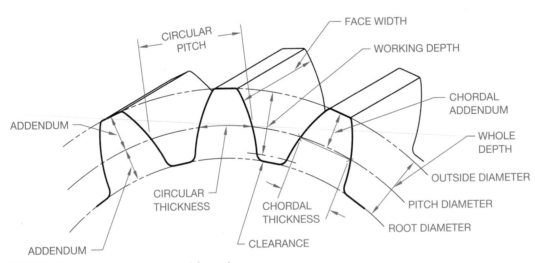

Figure 12–21 Spur gear terminology and formulas.

practice easier (see Figure 12–23) The simplified method shows the outside diameter and the root diameters as phantom lines and the pitch diameter as a centerline in the circular view. In addition, a side view is often required to show width dimensions and related features (see Figure 12–23(a)). If the gear construction has webs, spokes, or other items that require further clarification, a full section is normally used (see Figure 12–23(b)). Notice in the cross section that the gear tooth is left unsectioned and the pitch diameter is shown as a centerline.

When cluster gears are drawn, the circular view can show both sets of gear tooth representations in simplified form, or two circular views can be drawn (see Figure 12–24). When cluster gears are more complex than those shown here, multiple views and moved sections may be required.

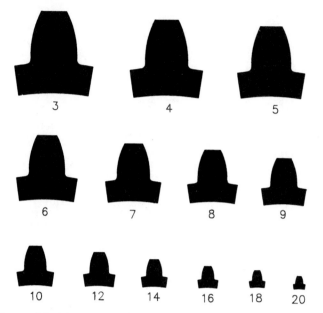

Figure 12–22 Diametral pitch.

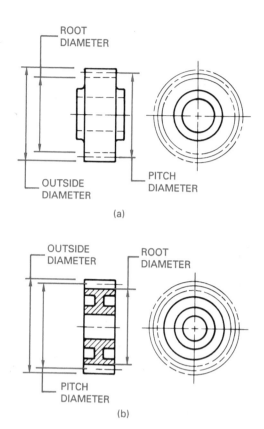

Figure 12–23 Typical spur gear drawings using simplified gear teeth representation. (a) Shown in Multiview. (b) Shown in Section.

The gear teeth, as mentioned earlier, are seldom drawn because it takes too much time to draw them. However, one or more teeth may be drawn for specific applications. For example, when a tooth must be in alignment with another feature of the gear, the tooth can be drawn, as shown in Figure 12–25.

Gear drawings typically contain a chart that shows the manufacturing information associated with the teeth and with related part detail dimensions placed on the specific views. Figure 12–26 shows all of the information in chart and dimensional form that is traditionally associated with a complete gear drawing.

For special presentations or when a computer-aided design system is used, the gear can be drawn with detailed teeth, as shown in Figure 12–27.

## CALCULATING GEAR TRAIN DATA

A gear train is an arrangement of two or more gears connecting driving and driven parts of a machine. Gear reducers and transmissions are examples of gear trains. The function of a gear train is to:

■ Transmit motion between shafts.

■ Decrease or increase the speed between shafts.

■ Change the direction of motion.

It is important for you to understand the relationship between two mating gears to calculate gear train data. When gears are designed, the end result is often a specific gear ration. Any two gears in mesh have a gear ratio. The gear ratio is expressed as a proportion, such as 2:1 or 4:1, between two similar values. The gear ratio between two gears is the relationship between the following characteristics:

■ Number of teeth

■ Pitch diameters

■ Revolutions per minute (RPM)

If you have Gear A (pinion) mating with Gear B, as shown in Figure 12–28, the gear ratio is calculated by dividing like values of the smaller gear into the larger gear as follows:

$$\frac{\text{Number Teeth}_{\text{Gear B}}}{\text{Number Teeth}_{\text{Gear A}}} = \text{Gear Ratio}$$

$$\frac{\text{Pitch Diameter}_{\text{Gear B}}}{\text{Pitch Diameter}_{\text{Gear A}}} = \text{Gear Ratio}$$

$$\frac{\text{RPM}_{\text{Gear B}}}{\text{RPM}_{\text{Gear A}}} = \text{Gear Ratio}$$

Now calculate the gear ratio for the two mating gears in Figure 12–29 if Gear A has 18 teeth, a 6-inch pitch diameter, and operates at 1200 RPM, and Gear B has 54 teeth, an 18-inch pitch diameter, and operates at 400 RPM:

| | | |
|---|---|---|
| Number of Teeth | = 54/18 | = 3:1 |
| Pitch Diameter | = 18/6 | = 3:1 |
| RPM | = 1200/400 | = 3:1 |

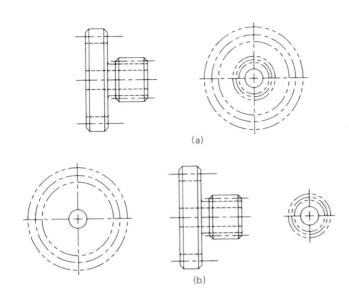

(a)

(b)

Figure 12–24 Cluster gear drawings.

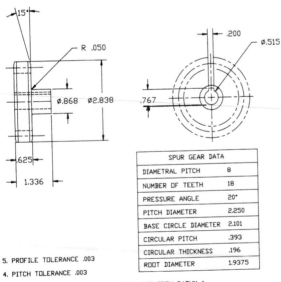

Ø .25 – CENTERLINE OF GEAR TOOTH MUST BE IN LINE WITH TOOLING HOLE

Figure 12–25 Showing the relationship of one gear tooth to another feature on the gear.

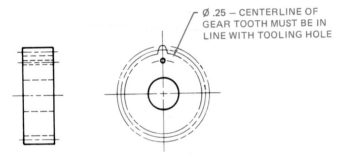

| SPUR GEAR DATA | |
|---|---|
| DIAMETRAL PITCH | 8 |
| NUMBER OF TEETH | 18 |
| PRESSURE ANGLE | 20° |
| PITCH DIAMETER | 2.250 |
| BASE CIRCLE DIAMETER | 2.101 |
| CIRCULAR PITCH | .393 |
| CIRCULAR THICKNESS | .196 |
| ROOT DIAMETER | 1.9375 |

5. PROFILE TOLERANCE .003

4. PITCH TOLERANCE .003

3. ALL TOOTH ELEMENT SPECIFICATIONS ARE FROM DATUM A.

2. INTERPRET GEAR DATA PER ANSI Y14.7.1-1979.

1. INTERPRET DIMENSIONS AND TOLERANCES PER ANSI Y14.5M-1994.

NOTES:

Figure 12–26 The complete CADD gear detail drawing with information in chart and dimensional format.

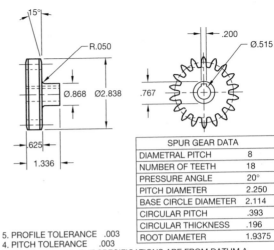

| SPUR GEAR DATA | |
|---|---|
| DIAMETRAL PITCH | 8 |
| NUMBER OF TEETH | 18 |
| PRESSURE ANGLE | 20° |
| PITCH DIAMETER | 2.250 |
| BASE CIRCLE DIAMETER | 2.114 |
| CIRCULAR PITCH | .393 |
| CIRCULAR THICKNESS | .196 |
| ROOT DIAMETER | 1.9375 |

5. PROFILE TOLERANCE .003
4. PITCH TOLERANCE .003
3. ALL TOOTH ELEMENT SPECIFICATIONS ARE FROM DATUM A
2. INTERPRET GEAR DATA PER ANSI Y14.7.1.
1. INTERPRET DIMENSIONS AND TOLERANCES PER ASME Y14.5M.
NOTES:

Figure 12–27 Gear detailed drawing using CADD to automatically draw teeth and gear data chart from given specifications.

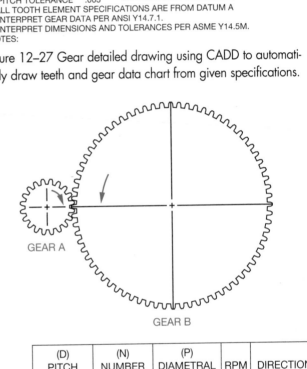

GEAR A

GEAR B

| | (D) PITCH DIAMETER | (N) NUMBER OF TEETH | (P) DIAMETRAL PITCH | RPM | DIRECTION |
|---|---|---|---|---|---|
| GEAR A | 6 | 18 | 3 | 1200 | C.WISE |
| GEAR B | 18 | 54 | 3 | 400 | C.C.WISE |

Figure 12–28 Calculating gear data.

You can solve for unknown values in the gear train if you know the gear ratio you want to achieve, the number of teeth and pitch diameter of one gear, and the input speed. For example, Gear A has 18 teeth, a pitch diameter of 6 inches, and an input speed of 1200 rpm, and the ratio between Gear A and gear B is 3:1. To keep this information well organized, it is recommended that you set up a chart similar to the one shown in Figure 12–29. The unknown values are shown in color for your reference. Determine the number of teeth for Gear B as follows:

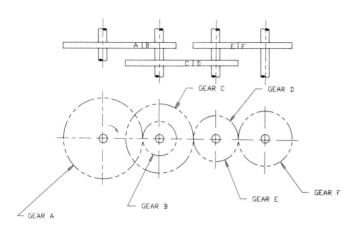

| | PITCH DIAMETER (D) | NUMBER OF TEETH (N) | DIAMETRAL PITCH (P) | RPM | DIRECTION | GEAR RATIO | CENTER DISTANCE |
|---|---|---|---|---|---|---|---|
| GEAR A | 7,000 | 28 | 4 | 300 | C.WISE | 2.33/1 | 5.000 |
| GEAR B | 3,000 | 12 | 4 | 699 | C.C.WISE | | |
| GEAR C | 6,000 | 18 | 3 | 699 | C.C.WISE | 1.5/1 | 5.000 |
| GEAR D | 4,000 | 12 | 3 | 1048.5 | C.WISE | | |
| GEAR E | 4,000 | 20 | 5 | 1048.5 | C.WISE | 1.2/1 | 4.400 |
| GEAR F | 4,800 | 24 | 5 | 1258.2 | C.C.WISE | | |

Figure 12–29 Calculate the unknown values for a given gear train.

$$\text{Teeth}_{\text{Gear A}} \times \text{Gear Ratio} = 18 \times 3 = \text{Teeth}_{\text{Gear B}} = 54$$

Or, if you know that Gear B has 54 teeth and the gear ratio is 3:1, then:

$$\frac{\text{Teeth}_{\text{Gear B}}}{\text{Gear Ratio}} \quad \frac{54}{3} = \text{Teeth}_{\text{Gear A}} = 18 \approx$$

Determine the RPM of gear B:

$$\frac{\text{RPM}_{\text{Gear A}}}{\text{Gear Ratio}} \quad \frac{1200}{3} = \text{RPM}_{\text{Gear B}} = 400$$

Determine the pitch diameter of gear B:

$$\text{Pitch Diameter} \times \text{Gear Ratio} = \text{Pitch Diameter}$$

| 6 inches | x | 3 | = | 18 inches |
|---|---|---|---|---|
| (Gear A) | | | | (Gear B) |

In some situations, you need to refer to the formulas given in Figure 12–21 to determine some unknown values. In this case, you must calculate the diametral pitch using the formula P = N/D, where P = diametral pitch, N = number of teeth, and D = pitch diameter:

$$P_{\text{Gear A}} = \frac{N}{D} = \frac{18}{6} = 3$$

Because the diametral pitch is tooth size and the teeth for mating gears must be the same size, the diametral pitch of gear B is also 3.

Keep in mind that the preceding example presents only one set of gear data. Other situations may be different. Always solve for unknown values based on information that you have and work with the standard formulas presented in this chapter. Following are some important points to keep in mind as you work with gear trains:

■ The RPM of the larger gear is always slower than the RPM of the smaller gear.

■ Mating gears always turn in opposite directions. For example, in Figure 12–29, Gear B turns counterclockwise, if Gear A turns clockwise.

■ Gears on the same shaft (cluster gears) always turn in the same direction and at the same speed (RPMs).

■ Mating gears have the same size teeth (diametral pitch).

■ The gear ratio between mating gears is a ratio between the number of teeth, the pitch diameters, and the RPMs.

■ The distance between the subscript shafts of mating gears is equal to $1/2\ D_{\text{Gear A}} + 1/2\ D_{\text{Gear B}}$. This distance is 3 + 9 = 12 inches between shafts in Figure 12–29.

Given the gear train shown in Figure 12–29, calculate the unknown values using the formulas and information discussed in this chapter. The unknown values are given in color to help you check your work.

## READING RACK AND PINION PRINTS

A rack is a straight bar with spur gear teeth used to convert rotary motion to reciprocating motion. Figure 12–30 shows a spur gear pinion mating with a rack. Notice the circular dimensions of the pinion become linear dimensions on the rack.

When reading a detailed drawing of the rack, the front view is normally shown with the profile of the first and last tooth drawn. Phantom lines are drawn to represent the top and root, and a centerline is used for the pitch line (see Figure 12–31). Related tooth and length dimensions are found on the front view and a depth dimension is placed on the side view. A chart is then placed in the field of the drawing to identify specific gear-cutting information.

## READING BEVEL GEAR PRINTS

Bevel gears are usually used to transmit power between intersecting shafts at 90°, although they may be used for any angle. Some gear terminology and formulas relate specifically to the manufacture of bevel gears. These formulas and a drawing of a bevel gear and pinion are shown in Figure 12–32. Most of the gear terms discussed for spur gears apply to bevel gears. In many cases information must be calculated using the formulas provided in the spur gear discussion shown in Figure 12–21.

The drawing for a bevel gear is similar to the drawing for a spur gear, but in many cases only one view is needed. This view is often a full section. Another view may be necessary to show the dimensions for a keyway. As with other gear drawings, a chart is used to specify gear-cutting data (see Figure 12–33). Notice, in this exam-

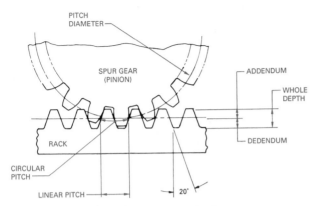

Figure 12–30 Rack and pinion terminology

ple, that only a partial side view is used to specify the bore and keyway.

# PLASTIC GEARS

The following is taken in part from *Plastic Gearing* by William McKinlay and Samuel D. Pierson, published by ABA/PGT Inc., Manchester, CT. Gears can be molded of many engineering plastics in various grades and in filled varieties. Filled plastics are those in which a material has been added to improve the mechanical properties. The additives normally used in gear plastics are glass, polytetrafluoroethylene (PTFE), silicones, and molybdenum disulphide. Glass fiber reinforcement can double the tensile strength and reduce the thermal expansion. Carbon fiber is often used to increase strength. Silicones, PTFE, and molybdenum disulphide are used to act as built-in lubricants and provide increased

wear resistance. Plastic gears are designed in the same manner as gears made from other materials. However, the physical characteristics of plastics make it necessary to follow gear design practices more closely than when designing gears that are machined from metals.

## Advantages of Molded Plastic Gears

Gears molded of plastic are replacing stamped and cut metal gears in a wide range of mechanisms. Designers are turning to molded plastic gears for one or more of the following reasons:

- ■ Reduced cost.
- ■ Increased efficiency.
- ■ Self-lubrication.
- ■ Increased tooth strength with nonstandard pressure angles.
- ■ Reduced weight.
- ■ Corrosion resistance.
- ■ Less noise.
- ■ Available in colors.

## Disadvantages of Molded Plastic Gears

Plastic gearing has the following limitations when compared with metal gearing:

- ■ Lower strength.
- ■ Greater thermal expansion and contraction.

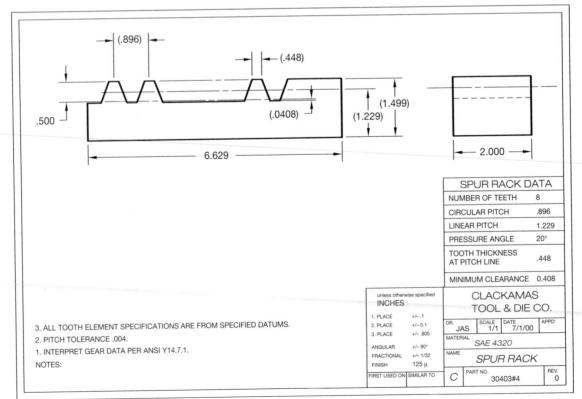

Figure 12–31 A detailed drawing of a rack.

| TERM | DESCRIPTION | FORMULA |
|---|---|---|
| Pitch Diameter (D) | The diameter of the base of the pitch cone. | $D = N/P$ |
| Pitch Cone | In Figure 14-64, the pitch come is identified as XYZ. | |
| Pitch Angle (Ø) | The angle between an element of a pitch cone and its axis. Pitch angles of mating gears depend on their relative diameters (gear a and gear b). | $\tan \varnothing_a = D_a/D_b$ $\tan \varnothing_b = D_b/D_a$ |
| Cone Distance (A) | Slant height of pitch cone. | $A = D/2 \sin \varnothing$ |
| Addendum Angle (δ) | The angle subtended by the addendum. It is the same for mating gears. | $\tan \delta = a/A$ |
| Dedendum Angle (Ω) | The angle subtended by the dedendum. It is the same for mating gears. | $\tan \Omega = b/A$ |
| Face Angle (Ø$_o$) | The angle between the top of the teeth and the gear axis. | $\varnothing_o = \varnothing - \delta$ |
| Root Angle (Ø$_r$) | The angle between the bottom of the tooth space and the gear axis. | $\varnothing_r = \varnothing - \Omega$ |
| Outside Diameter (D$_o$) | The diameter of the outside circle of gear. | $D_o = D + 2a(\cos \varnothing)$ |
| Crown Height (X) | The distance between the cone apex and the outer tip of the gear teeth. | $X = .5 (D_o)/\tan \varnothing_o$ |
| Crown Backing (Y) | The distance from the rear of the hub to the outer tip of the gear tooth, measured parallel to the axis of the gear. | |
| Face Width | The distance that should not exceed one-third of the come distance (A). | |

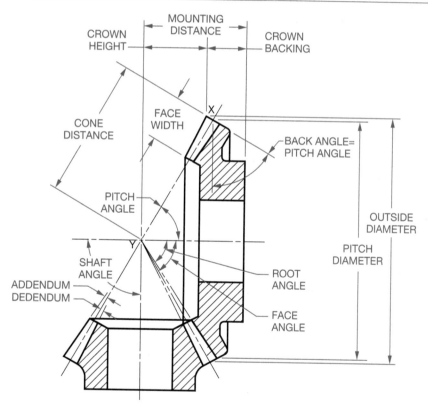

Figure 12–32 Bevel gear terminology and formulas.

- Limited heat resistance.
- Size change with moisture absorption.

## Accuracy of Molded Plastic Gears

Today's technology permits a very high degree of precision. In general, tooth-to-tooth composite tolerances can be economically held to .0005 or less for fine pitch gears. Total composite tolerance varies depending on configuration, evenness of product cross section, and the selection of the molding material.

## READING WORM GEAR PRINTS

Worm gears are used to transmit power between nonintersecting shafts. More importantly, they are used for large speed reductions in a small space as compared to other types of gears. Worm gears are also strong, move in either direction, and lock in place when

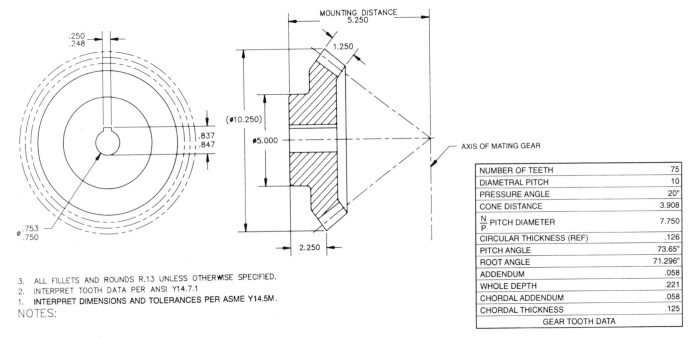

| NUMBER OF TEETH | 75 |
|---|---|
| DIAMETRAL PITCH | 10 |
| PRESSURE ANGLE | 20° |
| CONE DISTANCE | 3.908 |
| $\frac{N}{P}$ PITCH DIAMETER | 7.750 |
| CIRCULAR THICKNESS (REF) | .126 |
| PITCH ANGLE | 73.65° |
| ROOT ANGLE | 71.296° |
| ADDENDUM | .058 |
| WHOLE DEPTH | .221 |
| CHORDAL ADDENDUM | .058 |
| CHORDAL THICKNESS | .125 |
| GEAR TOOTH DATA | |

NOTES:
3. ALL FILLETS AND ROUNDS R.13 UNLESS OTHERWISE SPECIFIED.
2. INTERPRET TOOTH DATA PER ANSI Y14.7.1
1. INTERPRET DIMENSIONS AND TOLERANCES PER ASME Y14.5M.

Figure 12–33 A detailed drawing of a bevel gear.

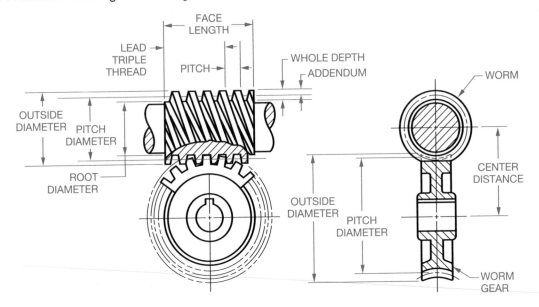

| TERM | DESCRIPTION | FORMULA |
|---|---|---|
| Pitch Diameter (worm) ($D_w$) | | $D_w = 2C - D_g$ |
| Pitch Diameter (gear) ($D_g$) | | $D_g = 2C - D_w$ |
| Pitch (P) | The distance from one tooth to the corresponding point on the next tooth measured parallel to the worm axis. It is equal to the circular pitch on the worm gear. | $P = L/T$ |
| Lead (L) | The distance the thread advances axially in one revolution of the worm. | $L = D_g/R$ |
| | | $L = P \times T$ |
| Threads (T) | Number of threads or starts on worm. | $T = L/P$ |
| Gear Teeth (N) | Number of teeth on worm gear. | $N = \pi D_g/P$ |
| Ratio (R) | Divide number of gear teeth by the number of worm threads. | $R = N/T$ |
| Addendum (a) | For single and double threads. | $a = .318P$ |
| Whole Depth (WD) | For single and double threads. | $WD = .686P$ |

Figure 12–34 Worm and worm gear terminology and formulas.

the machine is not in operation. A single lead worm advances one pitch with every revolution. A double lead worm advances two pitches with each revolution. As with bevel gears, the worm gear and worm have specific terminology and formulas that apply to their design. The representative worm gear and worm technology and design formulas are shown in Figure 12–34.

Figure 12–35 shows a worm gear drawing using a half-section method. A side view is provided to dimension the bore and keyway, and a chart is given for gear-cutting data.

A detailed drawing of the worm normally uses two views. The front view shows the first gear tooth on each end with phantom lines between for a simplified representation. The side view is the same as a spur gear drawing with the keyway specifications. The gear-cutting chart is then placed on the field of the drawing or over the title block. Figure 12–36 shows the worm drawing using CADD to provide a detailed representation of the worm teeth in the front view, rather than phantom lines, as in the simplified representation.

## BEARINGS

Bearings are mechanical devices used to reduce friction between two surfaces. They are divided into two large groups known as plain and rolling element bearings. Bearings are designed to accommodate either rotational or linear motion. Rotational bearings are used for radial loads, and linear bearings are designed for thrust loads. Radial loads are loads that are distributed around the shaft. Thrust loads are lateral. Thrust loads apply force to the end of the shaft. Figure 12–37 shows the relationship between rotational and linear motion.

### Plain Bearings

Plain bearings are often referred to as sleeve, journal bearings, or bushings. Their operation is based on a sliding action between mating parts. A clearance fit between the inside diameter of the bearing and the shaft is critical to ensure proper operation. Refer to fits between mating parts in Chapter 6 for more information. The

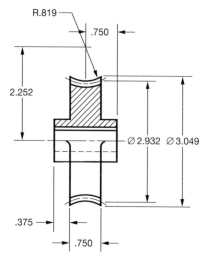

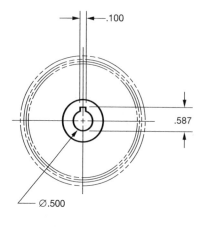

3. PITCH TOLERANCE ± .002.
2. INTERPRET GEAR DATA PER ANSI Y14.7.1.
1. INTERPRET DIMENSIONS AND TOLERANCES
   PER ANSI Y14.5M.
NOTES:

| WORM GEAR DATA | |
|---|---|
| NUMBER OF TEETH | 27 |
| PITCH DIAMETER | 2.933 |
| ADEDDUM | .057 |
| WHOLE DEPTH | .114 |
| TOOTH THICKNESS | .100 |

Figure 12–35 A detail drawing of a worm gear.

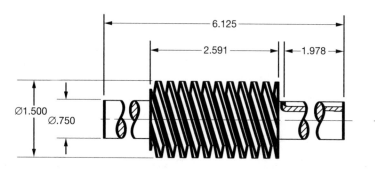

| WORM GEAR DATA | |
|---|---|
| PITCH DIAMETER | 1.251 |
| LEAD RIGHT OR LEFT | RIGHT |
| CENTER DISTANCE | 1.858 |
| WORKING DEPTH | .218 |
| CLEARANCE | .0312 |
| PRESSURE ANGLE | 20° |
| ADDENDUM | .125 |
| WHOLE DEPTH | 25 |
| CHORDAL THICKNESS | .125 |

2. PITCH TOLERANCE ± .002.
1. INTERPRET GEAR DATA PER ANSI Y14.7.1.
NOTES:

Figure 12–36 A detail drawing of a worm.

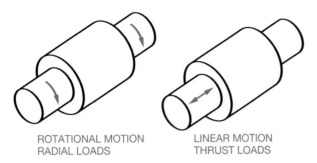

ROTATIONAL MOTION
RADIAL LOADS

LINEAR MOTION
THRUST LOADS

Figure 12–37 The relationship between rotational and linear motion.

bearing has an interference fit between the outside of the bearing and the housing or mounting device, as shown in Figure 12–38.

The material from which plain bearings are made is important. Most plain bearings are made from bronze or phosphor bronze. Bronze bearings are normally lubricated, while phosphor bronze bearings are commonly impregnated with oil and require no additional lubrication. Phosphor bronze is an excellent choice when anti-friction qualities are important and where resistance to wear and scuffing is needed.

## Rolling Element Bearings

Ball and roller bearings are the two classes of rolling element bearings. Ball bearings are the most commonly used rolling element bearings. In most cases, ball bearings have higher speed and lower load capabilities than roller bearings. Even so, ball bearings are manufactured for most uses. Ball bearings are constructed with two grooved rings, a set of balls placed radially around the rings, and a separator that keeps the balls spaced apart and aligned, as shown in Figure 12–39.

Single-row ball bearings are designed primarily for radial loads, but they can accept some thrust loads. Double-row ball bearings may be used where shaft alignment is important. Angular contact ball bearings support a heavy thrust load and a moderate radial load. Thrust bearings are designed for use in thrust load situations only. When both thrust and radial loads are necessary, both radial and thrust ball bearings are used together. Some typical ball bearings are shown in Figure 12–40. Ball bearings are available with shields and seals. A shield is a metal plate on one or both sides of the bearing. The shields act to keep the bearing clean and retain the lubricant. A sealed bearing has seals made of rubber, felt, or plastic placed on the outer and inner rings of the bearing. The sealed bearings are filled with special lubricant by the manufacturer. They require little or no maintenance in service. Figure 12–41 shows the shields and seals used on ball bearings.

Roller bearings are more effective than ball bearings for heavy loads. Cylindrical roller bearings have a high radial capacity and assist in shaft alignment. Needle roller bearings have small rollers and are designed for the highest load-carrying capacity of all rolling element bearings with shaft sizes under 10 inches. Tapered roller bearings are used in gear reducers, steering mechanisms, and machine tool spindles. Spherical roller bearings offer the best combination of high load capacity, tolerance to shock, and alignment,

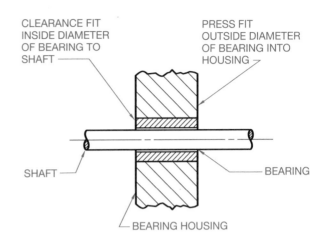

CLEARANCE FIT
INSIDE DIAMETER
OF BEARING TO
SHAFT

PRESS FIT
OUTSIDE DIAMETER
OF BEARING INTO
HOUSING

SHAFT

BEARING

BEARING HOUSING

Figure 12–38 Plain bearing terminology and fits.

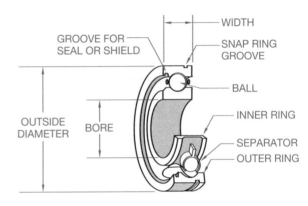

GROOVE FOR
SEAL OR SHIELD

WIDTH

SNAP RING
GROOVE

BALL

INNER RING

OUTSIDE
DIAMETER

BORE

SEPARATOR

OUTER RING

Figure 12–39 Ball bearing components.

**RADIAL BALL
BEARING**

**THRUST BALL
BEARING**

**ANGULAR CONTACT
BALL BEARING**

Figure 12–40 Typical ball bearings.

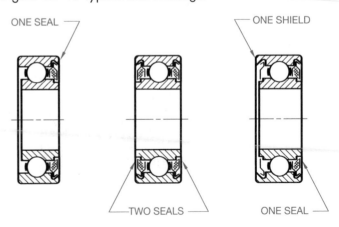

ONE SEAL

ONE SHIELD

TWO SEALS

ONE SEAL

Figure 12–41 Bearing seals and shields.

and are used on conveyors, transmissions, and heavy machinery. Some common roller bearings are displayed in Figure 12–42.

## BEARING SYMBOLS

Bearing symbols are standardized, as shown in Figure 12–43, to help the reader identify the type and location of bearings. Bearings are generally part of the total mechanism and are therefore normally found on the assembly drawing. As you will learn in Chapter 13, assemblies show each piece of the product put together and correlated to a parts list. The parts list clearly identifies the specific bearings used in the assembly. Bearings are also represented as symbols in bearing manufacturers' catalogs.

## BEARING CODES

Bearing manufacturers use similar coding systems for the identification and ordering of different bearing products. The bearing codes generally contain the following types of information:

- ■ Material
- ■ Bearing type
- ■ Bore size
- ■ Lubricant
- ■ Type of seals or shields

A sample bearing numbering system is shown in Figure 12–44.

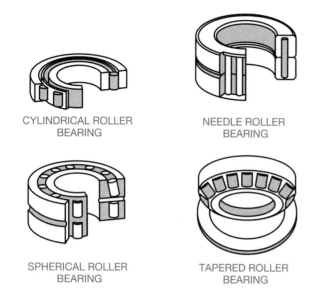

CYLINDRICAL ROLLER BEARING

NEEDLE ROLLER BEARING

SPHERICAL ROLLER BEARING

TAPERED ROLLER BEARING

Figure 12–42 Typical roller bearings.

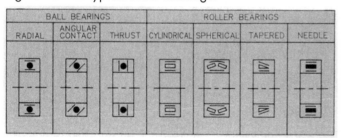

Figure 12–43 Bearing symbols drawn using CADD.

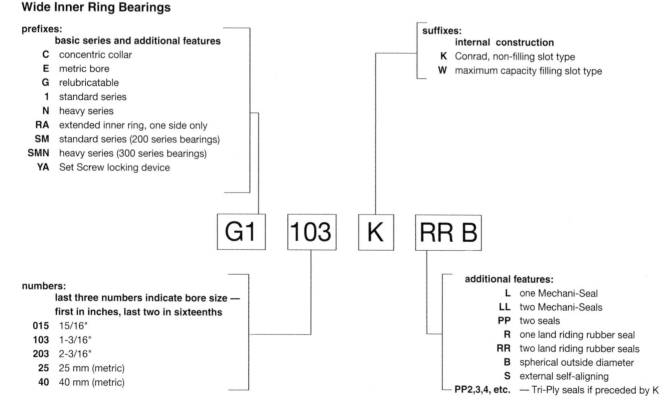

Figure 12–44 A sample bearing numbering system. Courtesy The Torrington Company.

# BEARING SELECTION

A variety of bearing types are available from manufacturers. Bearing design may differ depending on the use requirements. For example, a supplier may have light, medium, and heavy bearings available. Bearings can have specially designed outer and inner rings. Bearings are available open (without seals or shields) or with one or two shields or seals. Light bearings are generally designed to accommodate a wide range of applications, involving light to medium loads combined with relatively high speeds. Medium bearings have heavier construction than light bearings and provide a greater radial and thrust capacity. They are also able to withstand greater shock than light bearings. Heavy bearings are often designed for special service where extra-heavy shock loads are required. Bearings can also be designed to accommodate radial loads, thrust loads, or a combination of loading requirements.

## Bearing Bore, Outside Diameters, and Width

Bearings are dimensioned in relation to the bore diameter, outside diameter, and width. These dimensions are shown in Figure 12–45. After the loading requirements have been established, the bearing is selected in relationship to the shaft size. For example, if an approximate Ø 1.5-inch shaft size is required for a medium service bearing, a vendor's catalog chart, similar to the one shown in Figure 12–46, is used to select the bearing.

Referring to the chart shown in Figure 12–46, notice that the first column is the vendor's bearing number, followed by the bore size (B). To select a bearing for the approximate 1.5-inch shaft, go to the chart and pick the bore diameter of 1.5748, which is close to 1.5. This is the 308K bearing. The tolerance for this bore is specified in the chart as 1.5748 + .0000 and −.0005. Therefore, the limits dimension of the bore in this example is 1.5748/1.5743. The outside diameter is 3.5433 + .0000 − .0006 (3.5433/3.5427). The width of this bearing is .906 + .000 − .005 (.906/.901). The fillet

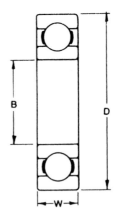

Figure 12–45 Bearing dimensions. *Courtesy The Torrington Company.*

radius is the maximum shaft or housing fillet radius in which the bearing corners will clear. The fillet radius for the 308K bearing is R.059. Notice that the dimensions are also given in millimeters.

## Shaft and Housing Fits

Shaft and housing fits are important, because tight fits may cause failure of the balls, rollers, or lubricant, or overheating. Loose fits can cause slippage of the bearings in the housing, resulting in overheating, vibration, or excessive wear.

### Shaft Fits

In general, for precision bearings it is recommended that the shaft diameter and tolerance be the same as the bearing bore diameter and tolerance. The shaft diameter used with the 308K bearing is dimensioned Ø1.5748/1.5743.

### Housing Fits

In most applications with rotating shafts, the outer ring is stationary and should be mounted with a push of the hand or light tapping. In general, the minimum housing diameter is .0001 larger

## DIMENSIONS — TOLERANCES

| Bearing Number | Bore B tolerance + .0000" + .000 mm to minus | | | | Outside Diameter D tolerance + .0000" + .000 mm to minus | | | | Width W +.000", -.005" +.00mm, -.13mm | | Fillet Radius (1) | | WL | | Static Load Rating C₀ | | External Dynamic Load Rating C | |
|---|---|---|---|---|---|---|---|---|---|---|---|---|---|---|---|---|---|---|
| | in. | mm | in. | mm | in. | mm | in. | mm | in. | mm | in. | mm | lbs. | kg | lbs | N | lbs. | N |
| 300K | .3937 | 10 | .0003 | .008 | 1.3780 | 35 | .0005 | .013 | .433 | 11 | .024 | .6 | .12 | .054 | 850 | 3750 | 2000 | 9000 |
| 301K | .4724 | 12 | .0003 | .008 | 1.4567 | 37 | .0005 | .013 | .472 | 12 | .039 | 1.0 | .14 | .064 | 850 | 3750 | 2080 | 9150 |
| 302K | .5906 | 15 | .0003 | .008 | 1.6535 | 42 | .0005 | .013 | .512 | 13 | .039 | 1.0 | .18 | .082 | 1270 | 5600 | 2900 | 13200 |
| 303K | .6693 | 17 | .0003 | .008 | 1.8504 | 47 | .0005 | .013 | .551 | 14 | .039 | 1.0 | .24 | .109 | 1460 | 6550 | 3350 | 15000 |
| 304K | .7874 | 20 | .0004 | .010 | 2.0472 | 52 | .0005 | .013 | .591 | 15 | .039 | 1.0 | .31 | .141 | 1760 | 7800 | 4000 | 17600 |
| 305K | .9843 | 25 | .0004 | .010 | 2.4409 | 62 | .0005 | .013 | .669 | 17 | .039 | 1.0 | .52 | .236 | 2750 | 12200 | 5850 | 26000 |
| 306K | 1.1811 | 30 | .0004 | .010 | 2.8346 | 72 | .0005 | .013 | .748 | 19 | .039 | 1.0 | .78 | .354 | 3550 | 15600 | 7500 | 33500 |
| 307K | 1.3780 | 35 | .0005 | .013 | 3.1496 | 80 | .0005 | .013 | .827 | 21 | .059 | 1.5 | 1.04 | .472 | 4500 | 20000 | 9150 | 40500 |
| 308K | 1.5748 | 40 | .0005 | .013 | 3.5433 | 90 | .0006 | .015 | .906 | 23 | .059 | 1.5 | 1.42 | .644 | 5600 | 24500 | 11000 | 49000 |
| 309K | 1.7717 | 45 | .0005 | .013 | 3.9370 | 100 | .0006 | .015 | .984 | 25 | .059 | 1.5 | 1.90 | .862 | 6700 | 3000 | 13200 | 58500 |
| 310K | 1.9685 | 50 | .0005 | .013 | 4.3307 | 110 | .0006 | .015 | 1.063 | 27 | .079 | 2.0 | 2.48 | 1.125 | 8000 | 35500 | 15300 | 68000 |
| 311K | 2.1654 | 55 | .0006 | .015 | 4.7244 | 120 | .0006 | .015 | 1.142 | 29 | .079 | 2.0 | 3.14 | 1.424 | 9500 | 41500 | 18000 | 80000 |
| 312K | 2.3622 | 60 | .0006 | .015 | 5.1181 | 130 | .0008 | .020 | 1.220 | 31 | .079 | 2.0 | 3.89 | 1.765 | 10800 | 48000 | 20400 | 90000 |

⁽¹⁾Maximum shaft or housing fillet radius which bearing corners will clear.

Figure 12–46 Bearing selection chart. *Courtesy The Torrington Company.*

Figure 12–47 Shaft shoulder and housing shoulder dimensions. *Courtesy The Torrington Company.*

than the maximum bearing outside diameter and the maximum housing diameter is .0003 larger than the minimum housing diameter. With this in mind, the housing diameter for the 308K bearing is 3.5433 + .0001 = 3.5434 and 3.5434 + .0003 = 3.5437. The housing diameter limits are 3.5437/3.5434.

**The Shaft Shoulder and Housing Shoulder Dimensions**

Next, you should size the shaft shoulder and housing shoulder diameters. The shaft shoulder and housing shoulder diameter dimensions are represented in Figure 12–47 as S and H. The shoulders should be large enough to rest flat on the face of the bearing and small enough to allow bearing removal. Refer to the chart in Figure 12–48 to determine the shaft shoulder and housing shoulder diameters for the 308K bearing selected in the preceding discussion. Find the basic bearing number 308 and determine the limits of the shaft shoulder and the housing shoulder. The shaft shoulder diameter is 2.00/1.93 and the housing shoulder diameter is 3.19/3.06. A partial detailed drawing of the shaft and housing for the 308K bearing is shown in Figure 12–49.

## Surface Finish of Shaft and Housing

The recommended surface finish for precision-bearing applications is 32 microinches (0.80 micrometer) for the shaft finish on shafts under 2 inches in diameter. For shafts over 2 inches in diameter, a 63-microinch (1.6 micrometer) finish is suggested. The housing diameter may have a 125-microinch (3.2 micrometer) finish for all applications.

## Bearing Lubrication

It is necessary to maintain a film of lubrication between the bearing surfaces. The factors to consider when selecting lubrication requirements include the:

- Type of operation, such as continuous or intermittent.
- Service speed in RPMs (revolutions per minute).
- Bearing load, such as light, medium, or heavy.

### Medium • 300, 7300WN Series — Shoulder Diameters

| Basic Bearing Number | shaft, S max | | shaft, S min | | housing, H max | | housing, H min | |
|---|---|---|---|---|---|---|---|---|
| | in. | mm | in. | mm | in. | mm | in. | mm |
| 300 | .59 | 15.0 | .50 | 12.7 | 1.18 | 30.0 | 1.15 | 29.2 |
| 301 | .69 | 17.5 | .63 | 16.0 | 1.22 | 31.0 | 1.21 | 30.7 |
| 302 | .81 | 20.6 | .75 | 19.0 | 1.42 | 36.1 | 1.40 | 35.6 |
| 303 | .91 | 23.1 | .83 | 21.1 | 1.61 | 40.9 | 1.60 | 40.6 |
| 304 | 1.06 | 26.9 | .94 | 23.9 | 1.77 | 45.0 | 1.75 | 44.4 |
| 305 | 1.31 | 33.3 | 1.14 | 29.0 | 2.17 | 55.1 | 2.09 | 53.1 |
| 306 | 1.56 | 39.6 | 1.34 | 34.0 | 2.56 | 65.0 | 2.44 | 62.0 |
| 307 | 1.78 | 45.2 | 1.69 | 42.9 | 2.80 | 71.1 | 2.72 | 69.1 |
| 308 | 2.00 | 50.8 | 1.93 | 49.0 | 3.19 | 81.0 | 3.06 | 77.7 |
| 309 | 2.28 | 57.9 | 2.13 | 54.1 | 3.58 | 90.9 | 3.41 | 86.6 |
| 310 | 2.50 | 63.5 | 2.36 | 59.9 | 3.94 | 100.1 | 3.75 | 95.2 |
| 311 | 2.75 | 69.8 | 2.56 | 65.0 | 4.33 | 110.0 | 4.13 | 104.9 |
| 312 | 2.94 | 74.7 | 2.84 | 72.1 | 4.65 | 118.1 | 4.44 | 112.8 |

### Light • 200, 7200WN Series — Shoulder Diameters

| Basic Bearing Number | shaft, S max | | shaft, S min | | housing, H max | | housing, H min | |
|---|---|---|---|---|---|---|---|---|
| | in. | mm | in. | mm | in. | mm | in. | mm |
| 200 | .56 | 14.2 | .50 | 12.7 | .98 | 24.9 | .97 | 24.6 |
| 201 | .64 | 16.3 | .58 | 14.7 | 1.06 | 26.9 | 1.05 | 26.7 |
| 202 | .75 | 19.0 | .69 | 17.5 | 1.18 | 30.0 | 1.15 | 29.2 |
| 203 | .84 | 21.3 | .77 | 19.6 | 1.34 | 34.0 | 1.31 | 33.3 |
| 204 | 1.00 | 25.4 | .94 | 23.9 | 1.61 | 40.9 | 1.58 | 40.1 |
| 205 | 1.22 | 31.0 | 1.14 | 29.0 | 1.81 | 46.0 | 1.78 | 45.2 |
| 206 | 1.47 | 37.3 | 1.34 | 34.0 | 2.21 | 56.1 | 2.16 | 54.9 |
| 207 | 1.72 | 43.7 | 1.53 | 38.9 | 2.56 | 65.0 | 2.47 | 62.7 |
| 208 | 1.94 | 49.3 | 1.73 | 43.9 | 2.87 | 72.9 | 2.78 | 70.6 |
| 209 | 2.13 | 54.1 | 1.94 | 49.3 | 3.07 | 78.0 | 2.97 | 75.4 |
| 210 | 2.34 | 59.4 | 2.13 | 54.1 | 3.27 | 83.1 | 3.17 | 80.5 |
| 211 | 2.54 | 64.5 | 2.41 | 61.2 | 3.68 | 93.5 | 3.56 | 90.4 |
| 212 | 2.81 | 71.4 | 2.67 | 67.8 | 3.98 | 101.1 | 3.87 | 98.3 |

### Extra-Light • 9100 Series — Shoulder Diameters

| Basic Bearing Number | shaft, S max | | shaft, S min | | housing, H max | | housing, H min | |
|---|---|---|---|---|---|---|---|---|
| | in. | mm | in. | mm | in. | mm | in. | mm |
| 9100 | .52 | 13.2 | .47 | 11.9 | .95 | 24.1 | .91 | 23.1 |
| 9101 | .71 | 18.0 | .55 | 14.0 | 1.02 | 25.9 | .97 | 24.6 |
| 9102 | .75 | 19.0 | .67 | 17.0 | 1.18 | 30.0 | 1.13 | 28.7 |
| 9103 | .81 | 20.6 | .75 | 19.0 | 1.30 | 33.0 | 1.25 | 31.8 |
| 9104 | .98 | 24.9 | .89 | 22.6 | 1.46 | 37.1 | 1.41 | 35.8 |
| 9105 | 1.18 | 30.0 | 1.08 | 27.4 | 1.65 | 41.9 | 1.60 | 40.6 |
| 9106 | 1.38 | 35.1 | 1.34 | 34.0 | 1.93 | 49.0 | 1.88 | 47.8 |
| 9107 | 1.63 | 41.4 | 1.53 | 38.9 | 2.21 | 56.1 | 2.15 | 54.6 |
| 9108 | 1.81 | 46.0 | 1.73 | 43.9 | 2.44 | 62.0 | 2.39 | 60.7 |
| 9109 | 2.03 | 51.6 | 1.94 | 49.3 | 2.72 | 69.1 | 2.67 | 67.8 |
| 9110 | 2.22 | 56.4 | 2.13 | 54.1 | 2.91 | 73.9 | 2.86 | 72.6 |
| 9111 | 2.48 | 63.0 | 2.33 | 59.2 | 3.27 | 83.1 | 3.22 | 81.8 |
| 9112 | 2.67 | 67.8 | 2.53 | 64.3 | 3.47 | 88.1 | 3.42 | 86.9 |

Figure 12–48 Shaft shoulder and housing shoulder dimensions selection chart. *Courtesy The Torrington Company.*

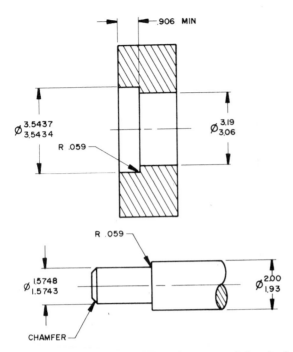

Figure 12-49 A partial detailed drawing of the shaft and housing for the 308K bearing.

Bearings may also be overlubricated, which can cause increased operating temperatures and early failure. Selection of the proper lubrication for the application should be determined by the manufacturer's recommendations. The ability of the lubricant is directly related, in part, to viscosity. Viscosity is the internal friction of a fluid, which makes it resist a tendency to flow. Fluids with low viscosity flow more freely than those with high viscosity. The chart in Figure 12-50 shows the selection of oil viscosity based on temperature ranges and speed factors.

## Oil Grooving of Bearings

In situations where bearings or bushings do not receive proper lubrication, it may be necessary to provide grooves for the proper flow of lubrication to the bearing surface. The bearing grooves help

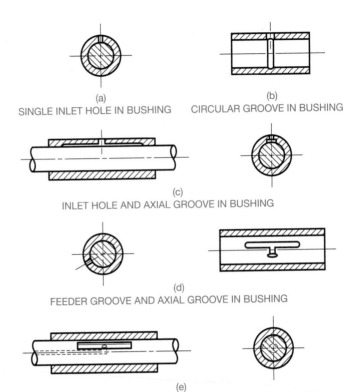

Figure 12-51 Methods of designing paths for lubrication to bearing surfaces.

provide the proper lubricant between the bearing surfaces and maintain adequate cooling. There are several methods of creating paths for the lubrication to the bearing surfaces, as shown in Figure 12-51.

## Sealing Methods

Machine designs normally include means for stopping leakage and keeping out dirt and other contaminants when lubricants are involved in the machine operation. This is accomplished using static or dynamic sealing devices. Static sealing refers to stationary devices that are held in place and stop leakage by applied pressure. Static seals, such as gaskets, do not come in contact with the moving parts of the mechanism. Dynamic seals are those that contact the moving parts of the machinery, such as packings.

Gaskets are made from materials that prevent leakage or access of dust contaminants into the machine cavity. Silicone rubber gasket materials are used in applications, such as water pumps, engine filter housing, and oil pans. Gasket tapes, ropes, and strips provide good cushioning properties for dampening vibration, and the adhesive sticks well to most materials. Nonstick gasket materials, such as paper, cork, and rubber, are available for certain applications. Figure 12-52 shows a typical gasket mounting.

Dynamic seals include packings and seals that fit tightly between the bearing or seal seat and the shaft.

The pressure applied by the seal seat or the pressure of the fluid causes the sealing effect. Molded lip packings that provide sealing

Oil Viscosities and Temperature Ranges for Ball Bearing Lubrication

| Maximum Temperature Range Degrees F | Optimum Temperature Range Degrees F | Speed Factor, S (inner race bore diameter (inches) x RPM) | |
|---|---|---|---|
| | | Under 1000 | Over 1000 |
| | | Viscosity | |
| −40 to +100 | −40 to −10 | 80 to 90 SSU (at 100° F) | 70 to 80 SSU (at 100° F) |
| −10 to +100 | −10 to +30 | 100 to 115 SSU (at 100° F) | 80 to 100 SSU (at 100° F) |
| +30 to +150 | +30 to +150 | SAE 20 | SAE 10 |
| +30 to +200 | +150 to +200 | SAE 40 | SAE 30 |
| +50 to +300 | +200 to +300 | SAE 70 | SAE 60 |

Figure 12-50 Selection of oil viscosity based on temperature ranges and speed factors.

as a result of the pressure generated by the machine fluid are available. Figure 12–53 shows examples of molded lip packings. Molded ring seals are placed in a groove and provide a positive seal between the shaft and bearing or bushing. Types of molded ring seals include labyrinth, O-ring, lobed ring, and others. A labyrinth, which means maze, refers to a seal that is made of a series of spaced strips that are connected to the seal seat, making it difficult for the lubrication to pass. Labyrinth seals are used in heavy machinery where some leakage is permissible. (See Figure 12–54.) The O-ring seal is the most commonly used seal because of its low cost, ease of application, and flexibility. The O-ring may be used for most situations involving rotating or oscillating motion. The O-ring is placed in a groove that is machined in either the shaft or the housing, as shown in Figure 12–55. The lobed ring has rounded lobes

that provide additional sealing forces over the standard O-ring seal. A typical lobed ring seal is shown in Figure 12–56.

Felt and wool seals are used where economical cost, lubricant absorption, filtration, low friction, and a polishing action are required. However, the ability to completely seal the machinery is not as positive as with the seals described earlier (see Figure 12–57).

### Bearing Mountings

There are a number of methods used for holding the bearing in place. Common techniques include a nut and lock washer, a nut and lock nut, or a retaining ring. Other methods may be designed to fit the specific application or requirements, such as a shoulder plate. Figure 12–58 shows some examples of mountings.

## GEAR AND BEARING ASSEMBLIES

Gear and bearing assemblies show the part of the complete mechanism as they appear assembled (see Figure 12–59). In some situations only a full sectional view is used to display all

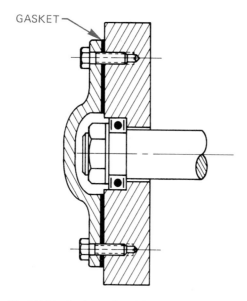

Figure 12–52 Typical Gasket Mounting

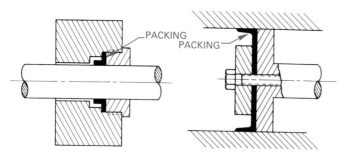

Figure 12–53 Molded lip packings.

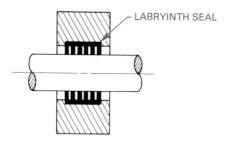

Figure 12–54 Labyrinth seal.

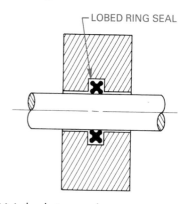

Figure 12–55 O-ring seals.

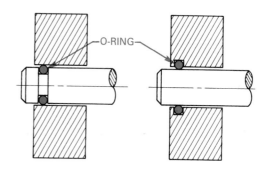

Figure 12–56 Lobed ring seal.

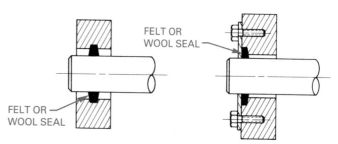

Figure 12–57 Felt and wool seals.

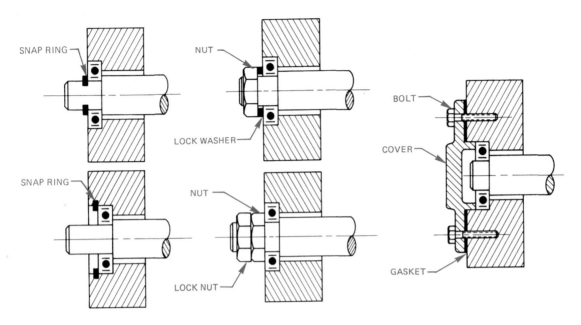

Figure 12–58 Typical bearing mountings.

the internal components. An exterior view, such as a front or top view, and a section is sometimes used. Dimensions are normally omitted from the assembly unless the dimensions are needed for assembly purposes. For example, when a specific dimension regarding the relationship of one part to another is require to properly assemble the parts, each part is identified with a number in a circle. This circle is referred to as a balloon. The balloons are connected to the part being identified with a leader. The balloon numbers correlate with a parts list. The parts list is normally placed on the drawing above or next to the title block, or on a separate sheet, as shown in Figure 12–60.

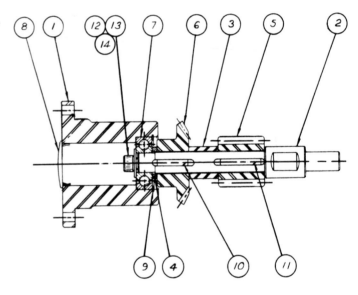

Figure 12–59 Assembly drawing with gear, bearing, bearing retainer, and bearing mounting. *Courtesy Curtis Associates.*

**ASSEMBLY** Cross Shaft Assembly

**USED ON**

**NUMBER OF UNITS**

**DATE**

| HAVE | NEED | P/Ø NO. W/Ø NO. | DET. NO. | PART NO. | DWG. | QTY. | PART NAME | DESCRIPTION | VENDOR |
|---|---|---|---|---|---|---|---|---|---|
| | | | 1 | | B | 1 | Bearing Retainer | Ø 3" C.D. Bar | |
| | | | 2 | | B | 1 | Cross Shaft | Ø 3/4" C.D. Bar | |
| | | | 3 | | A | 1 | Spacer | Ø 3/4 O.D. x 11GA Wall Tube x .738 Thick | |
| | | | 4 | | A | 1 | Spacer | Ø 3/4 O.D. x 11GA Wall Tube x .125 Thick | |
| | | | 5 | | - | 1 | Steel Worm 12 D.P. Single Thread | Boston Gear #H 1056 R.H. | |
| | | | 6 | | - | 1 | Bevel Gear 20° P.A. 2" P.D. | Boston Gear #HL 149 R.H. | |
| | | | 7 | | - | 1 | Ball Bearing .4724 Ø Bore | T.R.W. or Equivalent #MRC 201-S22 | |
| | | | 8 | | - | 1 | End Plug | | |
| | | | 9 | | - | 1 | Snap Ring | Waldes-Truarc #N 5000-125 | |
| | | | 10 | | - | 1 | Key Stock | 1/8 Sq. x 3/4 Lg. | |
| | | | 11 | | - | 1 | Key Stock | 1/8 Sq. x 1 Lg. | |
| | | | 12 | | - | 1 | Socket Head Cap Screw | 1/4 UNC x 3/4 Lg. | |
| | | | 13 | | - | 1 | Lockwasher | 1/4 Nominal | |
| | | | 14 | | - | 1 | Flat Washer | 1/4 Nominal | |

Figure 12–60 Parts list for the assembly drawing in Figure 12–59. *Courtesy Curtis Associates.*

# CHAPTER 12 TEST

Multiple choice: Respond to the following by selecting a, b, c, or d to best answer the question or complete the statement.

1. This is a mechanism that is used to convert rotary motion into a corresponding straight motion:
   a. cam
   b. gear
   c. follower
   d. sprocket

2. This common type of cam follower works well at high speeds, reduces friction and heat, and keeps wear to a minimum:
   a. flat-faced follower
   b. knife-edged follower
   c. roller follower
   d. high-speed follower

3. This follower is used for only low-speed and low-force applications, and has a low resistance to wear, but is very responsive and is used in situations that require abrupt changes in the cam profile:
   a. flat-faced follower
   b. knife-edged follower
   c. roller follower
   d. slow-speed follower

4. This cam follower is often used in situations where the cam profile has a steep rise or fall:
   a. flat-faced follower
   b. knife-edged follower
   c. roller follower
   d. round-faced follower

5. Information found on this type of drawing includes cam profile, hub dimensions with shaft hole and keyway, follower placement, and cam profile chart dimensions:
   a. cam layout drawing
   b. cam multiviews
   c. cam profile drawing
   d. cam manufacturing drawing

6. This type of cam is used when it is necessary for the follower to move in a path parallel to the axis of the cam:
   a. plate cam
   b. face cam
   c. cylindrical cam
   d. drum cam

7. The total movement of the cam follower in one complete revolution of the cam is called:
   a. dwell
   b. displacement
   c. rise
   d. fall

8. When the position of the cam follower remains constant for a period of rotation it is referred to as:
   a. dwell
   b. displacement
   c. rise
   d. fall

9. The most common and simplest form of gear is:
   a. bevel gear
   b. helical gear
   c. spur gear
   d. worm gear

10. When two or more spur gears are cut on a single shaft, they are referred to as:
    a. cluster gears
    b. gear train
    c. gear assembly
    d. group gears

11. When two gears are in mesh the larger is called the:
    a. driven gear
    b. driving gear
    c. pinion
    d. gear

12. When two gears are in mesh the smaller is called the:
    a. driven gear
    b. driving gear
    c. pinion
    d. gear

13. This mechanism exists when two or more gears are in combination for the purpose of transmitting power:
    a. cluster gears
    b. gear train
    c. gear assembly
    d. group gears

14. These are conical-shaped gears, allowing the shafts of the gear and pinion to intersect at any desired angle and provide for speed change between gear and pinion:
    a. bevel gear
    b. helical gear
    c. spur gear
    d. worm gear

15. These gears have their teeth cut at an angle which allows more than one tooth to be in contact, and they carry more load and are quieter than equally sized spur gears:
    a. bevel gear
    b. helical gear
    c. spur gear
    d. worm gear

16. This type of gear is commonly used when a large speed reduction is required in a small space:
    a. bevel gear
    b. helical gear
    c. spur gear
    d. worm gear

17. This is the direction of push transmitted from a tooth on one gear to a tooth on the mating gear or pinion:
    a. pitch diameter
    b. diametral pitch
    c. pressure angle
    d. circular pitch

18. This is the tooth size, and this must be the same on mating gears:
    a. pitch diameter
    b. diametral pitch
    c. pressure angle
    d. circular pitch

19. This is the most commonly used pressure angle because it provides a stronger tooth for quieter running and heavier load-carrying:
    a. 14.5°
    b. 20°
    c. 24°
    d. 24.5°

20. If you have a 2:1 gear ratio and the pinion has 14 teeth, how many teeth does the gear have?
    a. 7
    b. 28
    c. 56
    d. 21

21. If the gear in the train in Question 20 operates at 1200 RPM, how fast does the pinion operate?
    a. 600 RPM
    b. 2400 RPM
    c. 1200 RPM
    d. 4800 RPM

22. If you have a 4:1 gear ratio and the pinion has 18 teeth, how many teeth does the gear have?
    a. 9
    b. 36
    c. 54
    d. 72

23. Regarding the gear train in Question 22, what is the pitch diameter of the pinion if the gear has a 24" pitch diameter?
    a. 96"
    b. 48"
    c. 6"
    d. 24"

24. What is the distance between shafts of the gear train in Question 22?
    a. 12"
    b. 15"
    c. 24"
    d. 30"

25. What is the diametral pitch of a gear with a pitch diameter of 8 and 24 teeth?
    a. 3
    b. 4
    c. 1/3
    d. 12

26. These mechanical devices are used to reduce friction between two surfaces:
    a. cam
    b. gear
    c. bearing
    d. roller

27. These are often referred to as sleeve, journal bearings, or bushings and their use is based on a sliding action between parts.
    a. plain bearings
    b. rolling element bearings
    c. thrust bearings
    d. radial bearings

28. These are either ball or roller bearings:
    a. plain bearings
    b. rolling element bearings
    c. thrust bearings
    d. radial bearings

29. These are loads that are distributed around the shaft:
    a. circular loads
    b. cylindrical loads
    c. radial loads
    d. thrust loads

30. These loads apply force to the end of the shaft:
    a. length loads
    b. lineal loads
    c. radial loads
    d. thrust loads

31. These bearings are generally more effective for heavier loads:
    a. ball bearings
    b. roller bearings

32. The recommended surface finish for precision bearing applications on shafts under 2" in diameter is (in microinches):
    a. 16
    b. 32
    c. 63
    d. 125

33. The recommended surface finish for precision bearing applications on shafts over 2" in diameter is (in microinches):

    a. 16
    b. 32
    c. 63
    d. 125

34. It is necessary to maintain a film of lubrication between the bearing surfaces:

    a. true
    b. false

35. This type of seal does not come in contact with the moving parts:

    a. dynamic seal
    b. static seal

36. This type of seal does come in contact with the moving parts:

    a. dynamic seal
    b. static seal

37. This dynamic seal is most commonly used for most rotating or oscillating motion:

    a. packing
    b. labyrinth seal
    c. O-ring seal
    d. felt or wool seal

38. This type of seal is used for heavy machinery where some leakage is permissible:

    a. packing
    b. labyrinth seal
    c. O-ring seal
    d. felt or wool seal

39. This type of seal is used when low cost, lubricant absorption, filtration, low friction, and a polishing action is needed:

    a. packing
    b. labyrinth seal
    c. O-ring seal
    d. felt or wool seal

40. Where bearings or bushings do not receive proper lubrication, it may be necessary to provide these in the shaft or bushing:

    a. seals
    b. packing
    c. grooves
    d. grease bags

# CHAPTER 12 PROBLEMS

**PROBLEM 12–1** Answer the following questions based on the print shown on this page.

1. What type of cam is represented on this print?

2. Name the type of cam follower used with this cam.

3. Is the follower an in-line or offset follower?

4. What is the diameter for the cam shaft?

5. What is the diameter of the follower?

6. Give the cam plate thickness.

7. Provide the keyway dimensions.

8. Describe the feature that is used to hold the cam to the shaft at the keyway.

9. What is the tolerance of the Ø1.375 dimension?

10. What is the tolerance of angle A at each position?

11. What is the tolerance of the R dimensions at each position?

12. Give the radial displacement at each of the following angular displacements.

| Angular Displacement | Radial Displacement |
|---|---|
| 0° | _____ |
| 50° | _____ |
| 100° | _____ |
| 150° | _____ |
| 240° | _____ |
| 300° | _____ |

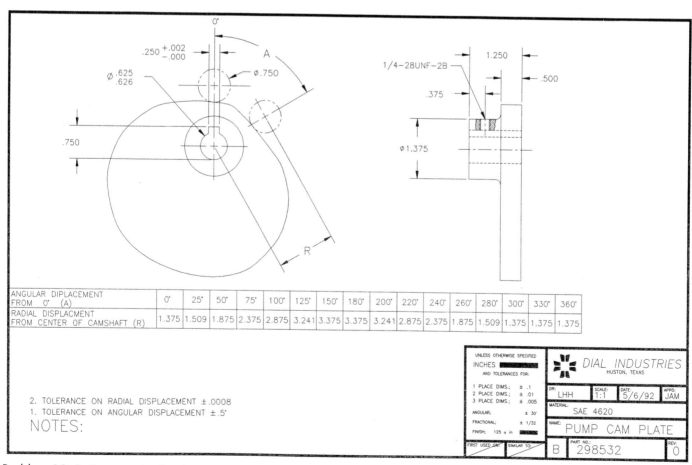

| ANGULAR DIPLACEMENT FROM 0° (A) | 0° | 25° | 50° | 75° | 100° | 125° | 150° | 180° | 200° | 220° | 240° | 260° | 280° | 300° | 330° | 360° |
|---|---|---|---|---|---|---|---|---|---|---|---|---|---|---|---|---|
| RADIAL DISPLACMENT FROM CENTER OF CAMSHAFT (R) | 1.375 | 1.509 | 1.875 | 2.375 | 2.875 | 3.241 | 3.375 | 3.375 | 3.241 | 2.875 | 2.375 | 1.875 | 1.509 | 1.375 | 1.375 | 1.375 |

NOTES:
1. TOLERANCE ON ANGULAR DISPLACEMENT ±.5°
2. TOLERANCE ON RADIAL DISPLACEMENT ±.0008

UNLESS OTHERWISE SPECIFIED
INCHES
AND TOLERANCES FOR:
1 PLACE DIMS.; ± .1
2 PLACE DIMS.; ± .01
3 PLACE DIMS.; ± .005
ANGULAR; ± 30'
FRACTIONAL; ± 1/32
FINISH; 125 u in

DIAL INDUSTRIES
HUSTON, TEXAS

DR: LHH | SCALE: 1:1 | DATE: 5/6/92 | APPD: JAM

MATERIAL: SAE 4620

NAME: PUMP CAM PLATE

FIRST USED ON | SIMILAR TO | B | PART NO.: 298532 | REV: 0

Problem 12–1 *Courtesy Dial Industries.*

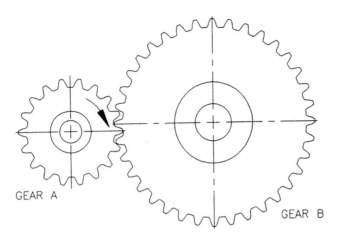

GEAR A

GEAR B

GEAR RATIO 2:1

|  | (D) PITCH DIAMETER | (N) NUMBER OF TEETH | (P) DIAMETRAL PITCH | RPM | DIRECTION |
|---|---|---|---|---|---|
| GEAR A | 6.000 | 18 |  | 1500 | CLOCKWISE |
| GEAR B |  |  |  |  |  |

GIVE THE DISTANCE BETWEEN SHAFTS _____

Problem 12–2

**PROBLEM 12–2** Given the gear train and chart shown above, calculate or determine the missing information to complete the chart.

**PROBLEM 12–3** Given the gear train and chart shown below, calculate or determine the missing information to complete the chart.

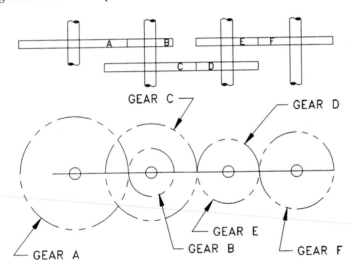

GEAR C

GEAR D

GEAR E

GEAR B

GEAR F

GEAR A

|  | PITCH DIAMETER (D) | NUMBER OF TEETH (N) | DIAMETRAL PITCH (P) | RPM | DIRECTION | GEAR RATIO | CENTER DISTANCE |
|---|---|---|---|---|---|---|---|
| GEAR A | 7.000 |  | 4 | 400 | C.WISE |  |  |
| GEAR B |  | 14 |  |  |  |  |  |
| GEAR C | 6.000 |  |  |  |  |  |  |
| GEAR D |  | 12 | 3 |  |  |  |  |
| GEAR E | 4.000 | 20 |  |  |  |  |  |
| GEAR F | 5.000 |  |  |  |  |  |  |

Problem 12–3

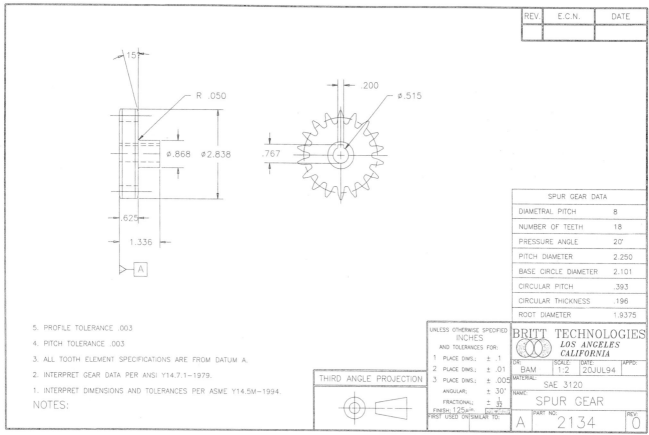

| SPUR GEAR DATA | |
| --- | --- |
| DIAMETRAL PITCH | 8 |
| NUMBER OF TEETH | 18 |
| PRESSURE ANGLE | 20° |
| PITCH DIAMETER | 2.250 |
| BASE CIRCLE DIAMETER | 2.101 |
| CIRCULAR PITCH | .393 |
| CIRCULAR THICKNESS | .196 |
| ROOT DIAMETER | 1.9375 |

NOTES:

5.  PROFILE TOLERANCE .003

4.  PITCH TOLERANCE .003

3.  ALL TOOTH ELEMENT SPECIFICATIONS ARE FROM DATUM A.

2.  INTERPRET GEAR DATA PER ANSI Y14.7.1–1979.

1.  INTERPRET DIMENSIONS AND TOLERANCES PER ASME Y14.5M–1994.

THIRD ANGLE PROJECTION

UNLESS OTHERWISE SPECIFIED
INCHES
AND TOLERANCES FOR:
1  PLACE DIMS.;  ± .1
2  PLACE DIMS.;  ± .01
3  PLACE DIMS.;  ± .005
ANGULAR;  ± 30'
FRACTIONAL;  ± $\frac{1}{32}$
FINISH; 125μin.
FIRST USED ON SIMILAR TO:

BRITT TECHNOLOGIES
*LOS ANGELES CALIFORNIA*
DR: BAM  SCALE: 1:2  DATE: 20JUL94  APPD:
MATERIAL: SAE 3120
NAME: SPUR GEAR
A  PART NO: 2134  REV: 0

| REV. | E.C.N. | DATE |
| --- | --- | --- |

Problem 12–4 *Courtesy David George.*

**PROBLEM 12–4** Answer the following questions based on the print shown on this page.

1.  Name the type of gear shown on this print.

2.  Describe the gear representation used on this print.

3.  How many teeth are there in this gear?

4.  What is the diametral pitch?

5.  What is the angle on the face of each gear tooth?

6.  Give the diameter of the hole provided for the shaft.

7.  Determine the pressure angle.

8.  What is the pitch tolerance?

9.  Give the outside diameter.

10. What is the hub diameter?

11. Determine the root diameter.

12. What is the pitch diameter?

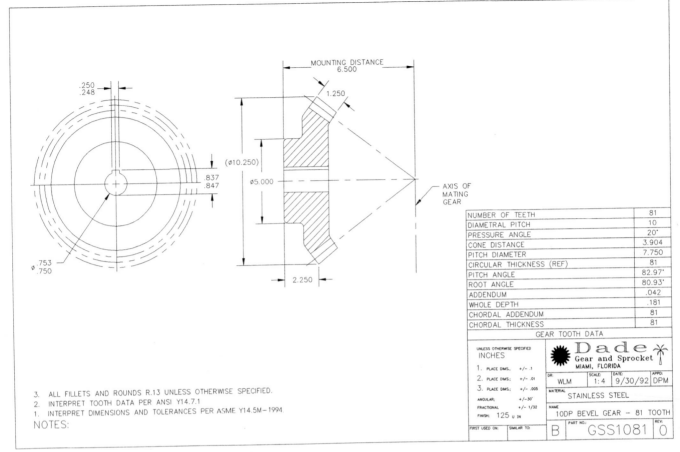

| GEAR TOOTH DATA | |
|---|---|
| NUMBER OF TEETH | 81 |
| DIAMETRAL PITCH | 10 |
| PRESSURE ANGLE | 20° |
| CONE DISTANCE | 3.904 |
| PITCH DIAMETER | 7.750 |
| CIRCULAR THICKNESS (REF) | 81 |
| PITCH ANGLE | 82.97° |
| ROOT ANGLE | 80.93° |
| ADDENDUM | .042 |
| WHOLE DEPTH | .181 |
| CHORDAL ADDENDUM | 81 |
| CHORDAL THICKNESS | 81 |

NOTES:
3. ALL FILLETS AND ROUNDS R.13 UNLESS OTHERWISE SPECIFIED.
2. INTERPRET TOOTH DATA PER ANSI Y14.7.1
1. INTERPRET DIMENSIONS AND TOLERANCES PER ASME Y14.5M – 1994.

UNLESS OTHERWISE SPECIFIED
INCHES
1. PLACE DIMS. +/– .1
2. PLACE DIMS.: +/– .01
3. PLACE DIMS.: +/– .005
ANGULAR: +/–30'
FRACTIONAL +/– 1/32
FINISH: 125 u IN

Dade
Gear and Sprocket
MIAMI, FLORIDA

| DR: WLM | SCALE: 1:4 | DATE: 9/30/92 | APPD: DPM |

MATERIAL  STAINLESS STEEL
NAME  10DP BEVEL GEAR – 81 TOOTH
FIRST USED ON: | SIMILAR TO: | B | PART NO.: GSS1081 | REV: 0

Problem 12–5 *Courtesy Dade Gear & Sprocket.*

**PROBLEM 12–5** Answer the following questions based on the print shown on this page.

1. Name the type of gear shown in this print.

2. Describe the type of gear representation used on this print.

3. What is the mounting distance?

4. How many teeth are there on this gear?

5. Give the keyway dimensions.

6. What is the diametral pitch?

7. Specify the pressure angle of this gear.

8. What is the pitch angle?

9. Give the whole depth.

10. What is the face width?

11. What do the parentheses mean on the Ø10.250 dimension?

12. Give the root angle.

| WORM GEAR DATA | |
|---|---|
| NUMBER OF TEETH | 5 |
| AXIAL PITCH | |
| PRESSURE ANGLE | 20° |
| PITCH DIAMETER | .750 |
| LEAD RIGHT HAND | .300 |
| LEAD ANGLE | 169° |
| ADDENDUM | .125 |
| WHOLE DEPTH | .250 |
| CHORDAL THICKNESS | .163 |
| WORM GEAR PART NUMBER | 1DT 1005 |

PART NAME: HIGH SPEED SHAFT
MATERIAL: SAE 4320
DRAWING NUMBER: 1DT 1005

## Problem 12–6

**PROBLEM 12–6** Answer the following questions based on the print shown on this page.

1. What type of gear is found on this print?
   _____

2. How many teeth are shown?
   _____

3. Give the total length of the gear.
   _____

4. What is the pressure angle?
   _____

5. Specify the outside diameter.
   _____

6. What is the root diameter?
   _____

7. Give the cordial thickness.
   _____

8. What is the pitch diameter?
   _____

9. What type of section is used on this print?
   _____

10. What is the dimension from the thread end to the start of the gear? (Use the specified dimensions.)
    _____

11. Give the thread specification.
    _____

12. What is the material?
    _____

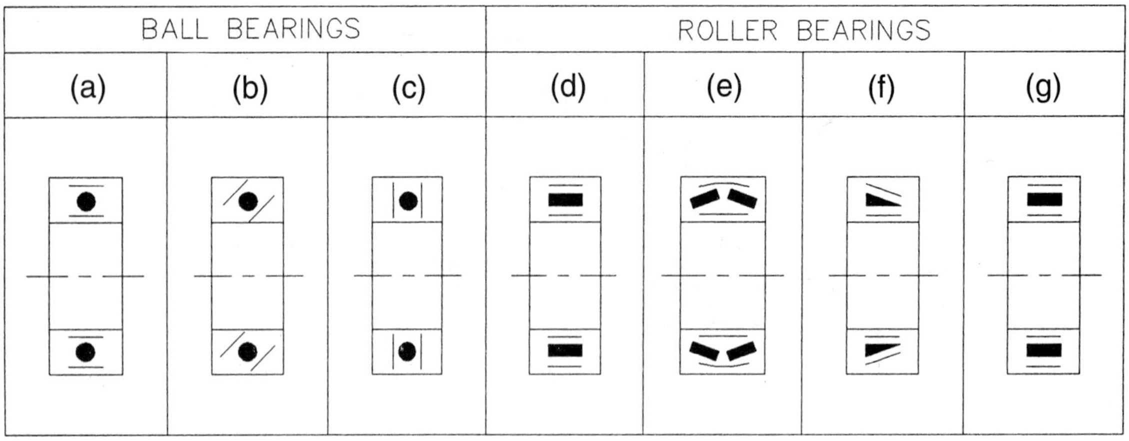

| BALL BEARINGS | | | ROLLER BEARINGS | | | |
|---|---|---|---|---|---|---|
| (a) | (b) | (c) | (d) | (e) | (f) | (g) |

Problem 12–7

**PROBLEM 12–7** Given the above bearing drawings, name each bearing type in the lines provided based on the letter in the box above it.

a. _____

b. _____

c. _____

d. _____

e. _____

f. _____

g. _____

# Reading Working Drawings

## LEARNING OBJECTIVES

After completing this chapter you will be able to:

- Read detail drawings.
- Read assembly drawings and parts lists.
- Identify types of assembly drawings.
- Answer questions related to engineering changes.

The drawings that are shown as examples or as print reading exercises in this text are a type of drawing, called detail drawings. When a product is designed and drawings are made for manufacturing, each part of the product must have a drawing. These drawings of individual parts are referred to as detail drawings. Component parts are assembled to create a final product, and the drawing that shows how the parts go together is called the assembly drawing. Associated with the assembly drawing and coordinated to the detail drawings is the parts list. When the detail drawings, assembly drawing, and parts list are combined, they are referred to as a complete set of working drawings. Working drawings, then, are a set of drawings that supply all the information necessary to manufacture any given product. A set of working drawings includes all the information and instructions needed for the purchase or construction of parts and the assembly of these parts into a product.

## DETAIL DRAWINGS

Detail drawings, used by workers in manufacturing, are drawings of each part contained in the assembly of a product. The only parts that may not have to be drawn are standard parts. Standard parts are items that can be purchased from an outside supplier more economically than they can be manufactured. Examples of standard, or purchased, parts are common bolts, screws, pins, keys, and any other product that can be purchased from a vendor. Standard parts do not have to be drawn because a written description clearly identifies the part, as shown in Figure 13–1.

1/2 - 13 UNC - 2 **x** 1.5 LONG, SOCKET HEAD CAP SCREW

Figure 13–1 Written description of standard or purchase part.

Detail drawings contain some or all of the following items:

1. Necessary multiviews.

2. Dimensional information.

3. Identification of the part, project name, and part number.

4. General notes and specific manufacturing information.

5. The material of which the part is made.

6. The assembly that the part fits (could be keyed to the part number).

7. Number of parts required per assembly.

8. Name of persons who worked on or with the drawing.

9. Engineering changes and related information.

In general, detail drawings have information that is classified into three groups:

1. Shape description, which shows or describes the shape of the part.

2. Size description, which shows the size and location of features on the part.

3. Specifications regarding items, such as material, finish, and heat treatment.

Figure 13–2 shows an example of a detail drawing.

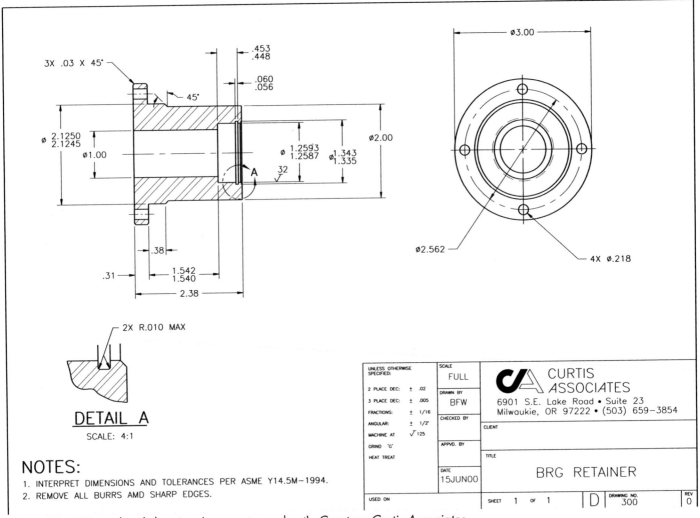

Figure 13–2 Monodetail drawing (one part per sheet). *Courtesy Curtis Associates.*

## Sheet Layout

Detail drawings may be prepared with one part per sheet, referred to as a monodetail drawing, as in Figure 13–2, or with several parts grouped on one sheet, which is called a multidetail drawing. The method of presentation depends on the choice of the individual company. The advantage of one detail per sheet is that each part stands alone so that the drawing of the part can be distributed to manufacturing without several other parts attached. Drawing sheet sizes, then, can vary depending on the part size, scale used, and information presented. This procedure requires that drawings be filed with numbers that allow the parts to be located in relation to the assembly. The advantage of a drawing with several details per sheet is one of economics. Several details per sheet saves paper and saves drafting time. The company may use one standard sheet size and place as many parts as possible on one sheet. When this practice is used, there may be a group of sheets with parts detailed for one assembly. The sheet numbers correlate the sheets to the assembly, and each part is keyed to the assembly. The sheets are given page numbers identifying the page number and the total number of pages in the set. For example, if there are three pages in a set, the first page is identified as 1 of 3, the second as 2 of 3, and the third as 3 of 3. Figure 13–3 shows an example of a multidetail drawing with several detail drawings on one sheet. Some companies may use both methods at different times depending on the purpose of the drawings and the type of product. For example, it is more common for the parts of a weldment to be drawn grouped on sheets, rather than one per sheet, because the parts are fabricated at one location in the shop.

## Detail Drawing Manufacturing Information

Detail drawings are drawn to suit the needs of the manufacturing processes. A detail drawing can have all the information necessary to completely manufacture the part; for example, casting and machining information on one drawing. In some situations a completely dimensioned machining drawing is sent to the pattern or die maker. The pattern or die is then made to accommodate extra material where machined surfaces are specified. When company standards require, two detail drawings can be prepared for each part. One detail gives views and dimensions that are necessary only for the casting or forging process. Another detail is drawn that does not give the previous casting or forging dimensions, but provides only the dimensions needed to perform the machining operations on the part.

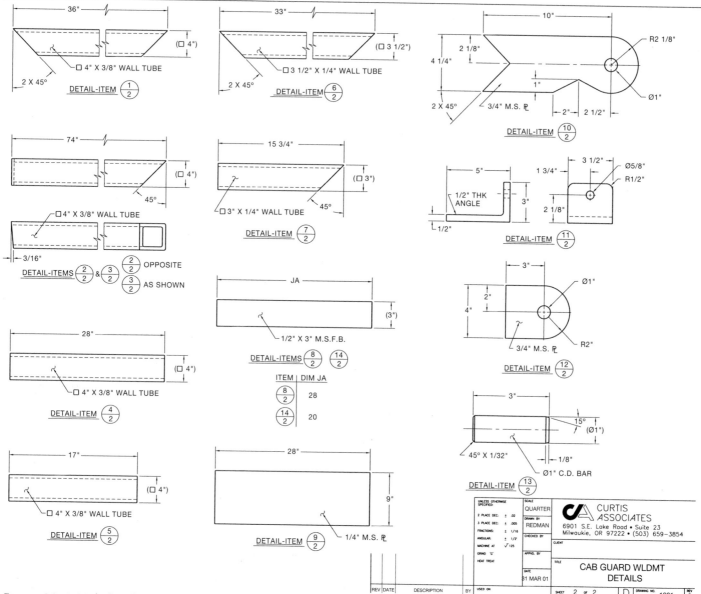

Figure 13–3 Multidetail drawing (several detail drawings on one sheet). *Courtesy Curtis Associates.*

## ASSEMBLY DRAWINGS

Most products are composed of several parts. A drawing showing how all of the parts fit together is called an assembly drawing. Assembly drawings can differ in the amount of information provided, and this decision often depends on the nature or complexity of the product. Assembly drawings are generally multiview drawings. In many cases a single front view is used for an assembly drawing (see Figure 13–4). Full sections are commonly associated with assembly drawings because the full section may expose the assembly of most or all of the internal features shown in Figure 13–5. If one section or view is enough to show how the parts fit together, a number of views or sections may be necessary. In some situations, a front view or group of views with broken-out sections is the best method of showing the external features, while exposing some of the internal features (see Figure 13–6). Another element of assembly drawings that makes them different from

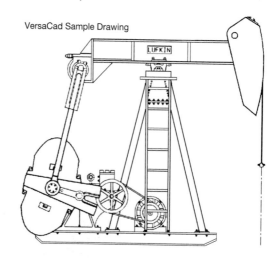

Figure 13–4 Assembly drawing with single front view. *Courtesy T & W Systems, Inc.*

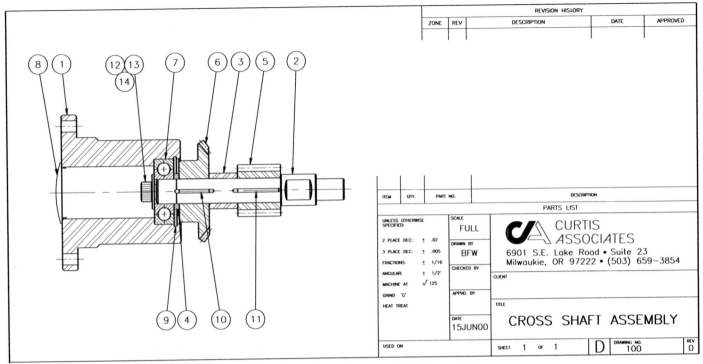

Figure 13–5 Assembly in full section. *Courtesy Curtis Associates.*

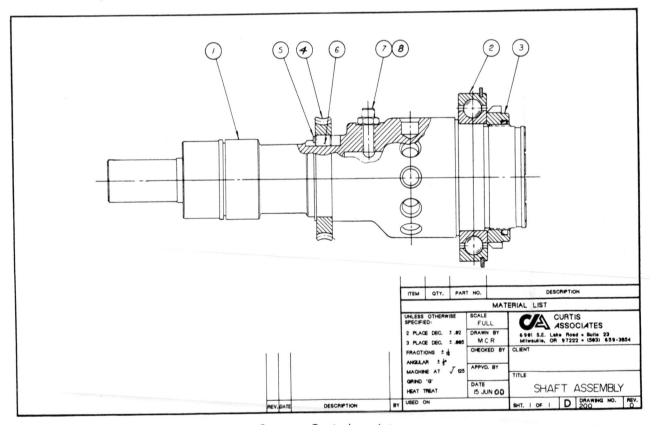

Figure 13–6 Assembly with broken-out sections. *Courtesy Curtis Associates.*

detail drawings is that they usually contain few or no hidden lines or dimensions. Dimensions serve no purpose on an assembly drawing unless the dimensions are used to show the assembly relationship of one part to another. These assembly dimensions are only necessary when a certain distance between parts must exist

before proper assembly can take place. Machining processes and other specifications are generally not given on an assembly drawing unless a machining operation must take place after two or more parts are assembled. Other assembly notes can include bolt-tightening specifications, assembly welds, or cleaning, painting, or decal

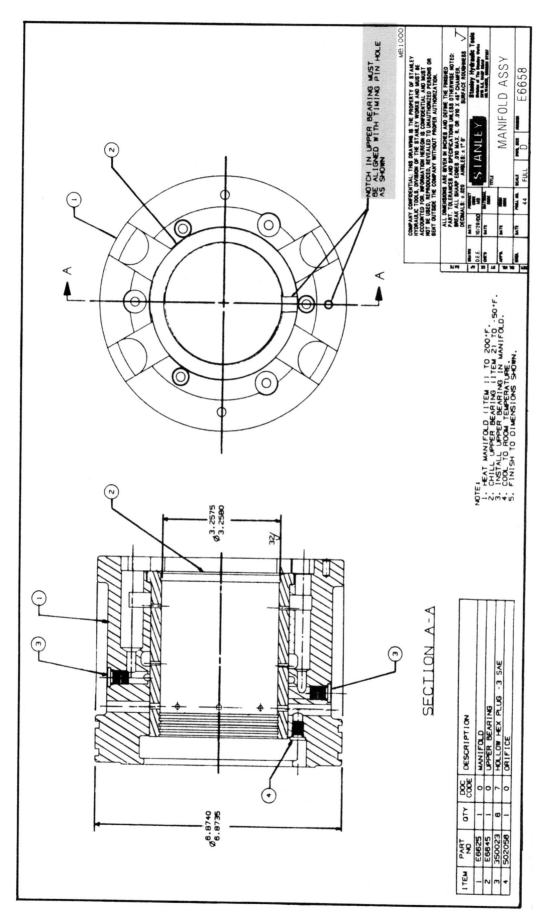

Figure 13–7 Process note on assembly drawing. Courtesy Stanley Hydraulic Tools, Division of the Stanley Works.

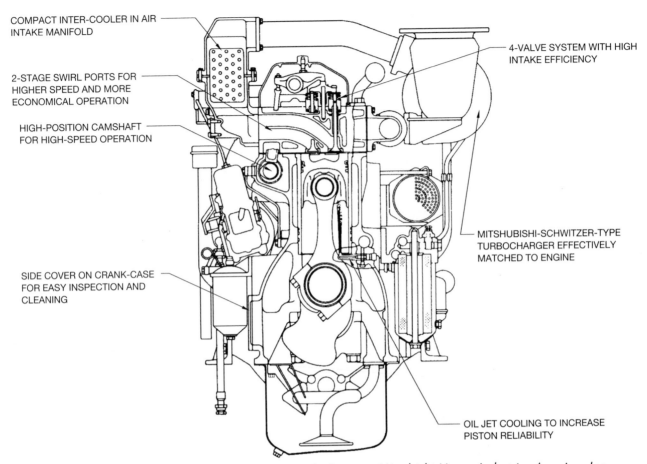

COMPACT INTER-COOLER IN AIR
INTAKE MANIFOLD

2-STAGE SWIRL PORTS FOR
HIGHER SPEED AND MORE
ECONOMICAL OPERATION

HIGH-POSITION CAMSHAFT
FOR HIGH-SPEED OPERATION

SIDE COVER ON CRANK-CASE
FOR EASY INSPECTION AND
CLEANING

4-VALVE SYSTEM WITH HIGH
INTAKE EFFICIENCY

MITSHUBISHI-SCHWITZER-TYPE
TURBOCHARGER EFFECTIVELY
MATCHED TO ENGINE

OIL JET COOLING TO INCREASE
PISTON RELIABILITY

**Figure 13–8** Full-section assembly with section lines omitted. *Courtesy Mitsubishi Heavy Industries America, Inc.*

placement that must take place after assembly. Figure 13–7 shows a process note applied to an assembly drawing. Some company standards or drawing presentations prefer that assemblies be drawn with sectioned parts shown without section lines. Figure 13–8 shows a full-section assembly with parts left unsectioned.

Assembly drawings may contain some or all of the following information:

- One or more views.
- Sections necessary to show internal features, function, and assembly.
- Any enlarged views necessary to show adequate detail.
- Arrangement of parts.
- Overall size and specific dimensions necessary for assembly.
- Manufacturing processes necessary for or during assembly.
- Reference or item numbers that key the assembly to a parts list and to the details.
- Parts list or bill of materials.

## TYPES OF ASSEMBLY DRAWINGS

There are several different types of assembly drawings used in industry:

1. Layout, or design, assembly

2. General assembly

3. Working-drawing or detail assembly

4. Erection assembly

5. Subassembly

6. Pictorial assembly

### Layout Assembly

Engineers and designers may prepare a design layout in the form of a sketch or as an informal drawing. These engineering design drawings are used to establish the relationship of parts in a product assembly. From the layout, the engineer prepares sketches or informal detail drawings for prototype construction. This research and development (R & D) is the first step in the process of taking a design from an idea to a manufactured product. Layout, or design, assemblies can take any form depending on the drafting ability of the engineer, the time frame for product implementation, the complexity of the product, or company procedures. Figure 13–9 shows a simple layout assembly of a product in the development stage. The limits of operation are shown in phantom lines.

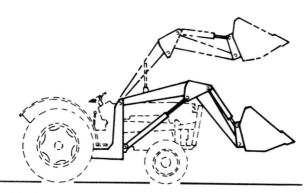

Figure 13-9 Layout assembly. The limits of operation are shown in phantom lines.

## General Assembly

General assemblies are the most common types of assemblies that are used in a complete set of working drawings. A set of working drawings contains three parts: detail drawings, an assembly drawing, and a parts list. The assembly drawing, as previously discussed, shows how all of the parts fit together. Figures 13–4 through 13–8 provide examples of general assembly drawings.

## Detail Assembly

When a drawing is created where details of parts are combined on the same sheet with an assembly of those parts, a detail assembly, also known as a working drawing assembly, is the result. While this practice is not as common as general assemblies, it is a practice at some companies. The use of working drawing assemblies may be a company standard, or this technique can be used in a specific situation even when it is not considered a normal procedure at a particular company. The detail assembly may be used when a particular end result requires that the details and assembly be combined on as few sheets as possible. An example may be a product with few parts that will be produced only once for a specific purpose (see Figure 13–10).

## Erection Assembly

Erection assemblies generally differ from general assemblies in that dimensions and fabrication specifications are commonly included. Typically associated with products that are made of structural steel, or cabinetry, erection assemblies are used for both fabrication and assembly. Figure 13–11 shows a CADD-produced erection assembly with multiviews, fabrication dimensions, and an isometric drawing that also helps display how the parts fit together.

## Subassembly

The complete assembly of a product can be made up of several component assemblies. These individual unit assemblies are called subassemblies. A complete set of working drawings can be made up of several subassemblies, each with its own detail drawings, and the general assembly. The general assembly of an automobile, for example, includes the subassemblies of the drive components, the engine components, and the steering column, just to name a few.

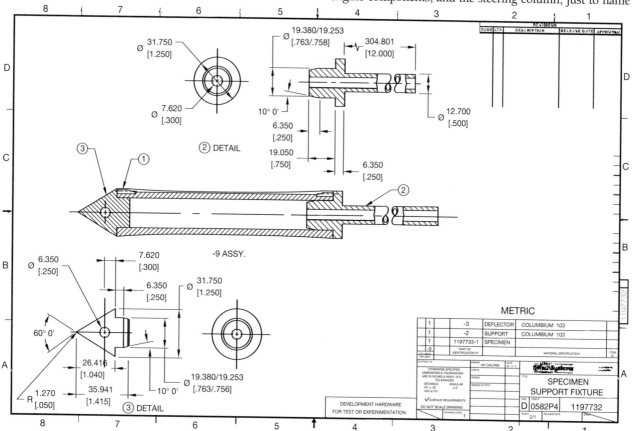

Figure 13–10 Working drawing, or detail, assembly. *Courtesy Aerojet Propulsion Division.*

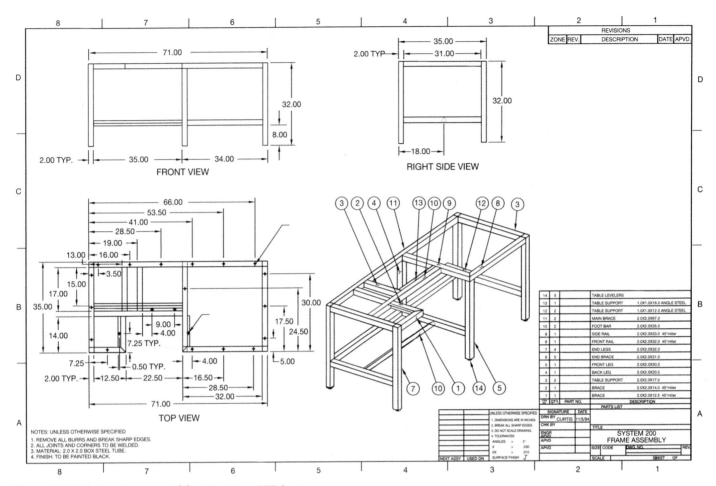

Figure 13–11 Erection assembly. *Courtesy EFT Systems.*

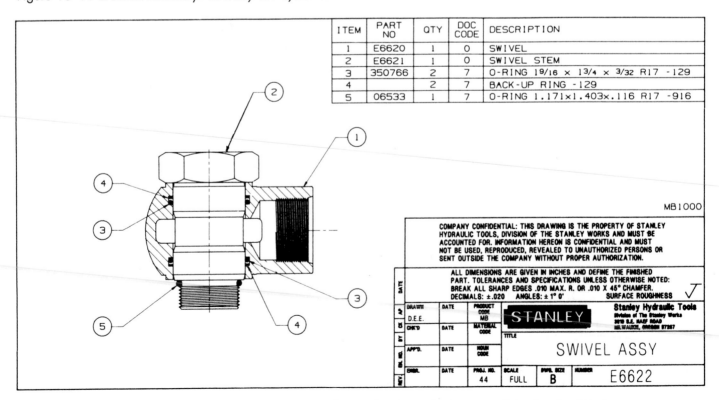

| ITEM | PART NO | QTY | DOC CODE | DESCRIPTION |
|------|---------|-----|----------|-------------|
| 1 | E6620 | 1 | 0 | SWIVEL |
| 2 | E6621 | 1 | 0 | SWIVEL STEM |
| 3 | 350766 | 2 | 7 | O-RING 19/16 × 13/4 × 3/32 R17 -129 |
| 4 |  | 2 | 7 | BACK-UP RING -129 |
| 5 | 06533 | 1 | 7 | O-RING 1.171×1.403×.116 R17 -916 |

Figure 13–12 Subassembly with parts list. *Courtesy Stanley Hydraulic Tools, Division of the Stanley Works.*

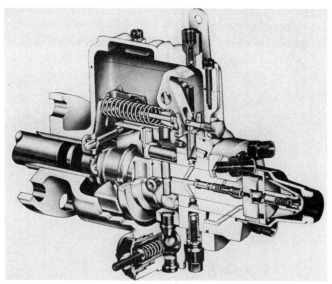

Figure 13–13 Pictorial assembly. *Courtesy Stanadyne Diesel Systems.*

A subassembly, such as an engine, may be made up of other subassemblies, such as the carburetor or the generator. Figure 13–12 shows a subassembly with a parts list.

## Pictorial Assembly

As the name suggests, pictorial assemblies are used to display a pictorial, rather than multiview, representation of the product, which may be used in other types of assembly drawings. Pictorial assemblies are made from photographs or artistic renderings. The pictorial assembly can be as simple as the isometric drawing in Figure 13–11, which was used to more clearly guide workers in the assembly of the product. Pictorial assemblies are commonly used in product catalogs or brochures. The purpose of these pictorial representations may be for sales promotion, customer self-assembly, or maintenance procedures (see Figure 13–13). Pictorial assemblies can also take the form of exploded technical illustrations, commonly known as illustrated parts breakdowns. These exploded multiview or isometric pictorials are used in vendors' catalogs and instruction manuals for maintenance or assembly. A pictorial assembly has been used by anyone that has put together a bicycle, model kit, or child's toy (see Figure 13–14).

## IDENTIFICATION NUMBERS

Identification or item numbers are used to key the parts from the assembly drawing to the parts list. Identification numbers are generally placed in balloons. Balloons are circles that are connected to the related part with a leader line. Several of the assembly drawings in this chapter show examples of identification numbers and balloons. Numbers in balloons are common, although some companies prefer to use identification letters (see Figure 13–15). In some situations when a particular group of parts are so closely related that individual identification is difficult, the identification balloons can be grouped adjacent to one another. For example, a cap screw, lock washer, and flat washer may require that the balloons be placed in a cluster or side-by-side, as shown at items 12 to 14 in Figure 13–5.

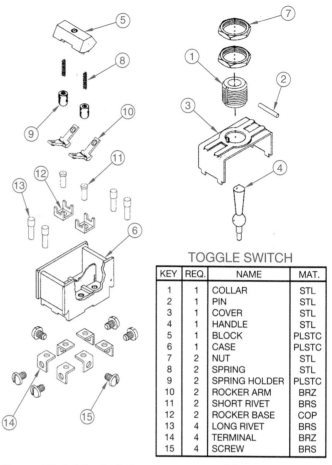

| KEY | REQ. | NAME | MAT. |
|-----|------|------|------|
| 1 | 1 | COLLAR | STL |
| 2 | 1 | PIN | STL |
| 3 | 1 | COVER | STL |
| 4 | 1 | HANDLE | STL |
| 5 | 1 | BLOCK | PLSTC |
| 6 | 1 | CASE | PLSTC |
| 7 | 2 | NUT | STL |
| 8 | 2 | SPRING | STL |
| 9 | 2 | SPRING HOLDER | PLSTC |
| 10 | 2 | ROCKER ARM | BRZ |
| 11 | 2 | SHORT RIVET | BRS |
| 12 | 2 | ROCKER BASE | COP |
| 13 | 4 | LONG RIVET | BRS |
| 14 | 4 | TERMINAL | BRZ |
| 15 | 4 | SCREW | BRS |

**TOGGLE SWITCH**

Figure 13–14 Exploded isometric assembly.

In some cases, the balloons not only key the detail drawings to the assembly and parts list, but they also key the assembly drawing and parts list to the page on which the detail drawing is found (see Figure 13–16). Figure 13–17 is an assembly drawing and parts list that is located on page one or a two-page set of working drawings. Notice how the balloons key the parts from the assembly and parts list to the detail drawings found on page 2. The page-two details are located in Figure 13–3.

## PARTS LISTS

The parts list is usually combined with the assembly drawing, yet remains one of the individual components of a complete set of working drawings. The information that is associated with the parts list generally includes:

1. Item number, from balloons

2. Quantity, the amount of that particular part needed for this assembly

3. Part or drawing number, which is a reference back to the detail drawing

4. Description, which is usually a part name or complete description of a purchase part or stock specification including sizes or dimensions

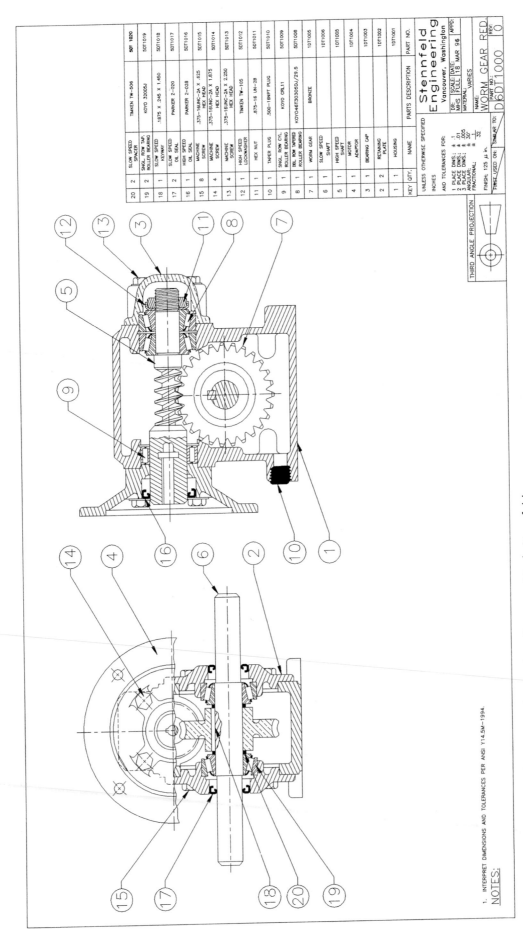

| KEY QTY. | | NAME | PARTS DESCRIPTION | PART NO. |
|---|---|---|---|---|
| 20 | 2 | SLOW SPEED SPACER | TIMKEN TW-506 | 50T1020 |
| 19 | 2 | SNGL. ROW TAP. ROLLER BEARING | KOYO 32005JJ | 50T1019 |
| 18 | 1 | SLOW SPEED KEYWAY | .1875 X .245 X 1.450 | 50T1018 |
| 17 | 1 | SLOW SPEED OIL SEAL | PARKER 2-020 | 50T1017 |
| 16 | 1 | HIGH SPEED OIL SEAL | PARKER 2-028 | 50T1016 |
| 15 | 8 | MACHINE SCREW | .375-16UNC-2A X .625 HEX HEAD | 50T1015 |
| 14 | 4 | MACHINE SCREW | .375-16UNC-2A X 1.875 HEX HEAD | 50T1014 |
| 13 | 4 | MACHINE SCREW | .375-16UNC-2A X 2.250 HEX HEAD | 50T1013 |
| 12 | 1 | HIGH SPEED LOCKWASHER | TIMKEN TW-105 | 50T1012 |
| 11 | 1 | HEX NUT | .875-16 UN-2B | 50T1011 |
| 10 | 1 | TAPER PLUG | .500-18NPT PLUG | 50T1010 |
| 9 | 1 | SNGL. ROW CYL. ROLLER BEARING | KOYO CRL11 | 50T1009 |
| 8 | 1 | DBL. ROW TAPERED ROLLER BEARING | KOYO46T303050J/29.5 | 50T1008 |
| 7 | 1 | WORM GEAR | BRONZE | 10T1005 |
| 6 | 1 | SLOW SPEED SHAFT | | 10T1006 |
| 5 | 1 | HIGH SPEED SHAFT | | 10T1005 |
| 4 | 1 | MOTOR ADAPTOR | | 10T1004 |
| 3 | 1 | BEARING CAP | | 10T1003 |
| 2 | 2 | RETAINING PLATE | | 10T1002 |
| 1 | 1 | HOUSING | | 10T1001 |

**Stennfeld Engineering**
Vancouver, Washington

| DR: MHS | SCALE: FULL | DATE: 18 MAR 96 | APPD: |
|---|---|---|---|

MATERIAL: VARIES

NAME: WORM GEAR RED.

| PART NO. D6DT1000 | REV: 0 |

UNLESS OTHERWISE SPECIFIED
INCHES
AND TOLERANCES FOR:

1 PLACE DIMS.: ± .1
2 PLACE DIMS.: ± .01
3 PLACE DIMS.: ± .005
ANGULAR: ± 30'
FRACTIONAL: ± $\frac{1}{32}$

FINISH: 125 $\mu$ in.

| USED ON: | SIMILAR TO: |

THIRD ANGLE PROJECTION

NOTES:
1. INTERPRET DIMENSIONS AND TOLERANCES PER ANSI Y14.5M—1994.

Figure 13–15 Assembly drawing and parts list. *Courtesy Mark Stennfeld.*

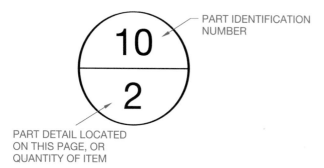

Figure 13–16 Balloon with page identification.

5. Material identification, the material that part is made of

6. Information about vendors for purchase parts

Parts lists may or may not contain all the above information, depending on company standards. Elements one through four in the list are the most common items. When all six elements are provided, the parts list is more appropriately called a list of materials.

You can draw the parts list on the assembly drawing, as in Figure 13–15 or Figure 13–17. When drawn on the assembly drawing, the parts list is located above the title block, in the upper-right or upper-left corner, or in a convenient location on the drawing field. The location depends on company standards, although the position over or near the title block is most common. The information on the parts list is usually presented with the first item number followed by consecutive item numbers. When the parts list is so extensive that the columns fill the page, a new group of columns is added next to the first. The reason that parts list data is provided from the bottom of the sheet upward or the top of the sheet downward is so that space on the parts list is available if additional parts are added to the assembly.

The parts list is not always placed on the assembly drawing. Some companies prefer to prepare parts lists on separate sheets for convenient filing. This method also enables the parts list to be computer-generated or typed separate from the drawings. Separate parts lists are usually prepared on a computer so information can be edited conveniently. Another option is to have parts lists typed on a standard parts list form, as shown in Figure 13–18.

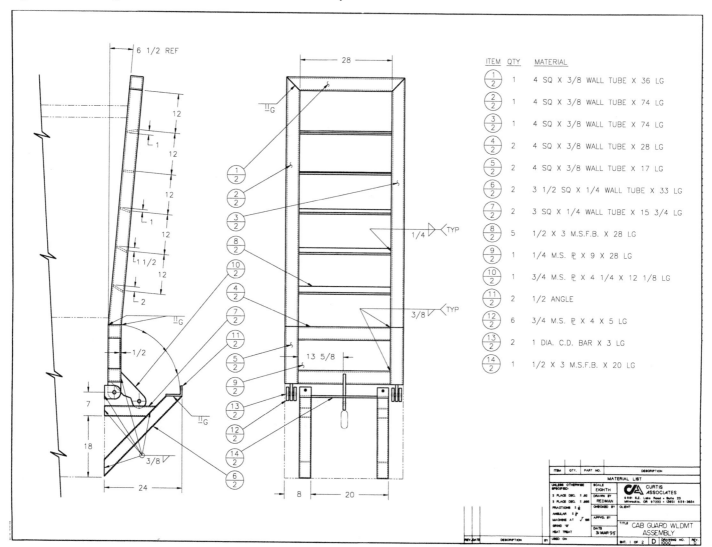

Figure 13–17 Assembly and parts list with page identification balloons. *Courtesy Curtis Associates.*

| HAVE | NEED | P/Ø NO. W/Ø NO. | DET. NO. | PART NO. | DWG. | QTY. | PART NAME | DESCRIPTION | VENDOR |
|------|------|------|------|------|------|------|-----------|-------------|--------|
| | | | 1 | | B | 1 | Bearing Retainer | Ø 3" C.D. Bar | |
| | | | 2 | | B | 1 | Cross Shaft | Ø 3/4" C.D. Bar | |
| | | | 3 | | A | 1 | Spacer | Ø 3/4 O.D × 11GA Wall Tube × .738 Thick | |
| | | | 4 | | A | 1 | Spacer | Ø 3/4 O.D × 11GA Wall Tube × .125 Thick | |

*ASSEMBLY* Cross Shaft Assembly   *USED ON* _____
*NUMBER OF UNITS* _____   *DATE* _____

Figure 13–18 Partial parts list separate from assembly drawing. *Courtesy Curtis Associates.*

## PURCHASE PARTS

Purchase or standard parts, as previously mentioned, are parts that are manufactured and available for purchase from an outside vendor. It is generally more economical for a company to buy these items than to make them. Parts that are available from suppliers can be found in the *Machinery's Handbook, Fastener's Handbook,* or vendors' catalogs. Purchase parts do not require detail drawings because a written description completely describes a part. For this reason, the purchase parts found in a given assembly are clearly and completely described in the parts list. Some companies have a purchase parts book that is used to record all purchase parts used in their product lines. The standard parts book gives a reference number for each part, which is placed on the parts list for convenient identification.

## ENGINEERING CHANGES

Engineering change documents are used to initiate and record changes to products in the manufacturing industry. Changes to engineering drawings can be requested from any branch of a company that deals directly with production and distribution of the product. For example, the engineering department may implement product changes as research and development results show a need for upgrading. Manufacturing departments request changes as problems arise in product fabrication or assembly. The sales staff can also initiate change proposals that stem from customer complaints.

### Engineering Change Request (ECR)

Before a drawing can be changed by the drafting department, an engineering change request (ECR) is needed. This ECR is the document that is used to initiate a change in a part or assembly. The ECR can come from any one of several sources in the company, such as engineering, manufacturing, or sales departments, for example. The ECR is usually attached to a print of the part affected. The print and the ECR show, by sketches and written descriptions, what changes are to be made. The ECR also contains a number that becomes the reference record of the change to be made. Figure 13–19 shows a sample ECR form.

## Engineering Change Notice (ECN)

Records of changes are kept so reference can be made between the existing and proposed product. To make sure that records of these changes are kept, special notations are made on the drawing and engineering change records. These records are commonly known as engineering change notices (ECN) or engineering change orders (ECO).

When a change is to be made, an ECR initiates the change to the drafting department. The drafters then alter the original drawing or computer program to reflect the change request. When the drawing of a part is changed, a revision letter or number is placed next to that change. For example, the first change is numbered R1, or A, the second change is numbered R2, or B, and so on. A circle drawn around the revision letter or number helps the identification stand out clearly from the rest of the other drawing. Figure 13–20 shows a part as it exists and also after a change has been made.

When the drafter chooses to make the change by not altering the drawing of the part, but only by changing the dimension, that dimension is labeled with a "not to scale" symbol. The method of making the change is the decision of the drafting department based on the extent of the change and the time required to make it. The not-to-scale symbol can be used to save time. Figure 13–21 shows a change made to a part with the new dimension identified as not to scale with a thick straight line placed under the new dimension.

After the part has been changed on the drawing and the proper R number or letter, placed next to the change, the change is recorded in the ECN column of the drawing. The location for the ECN column varies with different companies. This ECN identification can be found next to the title block or in a corner of the drawing.

*ANSI* The ANSI document, Drawing Sheet Sizes and Format, ANSI Y14.1 recommends that the ECN column be placed in the upper-right corner of a drawing. The revision block contains the revision number (number of times revised), ECN number (usually given on the ECR), and the date of the change. Some companies also have a column for a brief description of the change (as recommended by ANSI) and an approval. Figure 13–22(a) shows an expanded and condensed ECN column format. Notice that changes are added in alphabetic order from top to bottom.

Aerojet Liquid Rocket Company
## ENGINEERING CHANGE REQUEST & ANALYSIS

| DATE | ECRANG | PAGE |
|---|---|---|

| ORIGINATOR NAME | | DEPT | EXT | DATE | DOCUMENT NEED DATE | PROPOSED EFFECTIVITY |
|---|---|---|---|---|---|---|

| PART DOCUMENT NO | CURRENT REV TR | PART DOCUMENT NAME |
|---|---|---|

| USED ON NEXT ASSY NO | PROGRAMS AFFECTED | PROGRAMS AFFECTED | | |
|---|---|---|---|---|
| | | QTSS ITEM | YES ——— NO ——— |
| | | REQUAL REQD | YES ——— NO ——— |
| | | REVISE QTSS FORM | YES ——— NO ——— |

**DESCRIPTION OF CHANGE**

**JUSTIFICATION OF CHANGE**

| PROJECT ENGINEER SIGNATURE | DEPT | EXT | DATE | CCB REP SIGNATURE | DATE |
|---|---|---|---|---|---|

**DESIGN ENGINEERING TECHNICAL EVALUATION**

| DESIGN ENGINEER SIGNATURE | DEPT | EXT | DATE | CAUSING CONT OR WORK ORDER NO |
|---|---|---|---|---|

| ANY AFFECT ON | YES | NO | | YES | NO | | YES | NO |
|---|---|---|---|---|---|---|---|---|
| 1 PERFORMANCE | | | 5 WEIGHT | | | 9 OPERATIONAL COMPUTER PROGRAMS | | |
| 2 INTERCHANGEABILITY | | | 6 COST SCHEDULE | | | 10 RETROFIT | | |
| 3 RELIABILITY | | | 7 OTHER END ITEMS | | | 11 END ITEM IDENT | | |
| 4 INTERFACE | | | 8 SAFETY EMI | | | 12 VENDOR CHANGE CRITICAL ITEMS ONLY | | |

**CCB DECISION**

| SIGNATURES | DEPT | CON CUR | DIS SENT |
|---|---|---|---|
| QUALITY ASSURANCE | | | |
| MANUFACTURING | | | |
| ENGINEERING | | | |
| PRODUCT SUPPORT | | | |
| MATERIAL | | | |
| TEST OPERATION | | | |

| CCB CHAIRMAN SIGNATURE | DATE | CLASS I ☐ CLASS II ☐ |
|---|---|---|
| CUSTOMER SIGNATURE | DATE | EFFECTIVITY |

Figure 13–19 Sample engineering change request (ECR) form. *Courtesy Aerojet Propulsion Division.*

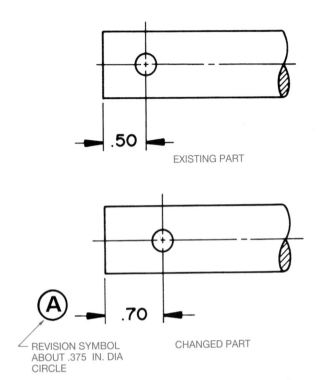

Figure 13–20 An existing part and the same part after a change has been made.

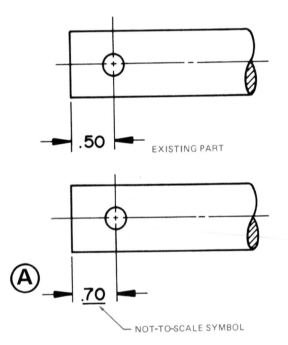

Figure 13–21 An existing part and the same part changed, using the not-to-scale symbol.

The condensed ECN Column in Figure 13–22(b) shows that the drawing has been changed twice. The first change was initiated by ECN number 2604 on September 10, 1993, and the second by ECN 2785 on November 18, 1994. Some companies may also identify the number of times a drawing has been changed by providing the letter of the current change in the title block, as shown in Figure 13–23.

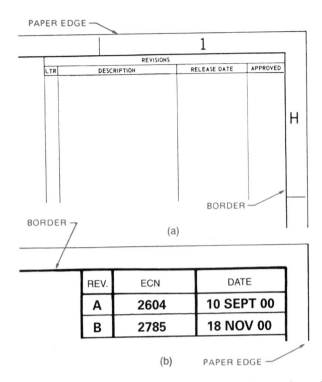

Figure 13–22 (a) Expanded revision columns. (b) Condensed revision columns. *Courtesy Aerojet Propulsion Division.*

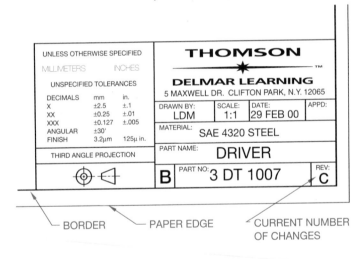

Figure 13–23 Title block displaying the current number of times a drawing has been changed.

When the drafter has made all drawing changes as specified on the ECR, an ECN is completed. The ECN ends the process and is filed for future reference. Usually the ECN thoroughly describes the part as it existed before the change and the change that was made as presented on the ECR. Figure 13–24 shows a typical ECN form. With a changed drawing and a filed ECN, anyone can verify what the part was before the change and the reason for the change. The ECN number is a reference for reviewing the change.

The general engineering change elements, terminology, and techniques are consistent among companies, although the actual format of engineering change documents may be con-

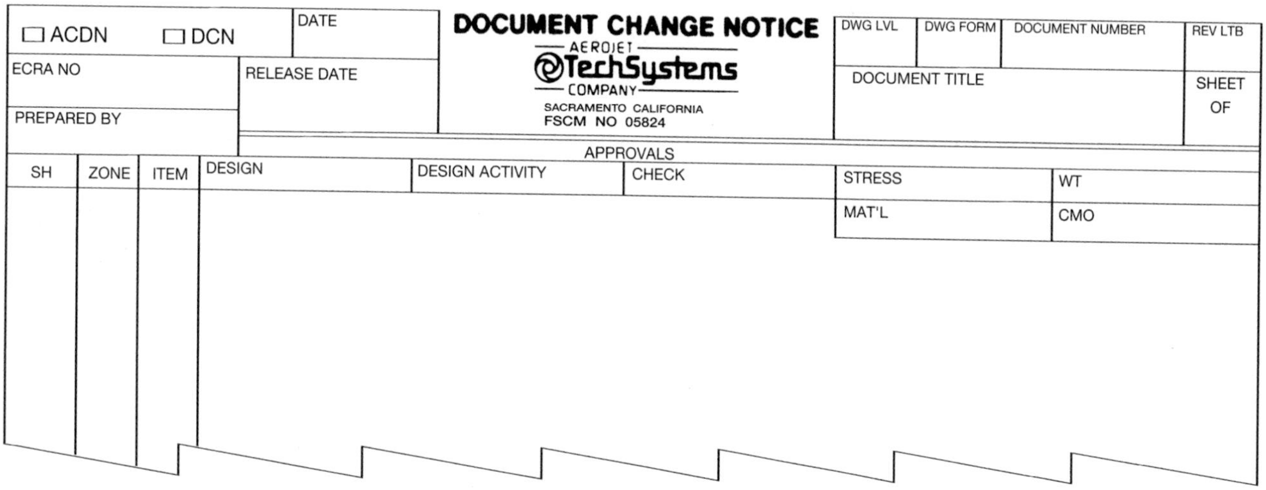

| | | | DATE | **DOCUMENT CHANGE NOTICE** | | DWG LVL | DWG FORM | DOCUMENT NUMBER | REV LTB |
|---|---|---|---|---|---|---|---|---|---|

☐ ACDN   ☐ DCN

—AEROJET—
@TechSystems
—COMPANY—
SACRAMENTO CALIFORNIA
FSCM NO 05824

| ECRA NO | | | RELEASE DATE | | | DOCUMENT TITLE | | | SHEET OF |
|---|---|---|---|---|---|---|---|---|---|
| PREPARED BY | | | | | | | | | |

APPROVALS

| SH | ZONE | ITEM | DESIGN | DESIGN ACTIVITY | CHECK | STRESS | | WT | |
|---|---|---|---|---|---|---|---|---|---|
| | | | | | | MAT'L | | CMO | |

Figure 13–24 Sample engineering change notice. *Courtesy Aerojet Propulsion Division.*

siderably different. Some formats are very simple, while others are much more detailed.

# CHAPTER 13 TEST

Multiple choice: Respond to the following by selecting a, b, c, or d to best answer the question or complete the statement.

1. The drawings that are used for the manufacture of individual parts are called:
   a. detail drawings
   b. assembly drawings
   c. working drawings
   d. manufacturing drawings

2. The drawings in a complete set of drawings, including a parts list, used to manufacture parts and assemble the product are called:
   a. detail drawings
   b. assembly drawings
   c. working drawings
   d. manufacturing drawings

3. The drawings that show how the component parts of a product fit together are called:
   a. detail drawings
   b. assembly drawings
   c. working drawings
   d. manufacturing drawings

4. The most common type of assembly that is used in a set of working drawings is:
   a. working-drawing assembly
   b. general assembly
   c. erection assembly
   d. layout assembly

5. This is usually an assembly created in the research and development stage of product design that is prepared by the engineer or designer to establish the relationship of parts in the product.
   a. working-drawing assembly
   b. general assembly
   c. erection assembly
   d. layout assembly

6. This type of assembly is where details of parts are combined on the same sheet with an assembly of these parts.
   a. working-drawing assembly
   b. general assembly
   c. erection assembly
   d. layout assembly

7. This type of assembly often includes dimensions and fabrication specifications and is typically associated with products that are made of structural steel.
   a. working-drawing assembly
   b. general assembly
   c. erection assembly
   d. layout assembly

8. Individual components of a complete assembly may, in turn, have an assembly. This component assembly is called:
   a. individual assembly
   b. element assembly
   c. fabrication assembly
   d. subassembly

9. This assembly of drawings includes an artistic representation, a photographic reproduction, a simple isometric drawing, or an exploded isometric drawing.
   a. pictorial assembly
   b. isometric assembly
   c. perspective assembly
   d. artistic assembly

10. These are circles that connect to the related parts of an assembly and contain numbers that correlate the parts to the parts list:
    a. circles
    b. balloons
    c. identification circles
    d. identification numbers

11. This is a list that identifies each part in the assembly and correlates the parts to the assembly. This list is usually on the assembly drawing, but may be separate:
    a. parts list
    b. list of materials
    c. description of parts
    d. description of materials

12. Also referred to as standard parts, these parts are manufactured and available from outside vendors:
    a. vendor parts
    b. manufactured parts
    c. purchase parts
    d. common parts

13. This written authorization is normally needed before a drawing can be changed by the drafting department:
    a. ECR
    b. ECN

14. Records of engineering changes are kept for reference. The record is called:
    a. ECR
    b. ECN

15. A straight, thick line placed under a dimension means the dimension is:
    a. reference
    b. a new dimension
    c. not to scale
    d. not for release

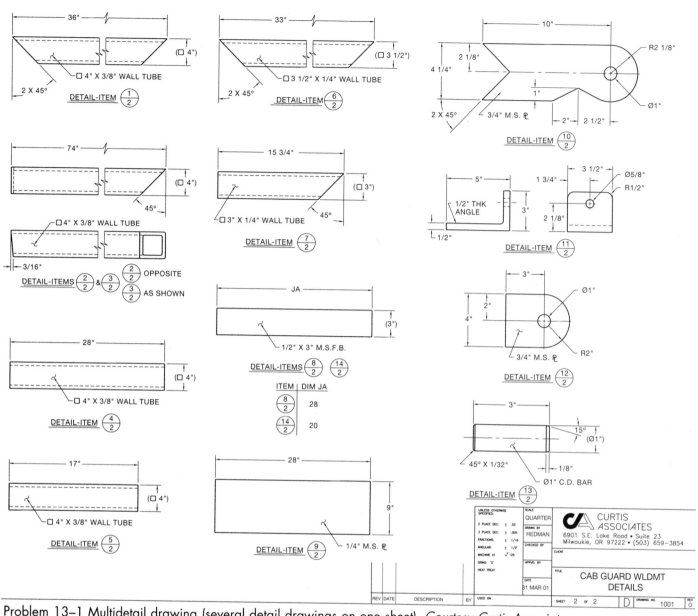

Problem 13–1 Multidetail drawing (several detail drawings on one sheet). *Courtesy Curtis Associates.*

## CHAPTER 13 PROBLEMS

**PROBLEM 13–1** Answer the following questions as you refer to the print shown on this page.

1. Is the drawing on this print a monodetail drawing or a multidetail drawing?

2. Explain the difference between a monodetail drawing and a multidetail drawing.

3. How many parts are detailed on this drawing?

4. What do the numbers in the circles next to the part names mean?

5. What is dimension A on part number 8/2?

6. Explain what the term OPPOSITE means as related to part number 2/2.

7. How many sheets are there in this set of drawings?

8. What is the title of this drawing?

9. Give the thickness of part number 10/2.

10. What does the note □4 x 3/8 WALL TUBE mean?

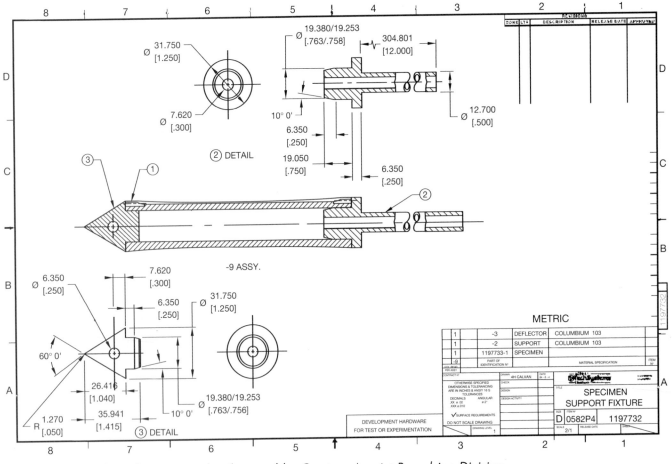

Problem 13–2 Working drawing, or detail, assembly. *Courtesy Aerojet Propulsion Division.*

**PROBLEM 13–2** Answer the following questions as you refer to the print shown on this page.

1. What type of drawing format is represented on this print?

2. How many parts make up the assembly?

3. When is an assembly or working drawing of this type used?

4. Why are dimensions given with two numbers where one number is in brackets?

5. What part is shown at zone C-D/3-6?

6. What is the total length, in millimeters, of the part at zone A-7?

7. Give the name of this assembly.

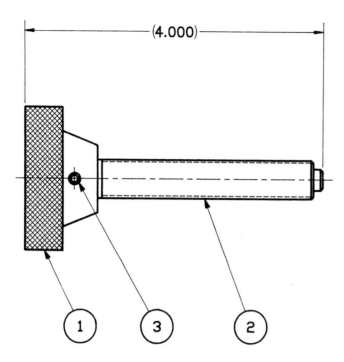

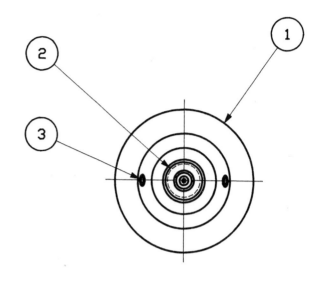

Problem 13–3 *Courtesy Stanley Hydraulic Tools.*

**PROBLEM 13–3** Answer the following questions as you refer to the print shown on this page.

1. What type of an assembly would you classify the drawing on this print as?

_____

2. How many parts are there in this assembly?

_____

3. Give the description of item 3.

_____
_____
_____

4. What is the overall assembled length of the assembly?

_____
_____

5. Give the name of the assembly.

_____

6. What is the part number for item 2?

_____

7. What is the general description of item 2?

_____
_____
_____

8. Explain how you would go about finding more specific information about the manufacturing information for item 2.

_____
_____
_____

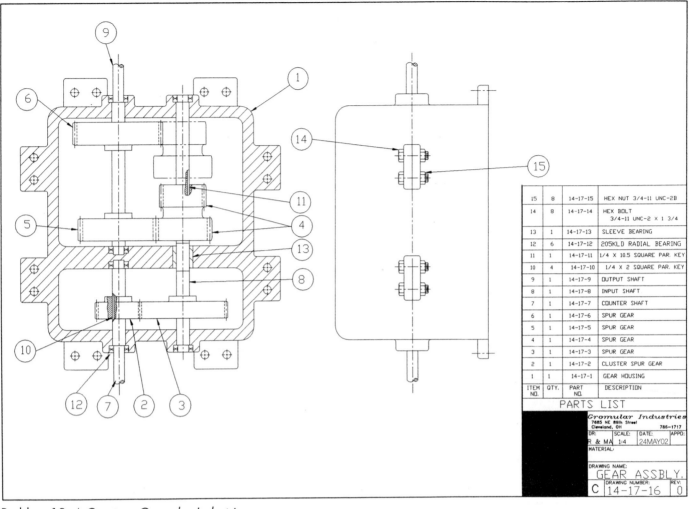

Parts List:

| 15 | 8 | 14-17-15 | HEX NUT 3/4-11 UNC-2B |
| 14 | 8 | 14-17-14 | HEX BOLT 3/4-11 UNC-2 X 1 3/4 |
| 13 | 1 | 14-17-13 | SLEEVE BEARING |
| 12 | 6 | 14-17-12 | 205KLD RADIAL BEARING |
| 11 | 1 | 14-17-11 | 1/4 X 10.5 SQUARE PAR. KEY |
| 10 | 4 | 14-17-10 | 1/4 X 2 SQUARE PAR. KEY |
| 9 | 1 | 14-17-9 | OUTPUT SHAFT |
| 8 | 1 | 14-17-8 | INPUT SHAFT |
| 7 | 1 | 14-17-7 | COUNTER SHAFT |
| 6 | 1 | 14-17-6 | SPUR GEAR |
| 5 | 1 | 14-17-5 | SPUR GEAR |
| 4 | 1 | 14-17-4 | SPUR GEAR |
| 3 | 1 | 14-17-3 | SPUR GEAR |
| 2 | 1 | 14-17-2 | CLUSTER SPUR GEAR |
| 1 | 1 | 14-17-1 | GEAR HOUSING |
| ITEM NO. | QTY. | PART NO. | DESCRIPTION |

PARTS LIST

Gromular Industries
7685 NE 86th Street
Cleveland, OH          786-1717

| DR: | SCALE: | DATE: | APPD: |
| R & MA | 1:4 | 24MAY02 | |

MATERIAL:

DRAWING NAME:
GEAR ASSBLY.

C | DRAWING NUMBER: 14-17-16 | REV: 0

Problem 13–4 *Courtesy Gromular Industries.*

**PROBLEM 13–4** Answer the following questions as you refer to the print shown on this page.

1. What type of an assembly would you classify the drawing on this print as?
   _____

2. How many parts are there in this assembly?
   _____

3. How many spur gears are there in this assembly?
   _____

4. Count the number of radial bearings in this assembly.
   _____

5. How many sleeve bearings are there in this assembly?
   _____

6. Fully describe the views given for this assembly.
   _____
   _____
   _____
   _____
   _____

7. How many hex bolts are used in this assembly?
   _____

8. Describe the hex bolts used in this assembly.
   _____

9. What is the name of the assembly?
   _____

10. Give the description of the radial bearings.
    _____

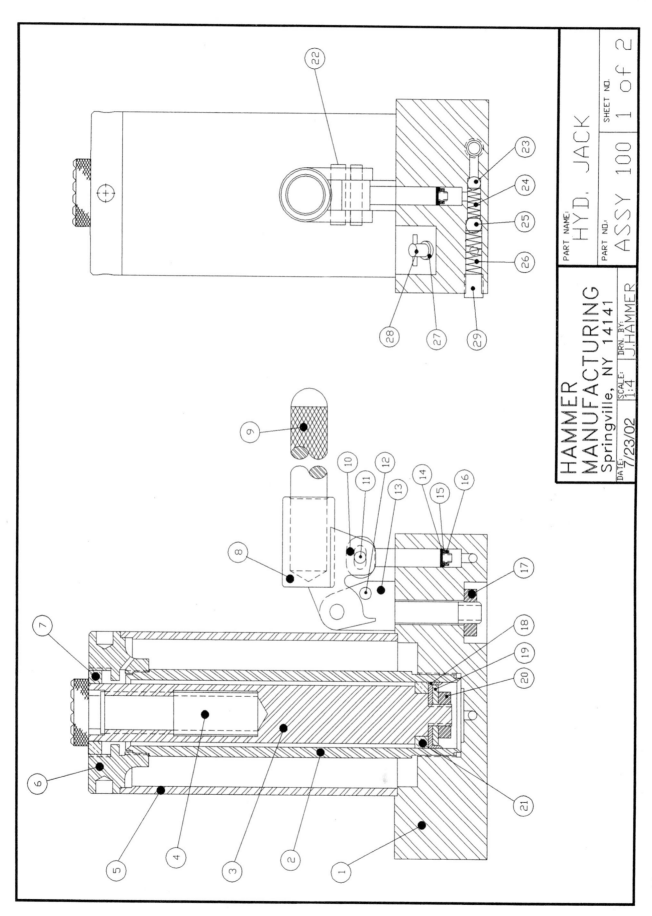

Problem 13–5 Courtesy Hammer Mfg. (continued on the next page)

PART NAME: HYD. JACK

PART NO.: ASSY 100    SHEET NO. 1 of 2

HAMMER MANUFACTURING
Springville, NY 14141

DATE: 7/23/02    SCALE: 1:4    DRN. BY: J.HAMMER

# HAMMER MANUFACTURING
Springville, NY 14141

ASSEMBLY__HYDRAULIC JACK_____     DATE:_7/23/92____
PRODUCT NUMBER:__100_____     UNITS:_25,000___

| ITEM | QTY | PART # | PART NAME | MATERIAL/DESCRIPTION |
|------|-----|--------|-----------|----------------------|
| 1 | 1 | 100–1 | BASE | CI |
| 2 | 1 | 100–2 | CYLINDER TUBE | FREE MACHINING LEADED CRMS |
| 3 | 1 | 100–3 | PISTON | HRMS |
| 4 | 1 | 100–4 | SCREW | MS |
| 5 | 1 | 100–5 | RESERVOIR TUBE | 3" PIPE STEEL SCH 40 |
| 6 | 1 | 100–6 | TOP CAP | MS |
| 7 | 1 | 100–7 | PACKING NUT | BRONZE |
| 8 | 1 | 100–8 | HANDLE SOCKET | HRMS |
| 9 | 1 | 100–9 | PUMP HANDLE | CRS |
| 10 | 1 | 100–10 | PUMP PLUNGER | HRS |
| 11 | 1 | 100–11 | DRIVE PIN | ⌀ 1/4 STL |
| 12 | 1 | 100–12 | STOP PIN | ⌀ 1/4 STL |
| 13 | 1 | 100–13 | PIVOT SUPPORT | CRS |
| 14 | 1 | 100–14 | CUP SEAL | ⌀ .445 NEOPRENE |
| 15 | 1 | 100–15 | ⌀ .125 WASHER | STEEL |
| 16 | 1 | 100–16 | NUT | 10–32 STL |
| 17 | 1 | 100–17 | NUT | 5/16–14UNC |
| 18 | 1 | 100–18 | LEATHER CUP | ⌀ 1.5 LEATHER |
| 19 | 1 | 100–19 | PISTON WASHER | SAE 1040 |
| 20 | 1 | 100–20 | NUT | 5/16–20UNF |
| 21 | 1 | 100–21 | GUIDE | BRONZE |
| 22 | 1 | 100–22 | PIVOT PIN | ⌀ .312 STL |
| 23 | 1 | 100–23 | BALL | ⌀ .25 BALL |
| 24 | 1 | 100–24 | .218 SPRING | 10 GA SPRING STL |
| 25 | 1 | 100–25 | BALL | ⌀ .3125 BALL |
| 26 | 1 | 100–26 | .343 SPRING | 10 GA SPRING STL |
| 27 | 1 | 100–27 | VALVE NUT | BRONZE |
| 28 | 1 | 100–28 | NEEDLE VALVE | MS |
| 29 | 1 | 100–29 | PLUG | 1/8–NPT |

Problem 13–5 continued.

**PROBLEM 13–5** Answer the following questions as you refer to the print and parts list shown on the previous two pages.

1. What type of an assembly would you classify the drawing on this print as?

2. How is this assembly different from the assembly in Problem 13–4?

3. Describe the view used in this assembly.

4. How many parts are there in this assembly?

5. List at least 12 items that could be purchased (standard) parts.

6. How many sheets are there in this assembly drawing?

7. What does MS mean?

8. Why are cylindrical breaks used on part 9?

9. Give the name of the assembly.

10. What is the product number?

11. What does CI mean?

12. Give the part number of the RESERVOIR TUBE.

# Reading Pictorial Drawings

After completing this chapter you will be able to:

- Identify types of pictorial drawings.
- Read pictorial drawing prints.

## PICTORIAL DRAWINGS

Most products are made from orthographic drawings that allow us to view an object with our line of sight perpendicular to the surface we are looking at. The one major shortcoming of this form of drawing is the lack of depth. Certain situations demand a single view of the object that provides a more realistic representation. This realistic single view is achieved with pictorial drawing. We commonly think of something that is pictorial as resembling a picture, which is much more true to life than an orthographic drawing.

The most common forms of pictorial drawings used in mechanical drafting are isometric and oblique. The most realistic type of pictorial illustration is perspective drawing. The use of vanishing points in the projection of these drawings gives them depth and distortion that we see with our eyes. These forms are discussed later.

Pictorial drawings are the most easily interpreted because they look realistic. In fact, pictorial views or drawings are sometimes included on two-dimensional (2D) multiview drawings for quick interpretation of the part or assembly. You may have also seen pictorial drawings used in assembly instructions that come with products you purchase. Pictorial drawings are used when a clear picture must be presented, especially for persons not trained to interpret 2D drawings. We naturally see things in three-dimensional form, and therefore drawings in this format are normally easy to read.

The presentation of pictorial drawings can be a simple oblique drawing or a complex exploded assembly that contains many parts, item numbers, and a parts list. Examples of these types of drawings are given in this chapter.

### Uses of Pictorial Drawings

Pictorial drawings are excellent aids in the design process, for they allow designers and engineers to view objects at various stages in development. Pictorial drawings are used in instruction manuals, parts catalogs, advertising literature, technical reports and presentations, and as aids in the assembly and construction of products. Pictorial drawings are often used as the base for technical illustrations, which can be shaded, rendered, airbrushed, or colored in a variety of techniques. Figure 14–1 is an example of a pictorial drawing, while Figure 14–2 shows a technical illustration.

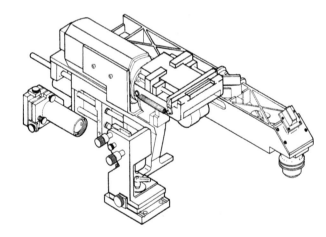

Figure 14–1 Pictorial drawing. *Courtesy Industrial Illustrators, Inc.*

## TYPES OF ISOMETRIC DRAWINGS

Isometric drawing is a form of pictorial drawing in which the receding axes are drawn at 30° from the horizontal, as shown in Figure 14–3. Three basic forms of isometric drawing include regular, reverse, and long-axis isometric.

Figure 14–2 Cutaway technical illustration. *Courtesy Industrial Illustrators, Inc.*

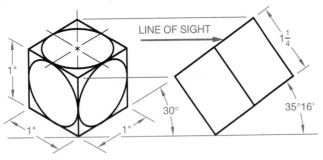

ISOMETRIC DRAWING
BASED ON TRUE MEASUREMENT IN ISOMETRIC VIEW
(PREFERRED METHOD)

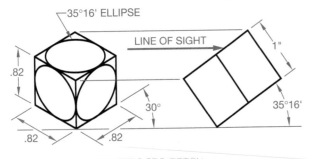

ISOMETRIC PROJECTION
BASED ON TRUE MEASUREMENT IN ORTHOGRAPHIC
(SELDOM USED)

Figure 14–3 The differences between isometric drawing and isometric projection.

## Regular Isometric

The top of an object can be seen in the regular isometric form of drawing. An example can be seen in Figure 14–4(a). This is the most common form of isometric drawing.

## Reverse Isometric

The only difference between reverse and regular isometric is that you can view the bottom of the part instead of the top. The 30° axis lines are drawn downward from the horizontal line instead of upward. Figure 14–4(b) shows an example of reverse isometric.

## Long-axis Isometric

The long-axis isometric drawing is normally used for objects that are long, such as shafts. Figure 14–4(c) shows an example of the long-axis form.

## Sections

Sections are views that show internal features of a part. A full section shows one half of the part cut away. A half section shows one quarter of the part cut away, and is used most often on circular or cylindrical parts that are symmetrical. The offset section has a jog in the cutting-plane line to cut through another feature that is "offset" from the first.

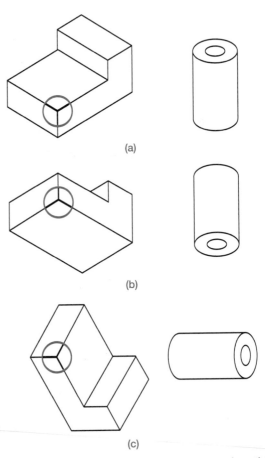

(a)

(b)

(c)

Figure 14–4 Isometric axis variations: (a) regular; (b) reverse; (c) long-axis.

Full and half sections are common in technical illustrations and are drawn along the isometric axes. The section lines in full sections are drawn in the same direction, while those on half sections appear in opposite directions. See the section lines illustrated in Figure 14–5(a) and (b). The section lines in offset sections change directions with each jog in the part, as seen in Figure 14–5(c).

## Isometric Dimensioning

It is not common for isometric drawings to be dimensioned; however, some isometric piping drawings rely heavily on dimensioning to get the message across. One technique that is used shows the

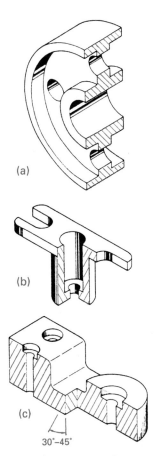

Figure 14–5 Isometric sections: (a) full; (b) half; (c) offset.

vertical strokes of the text parallel to extension lines, as shown in Figure 14–6((a) and (b)). This gives the appearance that the dimension is lying in the plane of the extension lines. Examples of unidirectional, aligned, and one-plane (horizontal) isometric dimensioning are shown in Figure 14–6(b).

## EXPLODED PICTORIAL DRAWINGS

Complicated parts and mechanisms are often illustrated as an exploded assembly to show the relationship of the parts in the most realistic manner. This type of drawing may be familiar to many readers who have seen exploded pictorial drawings in instruction manuals and parts catalogs (see Figure 14–7(a)).

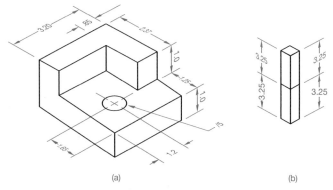

Figure 14–6 Isometric dimensioning: (a) dimensions parallel to extension lines; (b) various styles of dimensioning.

An exploded assembly is nothing more than a collection of parts, each drawn in the same pictorial method. Any of the pictorial drawing methods mentioned in this chapter are used to create exploded assemblies. Centerlines are used in exploded views to represent lines of explosion. These lines aid the eye in following a part to its position in or on another part. This is shown in Figure 14–7(b).

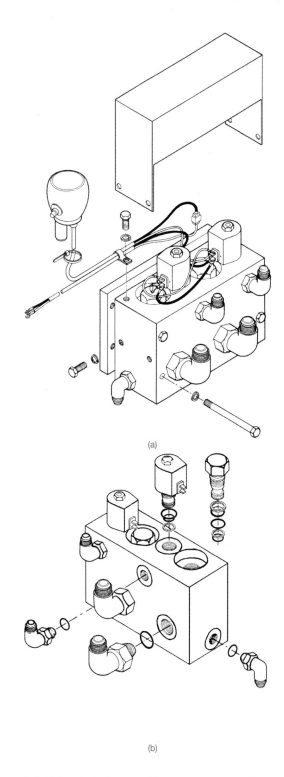

Figure 14–7 Exploded assemblies. *Courtesy Industrial Illustrators, Inc.*

A much more complex exploded assembly of a piece of electronic equipment is shown in Figure 14–8. It contains item numbers for quick identification of each part. Notice how important the lines of explosion are in this drawing. How easy would it be to determine where each part connected if the centerlines were omitted?

## PARTS LIBRARIES

Many companies create parts catalogs or libraries of the individual components they use or manufacture. These libraries may be printed with several parts to a page, many of which can look similar. If your job involves the use of parts libraries, it is important that you study each part carefully and check any component or identification number that is given with the part (see Figure 14–9).

## OBLIQUE DRAWING

Oblique drawing is a form of pictorial drawing in which the plane of projection is parallel to the front surface of the object. The lines of sight are at an angle to the plane of projection and are parallel to each other. This allows the viewer to see three faces of the object. The front face, and any surface parallel to it, is shown in true shape and size, while the other two faces are distorted in relation to the angle and scale used. Oblique drawings are useful if one face of an object needs to be shown without distortion.

There are three methods of oblique drawing: cavalier, cabinet, and general.

### Cavalier Oblique

The cavalier projection is one in which the receding lines are drawn true size, or full scale. This form of oblique drawing is usually drawn at an angle to a horizontal of 45° (see Figure 14–10).

### Cabinet Oblique

A cabinet oblique drawing is also drawn with a receding angle of 45°, but the scale along the receding axis is half size. Cabinet makers often draw their cabinet designs using this type of oblique drawing. A cabinet drawing is shown in Figure 14–11.

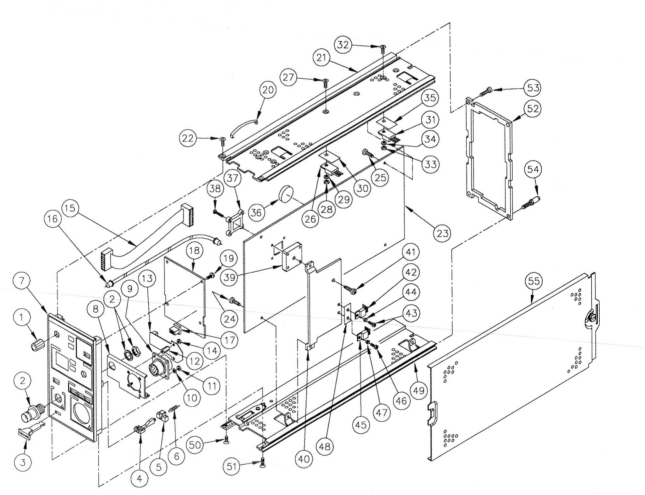

TEKTRONIX CURRENT PROBE AMPLIFIER

Figure 14–8 Exploded assembly with item numbers for part identification. *Courtesy Karen Miller, Technical Illustrator, Tektronix, Inc.*

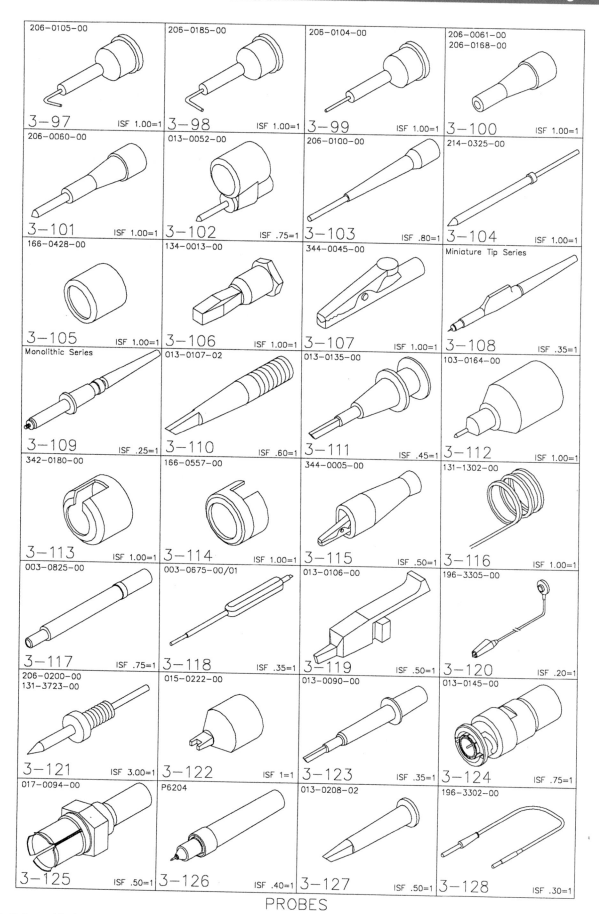

PROBES

Figure 14–9 Parts library page. *Courtesy Karen Miller, Technical Illustrator, Tektronix, Inc.*

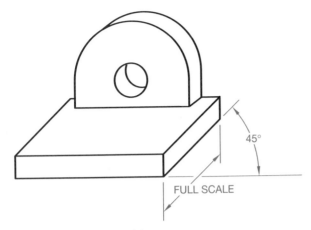

Figure 14–10 Cavalier oblique.

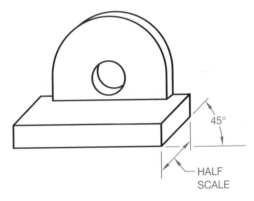

Figure 14–11 Cabinet oblique.

### General Oblique

The general oblique drawing is normally drawn at an angle other than 45°, and the scale on the receding axis is also different from those used in cavalier and cabinet. The most common angles for a general oblique drawing are 30°, 45°, and 60°. Any scale from half to full size can be used. A general oblique drawing is shown in Figure 14–12.

## PERSPECTIVE DRAWING

Perspective drawing is the most realistic form of pictorial illustration. It reflects the phenomenon of objects appearing smaller the farther away they are until they vanish at a point on the horizon. Two common types of perspective drawing techniques take their names from the number of vanishing points used in each. One-point, or parallel, perspective has one vanishing point and is used most often to show interiors of rooms. Two-point, or angular, perspective is the most popular and is used to illustrate exteriors of houses, small buildings, civil engineering projects, and occasionally, machine parts.

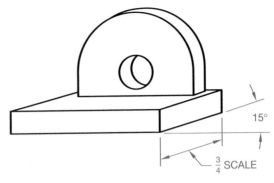

Figure 14–12 General oblique.

## ONE-POINT PERSPECTIVE

This form of perspective, also known as parallel perspective, has only one vanishing point. It is most often used by architects to show the interior of a single room. Lines in the perspective seem to project to a single vanishing point. Figure 14–13 is an example of a one-point perspective.

## TWO-POINT PERSPECTIVE

The two-point perspective method is also termed "angular perspective" because it is turned so that two of its principal planes are at an angle to the picture plane. This is the most popular form of perspective drawing. Architects use this form of perspective to illustrate the exterior of houses and commercial buildings. It provides the most realistic form of pictorial drawing, and is easy for people to visualize because it looks so real. Figure 14–14 and Figure 14–15 are examples of a two-point perspective.

Figure 14–13 One-point perspective. *Courtesy Lang Bates (Barrentine, Bates & Lee).*

Figure 14–14 Two-point perspective. *Courtesy Lang Bates (Barrentine, Bates & Lee).*

Figure 14–15 Two-point perspective. *Courtesy Lang Bates (Barrentine, Bates & Lee).*

# CHAPTER 14 TEST

Multiple choice: Respond to the following by selecting a, b, c, or d
to best answer the question or complete the statement.

1. The most common form of pictorial drawing is:
   a. perspective
   b. dimetric
   c. isometric
   d. orthographic

2. The most realistic form of pictorial illustration is:
   a. perspective
   b. dimetric
   c. isometric
   d. axonometric

3. Vanishing points are used in the _____ form of pictorial drawing:
   a. axonometric
   b. trimetric
   c. perspective
   d. oblique

4. Which of the following is not a good use of pictorial drawings?
   a. instruction manuals
   b. technical reports
   c. parts catalogs
   d. none of the above

5. How many sides of an object can be seen in an isometric?
   a. 2
   b. 4
   c. 3
   d. 5

6. Reverse isometric is used to principally show the _____ of the object:
   a. top
   b. left side
   c. right side
   d. bottom

7. Which of the following objects would best be shown in the long axis isometric format?
   a. shaft
   b. washer
   c. wedge
   d. box

8. Which section cuts away one half of the part?
   a. half
   b. quarter
   c. full
   d. removed

9. Which section cuts away one quarter of the part?
   a. half
   b. quarter
   c. full
   d. removed

10. Lines of explosion in exploded assembly drawings are also known as:
    a. hidden lines
    b. phantom lines
    c. center lines
    d. sight lines

11. What is the most common receding angle for oblique drawings?
    a. 45°
    b. 15°
    c. 30°
    d. 35°

12. Which form of perspective drawing is used to illustrate the exterior of a house?
    a. one point
    b. two point
    c. three point
    d. four point

13. Which form of perspective drawing is used to illustrate the interior of a single room?
    a. one point
    b. two point
    c. three point
    d. four point

14. One-point perspective is also known as _____ perspective:
    a. equal
    b. vanishing point
    c. converging
    d. parallel

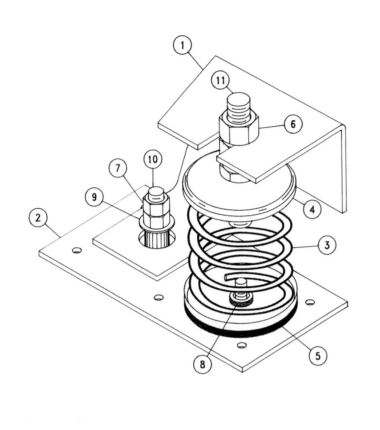

| MATERIAL LIST  4 REQ'D | | | | | |
|---|---|---|---|---|---|
| ITEM | TOTAL QTY. | DESCRIPTION | QTY. ONE | POWDER PAINT | WELD |
| 1 | 4 | SPRING ATTACHMENT CLIP | 1 | YES | NO |
| 2 | 4 | SPRING PAD MOUNTING PLATE | 1 | YES | NO |
| 3 | 4 | FZ-105 SPRING | 1 | YES | NO |
| 4 | 8 | FZ SPRING CUP WITH NUT | 2 | YES | NO |
| 5 | 4 | NEOPRENE PAD | 1 | YES | NO |
| 6 | 8 | 1" HEX NUT | 2 | YES | NO |
| 7 | 8 | 3/4" HEX NUT | 2 | YES | NO |
| 8 | 4 | GROMMET AN931-A8-20 | 1 | YES | NO |
| 9 | 4 | SHOULDER WASHER | 1 | YES | NO |
| 10 | 4 | 3/4" x 3" STUD | 1 | YES | YES |
| 11 | 4 | 1" DIA. THREADED ROD | 1 | YES | NO |

FILE NAME: EMT-A-02

FOR

**ISOLATION SPRING ASSEMBLY**

**PACE CLEAN-PAK®**
9420 S.E. LAWNFIELD
CLACKAMAS, OREGON 97015
PHONE: (503) 652-7488
FAX: (503) 652-7492

DRAWN BY GREG LANZ

DATE 3/5/92

REV.

JOB NUMBER: 435-045-14

PAGE    OF

Problem 14–1 *Courtesy PACE Co.*

## CHAPTER 14 PROBLEMS

**PROBLEM 14–1** Answer the following questions based on the print shown on this page.

1. What is this component?

2. What is the job number?

3. How many different parts are used for one assembly?

4. How many assemblies are required for this job?

5. How many total parts comprise the required assemblies?

6. What is part number 8?

7. What is part number 4?

8. Which part fits into part number 4?

9. Which part does item 7 fit into?

10. How many of item 7 are used for one assembly?

11. What is the file name of this drawing?

12. Which part number requires welding?

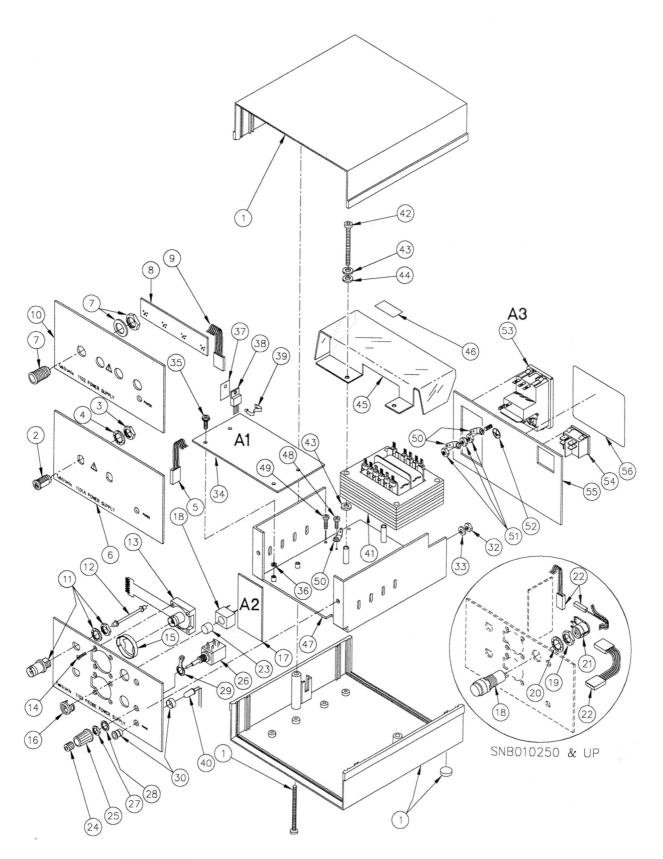

TEKTRONIX 1101A, 1102 & 1103 ACCESSORY POWER SUPPLIES

SNB010250 & UP

Problem 14–2 *Courtesy Karen Miller, Technical Illustrator, Tektronix, Inc.*

**PROBLEM 14–2** Answer the following questions based on the print shown on the previous page.

1. What type of drawing is this?

   _____

2. What form of pictorial drawing is used in this illustration?

   _____

3. How many different power supplies are indicated in this assembly?

   _____

4. What are the model numbers of these power supplies?

   _____

5. Item 1 is composed of how many different parts?

   _____

6. Which part does item 42 thread into?

   _____

7. Item 54 fits through a slot in which part?

   _____

8. Which two parts attach to 13?

   _____

9. Three side-by-side item 51s are bracketed by which part?

   _____

10. What part numbers are between item 35 and item 47?

    _____

11. How many parts is item 7 composed of?

    _____

12. Which part does item 3 thread onto?

    _____

13. How many of the following item numbers are used in this assembly?
    a. 50 _____
    b. 43 _____
    c. 51 _____

14. What is the purpose of the drawing inside the circle?

    _____

# Reading Precision Sheet Metal Drawings

## LEARNING OBJECTIVES

After completing this chapter you will be able to:

■ Read precision sheet metal prints.

■ Answer questions related to material bending, seams, and chassis layout.

■ Calculate bend allowance.

## PRECISION SHEET METAL DRAWINGS

Precision sheet metal drawings are found in industries where flat material is used to fabricate accurate product components. Common applications include automobile body panels, electronic chassis, and appliance housings. Precision sheet metal drawings are prepared showing the object in its final or bent form, as a flat pattern, or using a combination of bent and flat representations. The most common precision sheet metal print, especially for the electronics industry, is a flat pattern of the part using arrowless dimensioning, as shown in Figure 15–1. A flat pattern print is based on pattern development where the true size and shape of the object is laid out flat. When this flat pattern is bent along given bend lines the result is the desired sheet metal object. Computer-aided design and drafting (CADD) programs are available that calculate feature locations and allowances for bending material and send the information directly to computer-aided manufacturing (CAM) systems in the fabrication shop to run a computer numerical control metal-bending break or punch press. This type of CADD drawing is shown in Figure 15–1.

### Sheet Metal

Sheet metal is made by pressing metal between rollers to get the required thickness. Sheet metal thickness is determined by gaging systems. Actual material thickness varies from one gaging system to the next. For example, the thickness of Manufacturer's Standard gage for steel is different from Birmingham Gage for steel. The best way to ensure accuracy is to confirm the decimal equivalent of the desired gage size. Gage sizes decrease in thickness as the gage numbers increase. For example, a Manufacturer's Standard gage No. 4 is .2242 inch thick, while gage No. 20 is .0359 inch thick. Refer to the *Machinery's Handbook* for sheet metal gage sizes.

Aluminum, copper, and copper-base alloys are specified in fraction or decimal inches. Standard thicknesses below 1/4 inch are referred to by American Standard 20, 40, or 80 series decimal numbers. Refer to the *Machinery's Handbook* for a complete listing of these numbers.

Zinc sheet metal is normally ordered by decimal thickness, although zinc gage sizes exist and can be found in the *Machinery's Handbook*.

## MATERIAL BENDING AND CHASSIS LAYOUT

In precision sheet metal applications such as fabrication for electronics chassis components or sheet metal appliance body parts, the condition of material when bent must be taken into consideration.

### Seams

When sheet metal parts are bent and formed into the desired shape, a seam results where the ends of the pattern come together. The fastening method depends on the kind and thickness of material, on the fabrication processes available, and on the end use of the part. Sheet metal components that must hold gases or liquid, or are pressurized, may require soldering or welding. Other applications may use mechanical seams, which hold the part together by pressure-lapped metal, metal clips, or pop rivets. Some of the most common seams used in the sheet metal fabrication industry are shown in Figure 15–2.

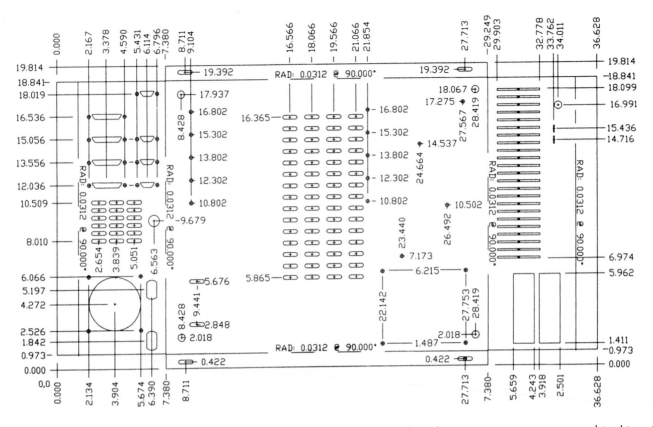

Figure 15–1 A typical precision sheet metal fabrication drawing uses arrowless dimensioning, as represented in this print. *Courtesy C.A.M. Systems, Inc.*

Hemmed edges are necessary when an exposed edge of a pattern must be strengthened. When hems are used, extra material is added to the pattern on the side of the hem. Figure 15–3 shows some common hems.

## Bend Allowance

Bend allowance is the amount of extra material needed for a bend to compensate for the compression during the bending process. Bend allowance is important when close tolerances must be held or when thick material must be bent or formed into desired shapes. The purpose of a bend allowance calculation is to determine the overall dimension of the flat pattern, so that, when bent, the desired final dimensions are achieved. You can use one of several slightly different methods to calculate bend allowance. There are different formulas in the *Machinery's Handbook*, ASME Handbook, and in most textbooks. Many companies use formulas taken from a proven method by individual experimentation. The amount of bend allowance depends on the material thickness, type of material, bending process, degree of bend, and bend radius. The grain of material, surface condition, and amount of lubrication used also influence bend allowance. The only exact way to establish bend allowance for a specific application is to experiment with the equipment and material to be used. Some companies have developed these tests and have created charts that give the bend allowance for the type of material, material thickness, and bend radius.

When material bends, there is compression on the inside of the bend and stretching on the outside. Somewhere between, there is

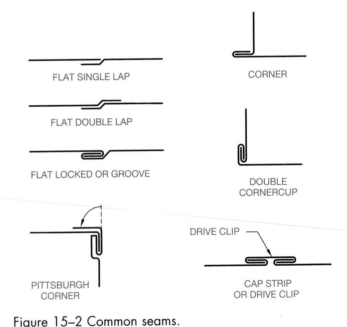

Figure 15–2 Common seams.

Figure 15–3 Common hems.

a neutral zone where neither stretching nor compression occurs; this is called the neutral axis. The neutral axis is approximately four-tenths of the thickness from inside the bend, but this depends on the material and other factors. Information related to calculating the bend allowance and length of the flat pattern is shown in Figure 15–4. The following formula may be used to calculate the length of the flat pattern (straight stock before bending):

Length of Flat Pattern = X + Y + Z

X = Horizontal Dimension to Bend = B – R – C

Y = Vertical Dimension to Bend = A – R – C

Z = Length of Neutral Axis

C = Material Thickness

R = Bend Radius

Z = 2(R + .4C) **x** p ÷ 4 (90° bend)

Given the sheet metal bend shown in Figure 15–5, determine the length of the flat pattern.

X = B – R – C = 12.250 – .125 – .125 = 12.000

Y = A –R – C = 4.875 – .125 – .125 = 4.625

Z = 2 (R + .4C) **x** p ÷ 4 = 2 (.125 + .4 **x** .125) **x** 3.14 ÷ 4 = .275

Length of Flat Pattern = X + Y + Z = 12.000 + 4.625 + .275 = 16.900

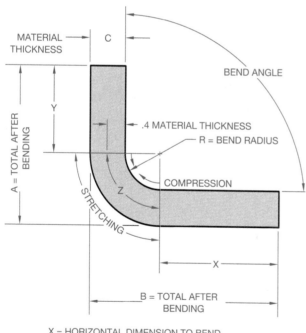

Figure 15–4 Bend allowance variables.

### Bend Relief

A corner with two edges bent in the same direction has internal stresses that can cause a crack at the corner. When necessary, some material at the corner can be cut away to help relieve this stress. This cutout at the corner is called bend relief (see Figure 15–6).

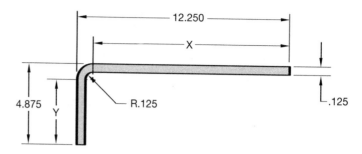

Figure 15–5 Sample bend allowance problem.

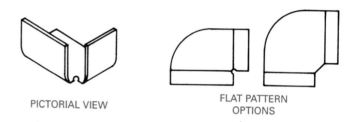

Figure 15–6 Bend relief.

## CHASSIS AND PRECISION SHEET METAL LAYOUT

Electronic assemblies often require shields, frames, panels, and chassis to be fabricated. Layout drawings are required for the fabrication of these items. The method of dimensioning these parts includes standard unidirectional, arrowless, or tabular dimensioning systems. (Refer to Chapter 7, Reading Dimensions.) A precision sheet metal pattern can be drawn as a flat pattern, as a finished product, or with the finished part shown and the flat pattern drawn using phantom lines.

### Flat Pattern Layout

As previously discussed, the most common precision sheet metal print is given as a flat pattern layout. When this is done, the bend allowance is already calculated based on the type of material, material thickness, bend radius, and bend angle. Figure 15–7 shows a simple chassis fabrication print, given as a flat layout, using arrowless tabular dimensioning.

### Formed View Drawing

Some companies prepare sheet metal fabrication prints using the formed view or representation of the object in its final bent shape, as shown in Figure 15–8. A combination of formed view and flat pattern may also be found on precision sheet metal prints, as shown in Figure 15–9.

### Flat Pattern Reference

A method used by some companies combines the formed views of the part in its final shape and the flat pattern reference given using phantom lines. Dimensions to the flat pattern reference, shown in Figure 15–10, include calculations for the bend allowance.

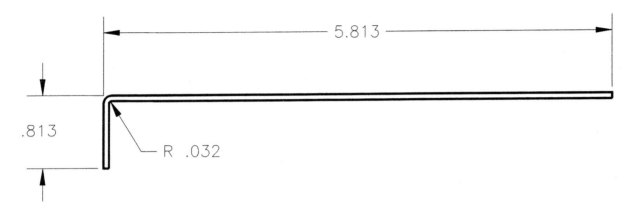

MATERIAL: .048 CRS

Figure 15–7 Flat pattern CADD drawing using arrowless tabular dimensioning.

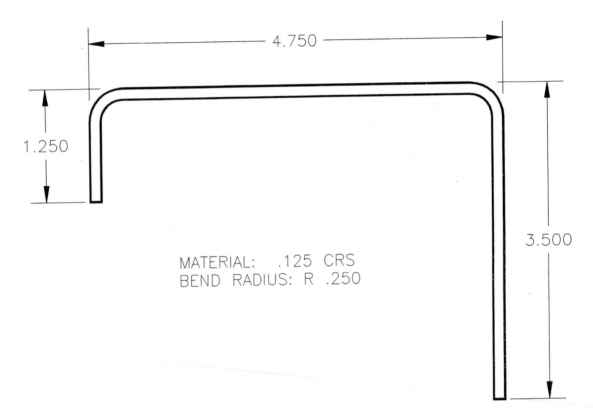

MATERIAL: .125 CRS
BEND RADIUS: R .250

Figure 15–8 CADD drawing showing formed views. *Courtesy TEMCO.*

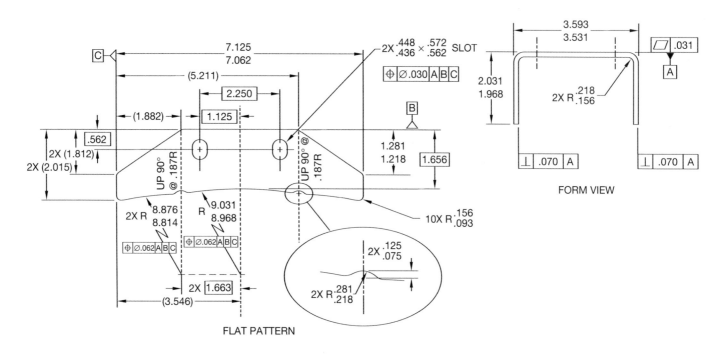

Figure 15–9 CADD drawing showing formed view and flat pattern. *Courtesy TEMCO.*

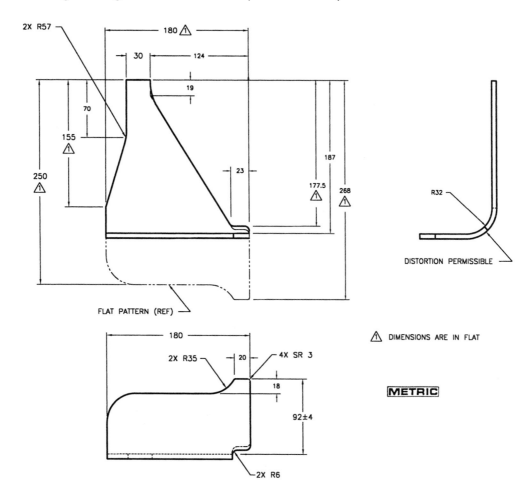

Figure 15–10 Drawing with flat pattern shown using phantom lines. *Courtesy Hyster Company.*

# CHAPTER 15 TEST

Multiple choice: Respond to the following by selecting a, b, c, or d to best answer the question or complete the statement.

1. This type of material is made by pressing metal between rollers to get the required thickness:
   a. wire
   b. sheet metal
   c. flat metal
   d. gage metal

2. The best way to ensure accuracy of sheet metal thickness is to use the decimal equivalent:
   a. true
   b. false

3. This is used to strengthen an exposed edge:
   a. bend allowance
   b. seam
   c. hem
   d. bend relief

4. This cutout at the corner where two edges bend is used to relieve stress:
   a. bend allowance
   b. seam
   c. hem
   d. bend relief

5. This is the amount of extra material needed for a bend to compensate for the compression during the bending process:
   a. bend allowance
   b. seam
   c. hem
   d. bend relief

6. This results when sheet metal parts are bent and formed into the desired shape, where the ends of the pattern come together:
   a. bend allowance
   b. seam
   c. hem
   d. bend relief

7. When material bends, this occurs on the inside of the bend:
   a. compression
   b. stretching

8. When material bends, this occurs on the outside of the bend:
   a. compression
   b. stretching

9. When material bends, there is compression and stretching, and somewhere between there is a zone called:
   a. ozone
   b. neutral zone
   c. neutral axis
   d. central zone

10. A precision sheet metal pattern can be drawn as:
    a. a flat pattern
    b. a finished product
    c. a finished part with a flat pattern shown in phantom lines
    d. all of the above

11. What is the length of the flat pattern if the overall finished dimensions are 12.250 ✗ 4.875", the bend is 90°, the bend radius is .125, and the material thickness is .125?
    a. 16.900
    b. 17.125
    c. 16.875
    d. 17.000

12. One of the most common methods of dimensioning precision sheet metal prints, especially for the electronics industry, is:
    a. chain dimensioning
    b. point-to-point dimensioning
    c. arrowless dimensioning
    d. precision dimensioning

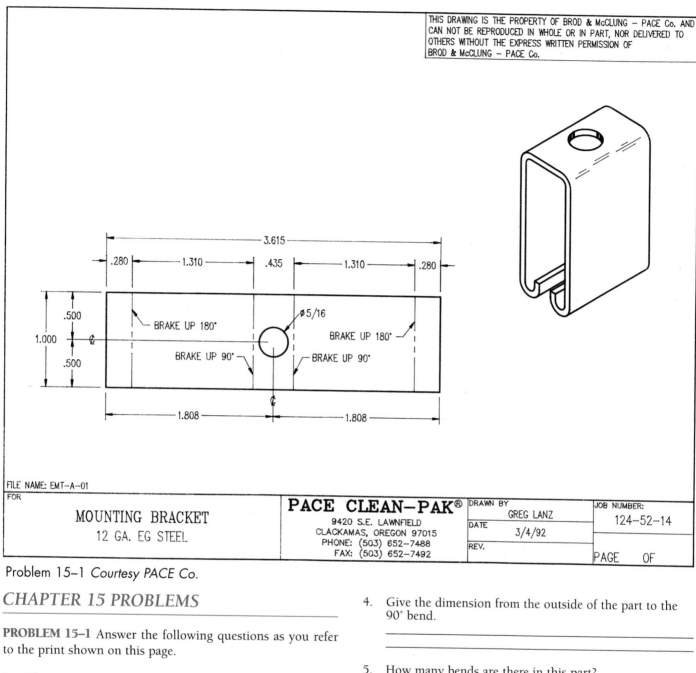

FILE NAME: EMT–A–01

FOR

MOUNTING BRACKET
12 GA. EG STEEL

PACE CLEAN–PAK®
9420 S.E. LAWNFIELD
CLACKAMAS, OREGON 97015
PHONE: (503) 652–7488
FAX: (503) 652–7492

DRAWN BY GREG LANZ
DATE 3/4/92
REV.

JOB NUMBER: 124–52–14

PAGE    OF

Problem 15–1 *Courtesy PACE Co.*

## CHAPTER 15 PROBLEMS

**PROBLEM 15–1** Answer the following questions as you refer to the print shown on this page.

1. There were several methods, described in this chapter, used to show precision sheet metal fabrication drawing representations. Describe the method used on this print.

2. Give the overall dimensions of the flat pattern.

3. What is the material?

4. Give the dimension from the outside of the part to the 90° bend.

5. How many bends are there in this part?

6. Give the location dimensions to the Ø5/16 hole.

7. List the dimensions of the bend lines.

8. Give the angle of the bends.

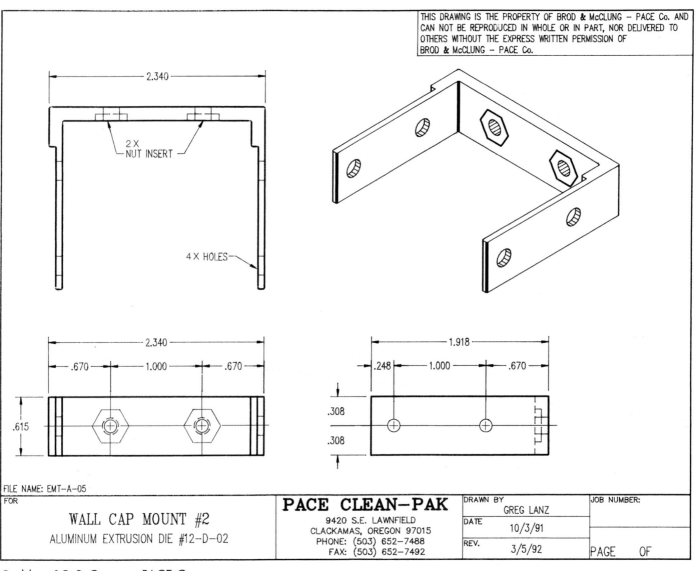

THIS DRAWING IS THE PROPERTY OF BROD & McCLUNG – PACE Co. AND CAN NOT BE REPRODUCED IN WHOLE OR IN PART, NOR DELIVERED TO OTHERS WITHOUT THE EXPRESS WRITTEN PERMISSION OF BROD & McCLUNG – PACE Co.

2 X
NUT INSERT

4 X HOLES

2.340

2.340

.670  1.000  .670

.615

1.918

.248  1.000  .670

.308

.308

FILE NAME: EMT-A-05

| FOR | | DRAWN BY | GREG LANZ | JOB NUMBER: |
| WALL CAP MOUNT #2 | **PACE CLEAN–PAK** | DATE | 10/3/91 | |
| ALUMINUM EXTRUSION DIE #12-D-02 | 9420 S.E. LAWNFIELD CLACKAMAS, OREGON 97015 PHONE: (503) 652-7488 FAX: (503) 652-7492 | REV. | 3/5/92 | PAGE    OF |

Problem 15–2 *Courtesy PACE Co.*

**PROBLEM 15–2** Answer the following questions as you refer to the print shown on this page.

1. There are several methods, described in this chapter, used to show precision sheet metal fabrication drawing representations. Describe the method used on this print.

2. Give the overall finish dimensions.

3. Give the location dimensions to the side holes from the top edge of the part.

4. How many side holes are there?

5. How many nut inserts are there?

**BONUS QUESTION** What do you think is the purpose of the pictorial drawing in the upper-right corner of this print?

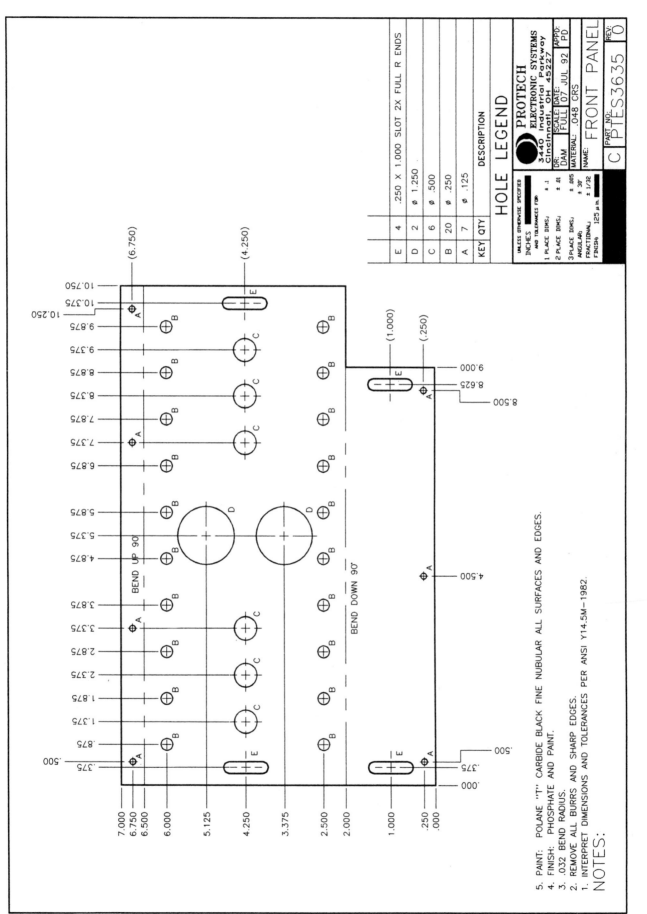

Problem 15-3 Courtesy Protech Electronic Systems.

HOLE LEGEND

| KEY | QTY | DESCRIPTION |
|-----|-----|-------------|
| E | 4 | .250 X 1.000 SLOT 2X FULL R ENDS |
| D | 2 | Ø 1.250 |
| C | 6 | Ø .500 |
| B | 20 | Ø .250 |
| A | 7 | Ø .125 |

PROTECH
ELECTRONIC SYSTEMS
3440 Industrial Parkway
Cincinnati, OH 45227

| DR: DAM | SCALE: FULL | DATE: 07 JUL 92 | APPD: PD |
|---------|-------------|-----------------|----------|

MATERIAL: .048 CRS

NAME: FRONT PANEL

PART NO.: C PTES3635 REV: 0

UNLESS OTHERWISE SPECIFIED
INCHES
AND TOLERANCES FOR:
1 PLACE DIMS. ± .1
2 PLACE DIMS. ± .01
3 PLACE DIMS. ± .005
ANGULAR: ± 30'
FRACTIONAL: ± 1/32
FINISH: 125 μ in.

BEND UP 90°

BEND DOWN 90°

NOTES:
1. INTERPRET DIMENSIONS AND TOLERANCES PER ANSI Y14.5M-1982.
2. REMOVE ALL BURRS AND SHARP EDGES.
3. .032 BEND RADIUS.
4. FINISH: PHOSPHATE AND PAINT.
5. PAINT: POLANE "T" CARBIDE BLACK FINE NUBULAR ALL SURFACES AND EDGES.

**PROBLEM 15–3** Answer the following questions as you refer to the print shown on this page.

1. There are several methods, described in this chapter, used to show precision sheet metal fabrication drawing representations. Describe the method used on this print.

   _____

2. How many different holes are there in this part?

   _____

3. Give the overall length, width, and thickness.

   _____

4. Give the quantity and specifications of the slots.

   _____

5. What is the tolerance of the 5.875 dimension?

   _____

6. Explain the meaning of the parentheses on the (4.250) dimension.

   _____

7. How many "B" holes are there?

   _____

8. What is the diameter of the "B" holes?

   _____

9. Give the full name of the material used to make this part.

   _____

10. What is the bend radius?

    _____

11. Give the finish specification.

    _____

12. Give the paint specification.

    _____

13. Give the X coordinate location of the two "D" holes.

    _____

14. Give the Y coordinate location of the two "D" holes.

    _____

15. Give the X coordinate location of the upper right "A" feature.

    _____

16. Give the Y coordinate location of the upper right "A" feature.

    _____

17. Give the dimensions to the bend lines.

    _____

18. What is to be done with sharp edges?

    _____

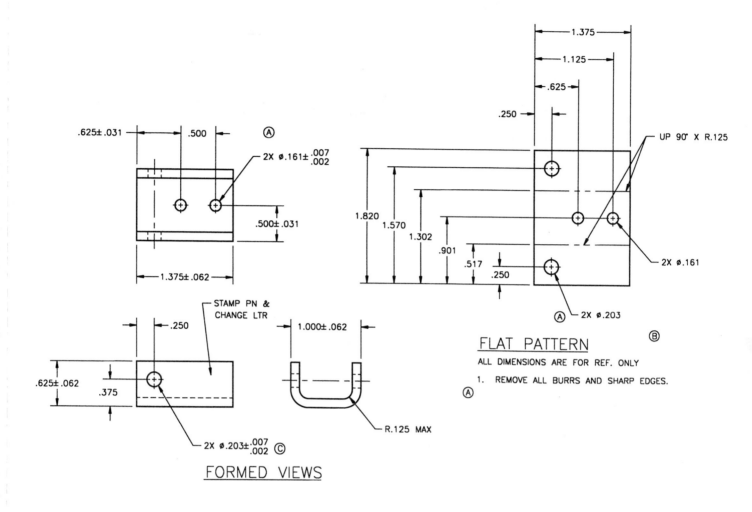

FLAT PATTERN

ALL DIMENSIONS ARE FOR REF. ONLY

1.  REMOVE ALL BURRS AND SHARP EDGES.

FORMED VIEWS

Problem 15–4

**PROBLEM 15–4** Answer the following questions as you refer to the print shown on this page.

1.  There are several methods, described in this chapter, used to show precision sheet metal fabrication drawing representations. Describe the method used on this print.

2.  Give the overall dimensions of the part in its final condition.

3.  What is the length and width of the flat pattern?

4.  Give the material thickness.

5.  What is the material specification?

6.  What is the bend radius?

7.  Give the quantity and size of each hole in this part.

8.  What is the tolerance of the 1.375 dimension?

9.  What is the tolerance of the Ø.203 feature?

10. Give the location dimensions to the bend lines.

11. What is the bend angle?

12. Identify the information given about stamping the part number and change letter.

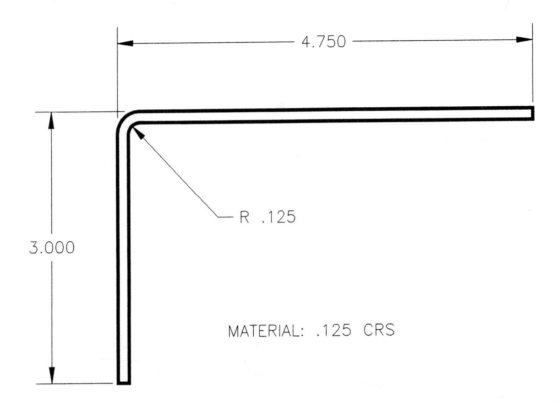

Problem 15–5

**PROBLEM 15–5** Given the sheet metal bend shown in the above drawing, determine the length of the flat pattern. Show all your formulas and calculations in the space provided.

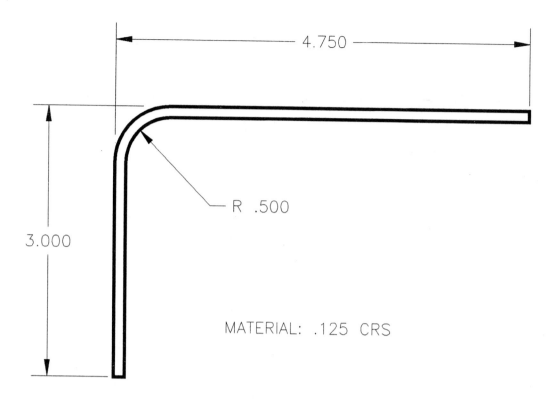

4.750

3.000

R .500

MATERIAL: .125 CRS

Problem 15–6

**PROBLEM 15–6** Given the sheet metal bend shown in the above drawing, determine the length of the flat pattern. Show all of your formulas and calculations in the space provided.

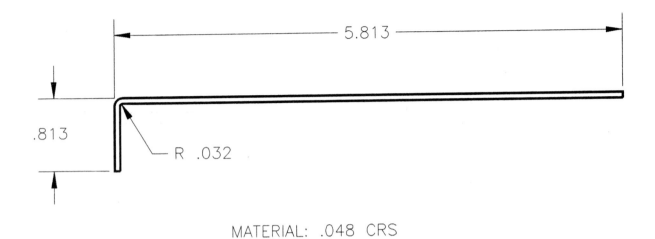

MATERIAL: .048 CRS

Problem 15–7

**PROBLEM 15–7** Given the sheet metal bend shown in the above drawing, determine the length of the flat pattern. Show all your formulas and calculations in the space provided.

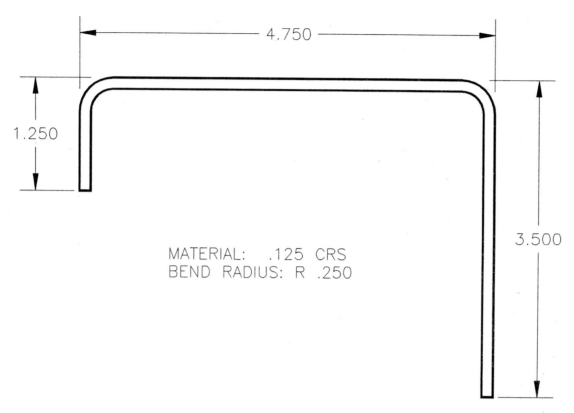

Problem 15–8

**PROBLEM 15–8** Given the sheet metal bend shown in the above drawing, determine the length of the flat pattern. Show all your formulas and calculations in the space provided.

# Reading Electrical Diagrams and Schematics

After completing this chapter you will be able to:

- Identify the ANSI standard for electrical drawings.
- Answer questions related to the different types of electrical diagrams.
- Read the identification system of highway diagrams.
- Read cable diagrams and assemblies.
- Read pictorial and semipictorial drawings and parts lists.

## ELECTRICAL AND ELECTRONIC STANDARDS

*ANSI* A key to effective communication on electrical drawings is the use of standardized symbols so anyone reading the diagram interprets it the same way. To ensure proper standardization, these engineering drawings and related documents must be prepared in accordance with the American National Standards Institute publication ANSI Y32.2.

## FUNDAMENTALS OF ELECTRICAL DIAGRAMS

The purpose of electrical diagrams is to communicate information about the electrical system or circuit in a simple, easy-to-understand format of lines and symbols. Electrical circuits provide the path for electrical flow from the source of electricity through system components and connections and back to the source. Electrical diagrams are generally not drawn to scale.

### Pictorial Diagram

The easiest drawing for most people to understand is the pictorial diagram. Pictorial diagrams represent the electrical circuit as a three-dimensional drawing. This type of diagram provides the most realistic and easy-to-understand representation and may commonly be used in sales brochures, catalogs, service manuals, or assembly drawings. Figure 16–1 shows the pictorial drawing of a simple doorbell circuit.

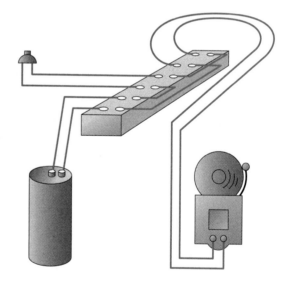

Figure 16–1 Pictorial drawing of a simple doorbell circuit.

### Schematic Diagram

Schematic diagrams are drawn as a series of lines and symbols that represent the electrical current path and the components of the circuit. These drawings result in a clear, simplified layout of the circuit. Schematic diagrams may not show the individual wires or connections of an electrical circuit, but they do show the electrical relationship of components. Figure 16–2 shows a schematic diagram of the doorbell circuit.

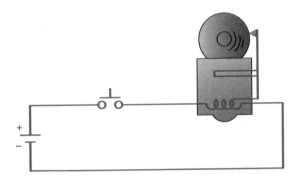

Figure 16–2 Schematic diagram of the doorbell circuit shown in Figure 16–1.

## One-line Diagram

The one-line diagram is a simplified version of the schematic diagram. The difference is that the one-line diagram does not show the complete circuit, although the circuit is assumed. A single line represents the flow of electricity or information from one component to the next, using simplified symbols or labeled blocks. Simplified symbols exhibit a minimum of detail of the component and generally no connections at individual terminals, as shown in Figure 16–3.

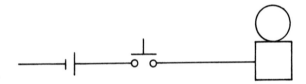

Figure 16–3 One-line diagram of the doorbell circuit shown in Figure 16–1.

## Wiring Diagram

The wiring diagram is a type of schematic that shows all the interconnections of the system components. These diagrams are often referred to as point-to-point, interconnecting wiring diagrams. The wiring diagram is much more detailed than a standard schematic diagram because it shows the layout of individual wire runs. Figure 16–4 shows a wiring diagram of the doorbell circuit.

## Schematic Wiring Diagram

A schematic wiring diagram combines the simplicity of a schematic diagram and the completeness of a wiring diagram. The complete circuit is drawn as a series of lines and symbols that represent the electrical current path and the components of the circuit. Also, in the schematic wiring diagram, the connection terminals are shown in their proper locations along the circuit (see Figure 16–5).

## Highway Wiring Diagram

Also known as highway diagrams, these drawings are used for fabrication, quality control, and trouble-shooting of the wiring of elec-

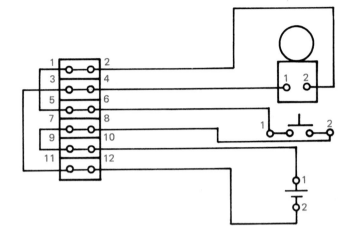

Figure 16–4 Wiring diagram of the doorbell circuit shown in Figure 16–1.

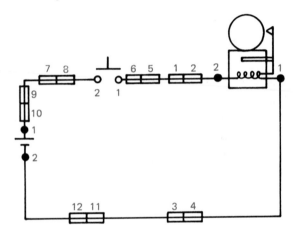

Figure 16–5 Schematic wiring diagram of the doorbell circuit shown in Figure 16–1.

trical circuits and systems. A highway wiring diagram is a simplified or condensed representation of a point-to-point, interconnecting wiring diagram. Highway wiring diagrams are used when it's difficult to draw individual connection lines because of diagram complexity, or when it's not necessary to show all the wires between terminal blocks.

The wiring lines are merged at convenient locations into main trunk lines, or highways, that run horizontally or vertically between component symbols. The lines that run from the component to the trunk lines are called feed lines. The feed lines are identified by code letter, number, or a combination letter and number at the point where each line leaves the component. By reducing the number of lines, the writing diagram is easier to interpret. Figure 16–6 shows an interconnecting wiring diagram and the same electricity system converted to a highway diagram.

A complete highway diagram has an identification system that guides the reader through the system, with a code at one terminal related to a corresponding terminal on another component. An identification system recommended by the American Standards Association (ASA) is a code made up of the wire destination, ter-

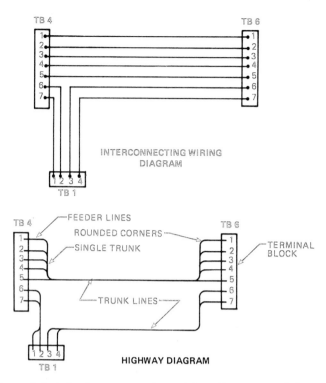

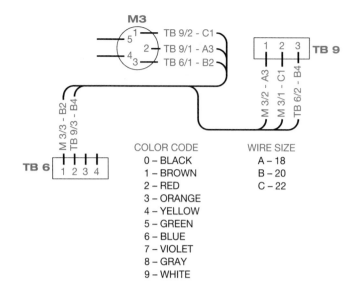

**Figure 16–6** An interconnecting wiring diagram and the same electrical system converted to a highway diagram.

**Figure 16–7** Highway wiring diagram.

minal number at the destination, wire size, and wire-covering color. The individual letters and numbers that make up the electrical system are determined by the company standards and predetermined codes for specific items. Look at Figure 16–7 as you interpret the following code:

| Component-M3 | Component-TB6 |
|---|---|
| TB6 / 1-B2 | M3/3-B2 |
| TB6 = Destination | M3 = Destination |
| 1 = Terminal at destination | 3 = Terminal at destination |
| B = Wire size | B = Wire size |
| 2 = Color of wire covering | 2 = Color of wire covering |

## Wireless Diagram

Wireless diagrams are similar to highway diagrams, except that interconnecting lines are omitted. Components in a wireless diagram are shown in the locations they occupy related to the complete system. The interconnection of terminals is provided by coding as previously described in the discussion about highway diagrams. A wireless diagram is shown in Figure 16–8.

## Cable Diagram and Assemblies

Cable diagrams are associated with multiconductor systems. There may be no difference between a cable diagram and highway or wireless diagrams if the system incorporates multiconductors. A multiconductor is a cable, or group of insulated wires put together

in one sealed assembly. (Cable in another application refers to a stranded wire.) For example, the trunk line, shown in the highway diagram in Figure 16–6 or 16–7, could be a multiconductor. Cable systems are made up of insulated conductor wires, a protective outer jacket or some other means of holding the wires together, and connectors at one end or both ends. Cables are used to connect components, equipment assemblies, or systems together. Cable diagrams usually provide circuit destination, conductor size, number of leads, conductor type, and power rating. A cable diagram is shown in Figure 16–9.

Cable assemblies are the components of the cables put together into one package. The difference between cable assemblies is the number of individual insulated wires in the assembly and the type of end connectors or end terminations. The information that is typically provided on cable assembly drawings includes:

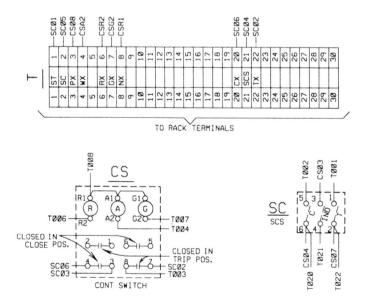

**Figure 16–8** CADD-generated wireless diagram. *Courtesy Bonneville Power Administration.*

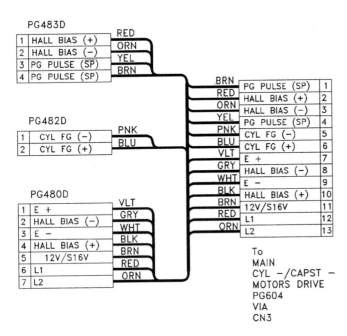

Figure 16-9 Cable diagram. *Reproduced by permission of RCA Consumer Electronics.*

1. Overall length of finished cable.

2. Type of end terminations.

3. Connecting diagram for multiconductor cables, when used.

4. Method of assembly.

5. Wire list for multiconductor cables, specifying wire type, gauge, color, insulation, and length for purchase purposes.

6. Type, gauge, color, and length of outer covering jacket material.

7. Marking or identification, if required.

Cable assemblies are drawn to scale and include a parts list that is coordinated with the drawing by identification balloons, as shown in Figure 16–10.

## PICTORIAL DRAWINGS

Pictorial diagrams were introduced earlier in this chapter. Three-dimensional drawings are frequently made to show pictorial representations of products for display in vendors' catalogues or instruction manuals. Pictorial assembly drawings are often used to show the physical arrangement of components so components are properly placed by factory workers during assembly. These drawings are also helpful for maintenance, as they provide a realistic representation of the product.

One of the most effective uses of photodrafting is found in the electronics field. This is where a photograph is taken of a complete assembly or product and drafting is used to label individual components. A common method of pictorial representation is the exploded technical illustration often used for parts identification and location, as shown in Figure 16–11.

Another type of pictorial diagram, known as a semipictorial wiring diagram, uses two-dimensional images of components in a diagram form for use in electrical or electronic applications. The idea is to show components as features that anyone can recognize. This type of pictorial schematic is commonly used in automobile operations manuals, as shown in Figure 16–12.

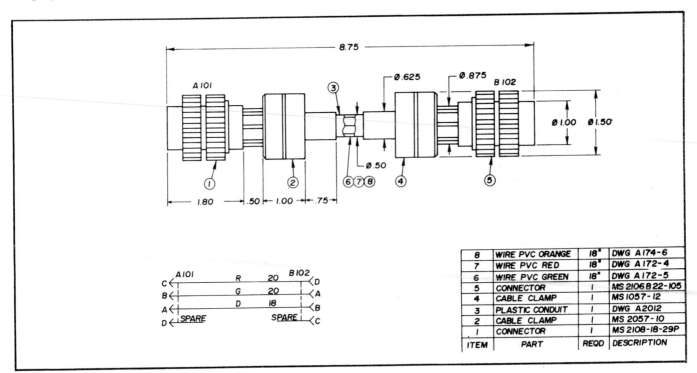

Figure 16–10 Cable assembly.

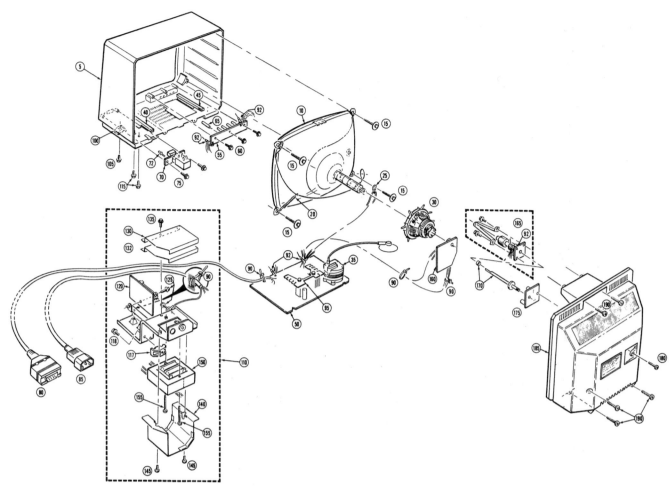

Figure 16–11 Parts pictorial, exploded technical illustration. *Reprinted by permission of Heath Company.*

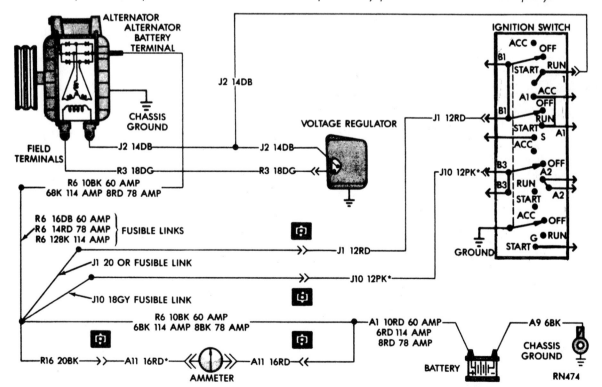

Figure 16–12 Semipictorial wiring diagram. *Courtesy Chrysler Corporation.*

# CHAPTER 16 TEST

Multiple choice: Respond to the following by selecting a, b, c, or d to best answer the question or complete the statement.

1. Electrical diagrams are generally drawn at this scale:
   a. 1:1
   b. 1:2
   c. 2:1
   d. no scale

2. This is a simplified electrical diagram where the complete circuit is not shown, and a single line represents the flow of electricity from one component to the next using simplified symbols or labeled blocks:
   a. pictorial diagram
   b. schematic diagram
   c. one-line diagram
   d. wiring diagram

3. This is the easiest drawing for most people to understand and represents the electrical circuit as a three-dimensional drawing:
   a. pictorial diagram
   b. schematic diagram
   c. one-line diagram
   d. wiring diagram

4. This is a type of electrical drawing that shows all of the interconnections of the system components, and achieves quite a lot of detail by showing the individual wire runs:
   a. pictorial diagram
   b. schematic diagram
   c. one-line diagram
   d. wiring diagram

5. These diagrams are drawn as a series of lines and symbols that represent the electrical current path and the components of the circuit, but they may not show the individual wires or connections of an electrical circuit. These diagrams, however, show the electrical relationship of components:
   a. pictorial diagram
   b. schematic diagram
   c. one-line diagram
   d. wiring diagram

6. This type of diagram combines the simplicity of a schematic diagram and the completeness of a wiring diagram. It contains the complete circuit drawn as a series of lines and symbols that represent the electrical current path and the components of the circuit, including the connection terminals in their proper locations:
   a. highway wiring diagram
   c. wireless diagram
   b. schematic wiring diagram
   d. cable diagram

7. This type of diagram is associated with multiconductor systems and usually provides circuit designation, conductor size, number of leads, conductor type, and power rating:
   a. highway wiring diagram
   b. schematic wiring diagram
   c. wireless diagram
   d. cable diagram

8. These drawings are used for fabrication, quality control, and troubleshooting of the wiring of electrical circuits and systems, and represent a simplified version of a point-to-point, interconnecting wiring diagram:
   a. highway wiring diagram
   b. schematic wiring diagram
   c. wireless diagram
   d. cable diagram

9. These diagrams omit the interconnecting lines and show components in their locations related to the complete system:
   a. highway wiring diagram
   b. schematic wiring diagram
   c. wireless diagram
   d. cable diagram

10. This is a multiconductor or group of insulated wires put together in one sealed assembly:
    a. harness
    b. cable
    c. package
    d. wire assembly

Problem 16–1 Courtesy David George.

## CHAPTER 16 PROBLEMS

**PROBLEM 16–1** Answer the following questions as you refer to the print shown on the previous page.

1. What type of electrical drawing is displayed on this print?
   _____

2. Wiring lines merge at thick lines. Give two names that these thick lines are called.
   _____
   _____

3. What are the lines called that run from the component to the thick lines described in question 2?
   _____
   _____

4. Explain the difference between the electrical diagram shown on this print and an interconnecting wiring diagram.
   _____
   _____
   _____
   _____
   _____
   _____

5. What do the codes M1, M2, XDS1, XDS5, S1, and S4 have in common?
   _____
   _____

6. Explain the elements of the code XDS5-2.
   _____
   _____
   _____

7. What does the underline on the E in the code P1-E mean?
   _____
   _____

8. Where does the wire connect from component P1 pin Z?
   _____

9. Give the part number of this electrical circuit.
   _____

10. How are wire lengths determined?
   _____

**PROBLEM 16–2** Answer the following questions as you refer to the print shown on the following page.

1. What type of electrical drawing is displayed on this print?
   _____
   _____

2. Explain the difference between this type of electrical diagram and a highway wiring diagram.
   _____
   _____
   _____
   _____

3. What does this note mean:
   FROM MARKER CONTROL PANEL B1-31-27?
   _____
   _____
   _____

4. Where do pins 1 through 6 connect to from the BALANCE CONTROL PANEL?
   _____
   _____
   _____

5. Give the part number for this electrical system.
   _____

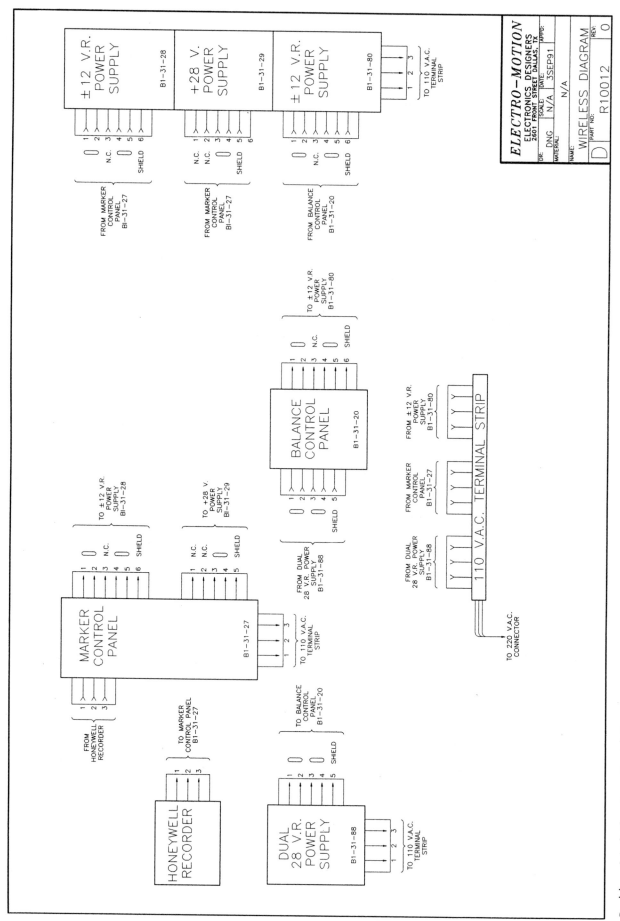

Problem 16–2 Courtesy David George.

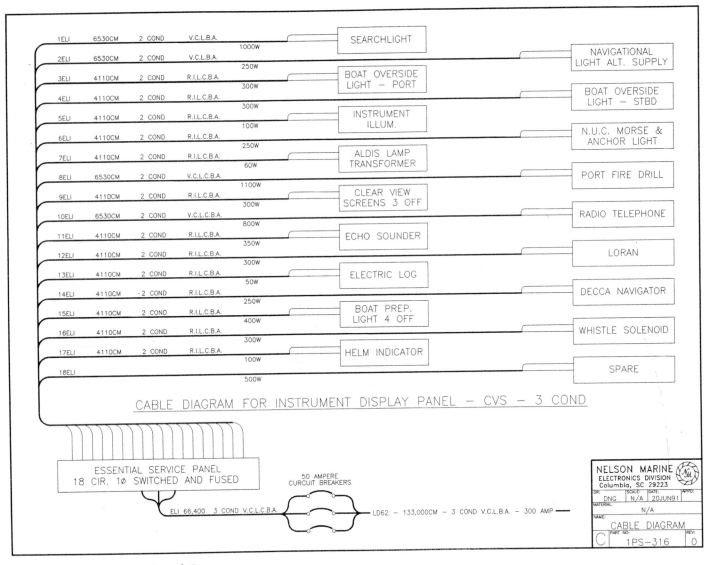

CABLE DIAGRAM FOR INSTRUMENT DISPLAY PANEL — CVS — 3 COND

Problem 16–3 *Courtesy David George.*

**PROBLEM 16–3** Answer the following questions as you refer to the print shown on this page.

1. What type of electrical drawing is displayed on this print?

2. What is the purpose of this type of electrical diagram?

3. Give the name of the system that this electrical diagram represents.

4. Give the size of the three circuit breakers.

5. Give the specifications listed on the multiconductor for the RADIO TELEPHONE circuit.

6. Give the specifications listed on the multiconductor for the WHISTLE SOLENOID circuit.

7. How many circuits are there in the SERVICE PANEL?

8. Give the part number of this electrical system.

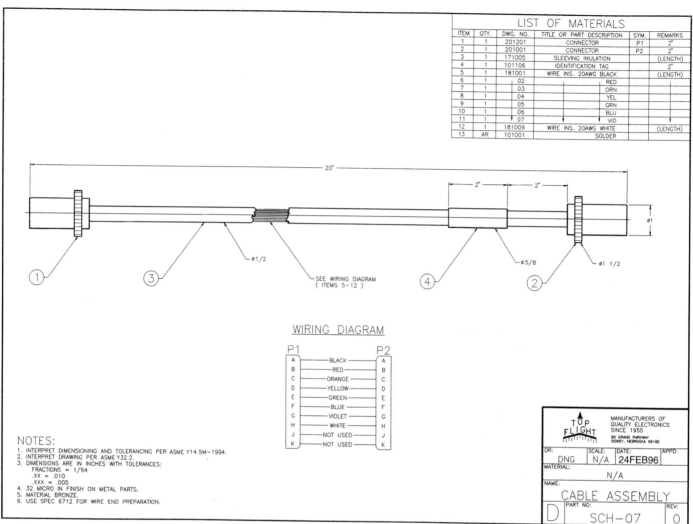

LIST OF MATERIALS

| ITEM | QTY. | DWG. NO. | TITLE OR PART DESCRIPTION | SYM. | REMARKS |
|------|------|----------|---------------------------|------|---------|
| 1 | 1 | 201201 | CONNECTOR | P1 | 2" |
| 2 | 1 | 201001 | CONNECTOR | P2 | 2" |
| 3 | 1 | 171005 | SLEEVING INULATION | | (LENGTH) |
| 4 | 1 | 101106 | IDENTIFICATION TAG | | 2" |
| 5 | 1 | 181001 | WIRE INS. 20AWG BLACK | | (LENGTH) |
| 6 | 1 | 02 | RED | | |
| 7 | 1 | 03 | ORN | | |
| 8 | 1 | 04 | YEL | | |
| 9 | 1 | 05 | GRN | | |
| 10 | 1 | 06 | BLU | | |
| 11 | 1 | 07 | VIO | | |
| 12 | 1 | 181009 | WIRE INS. 20AWG WHITE | | (LENGTH) |
| 13 | AR | 101001 | SOLDER | | |

WIRING DIAGRAM

P1                    P2
A —— BLACK —— A
B —— RED —— B
C —— ORANGE —— C
D —— YELLOW —— D
E —— GREEN —— E
F —— BLUE —— F
G —— VIOLET —— G
H —— WHITE —— H
J —— NOT USED —— J
K —— NOT USED —— K

TOP FLIGHT
MANUFACTURERS OF QUALITY ELECTRONICS SINCE 1955
86 GRAND PARKWAY
SIDNEY, NEBRASKA 69160

DR: DNG    SCALE: N/A    DATE: 24FEB96    APPD:
MATERIAL: N/A
NAME: CABLE ASSEMBLY
D    PART NO: SCH—07    REV: 0

NOTES:
1. INTERPRET DIMENSIONING AND TOLERANCING PER ASME Y14.5M—1994.
2. INTERPRET DRAWING PER ASME Y32.2.
3. DIMENSIONS ARE IN INCHES WITH TOLERANCES:
     FRACTIONS = 1/64
     .XX = .010
     .XXX = .005
4. 32 MICRO IN FINISH ON METAL PARTS.
5. MATERIAL BRONZE.
6. USE SPEC 6712 FOR WIRE END PREPARATION.

Problem 16–4 *Courtesy David George.*

**PROBLEM 16–4** Answer the following questions as you refer to the print shown on this page.

1. What type of electrical drawing is displayed on this print?

2. Explain the purpose of this type of electrical drawing.

3. How many individual insulated wires are used in this part?

4. Give the drawing numbers of the connectors.

5. What is the total length of the part shown on this print?

6. What is the color of the insulated wire connecting to the E terminal?

7. Give the tolerance of the Ø5/8" dimension.

8. Give the description of the RED wire.

9. What is the material and finish of the metal parts?

10. What is the length of item 2?

11. Name the specification used for the wire end preparation.

12. What is item 13?

# GLOSSARY

**Acme**  A thread system used especially for feed mechanisms.

**Addendum (Spur Gear)**  The radial distance from the pitch circle to the top of the tooth.

**Addendum Angle (Bevel Gear)**  The angle subtended by the addendum.

**Aligned Section**  The cutting plane is staggered to pass through offset features of an object.

**Allowance**  The tightest possible fit between two mating parts.

**Alloys**  A mixture of two or more metals.

**Amplifier (AMP)**  A device that allows an input signal to control power, capable of having an output signal greater than the input signal.

**Annealing**  Under certain heating and cooling conditions and techniques, steel may be softened.

**Apparent Intersection**  This is a condition where lines or planes look like they may be intersecting, but in reality they may not be intersecting.

**Auxiliary View**  A view that is required when a surface is not parallel to one of the principal planes of projection; the auxiliary projection plane is parallel to the inclined surface so that the surface may be viewed in its true size and shape.

**Axis**  The centerline of a cylindrical feature.

**Ball Bearing**  A friction-reducer where balls roll in two grooved rings.

**Base Circle (CAM)**  The smallest circle tangent to the CAM follower at the bottom of displacement.

**Base Circle Diameter (Spur Gear)**  The diameter of a circle from which the involute tooth is generated.

**Basic Dimension**  A numerical value used to describe the theoretically exact size, profile, orientation, or location of a feature or datum target. It is the basis from which permissible variations are established by tolerances on other dimensions, in notes, or in feature control frames.

**Bearing (Mechanical)**  A mechanical device that reduces friction between two surfaces.

**Bearing Seal**  A rubber, felt, or plastic seal on the outer and inner rings of a bearing. Generally, it is filled with a special lubricant by the manufacturer.

**Bearing Shield**  A metal plate on one or both sides of the bearing; serves to retain the lubricant and keep the bearing clean.

**Bellcrank**  A link, pivoted near the center, that oscillates through an angle.

**Bend Allowance**  The amount of extra material needed for a bend to compensate for compression during the bending process.

**Bend Relief**  Cutting away material at a corner to help relieve stress.

**Bevel Gear**  Used to transmit power between intersecting shafts; takes the shape of a frustum of a cone.

**Bias**  The voltage applied to a circuit element to control the mode of operation.

**Bilateral Tolerance**  A tolerance in which variation is permitted in both directions from the specified dimension.

**Bit**  Binary digit.

**Bolt Circle**  Holes located in a circular pattern.

**Bore**  To enlarge a hole with a single pointed machine tool in a lathe, drill press, or boring mill.

**Boss**  A cylindrical projection on the surface of a casting for forging.

**Broken Out Section**  A portion of a part is broken away to clarify an interior feature; there is no associated cutting-plane line.

**Bus**  An aluminum or copper plate or tubing that carries the electrical current.

**Bushing**  A replaceable lining or sleeve used as a bearing surface.

**Butt Weld** A form of pipe manufacture in which the seam of the pipe is a welded flat-faced joint. Also, a form of welding in which two pieces of material are "butted" against each other and welded.

**Cabinet Oblique Drawing** A form of oblique drawing in which the receding lines are drawn at half scale, and usually at a 45° angle from horizontal.

**CAD** Computer-aided design.

**CAD/CAM** Computer-aided design/computer-aided manufacturing.

**CADD** Computer-aided design and drafting.

**CAE** Computer-aided engineering.

**Cam** A machine part used to convert constant rotary motion into timed irregular motion.

**Cam Motion** The base point from which to begin cam design. There are four basic types of motion: simple harmonic, constant velocity, uniform accelerated, and cycloidal.

**Capacitor** An electronic component that opposes a change in voltage and storage of electronic energy.

**Carburization** A process where carbon is introduced into metal by heating to a specified temperature range while in contact with a solid, liquid, or gas material consisting of carbon.

**Cartesian Coordinate System** A measurement system based on rectangular grids to measure width, height, and depth (x, y, and z).

**Casting** An object or part produced by pouring molten metal into a mold.

**Cavalier Oblique Drawing** A form of oblique drawing in which the receding lines are drawn true size, or full scale. Usually drawn at an angle of 45 horizontal degrees.

**Central Processing Unit (CPU)** The processor and main memory chips in a computer. Specifically, the CPU is just the processor, but generally it refers to the computer.

**Chain Dimensioning** Also known as point-to-point dimensioning when dimensions are established from one point to the next.

**Chamfer** A slight surface angle used to relieve a sharp corner.

**Chordal Addendum (Spur Gear)** The height from the top of the tooth to the line of the chordal thickness.

**Chordal Thickness (Spur Gear)** The straight-line thickness of a gear tooth on the pitch circle.

**CIM** Computer-integrated manufacturing that combines CADD, CAM, and CAE into a controlled system.

**Circular Pitch (Spur Gear)** The distance from a point on one tooth to the corresponding point on the adjacent tooth, measured on the pitch circle.

**Circular Thickness (Spur Gear)** The length of an arc between the two sides of a gear tooth on the pitch circle.

**Clearance (Spur Gear)** The radial distance between the top of a tooth and the bottom of the mating tooth space.

**Coil or Inductor** A conductor wound on a form or in a spiral; contains inductance.

**Cold Rolled Steel (CRS)** The additional cold forming of steel after initial hot rolling; cleans up hot formed steel.

**Compressive** Pushing toward the point of currency, as in forces that are compressed.

**Concentric** Two or more circles sharing the same center.

**Cone Distance (Bevel Gear)** The slant height of the pitch cone.

**Counterbore** To cylindrically enlarge a hole; generally to allow the head of a screw or bolt to be recessed below the surface of an object.

**Counterdrill** A machined hole that looks similar to a countersink/counterbore combination.

**Countersink** Used to recess the tapered head of a fastener below the surface of an object.

**Crank** A link, usually a rod or bar, that makes a complete revolution about a fixed point.

**Crown Backing (Bevel Gear)** The distance between the cone apex and the outer tip of the gear teeth.

**CRT** Cathode ray tube.

**Datum** A theoretically exact point, axis, or plane derived from the true geometric counterpart of a specified datum feature. The original from which the location or geometric characteristics of features of a part are established.

**Datum Dimensioning** A dimensioning system where each dimension originates from a common surface, plane, or axis.

**Dedendum (Spur Gear)** The radial distance from the pitch circle to the bottom of the tooth.

**Dedendum Angle (Bevel Gear)** The angle subtended by the dedendum.

**Detail** A drawing of an individual part that contains all of the views, dimensions, and specifications necessary to manufacture the part.

**Diametral Pitch** A ratio equal to the number of teeth on a gear per inch of pitch diameter.

**Diazo** A printing process that produces blue, black, or brown lines on various media (other resultant colors are also produced with certain special products). The print process is a combination of exposing an original in contact with a sensitized material exposed to an ultraviolet light, and then running the exposed material through an ammonia chamber to activate the remaining sensitized image to form the desired print. This is a fast and economical method of making prints commonly used in drafting.

**Dimetric Drawing** A pictorial drawing in which two axes form equal angles with the plane of projection. These can be greater than 90° but less than 180° and cannot have an angle of 120°. The third axis may have an angle less or greater than the two equal axes.

**Displacement Diagram** A graph; the curve on the diagram is a graph of the path of the cam follower. In the case of a drum cam displacement diagram, the diagram is actually the developed cylindrical surface of the cam.

**Documentation** Instruction manuals, guides, and tutorials provided with any computer hardware/software system.

**Dowel Pin** A cylindrical fastener used to retain parts in a fixed position or to keep parts aligned.

**Draft** The taper on the surface of a pattern for castings of the die for forgings, designed to help facilitate removal of the pattern from the mold or the part from the die. Draft is often 7–10° but depends on the material and the process.

**Drilling Drawing** Used to provide size and location dimensions for trimming the printed circuit board.

**Drum Cam** A drum, or cylindrical, cam is a cylinder with a groove in its surface. As the cam rotates, the follower moves through the groove, producing a reciprocating motion parallel to the axis of the camshaft.

**Duct** Sheet metal, plastic, or other material pipe designed as the passageway for conveying air from the HVAC equipment to the source.

**Ductility** The ability to be stretched, drawn, or hammered without breaking.

**Eccentric Circle** Not having the same center.

**Electrical Relays** Magnetic switching devices.

**Electro-Discharge Machining (EDM)** A process where material to be machined and an electrode are submerged in a fluid that does not conduct electricity, forming a barrier between the part and the electrode. A high-current, short-duration electrical charge is then used to remove material.

**Electroless** The depositing of metal on another material through the action of an electric current.

**Electron Beam (EB)** Generated by a heated tungsten filament used to cut or machine very accurate features in a part.

**Elementary Diagrams** Diagrams that provide the detail necessary for engineering analysis and operation or maintenance of substation equipment.

**Engineering Change Documents** Documents used to initiate and implement a change to a production drawing; engineering change request (ECR) and engineering change notice (ECN) are examples.

**Exploded Assembly** A pictorial assembly showing all parts removed from each other and aligned along axis lines.

**Face Angle (Bevel Angle)** The angle between the top of the teeth and the gear axis.

**Fault Condition** A short circuit that is a zero-resistance path for electrical current flow.

**Fillet** A curve formed at the interior intersection between two or more surfaces.

**Fixture** A device for holding work in a machine tool.

**Flange** A thin rim around a part.

**Fold Lines** The reference line of intersection between two reference planes in orthographic projection.

**Follower** The cam follower is a reciprocating device whose motion is produced by contact with the cam surface.

**Foreshortened Line** A line that appears shorter than its actual length because it is at an angle to the line of sight.

**Four-Bar Linkage** The most commonly used linkage mechanism. It contains four links: a fixed link called the ground link, a pivoting link called a driver, another pivoted link called a follower, and a link between the driver and follower called a coupler.

**Full Section** The cutting plane extends completely through the object.

**Gate** The part of an electronic system that makes the electronic circuit operate; permits an output only when a predetermined set of input conditions are met.

**Gear** A cylinder or cone with teeth on its contact surface; used to transmit motion and power from one shaft to another.

**Gear Ratio** Any two mating gears have a relationship to each other called a gear ratio. This relationship is the same between any of the following RPMs, number of teeth, and pitch diameters of the gears.

**Gear Train** Formed when two or more gears are in contact.

**Graphical Kinematic Analysis** The process of drawing a particular mechanism in several phases of a full cycle to determine various characteristics of the mechanism.

**Half Section** Used typically for symmetrical objects; the cutting-plane line actually cuts through one quarter of the part. The sectional view shows half of the interior and half of the exterior at the same time.

**Hone** A method of finishing a hole or other surface to a desired close tolerance and fine surface finish using an abrasive.

**Inductance** The property in an electronic circuit that opposes a change in current flow or where energy may be stored in a magnetic field, as in a transformer.

**Integrated Circuit (IC)** When all of the components in a schematic are made up of one piece of semiconductor material.

**Intersecting Lines** When lines are intersecting, the point of intersection is a point that lies on both lines.

**Isometric Drawing** A form of pictorial drawing in which all three drawing axes form equal angles (120°) with the plane of projection.

**???** A device used for guiding a machine tool in the matching of a part or feature.

**Joint** The connection point between two links.

**Kerf** A groove created by the cut of a saw.

**Kinematics** The study of motion without regard to the forces causing the motion.

**Lap Weld** A form of pipe manufacture in which the seam of the pipe is an angular "lap."

**Large-Scale Integration (LSI)** More circuits on a single small IC chip.

**Laser (Light Amplifcation by Stimulated Emission of Radiation)** A device that amplifies focused light waves and concentrates them in a narrow, very intense beam.

**Lay**  Describes the basic direction or configuration of the predominant surface pattern in a surface finish.

**Lead (Worm Thread)**  The distance that the thread advances axially in one revolution of the worm or thread.

**Lever**  A link that moves back and forth through an angle; also known as a rocker.

**Line of Sight**  An imaginary straight line from the eye of the observer to a point on the object being observed. All lines of sight for a particular view are assumed to be parallel and are perpendicular to the projection plane involved.

**Logic Diagrams**  A type of schematic that is used to show the logical sequence in an electronic system.

**Magnetic Declination**  The degree difference between magnetic azimuth and true azimuth.

**Malleable**  The ability to be hammered or pressed into shape without breaking.

**Master Pattern**  A 1:1 scale circuit pattern that is used to produce a printed circuit board.

**Mechanism**  A combination of two or more machine members that work together to perform a specific motion.

**Multiview Projection**  The views of an object as projected upon two or more picture planes in orthographic projection.

**Neck**  A groove around a cylindrical part.

**Nominal Size**  The designation of the size for a commercial product.

**Nondestructive Testing (NDS)**  Tests for potential defects in welds; they do not destroy or damage the weld or the part.

**Normal Plane**  A plane surface that is parallel to any of the primary projection planes.

**Normalizing**  A process of heating steel to a specific temperature and then allowing the material to cool slowly by air, bringing the steel to a normal state.

**Numerical Control (NC)**  A system of controlling a machine tool by means of numeric codes that direct the commands for the machine movements; computer numerical control (CNC) is a computer command control of the machine movement.

**Oblique Drawing**  A form of pictorial drawing in which the plane of projection is parallel to the front surface of the object and the receding angle is normally 45°.

**Oblique Line**  A straight line that is not parallel to any of the six principal planes.

**Oblique Plane**  Inclined to all of the principal projection planes.

**Offset Section**  The cutting plane is offset through staggered interior features of an object to show those features in section as if they were in the same plane.

**Operational Amplifier (OPAMP)**  A high-gain amplifier created from an integrated circuit.

**Outside Diameter (Spur Gear)**  The overall diameter of the gear; equal to the pitch diameter, plus two addendum.

**Pads**  Or lands; the circuit-termination locations where the electronic devices are attached.

**Pattern Development**  Based on laying out geometric forms in true size and shape flat patterns.

**Perspective Drawing**  A form of pictorial drawing in which vanishing points are used to provide the depth and distortion that is seen with the human eye. Perspective drawings can be drawn using one, two, and three vanishing points.

**Photodrafting**  A combination of a photograph or photographs with line work and lettering on a drawing.

**Pictorial Drawing**  A form of drawing that shows an object's depth. Three sides of the object can be seen in one view.

**Pie Charts**  Used for presentation purposes where portions of a circle represent quantity.

**Pinion Gear**  When two gears are mating, the pinion gear is the smaller, usually the driving gear.

**Pitch**  A distance of uniform measure determined at a point on one unit to the same corresponding point on the next unit; used in threads, springs, and other machine parts.

**Pitch (Worm)**  The distance from one tooth to the corresponding point on the next tooth measured parallel to the worm axis; equal to the circular pitch on the worm gear.

**Pitch Angle (Bevel Gear)**  The angle between an element of a pitch cone and its axis.

**Pitch Diameter (Bevel Gear)**  The diameter of the base of the pitch cone.

**Pitch Diameter (Spur Gear)**  The diameter of an imaginary pitch circle on which a gear tooth is designed. Pitch circles of two spur gears are tangent.

**Plain Bearing**  Based on a sliding action between the mating parts; also called sleeve or journal bearings.

**Plane**  A surface that is not curved or warped. It is a surface in which any two points may be connected by a straight line, and the straight line will always lie completely within the surface.

**Plat**  A tract of land showing building lots.

**Plate Cam**  A cam in the shape of a plate or disk. The motion of the follower is in a plane perpendicular to the axis of the camshaft.

**Polyester Drafting Film**  A high-quality drafting material with excellent reproduction, durability, and dimensional stability; also known by the trade name Mylar®.

**Polygons**  Enclosed figures, such as triangles, squares, rectangles, parallelograms, and hexagons.

**Pressure Angle**  The direction of pressure between contacting gear teeth. It determines the size of the base circle and the shape of the involute spur gear tooth, commonly 20°.

**Prime Circle (CAM)**  A circle with a radius equal to the sum of the base circle radius and the roller follower radius.

**Printed Circuits (PC)**  Electronic circuits printed on a board that form the interconnection between electronic devices.

**Printer** A device that receives data from the computer and converts it into alphanumeric or graphic printed images.

**Profile Line** A line that is parallel to the profile projection plane; its projection appears in true length in the profile view.

**Projection Line** A straight line at 90° to the fold line, which connects the projection of a point in a view to the projection of the same point in the adjacent view.

**Projection Plane** An imaginary surface on which the view of the object is projected and drawn. This surface is imagined to exist between the object and the observer.

**Quench** To cool suddenly by plunging into water, oil, or other liquid.

**Rack** Basically a straight bar with teeth on it. Theoretically, it is a spur gear with an infinite pitch diameter.

**Radial Motion** Exists when the path of the motion forms a circle, the diameter of which is perpendicular to the center of the shaft; also known as rotational motion.

**Ream** To enlarge a hole slightly with a machine tool called a reamer to produce greater accuracy.

**Relief** A slight groove between perpendicular surfaces to provide clearance between the surfaces for machining.

**Removed Section** A sectional view taken from the location of the section cutting plane and placed in any convenient location of the drawing, generally labeled in relation to the cutting plane.

**Resistors** Components that contain resistance to the flow of electric current.

**Revolution** An alternate method for solving descriptive geometry problems in which the observer remains stationary and the object is rotated to obtain various views.

**Revolved Section** A sectional view established by revolving 90° in place into a plane perpendicular to the line of sight, generally used to show the cross section of a part or feature that has consistent shape throughout the length.

**Rib** A thin metal section between parts to reinforce while reducing weight in a part.

**Right Angle** An angle of 90°.

**Rocker** A link that moves back and forth through an angle; also known as a lever.

**Roller Bearings** A bearing composed of two grooved rings and a set of rollers. The rollers are the friction-reducing element.

**Root Diameter (Spur Gear)** The diameter of a circle coinciding with the bottom of the tooth spaces.

**Round** Two or more exterior surfaces rounded at their intersection.

**RPM** Revolutions per minute.

**Runouts** Characteristics of intersecting features, determined by locating the line of intersection between the mating parts.

**Schematic Diagrams** Drawn as a series of lines and symbols that represent the electrical current path and the components of the circuit. Provides the basic circuit connection information for electronic products.

**Semiconductors** Devices that provide a degree of resistance in an electronic circuit. Types include diodes and transistors.

**Slider** A link that moves back and forth in a straight line.

**Slope Angle** The angle in degrees that the line makes with the horizontal plane.

**Solder** An alloy of tin and lead.

**Solder Mask** A polymer coating to prevent the bridging of solder between pads or conductor traces on a printed circuit board.

**Solids Modeling** A design and engineering process in which a 3-D model of the actual part is created on the screen as a solid part showing no hidden features.

**Spline** One of a series of keyways cut around a shaft and mating hole; generally used to transfer power from a shaft to a hub while allowing a sliding action between the parts.

**Spur Gear** The simplest, most common type of gear used for transmitting motion between parallel shafts. Its teeth are straight and parallel to the shaft axis.

**Stretch-Out Line** Typically, the beginning line upon which measurements are made and the pattern development is established.

**Surface Finish** Refers to the roughness, waviness, lay, and flaws of a machine surface.

**Surface Mount Technology (SMT)** The traditional component lead through is replaced with a solder paste to hold the electronic components in place on the surface of the printed circuit board, and taking up to less than one-third of the space of conventional PC boards.

**Tangent** A straight or curved line that intersects a circle or arc at one point only; is always 90° relative to the center.

**Taper** A conical shape on a shaft or hole, or the slope of a plane surface.

**Tempering** A process of reheating normalized or hardened steel to a specified temperature, followed by cooling at a predetermined rate to achieve certain hardening characteristics.

**Tensile Forces** Forces that pull away.

**Tensile Strength** Ability to be stretched.

**Thermoplastic** Plastic material may be heated and formed by pressure. Upon reheating, the shape can be changed.

**Thermoset** Plastics are formed into permanent shape by heat and pressure and may not be altered after curing.

**Thermostat** An automatic mechanism for controlling the amount of heating or cooling given by a central or zoned heating or cooling system.

**Tolerance** The total permissible variation in a size or location dimension.

**Transistors** Semiconductor devices in that they are conductors of electricity with resistance to electron flow applied and are used to transfer or amplify an electronic signal.

**Transition Piece** A duct component that provides a change from square or rectangular to round; also known as a "square-to-round."

**Translational Motion** Linear motion.

**Triangulation** A technique used to lay out the true size and shape of a triangle with the true lengths of the sides; used in pattern development on objects such as the transition piece.

**Trimetric Drawing** A type of pictorial drawing in which all three of the principal axes do not make equal angles with the plane of projection.

**True Length or True Size and Shape** When the line of sight is perpendicular to a line, surface, or feature.

**True Position** The theoretically exact location of a feature established by basic dimensions.

**Ultrasonic Machining** A process where a high-frequency mechanical vibration is maintained in a tool designed to a desired shape; also known as impact grinding.

**Undercut** A groove cut on the inside of a cylindrical hole.

**Unilateral Tolerance** A tolerance in which variation is permitted in only one direction from the specified dimension.

**Upset** A forging metal used to form a head or enlarged end on a shaft by pressure or hammering between dies.

**Valve** Any mechanism, such as a gate, ball, flapper, or diaphragm, used to regulate the flow of fluids through a pipe.

**Vellum** A drafting paper with translucent properties.

**Viewing-Plane Line** Represents the location of where a view is established.

**Visualization** The process of recreating a three-dimensional image of an object in a person's mind.

**Web** See Rib.

**Whole Depth (Spur Gear)** The full height of the tooth. It is equal to the sum of the addendum and the dedendum.

**Wire Form** A three-dimensional form in which all edges and features show as lines, thus appearing to be constructed of wire.

**Wireless Diagram** Similar to highway diagrams except that interconnecting lines are omitted. The interconnection of terminals is provided by coding.

**Wiring Diagram** A type of schematic that shows all of the interconnections of the system components, also referred to as a point-to-point interconnecting wiring diagram.

**Working Depth (Spur Gear)** The distance that a tooth occupies in the mating space. It is equal to two times the addendum.

**Worm Gears** Used to transmit power between nonintersecting shafts. The worm is like a screw and has teeth similar to the teeth on a rack. The teeth on the worm gear are similar to the spur gear teeth, but they are curved to form the teeth on the worm.

**Zoning** A system of numbers along the top and bottom and letters along the left and right margins of a drawing used for ease of reading and locating items.

# INDEX